Lecture Notes in Mathematics

A collection of informal reports and seminars
Edited by A. Dold, Heidelberg and B. Eckmann, Zürich

186

Recent Trends in Graph Theory

Proceedings of the First New York City
Graph Theory Conference
held on June 11, 12, and 13, 1970

Sponsored by St. John's University, Jamaica, New York

Edited by M. Capobianco, J. B. Frechen and M. Krolik
St. John's University, Jamaica, New York

Springer-Verlag
Berlin · Heidelberg · New York 1971

AMS Subject Classifications (1970): 05C99

ISBN 3-540-5386-7 Springer-Verlag Berlin · Heidelberg · New York
ISBN 0-387-5386-7 Springer-Verlag New York · Heidelberg · Berlin

Offsetdruck: Julius Beltz, Weinheim/Bergstr.

FOREWORD

These are the proceedings of the first graph theory conference held in New York City. In spirit and scope it followed the graph theory conferences recently held in Ann Arbor and Kalamazoo: the contributors and participants were persons from Canada and the United States who are interested and very active in graph theory and its applications.

We thank St. John's University which, in connection with its Centennial Year celebration, sponsored the Conference financially, and lent the use of some of its facilities. In particular, we thank the Rev. Joseph I. Dirvin, C.M., through whose efforts the funds were made available.

We also wish to extend our gratitude to Bruce Monte, John M. O'Shaughnessy, and their staff for their work in helping to organize the Conference.

Special acknowledgment goes to our Mathematics Department secretary, Mary O'Shea, and her assistant, Marsha Davis, who tended to so many of the details in the preparation of the Conference and the manuscript for these Proceedings.

October, 1970 Joseph B. Frechen

TABLE OF CONTENTS

THE CONNECTIVITY AND LINE-CONNECTIVITY OF COMPLEMENTARY GRAPHS

Yousef Alavi, Western Michigan University, Kalamazoo, MI 49001
John Mitchem, San Jose State College, San Jose, CA 95114

Introduction. The complement $\overline{G}$ of a graph G is the graph whose point set is that of G and such that two points are adjacent in $\overline{G}$ if and only if they are not adjacent in G. In [4] Nordhaus and Gaddum showed that for any graph G of order p the following inequalities hold:

$$\{2\sqrt{p}\} \le \chi(G) + \chi(\overline{G}) \le p + 1 \ ;$$

$$p \le \chi(G) \cdot \chi(\overline{G}) \le \left[\left(\frac{p+1}{2}\right)^2\right] ,$$

where $\chi(G)$ is the chromatic number of G. Finck [2] and Stewart [5] showed that the bounds are sharp for every positive integer p. In this paper we find the corresponding inequalities for connectivity and line-connectivity.

Definitions and Known Results. The connectivity $\kappa(G)$ of a graph G is the minimum number of points whose removal results in a disconnected or trivial graph. The line-connectivity $\lambda(G)$ is defined analogously. For any real number r, $[r]$ and $\{r\}$ denote the greatest integer not exceeding r and the least integer not less than r, respectively. The following results and observations will prove to be useful:

(1) If G is disconnected, then $\overline{G}$ is connected.

(2) For any graph G, $\kappa(G) \le \lambda(G) \le \delta(G)$, where $\delta(G)$ is the minimum degree among the points of G.

(3) Chartrand [1] showed that if $\delta(G) \ge \frac{p-1}{2}$ for a graph G of order p, then $\delta(G) = \lambda(G)$.

(4) Let p and n be integers such that $0 \le n < p$. Harary has shown in [3] that in the case where p and n are not both odd, there exists a regular graph G of order p for which $\delta(G) = \lambda(G) = \kappa(G) = n$.

Main Results. We first state and prove the following lemma.

Lemma. For any graph G of order p, we have

$$(5) \quad \delta(G)\cdot\delta(\overline{G}) \le \left[\frac{p-1}{2}\right]\left\{\frac{p-1}{2}\right\} , \quad \text{if } p \equiv 0, 1, 2, \pmod 4 .$$

and

$$(6) \quad \delta(G)\cdot\delta(\overline{G}) \le \left(\frac{p-3}{3}\right)\left(\frac{p+1}{2}\right) \quad \text{if } p \equiv 3 \pmod 4 .$$

Proof. It is clear that $\delta(G) + \delta(\overline{G}) \le p - 1$. If the sum of two numbers is $p - 1$, then their product is maximum when each number is $(p-1)/2$. If $p \equiv 1 \pmod 4$, then $(p-1)/2$ is an integer and

$$\delta(G)\cdot\delta(\overline{G}) \le \left(\frac{p-1}{2}\right)^2 ,$$

so that (5) is obtained. On the other hand, when $p \equiv 0$ or $2 \pmod 4$, the number $(p-1)/2$ is not an integer. However,

$$\left(\frac{p-1}{2}\right)^2 = \frac{p}{2}\left(\frac{p-2}{2}\right) + \frac{1}{4},$$

where $p/2$ and $(p-2)/2$ are integers, i.e., if $p \equiv 0$ or $2 \pmod 4$, the maximum product of two integers (whose sum is at most $p - 1$) is $(p^2-2p)/4$. Hence (5) is verified here also.

In the case where $p \equiv 3 \pmod 4$, the maximum occurs at the odd integer $(p-1)/2$. However, if $\delta(G) = \delta(\overline{G}) = (p-1)/2$, then each of G and $\overline{G}$ is a regular graph of odd degree and odd order, which is impossible. Therefore,

$$\delta(G)\cdot\delta(\overline{G}) < \left(\frac{p-1}{2}\right)^2 = \left(\frac{p-3}{2}\right)\left(\frac{p+1}{2}\right) + 1,$$

which gives us (6).

We now prove our main result.

Theorem 1. For any graph G of order $p \geq 2$, we have

(7) $1 \leq \lambda(G) + \lambda(\overline{G}) \leq p - 1$,

(8) $0 \leq \lambda(G) \cdot \lambda(\overline{G}) \leq M(p)$,

where

$$M(p) = \left[\frac{p-1}{2}\right]\left\{\frac{p-1}{2}\right\} \quad \text{if} \quad p \equiv 0, 1, 2 \pmod 4,$$

and

$$M(p) = \left(\frac{p-3}{2}\right)\left(\frac{p+1}{2}\right) \quad \text{if} \quad p \equiv 3 \pmod 4.$$

Proof. The lower bound of (8) is clear and the one in (7) follows from (1). From (2) we have $\lambda(G) + \lambda(\overline{G}) \leq \delta(G) + \delta(\overline{G})$. This together with the fact that $\delta(G) + \delta(\overline{G}) \leq p - 1$ implies the upper bound of (7). The upper bound of (8) follows from the Lemma and the fact that $\lambda(G) \leq \delta(G)$ for any graph G.

In the following theorem we show that the bounds in (7) and (8) are sharp for every integer $p \geq 2$.

Theorem 2. For $p \geq 2$ and any of the bounds given in Theorem 1, there exists a graph G of order p such that the equality holds for that bound.

Proof. The complete graph with p points shows that the upper bound of (7) and the lower bound of (8) are attained. To see that the lower bound of (7) is sharp it suffices to let G be the complete bipartite graph $K(1,p-1)$. To show that the remaining bound is sharp, let G be a regular graph of order p such that $\lambda(G) = \delta(G)$, where

$$\delta(G) = \left[\frac{p-1}{2}\right] \quad \text{if} \quad p \equiv 0, 1, 2 \pmod 4,$$

$$= \frac{p-3}{2} \quad \text{if} \quad p \equiv 3 \pmod 4.$$

From (4) we know such graphs exist. It then follows that $\overline{G}$ is regular and

$$\delta(\overline{G}) = \left\{\frac{p-1}{2}\right\} \quad \text{for} \quad p \equiv 0,\ 1,\ 2 \pmod 4 ,$$

$$= \frac{p+1}{2} \quad \text{for} \quad p \equiv 3 \pmod 4 .$$

Now since $\delta(\overline{G}) \geq \frac{p-1}{2}$, we conclude from (3) that $\lambda(\overline{G}) = \delta(\overline{G})$. Thus the upper bound of (8) is attained.

The same bounds as those presented for line-connectivity can be established for the connectivity of complementary graphs. These results are stated in the following theorem, where the upper bounds follow from (2) and Theorem 1, and the lower bounds are immediate .

Theorem 3. For any graph G of order $p \geq 2$, the following inequalities hold:

(9) $1 \leq \kappa(G) + \kappa(\overline{G}) \leq p - 1$,

(10) $0 \leq \kappa(G) \cdot \kappa(\overline{G}) \leq M(p)$,

where $M(p)$ is the same as given in Theorem 1.

REFERENCES

1. G. Chartrand, "A Graph-Theoretic Approach to a Communcations Problem", J. SIAM Appl. Math., 14 (1966), 778-781.

2. H.J. Finck, "On the Chromatic Numbers of a Graph and its Complement", Theory of Graphs, (P. Erdos and G. Katona eds.) Academic Press (1968), 99-113.

3. F. Harary, "The Maximum Connectivity of a Graph", Proc. Nat. Acad. Sci. USA, 48 (1962), 1142-1146.

4. E.A. Nordhaus and J.W. Gaddum, "On Complementary Graphs", Amer. Math. Monthly, 63 (1956), 175-177.

5. B.M. Stewart, "On a Theorem of Nordhaus and Gaddum J. Combinatorial Theory, 6 (1969), 217-218.

MAXIMAL m-GONS IN 4-REGULAR MAJOR MAPS

Ruth A. Bari
George Washington University, Washington, DC 20006

Let P_n be a map with n regions on a sphere. Then P_n is said to be regular if it is a cubic map of simply connected regions without proper 2-rings or proper 3-rings.

P_n is said to be a major map if it has no regions with fewer than 5 sides. Furthermore, if P_n is a regular major map which has no proper 4-rings, P_n is called a 4-regular major map.

An attempt to enumerate all 4-regular major maps with a given number of regions led to the following questions:

1. What is the largest number m such that P_n may contain at least one m-gon?
2. Every m-gon in a regular major map is surrounded by a proper m-ring C. What is the largest number k such that the ring C_1 of regions which meet C has m + k regions?

Definition. Let C be a proper ring of a map P_n. Then the strict interior map C^{Int} is the submap of P_n consisting only of the regions of P_n which are inside C.

The augmented strict interior map C^{Int+} is the submap of P_n obtained from P_n by replacing the regions of C and the regions outside C by a single region.

Theorem 1. Let P_n be a 4-regular major map of n regions, R an m-gon of P_n. If the m-ring C of regions surrounding R in P_n consists of m pentagons, there is at least one region of C^{Int} which does not meet C.

Proof. Suppose that every region of C^{Int} meets at least one pentagon of C. Then every region of C^{Int} meets at least two pentagons of C, for otherwise P_n would not be 4-regular. Hence there are exactly m regions in C^{Int}, each of which meets the outside boundary of C^{Int} exactly once.

Let R' be the outside region of the augmented strict interior map C^{Int+}. In this case, we now prove that there is at least one triangle in C^{Int} which meets R' in C^{Int+}.

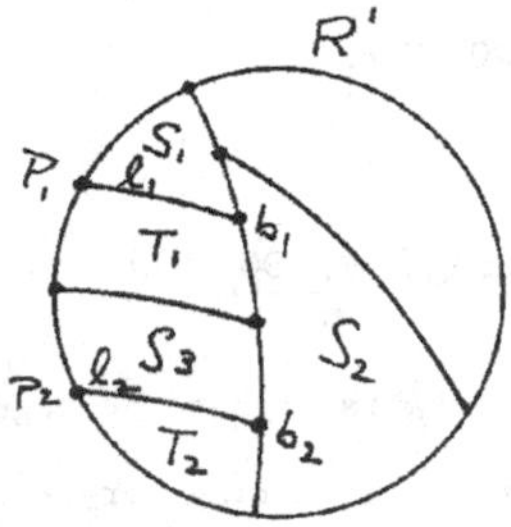

Figure 1

Let p_1 be a vertex of R' in C^{Int+}. Then there is a boundary ℓ_1 of C^{Int} which meets p_1. If b_1 is the second vertex of ℓ_1, b_1 is not a vertex of R', but is a vertex of a boundary which meets R' in C^{Int+}.

Let S_1 and T_1 be the two regions of C^{Int+} which have the boundary ℓ_1 in common. Let S_2 be the third region which meets b_1. Since S_2 is in C^{Int}, S_2 meets R' exactly once in C^{Int+}. Then there is a (not necessarily proper) 3-ring C_1 with defining curve c_1 passing successively from R' to S_1 to S_2 and back to R' with T_1 either strictly inside or strictly outside c_1. Since the two situations are completely analogous, we may consider the case when T_1 is inside c_1.

If T_1 is the only region inside the 3-ring C_1, T_1 is a triangle, and there is nothing more to prove. We may therefore suppose that ${C_1}^{Int}$ has more than one region. Since ℓ_1 is the complete boundary of S_1 inside c_1, there is a vertex other than b_1 on the boundary of S_2 inside c_1, for otherwise T_1 would meet R' more than once, contrary to hypothesis. Let b_2 be the vertex on the boundary of S_2 inside C_1 such that there is a boundary ℓ_2 which meets R' at p_2 and has b_2 as its second vertex. Let T_2 be the other region which has ℓ_2 as its boundary, and let S_3 be the third region which meets T_2 and S_3 at b_2. Since S_3 is in C^{Int}, S_3 also meets R' exactly once. Then there is a ring C_2 with defining curve c_2 passing from R' to S_2 to S_3, and back to R'. The number of regions inside c_2 is obviously smaller than the number of regions inside c_1, since S_3 is completely inside c_1 but is not completely inside c_2. If the ring C_2 contains only T_2, T_2 is again a triangle. If not, we repeat the process, and obtain a ring C_3 with fewer regions inside it than C_2.

Since there are only m regions inside C, in a finite number of steps, we must find a ring C_k which has only one region T_k in its interior, and T_k will be a triangle.

But in this case, since T_k meets only one interior ring vertex of C in P_n, and no other vertex it did not meet in C^{Int+}, P_k is a quadrilateral in P_n, and P_n is not

4-regular.

Hence the assumption that every region of C^{Int} meets a region of C leads to a contradiction, and at least one region of C^{Int} fails to meet C.

<u>Corollary</u>. Let P_n be a 4-regular major map, R an m-gon of P_n. If the m-ring C_1 of regions surrounding R in P_n consists of m pentagons, with R outside C, there is at least one proper ring C_1 inside C whose interior is a region R_1.

<u>Proof</u>. Let R_1 be a region inside C which does not meet C, which exists by theorem 1. Then R_1 is surrounded by a proper ring C_1. If one of the regions of C_1 is also a region of C, R_1 meets C_1, contrary to hypothesis.

Hence the ring C_1 is inside C.

<u>Theorem 2</u>. If P_n is a 4-regular major map of n regions, the largest possible m-gon in P_n is one with $\frac{n-2}{2}$ sides.

<u>Proof</u>. Suppose, on the contrary, that there is a region R_1 in P_n with $m > \frac{n-2}{2}$ sides.

Since P_n is a 4-regular major map, there is a proper m-ring C_1 around R_1, and a ring C_2 of at least m distinct regions around C_1. C_2 has exactly m distinct regions only if every region in C_1 is a pentagon. In this case, there is a region R_2 inside C_1 which does not meet C_1. Hence R_2 is not in C_2.

Thus if every region in C_1 is a pentagon, there are at least $2m + 2$ regions in P_n, m regions in C_1, m regions in C_2, and R_1 and R_2.

If a region of C_1 has 6 or more sides, C_2 has at least $m + 1$ regions, C_1 has m regions, and these together with R_1 and R_2 make $n \geq 2m + 2$.

But it was assumed that $m > \frac{n-2}{2}$. Hence $n - 2 < 2m$, and $n < 2m + 2$.

This contradicts $n \geq 2m + 2$, so we must have $m \leq \frac{n-2}{2}$.

<u>Corollary</u>. There is no regular major map P_{13}, of 13 regions.

<u>Proof</u>. If f_i stands for the number of i-sided regions in P_n, as a consequence of the Euler polyhedral formula, we have $4f_2 + 3f_3 + 2f_4 + f_5 = 12 + f_7 + 2f_8 + \dots$. Hence if P_{13} exists, it must satisfy the equations $f_5 = 12 + f_7 + 2f_8 + \dots$ and $f_5 + f_6 + f_7 + \dots = 13$.

The only simultaneous solution of these equations for non-negative f_i is $f_5 = 12$, $f_6 = 1$, $f_7 = f_8 = \dots = 0$.

Since every regular major map with fewer than 20 regions is 4-regular, the largest m-gon in P_{13} has $m \le \frac{13-2}{2} = \frac{11}{2} = 5\frac{1}{2} < 6$.

Hence there is no regular major map of 13 regions containing a hexagon, and therefore no regular major map of 13 regions.

Theorem 3. Let P_n be a 4-regular major map of n regions, R an m-gon of P_n, C a proper m-ring around R. If C_1 is a proper $(m + k)$-ring which meets C, $k > 0$, then $k < n - 2m - 2$.

Proof. Suppose $k \geq n - 2m - 2$. Since R, C, and C_1 have $1 + m + (m + k) = 1 + 2m + k$ regions, it follows that $1 + 2m + k \geq 1 + 2m + (n - 2m - 2)$, or $1 + 2m + k \geq n - 1$.

But if R, C, and C_1 have at least n-1 regions, there is at most one region inside C_1. Since C_1 is a proper ring, there is at least one region inside C_1, and therefore exactly one region inside C_1.

Now since $k > 0$, at least k vertices of C_1^{Int} must meet regions of C_1 in P_n, or C_1 will have k quadrilaterals and will not be a major map. But if C_1 is a single region, no vertices of C_1^{Int} can meet the regions of C_1.

This is a contradiction, so $k < n - 2m - 2$.

References:

1. Bari, Ruth A. Absolute Reducibility of Maps of at Most 19 Regions. Ph.D Thesis, The Johns Hopkins Univ. (1966)

2. Birkhoff, George D. and Lewis, Daniel C. Chromatic Polynomials. Transactions of the American Mathematical Society. 60, 355-451 (1946)

ON A THEOREM OF TUTTE,

A 4-COLOR THEOREM AND B-SETS

David W. Barnette[1] and Sherman K. Stein
University of California, Davis, CA 95616

1. Introduction. Let G be a graph in the sphere and let C be a circuit in G. Let B be a maximal subgraph of G such that given any two vertices of B then there is a path P joining these vertices which misses C except possibly at the endpoints of P. Such a subgraph is called a <u>bridge</u> of C. The vertices of B that are also in C are called the <u>vertices of attachment of</u> B. A Theorem of Tutte [2] states the following:

<u>Any planar 2-connected graph has a circuit</u> C <u>such that all bridges of</u> C <u>have at most three vertices of attachment</u>.

Such a circuit will be called a <u>Tutte circuit</u>. We shall use this theorem to prove

Theorem 1. <u>Any map on the sphere without non-trivial 2-cycles can be colored with four colors such that no country meets more than one other country of the same color.</u> Here, an n-<u>cycle</u> is defined to be a sequence of countries, $C_1, \cdots, C_n$ such that C_i meets C_{i-1} on an edge and C_n meets C_1 on an edge.

A family of sets $\mathfrak{F}$ is said to have a B-set $\mathfrak{B}$ if $\mathfrak{B}$ meets each member of $\mathfrak{F}$ and contains none of them. We shall show that Tutte's theorem can be translated into a theorem on B-sets which is also very close to the 4-color conjecture. For more on map coloring and B-sets see [1].

2. Proof of Theorem 1.

Lemma 1. <u>Let</u> $\mathfrak{M}$ <u>be a regular map on the sphere with no non-trivial 2-or 3-cycles. Then, there exists a coloring of the map with 4 colors so that no country meets more than one other country of the same color. Moreover, given any vertex of the map we may prescribe the colors of the countries meeting that vertex.</u>

---
[1] Research supported by NSF Grant #GP-8470.

Proof. As is often the case in inductive proofs, Tutte proved a stronger form of the theorem that we quoted. He proved that not only is there a Tutte circuit, but that any two edges of any given country in the map may be chosen to be edges of the circuit. We shall need this strengthening in the proof of our theorem.

Let v be a given vertex of $\mathfrak{M}$ and let C_1, C_2 and C_3 be the countries meeting v. We consider the two cases:

Case I. C_1 and C_2 are to have the same color in the 4-coloring. Let e_1 and e_2 be the edges between C_1 and C_3, and C_2 and C_3 and let Γ be a Tutte circuit containing e_1 and e_2. Let B be any bridge of Γ that contains a vertex not on Γ. Since $\mathfrak{M}$ is 3-connected, B must have exactly three vertices of attachment. These three vertices of attachment meet a 3-cycle that separates the vertices of B from the rest of the graph. Since there are no non-trivial 3-cycles, B must consist of one vertex and three edges joining it to Γ.

Now that we have this characterization of Bridges of Γ we can now examine the way that the countries inside Γ are arranged. To do this, we shall construct a "dual" graph G as follows: the vertices of G will correspond to countries inside Γ. Two vertices of G are joined by an edge if and only if the corresponding countries meet on an edge. The graph G now has the following properties

(i) It is connected.

(ii) If G is embedded in the sphere then the circuits of G that bound countries of G will correspond to vertices of $\mathfrak{M}$, thus these circuits will be of length 3.

(iii) By our characterization of Bridges, we see that no two vertices inside Γ are joined by an edge, thus all circuits in G are of length 3.

We now show, by induction on the number of edges, that the vertices of G can be colored with two colors so that no vertex is joined by an edge to more than one other of the same color, and that we may prescribe the colors of two vertices which are joined by an edge. This is clearly true if G has at most 3 edges. If G does not consist of a single edge or a single 3-cycle, it is easy to show that we can find two subgraphs of G with one vertex of attachment, where these subgraphs are either edges or 3-cycles. We choose such a subgraph G', which does not contain both vertices that are to be prescribed and we consider the graph $G \sim G'$. This graph also satisfies (i), (ii) and (iii), thus by induction we can find the desired 2-coloring of $G \sim G'$. We now use this coloring in G and we give the vertices of G that are not in $G \sim G'$, a different color than the vertex of attachment.

This coloring now gives us a way of coloring the countries inside Γ in the desired way, while prescribing the color of two countries. We can do the same thing outside Γ with two different colors. By the choice of edges e_1 and e_2 on Γ we have that C_1 and C_2 are either both outside or both inside Γ so that we may require that they have the same color.

Case II. C_1, C_2 and C_3 have different colors. We choose any two edges meeting v to be on Γ and repeat the above process. This time two of the 3 countries will lie outside or inside Γ and we prescribe different colors for them.

If $\mathfrak{M}$ is a map on the sphere and C is a non-trivial 3-cycle in $\mathfrak{M}$ then C breaks up the sphere into two simply-connected components. These components will be called the <u>regions of</u> C.

Lemma 2. <u>If $\mathfrak{M}$ is a regular map in the sphere without 2-cycles but with non-trivial 3-cycles then there is a 3-cycle with the property that if one of the regions of the 3-cycle is shrunk to a point the resulting map $\mathfrak{M}'$ has no non-trivial 3-cycles.</u>

Proof. Among all non-trivial 3-cycles choose one, C, such that the number of countries in one of its regions is minimal. If we shrink the other region to a point we will have a map $\mathfrak{M}'$ with no non-trivial 3-cycles. To see this, observe that in $\mathfrak{M}'$ if two countries meet on an edge then they meet on an edge in $\mathfrak{M}$ thus any non-trivial 3-cycle in $\mathfrak{M}'$ was a non-trivial 3-cycle in $\mathfrak{M}$. This non-trivial 3-cycle must have been in the union of C and its minimal region, but this implies that the region was not minimal.

Theorem 2. <u>Let $\mathfrak{M}$ be a regular map on the sphere with no 2-cycles, then $\mathfrak{M}$ can be 4-colored so that no country meets more than one other country of the same color.</u>

Proof. The proof is by induction on the number of countries. We may use the maps without non-trivial 3-cycles to start the induction. If $\mathfrak{M}$ has a non-trivial 3-cycle, we choose one, call it C, such that if we shrink one of its regions R_1 to a point p_1 the resulting map $\mathfrak{M}_1$ has no non-trivial 3-cycles. Let R_2 be the other region and let $\mathfrak{M}_2$ be the map obtained by shrinking R_2 to a point p_2. By induction $\mathfrak{M}_2$ can be given the desired 4-coloring. What we must now do is 4-color $\mathfrak{M}_1$ so that the colors meeting p_1 match the colors meeting p_2, so that we can induce a coloring in the original map where the colors of the countries meeting p_1 and p_2 will be the colors of the 3-cycle C. This, however, can be done by Lemma 1.

Theorem 1 now can be obtained by making the usual reduction, changing any map to a regular map by adding countries at vertices and then observing that the coloring of the regular map induces the desired coloring in the original map.

It is interesting that if we are restricted to three colors then the "pieces" of the map of any given color may get arbitrarily large, as this next theorem shows.

Theorem 3. Given any n there exists a map such that in any coloring of the vertices of the map with 3 colors a, b and c, there will be a path of at least n vertices of one color.

Proof. We construct the family of maps $\{\mathfrak{M}_i\}$ as follows: $\mathfrak{M}_1$ is the map in Fig. 1, $\mathfrak{M}_i$ is obtained from $\mathfrak{M}_{i-1}$ by dividing each bounded region into three regions (see Fig. 2).

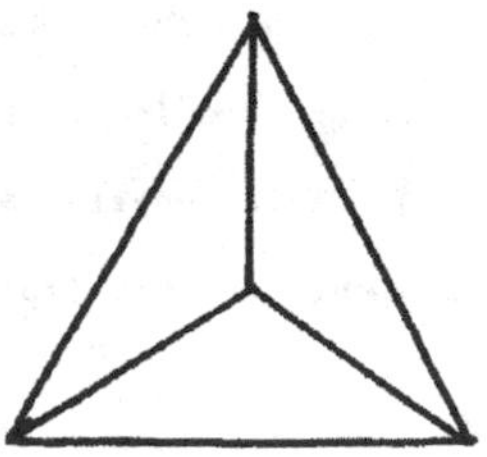

$\mathfrak{M}_1$

Fig. 1

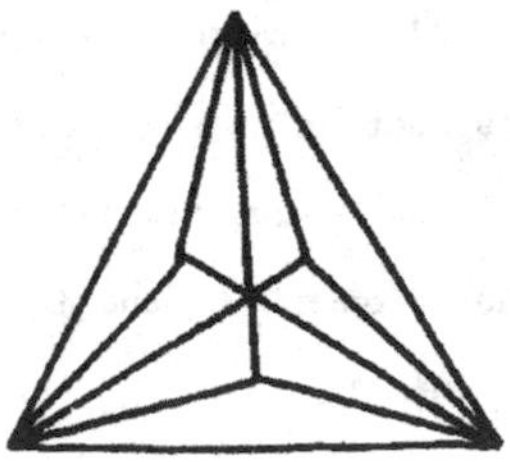

$\mathfrak{M}_2$

Fig. 2

If $i > n$ then at least two colors must be used on $\mathfrak{M}_i$, for otherwise $\mathfrak{M}_i$ would obviously contain a path of length n of one color.

Next we shall show that if $i > 3n$ then either the map has a triangle whose vertices are colored a, b and c, or there is a path of one color of length at least n (such a triangle will be called a complete triangle).

Suppose no such complete triangle exists in $\mathfrak{M}_i$, $i > 3n$. We consider $\mathfrak{M}_n$ which is a subgraph of $\mathfrak{M}_i$. We can assume $\mathfrak{M}_n$ has a country C_1 with vertices labeled a and b. In constructing $\mathfrak{M}_{n+1}$ we add a vertex v_1 inside C_1 and it is colored either a or b. This vertex together with two vertices of C determines a triangle C_2 with vertices labeled a and b. In $\mathfrak{M}_{n+2}$ there is a vertex v_2 inside C_2 that is labeled either a or b and it, together with two vertices of C_2, determines a triangle C_3 that is labeled a and b. In this way we have a sequence $v_1, v_2, \cdots, v_{2n}$ of vertices colored either a or b. The a vertices all lie on a path and so do the b vertices, thus we have a path of one color of length at least n.

Now we show that if $i > 6n$ then there is a path of one color of length at least n. We may now assume that for some $j \leq 3n$, $\mathfrak{M}_j$ has a country C_1 that is a complete triangle. To construct $\mathfrak{M}_{j+1}$ we must add a vertex v_1 inside C_1. This vertex together with two vertices of C_1 determines a complete triangle C_2. Similarly a vertex v_2 in C_2 together with two vertices of C_2 determines a complete triangle C_3. This gives us a sequence $v_1, v_2, \cdots, v_{i-j}$ with at least $3n$ elements. The vertices of any given color in this sequence lie on a path, thus one of these paths has length at least n.

3. B-sets. The next theorem relates Tutte's theorem to the existence of a B-set for a certain family.

Theorem 4. Let $\mathfrak{M}$ be a regular map in the plane with no non-trivial 2- or 3-cycles. Then there is a Tutte-circuit for $\mathfrak{M}$ if and only if the family of all cycles of length at least four has a B-set.

Proof. Assume that $\mathfrak{M}$ has the Tutte-circuit Γ. Let $\mathfrak{B}$ consist of the countries inside Γ. If $\mathfrak{B}$ contained a cycle, γ, of length at least four then γ would separate some vertex v from Γ. The bridge to which v belongs would not consist of v and the three edges incident to v. Thus neither $\mathfrak{B}$ nor the complement of $\mathfrak{B}$ contains a cycle of length at least four. Thus $\mathfrak{B}$ is a B-set for the cycles of length at least four.

Conversely, assume that the family of cycles of length at least four has the B-set $\mathcal{B}$. $\mathcal{B}$ is connected; for, if not, the complement $\mathcal{B}$ would contain a cycle that disconnected the sphere. Such a cycle would have length at least four (since the 2- and 3-cycles are degenerate). Thus $\mathcal{B}$ and its complement are connected and simply connected, hence cells.

Let the common border of $\mathcal{B}$ and its complement be the circuit Γ. We assert that Γ is a Tutte-circuit. For if some bridge had $k \geq 4$ vertices of attachment, there would be a cycle of k countries in $\mathcal{B}$ (or in its complement). This would contradict our choice of $\mathcal{B}$.

We should point out that the set $\mathcal{B}$ in the preceding theorem comes "fairly near" being a B-set for all cycles of length at least 3. To see this, note that a Tutte-circuit Γ passes through at least three-quarters of the vertices. The interior of Γ is a B-set for the family of three-cycles around the vertices through which Γ passes. As is pointed out in [1], the family of all cycles of length at least three has a B-set if and only if the edge graph has a Hamiltonian circuit.

4. Remarks.

(i) Theorem 1 may be proved for all maps by modifying Lemma 2 and Theorem 2.

(ii) We pose the following question: Does the family of all 3-cycles around vertices and 5-cycles around pentagons have a B-set?

References

1. S. K. Stein, B-sets and coloring problems, Bull. Amer. Math. Soc., 1970, pp. 304-305.

2. W. T. Tutte, A theorem on planar graphs, Trans. Amer. Math. Soc., 82 (1956), pp. 99-116.

DERIVED GRAPHS WITH DERIVED COMPLEMENTS

Lowell W. Beineke
Purdue University at Fort Wayne
Fort Wayne, Indiana 46805

The operation of forming the derived graph of a graph is probably the most interesting operation by which one graph is obtained from another. Furthermore, derived graphs are among the most interesting classes of graphs. They have been rediscovered many times and are known by many other names: line graph, interchange graph, adjoint, and edge-to-vertex dual. Characterizations and isomorphism theorems have provided much of the existing theory, and before studying those graphs which are derived and have derived complements, we give some of this background material.

DERIVED GRAPHS

The <u>derived graph</u> G' of a graph G has as its vertices the edges of G, with two vertices being adjacent if the corresponding edges are adjacent in G.

Figure 1 shows two graphs, the second of which is the derived graph of both. The following theorem of Whitney [6] shows that these graphs are unique in that respect.

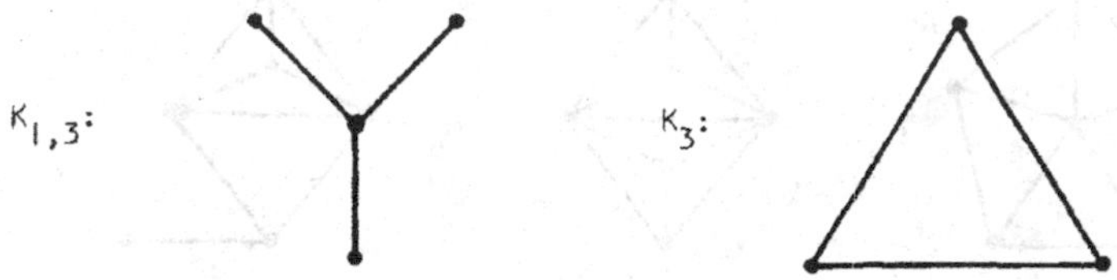

Figure 1

<u>Theorem 1</u>. The only different connected graphs with isomorphic derived graphs are $K_{1,3}$ and K_3, shown in Figure 1.

The next result has been rediscovered many times.

<u>Theorem 2</u>. A connected graph is isomorphic to its line graph if and only if it is a cycle.

The third theorem gives several characterizations of derived graphs; the first is due to Krausz [4], the second to van Rooij and Wilf [5], and the last to Beineke [2] and Robertson (unpublished). One definition is required: A triangle in a graph is <u>odd</u> if some other vertex is adjacent to an odd number of its vertices.

<u>Theorem 3</u>. The following are equivalent for a graph G:

(1) G is a derived graph.

(2) The edges of G can be partitioned into complete subgraphs in such a way that no vertex appears in more than two.

(3) G has no induced subgraph isomorphic to G_1 of Figure 2; and if uvw and vwx are distinct odd triangles, then u is adjacent to x.

(4) G has no induced subgraph isomorphic to any of the graphs in Figure 2.

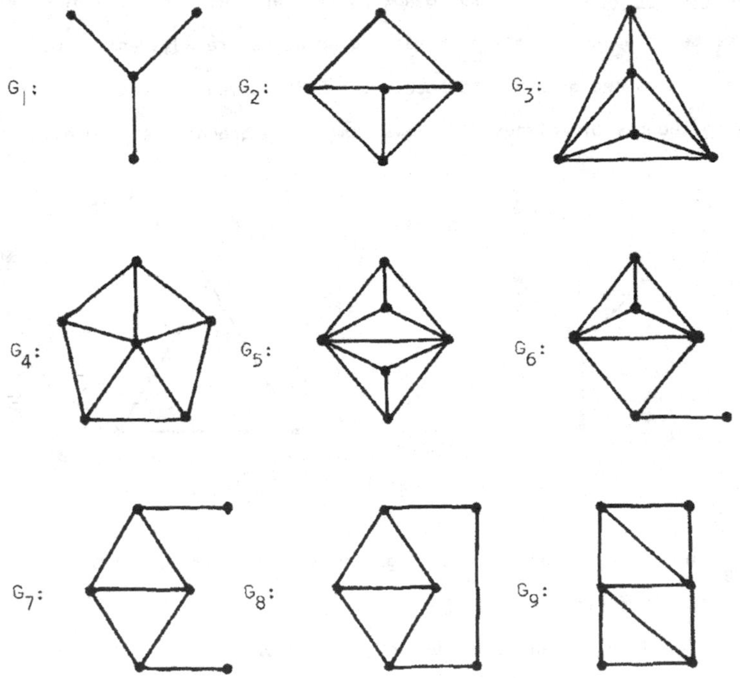

Figure 2

Of the many other results on derived graphs, there is only one which we want to mention now; it is due to Aigner [1] and is of the same nature as our results.

Theorem 4. If G is a graph with its derived graph isomorphic to its complement, then it is one of the two graphs in Figure 3.

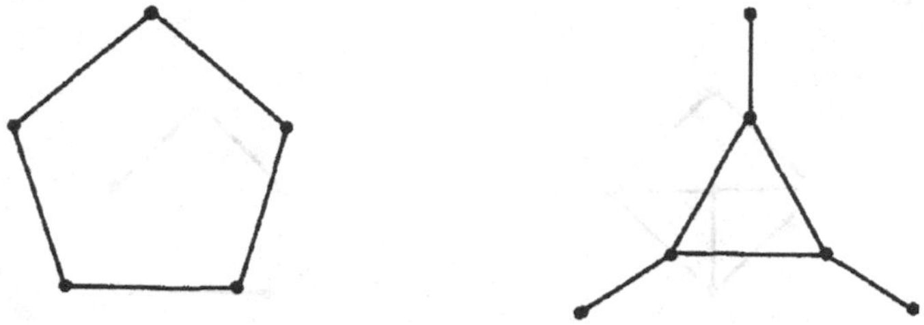

Figure 3

CODERIVED GRAPHS

We call a graph coderived if both it and its complement are derived graphs. Clearly both the complete graph K_p on p vertices and its complement the null graph with p vertices are derived. For, the star $K_{1,p}$ has K_p as its derived graph, and the graph with p components, each a copy of K_2, has $\overline{K}_p$. Thus K_p and $\overline{K}_p$ are coderived. We will show that there are only a finite number of other coderived graphs.

We observe that every induced subgraph of a derived graph must be derived, namely, from the subgraph with the corresponding edges of the original graph. As a consequence of Theorem 3, one can obtain the following "excluded subgraph" characterization of coderived graphs.

Theorem 5. A graph is coderived if and only if it has none of the ten graphs of Figure 4 as an induced subgraph.

Proof. From conditions (1) and (4) of Theorem 3, it follows that a graph is coderived if and only if neither it nor its complement has any of the nine graphs of Figure 2 as an induced subgraph. Equivalently, it is coderived if and only if it contains none of the same nine graphs or their complements as a derived subgraph. However, this set of eighteen graphs is not minimal. For, the complements of G_6,

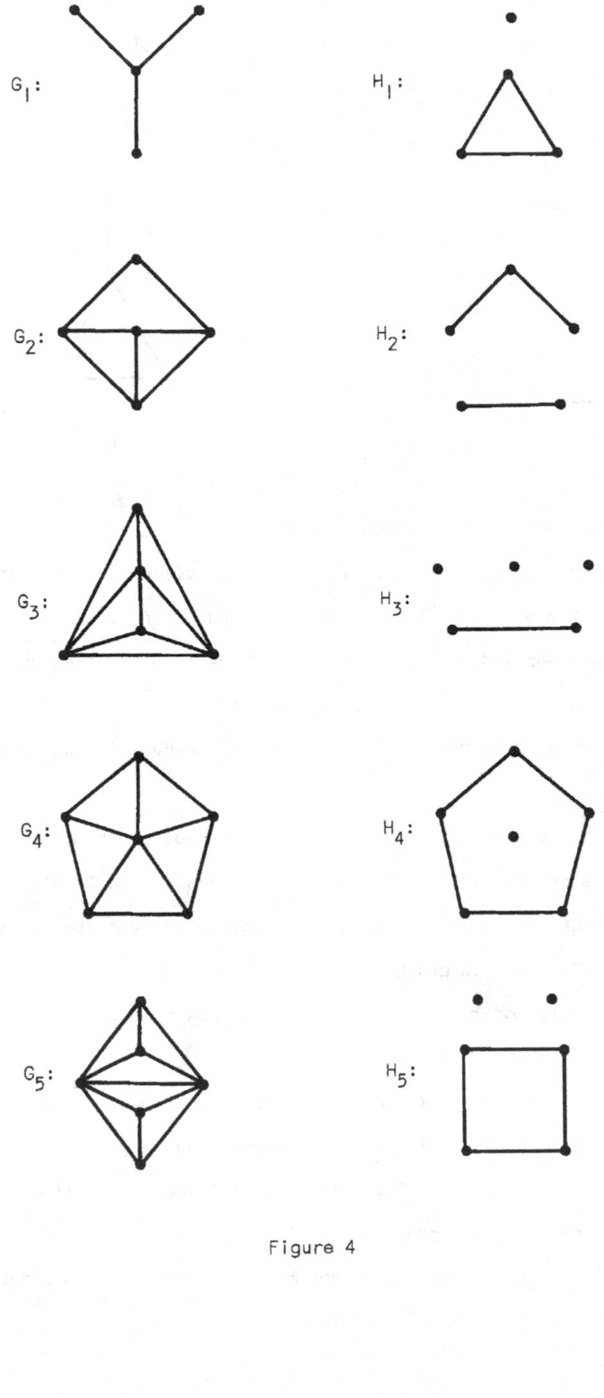

Figure 4

G_7, G_8, and G_9 all contain G_1 as an induced subgraph, and of course their complements similarly contain the complement of G_1. This leaves only the ten graphs, which completes the proof.

We shall now determine precisely which graphs are coderived. Clearly, for graphs with p vertices one need only consider those with up to one-half the possible number of edges; that is, at most $\frac{1}{4}p(p-1)$ edges. As mentioned earlier, all complete and null graphs are coderived, and so we shall only consider others here. There are of course no others with 1 or 2 vertices.

<u>3 to 6 vertices</u>

The coderived graphs with from 3 to 6 vertices are shown in Figure 5, in complementary pairs (a few are self-complementary). That these constitute all coderived graphs with up to 6 vertices is most easily seen by using Theorem 5 and exhaustively eliminating all others (with up to half the possible number of edges) from a complete list of graphs, as found in Harary [3, 215 ff.].

Since none of the ten graphs in Figure 4 has more than six vertices, we get a new criterion for coderived graphs with more vertices: A graph with p vertices, $p \geq 7$, is coderived if and only if each of its induced subgraphs with p-1 vertices is coderived.

<u>7 vertices</u>

Here we consider graphs with from 1 to 10 edges. Let G be such a coderived graph. Because H_3 (of Figure 4) is an excluded subgraph, there can be no coderived graph with 4, 5, or 6 components.

Suppose G has 3 components. Because of H_1, there can be no triangle. Since one component must have at least 3 vertices, neither of the others can have an edge, because of H_2. Thus, two of the components must consist of isolated vertices, and the third must be a connected coderived 5-vertex graph with no triangles. From Figure 5, we see there are only two possibilities: a path and a cycle. But then both of the corresponding 7-vertex graphs contain H_3, so there cannot be three components.

Now assume G has 2 components. Again, because of H_1 and H_2, there can be no triangles and one component must be a single vertex. From the connected coderived

6-vertex graphs in Figure 5, we see there is only one possibility: the other component must be a cycle of length 6. Such a graph is coderived; see Figure 6.

Now assume G is a connected graph. If it contains a 6-vertex cycle as an induced subgraph, then the other vertex must be adjacent to two consecutive vertices, because of G_1. But since that is the only connected coderived 6-vertex graph without a triangle, G itself must contain a triangle. Furthermore, because of H_1, every vertex must be adjacent to some vertex on each triangle, a very important observation.

Suppose G has two triangles abc and bcd with a common edge. If their other two vertices a and d are adjacent (that is, K_4 is present), then each of the other three vertices of G must be adjacent to at least two of those four vertices. But this implies the existence of at least twelve edges, so a and d cannot be adjacent. If one of the other three vertices were not adjacent to a or d, it would be adjacent to b or c and G_1 would be present. Hence, each of the other three must be adjacent to both a and d (a point of each triangle) and again there are too many edges. Therefore, there cannot be two triangles with a common edge.

Let abc denote a triangle of G. Two of the other four vertices must be adjacent to the same vertex of the triangle, and because of G_1, they must be adjacent to each other. Hence, there are triangles abc and cde. Because no two triangles can have a common edge, there are no other edges between them and no other edge at c (because of G_1). Therefore each of the other vertices must be adjacent to either a or b and to either d or e. Since this implies the existence of 10 edges, there can be no others, and hence the other two vertices cannot be adjacent to the same vertex. Thus a unique graph is determined; it is coderived and shown in Figure 6 with its complement.

8 vertices

Let G be a coderived graph with 8 vertices and at most 14 edges. As for the 7-vertex case, G cannot have more than two components, so suppose G has two. Then as before, neither can have a triangle and one must be an isolated vertex. It is clear from the 7-vertex coderived graphs that no such graph exists.

Suppose now G is connected. The disconnected 7-vertex coderived graph cannot be an induced subgraph since there would then be a vertex of degree 1 and none of the eligible 7-vertex graphs have such a vertex. Hence each of the 7-vertex subgraphs must be one of those with 10 or 11 edges. Neither of these has two triangles with a common edge, so G cannot either. Furthermore G cannot have a vertex of degree 2 or 5 since no 7-vertex coderived graph has one of degree 5. Therefore, if G contains the 7-vertex 10-edge graph as a subgraph, the 8'th vertex must be adjacent to the two vertices of degree 2. Because of G_1, it must also be adjacent to vertices adjacent to each of these. However, two of these cannot be adjacent because of the triangle property mentioned above. A unique 14 edge graph is therefore obtained. It is coderived and self-complementary; each of the two eligible 7-vertex graphs appear as subgraphs 4 times. It is shown in Figure 6.

<u>9 vertices</u>

All of its 8-vertex induced subgraphs must be isomorphic. The only way to achieve this is to add a new vertex adjacent to the four vertices of degree 3 in the 8-vertex coderived graph. This graph too is self-complementary, and is shown in Figure 6.

<u>Summary</u>

It is apparent that no new coderived graph with ten vertices can exist, for if one did, some vertex would have to have degree 5 in a coderived graph with 9 vertices. Therefore, we have the following theorem.

<u>Theorem 6</u>. A graph is coderived if and only if it is complete, is null, or is one of the graphs in Figures 5 or 6.

From the theorem and figures, one can make a number of interesting observations. There are precisely 37 coderived graphs other than the obvious classes of complete and null graphs. Of these, 26 are connected. There are just 6 self-complementary derived graphs, two with 5 vertices, and one each with 1, 4, 8, and 9 vertices.

As mentioned earlier, complete bipartite graphs $K_{1,p}$ and the graphs pK_2 consisting of p copies of K_2 have derived graphs which are coderived. The complete graph K_3 also has this property. We leave to the reader the construction of the 37 other graphs (unique up to isolates) having derived graphs which are coderived.

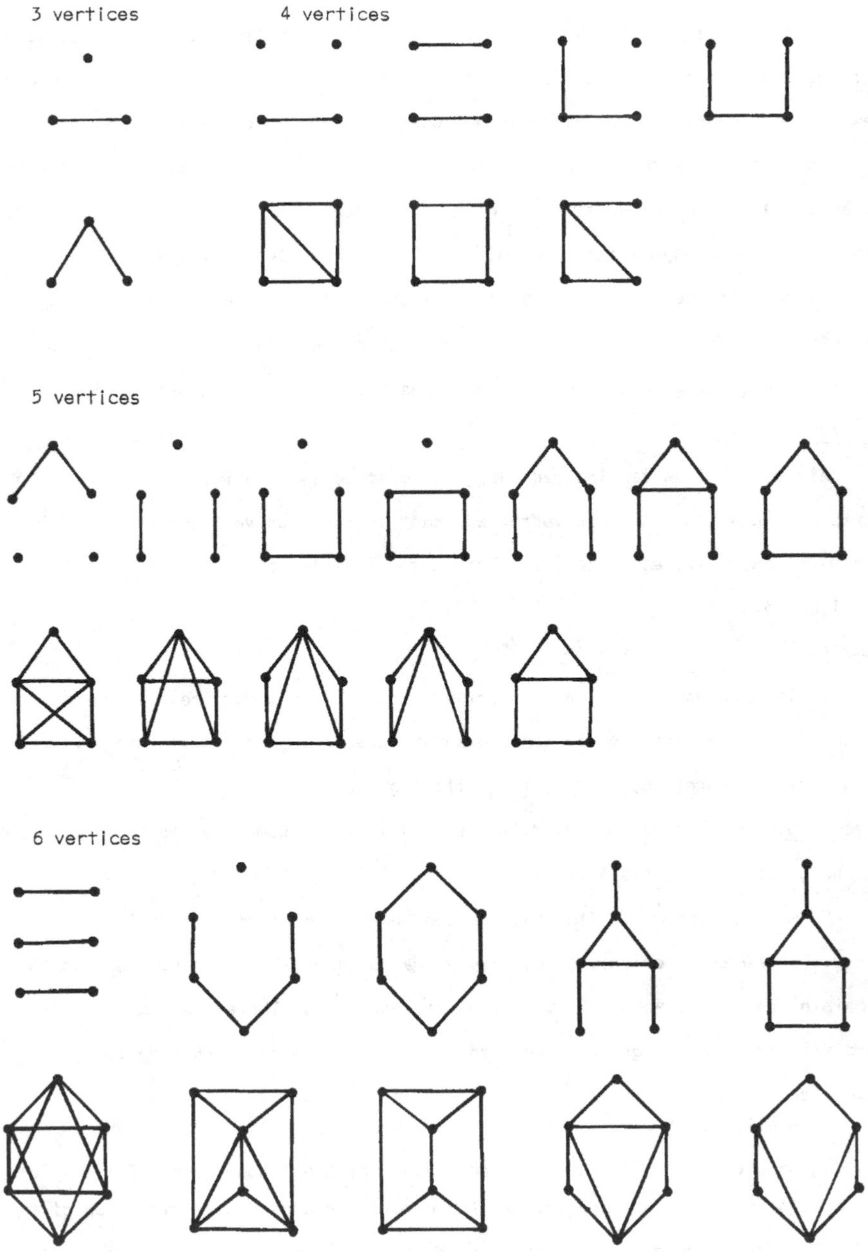

Figure 5

7 vertices

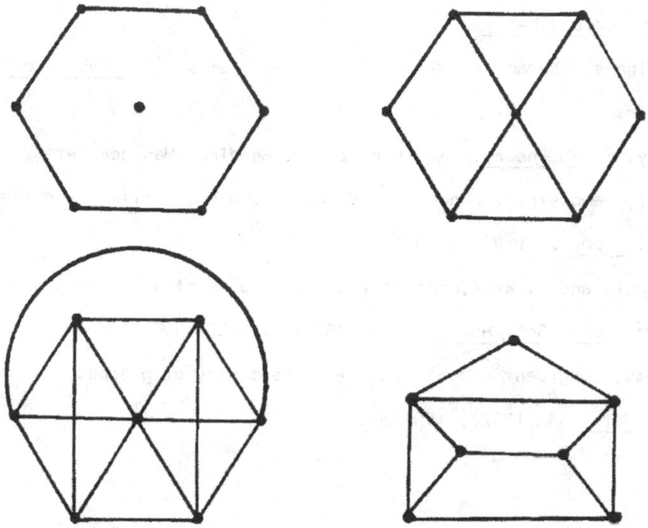

8 vertices

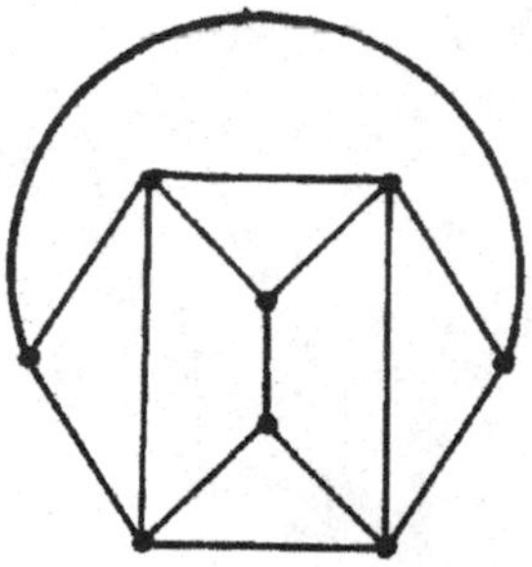

9 vertices

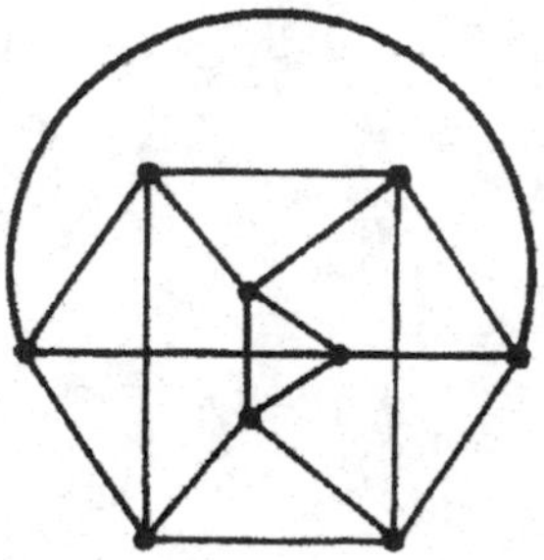

Figure 6

BIBLIOGRAPHY

1. M. Aigner, Graphs whose complement and line graph are isomorphic. J. Combinatorial Theory 7 (1969), 273-275.
2. L. W. Beineke, Characterizations of derived graphs. J. Combinatorial Theory (to appear).
3. F. Harary, Graph Theory. Addison-Wesley, Reading, Massachusetts, 1969.
4. J. Krausz, Demonstration nouvelle d'une theoreme de Whitney sur les reseaux. Mat. Fiz. Lapok 50 (1943), 75-89.
5. A. van Rooij and H. Wilf, the interchange graphs of a finite graph. Acta Math. Acad. Sci. Hungar. 16 (1965), 263-269.
6. H. Whitney, Congruent graphs and the connectivity of graphs. Amer. J. Math. 54 (1932), 150-168.

BINARY RELATIONS

Kim Ki-Hang Butler

Pembroke State University
Pembroke, North Carolina 28372

Binary relations on a finite set X are investigated with respect to the Green's equivalences on the semigroup B_X. Some counting results are obtained.

A binary relation on a set X is a subset X x X, and set of all such relations on X is denoted by B_X. Now B_X is a semigroup under the operation

$$AB = \{(a,b) \in X \times X: (a,c) \in A,\ (c,b) \in B \text{ for some } c \in X\}.$$

Let $X = \{x_1, x_2, \ldots, x_n\}, n > 1$. Then the map

$$A \longrightarrow (a_{ij}) = \begin{cases} 1 \text{ if } (x_i, x_j) \in A \\ 0 \text{ otherwise} \end{cases}$$

is an isomorphism of B_X onto the semigroup of all n x n matrices over the Boolean algebra on $\{0, 1\}$, under matrix multiplication. In this way every binary relation on X determines and is determined by an incidence matrix for a graph on X. In particular, we shall use the term graph and relation interchangeably. We use the notation and terminology of Clifford and Preston [4] unless stated otherwise. Let $R = \{0, 1\}$ be the set with two binary operations + and . on R as follows: $0 + 0 = 0$, $1 + 1 = 1 + 0 = 0 + 1 = 1$, $0 \cdot 0 = 1 \cdot 0 = 0 \cdot 1 = 0$, $1 \cdot 1 = 1$. Then $\{R, +, .\}$ is a binary Boolean semiring. A nonempty set V is said to be a semivector space over R if V is an abelian semigroup under an operation we denote +, and it must also satisfy the laws of scalar multiplication. Let $V_n(R)$ denote the set of all n-tuples over R. Then $V_n(R)$ is a semivector space. A set X in $V_n(R)$ is said to form a subspace of $V_n(R)$ if X is a semivector space. An element x in $V_n(R)$ is said to be linearly dependent on a set X in $V_n(R)$ if $x = x_1 + x_2 + \ldots + x_k$ for some $x_i \in X$; otherwise, x is called linearly independent.

A set X in $V_n(R)$ is said to be linearly dependent if there exists an element x in X which is linearly dependent on $X - x$; otherwise, X is called linearly independent. There is a set of vectors which forms a basis, i.e., they are linearly independent and span the space. If x in $V_n(R)$ then x ($x \neq 0$) takes the form $x = e_i + e_j + \dots + e_m$, $1 \leq i, j, \dots, m \leq n$, where e_i denotes the n-tuples in which the ith coordinates is 1 and all other coordinates are 0. We define $\rho(x)$ by $\rho(x) = \rho(e_i + e_j + \dots + e_m) = k$. If $|\{i, j, \dots, m\}| = k$, then $\rho(x)$ is called the rank of x. $|X|$ denotes the cardinality of a set X. We need the following theorem.

THEOREM 1. Let X be a subspace of $V_n(R)$. If X has bases B_1 and B_2, then $B_1 = B_2$.

Proof. Before proceeding we define $N = \{1, 2, \dots, n\}$. Let $a, a_i \in V_n(R)$ such that $a = a_1 + a_2 + \dots + a_n$. Then we claim

(1) $\rho(a) = 0 \iff a = 0$,

(2) $\rho(a) \geq \rho(a_i)$ for every i,

(3) $\exists\, i_0 \in N \ni \rho(a) = \rho(a_{i_0}) \Rightarrow a = a_{i_0}$.

The proof of (1) is trivial. To prove (2), we will use mathematical induction on i. We shall show that $\rho(a_1 + a_2) \geq \rho(a_i)$, $i = 1, 2$. Let $a_i = (a_{1i}, a_{2i}, \dots, a_{ni}) \in V_n(R)$, $i = 1, 2$. Then $a_1 + a_2 = (a_{11} + a_{12}, a_{21} + a_{22}, \dots, a_{n1} + a_{n2})$. It follows that if we set $a_{ki} = 1$ then $a_{k1} + a_{k2} = 1$, $k \in N$. Therefore, we have $\rho(a_1 + a_2) \geq \rho(a_i)$, $i = 1, 2$. To prove (3), let $a = a_1 + a_2$. Then $\rho(a) = \rho(a_1)$. Set $a = (c_1, c_2, \dots, c_n) \in V_n(R)$. Then $c_i = a_{i1} + a_{i2} \Rightarrow a_{i1} = 1 \Rightarrow c_i = 1$. Furthermore, $\rho(a) = \rho(a_1)$ $\iff a_{i1} = 1 \iff c_i = 1 \iff a = a_1$. We define

$$m.r.X = \min \{\rho(a) : a \in X\},$$

$$m.v.X = \{a \in X : \rho(a) = m.r.X\},$$

$$B^r = \{a \in B : \rho(a) \leq r,\ B = V_n(R)\}.$$

Then we claim (4) $m.r.B_1 = r = m.r.B_2$, (5) $B_1^r = B_2^r$, (6) $B_1^k = B_2^k$ for

for all $k \in N$. To prove (4), let $m.r.B_1 < m.r.B_2$. Let $u \in B_1$ such that $\rho(u) = m.r.B_1$ but $u = b_1 + b_2 + \dots + b_k$ for every $b_j \in B_2$. Then by (2) we have $m.r.B_1 = \rho(u) \geqq \rho(b_j) \geqq m.r.B_2$ for all j. Which is a contradiction. To prove (5), let $u \in B_1^r$. Then $u = b_1 + b_2 + \dots + b_k$ for every $b_j \in B_2 \Rightarrow r = m.r.B_1 = \rho(u) \geqq \rho(b_j) \geqq m.r.B_2 = r \Rightarrow \rho(u) = \rho(b_j) \Rightarrow u = b_1 = b_2 = \dots = b_k$. Hence $B_1^r \subseteq B_2^r$. Similarly, we have $B_2^r \subseteq B_1^r$. To prove (6), we assume true for $k = p$ and $B_1^{p+1} - B_1^p \neq \phi$ (if $B_1^{p+1} - B_1^p = \phi$ then by the following argument $B_2^{p+1} - B_2^p \neq \phi$). Let $v \in B_1^{p+1} - B_1^p$ and $v = b_1 + b_2 + \dots + b_k$. Then $(p + 1) = \rho(v) \geqq \rho(b_j)$ for every j. Lastly, we shall show that there exists $j_o \in N$ such that $\rho(b_{j_o}) = (p + 1)$. We have $\rho(b_j) \leqq p$ for all $j \Rightarrow b_j \in B_2^p$ for all $j \Rightarrow b_j \in B_1^p \subseteq B_1$ for all $j \Rightarrow v$ is linearly dependent on $B_1 - v$ $\Rightarrow B_1$ is not a basis. Which is a contradiction. Hence there exists $j_o \in N$ such that $\rho(b_{j_o}) = (p + 1) = \rho(v)$, which in turn implies that $v = b_{j_o}$ by (3). Hence $B_1^{p+1} \subseteq B_2^{p+1}$. Similarly, we have $B_2^{p+1} \subseteq B_1^{p+1}$, and therefore $B_1^{p+1} = B_2^{p+1}$. We immediately conclude that $B_1 = B_2$, and $B_i = B_i^n$ for all i and $n \in N$. This proves the theorem. The following is a dual of Theorem 1. We define $V^n(R) = \{x^t : x \in V_n(R)\}$, where x^t denotes the transpose of u.

THEOREM 1′. Let X' be a subspace of $V^n(R)$. If X' has two bases B_1' and B_2', then $B_1' = B_2'$.

Proof. The proof is similar to that of Theorem 1.

Let $M_{m,n}(R)$ denote the set of all $m \times n$ matrices over R. In case $m = n$, denoted by $M_n(R)$. Then $M_n(R)$ is a (0,1)-matrix semigroup under matrix multiplication. Let $A \in M_{m,n}(R)$. With A we associate the following sets: (i) the row space $R(A) = \{xA : x \in V_m(R)\}$, (ii) the column space $C(A) = \{Ay : y \in V^n(R)\}$. We give some properties of the row and column spaces. Let A_{i*} (A_{*i}) denote the ith row (column) of A.

LEMMA 2. Let $A \in M_n(R)$. If $u_iA = w_i$ for $u_i \in V_n(R)$ $(i = 1, 2)$,

<u>then</u> $(u_1 + u_2)A = w_1 + w_2$.

Proof. It suffices to show that for any pair e_i and e_j of $V_n(R)$ we have $(e_i + e_j)A = e_iA + e_jA = A_{i*} + A_{j*}$. Dual proof holds for the column space.

<u>LEMMA 2'</u>. <u>Let</u> $A \varepsilon M_n(R)$. <u>If</u> $Au_i' = w_i'$ <u>for</u> $u_i' \varepsilon V^n(R)$ $(i = 1, 2)$, <u>then</u> $A(u_1' + u_2') = w_1' + w_2'$.

<u>LEMMA 3</u>. <u>If</u> $A \varepsilon M_n(R)$, <u>then</u> $R(A)$ <u>is a subspace of</u> $V_n(R)$.

Proof. Let $x, y \varepsilon R(A)$ such that $x = uA$ and $y = vA$ for $u, v \varepsilon V_n(R)$. Then by Lemma 2, $x + y = uA + vA = (u + v)A = wA \varepsilon R(A)$ $(w \varepsilon V_n(R))$. Hence $R(A)$ is a subspace of $V_n(R)$. Dual proof holds for the column space.

<u>LEMMA 3'</u>. <u>If</u> $A \varepsilon M_n(R)$, <u>then</u> $C(A)$ <u>is a subspace of</u> $V^n(R)$.

We shall be making extensive and constant use of the following two lemmas.

<u>LEMMA 4</u>. <u>If</u> $A \varepsilon M_n(R)$ <u>and</u> $B \varepsilon M_n(R)$, <u>then</u> $R(AB) \subseteq R(B)$.

Proof. Since the rows of AB are linear combinations of the rows of B, $uAB = (uA)B$ for every nonzero vector $u \varepsilon V_n(R)$. Dual proof holds for the column space.

<u>LEMMA 4'</u>. <u>If</u> $A \varepsilon M_n(R)$ <u>and</u> $B \varepsilon M_n(R)$, <u>then</u> $C(AB) \subseteq C(A)$.

An immediate consequence of Theorem 1 is that the row space $R(A)$ of a matrix $A \varepsilon M_n(R)$ has a unique basis called the row basis of A and is denoted by $B_r(A)$. The cardinality of this basis is called the row rank of A, denoted by $\rho_r(A)$. The definition of the column basis and column rank are dual. $B_c(A)$ ($\rho_c(A)$) denotes column basis (column rank) of A. If $\rho_r(A) = \rho_c(A)$, then the common value of the row and column rank is called the rank of A and denoted by $\rho(A)$.

We remark here that $A \varepsilon M_n(R)$ need not have $\rho_r(A) = \rho_c(A)$. For example, if

$$A = \begin{bmatrix} 1 & 0 & 1 & 0 \\ 1 & 1 & 0 & 0 \\ 0 & 1 & 0 & 0 \\ 0 & 0 & 1 & 0 \end{bmatrix},$$

then $\rho_r(A) = 4$ and $\rho_c(A) = 3$.

<u>THEOREM 5</u>. <u>Let</u> $A \in M_n(R)$. (i) $\rho_r(A) = 1 \iff \rho_c(A) = 1$, (ii) $\rho_r(A) = 2 \iff \rho_c(A) = 2$, (iii) $\rho_r(A) = 3 \Rightarrow 3 \leq \rho_c(A) \leq 4$, $\rho_c(A) = 3 \Rightarrow 3 \leq \rho_r(A) \leq 4$.

Proof. (i) Trivial. (ii) We define binary operation $\otimes$ on $V_n(R)$ as follows: $x \otimes y = (x_1y_1, x_2y_2, \ldots, x_ny_n) \in V_n(R)$.

<u>Necessity:</u> Let $B_r(A) = \{v_1, v_2\}$. Let $M_o = \{h \in N: A_{h*} = 0\}$, $M_1 = \{h \in N: A_{h*} = v_1\}$, $M_2 = \{h \in N: A_{h*} = v_2\}$, and $M_3 = \{h \in N: A_{h*} = v_1 + v_2\}$. Let $N_1 = \{h \in N: e_h \otimes v_1 = e_h\}$, $N_2 = \{h \in N: e_h \otimes v_2 = e_h\}$, and $N_3 = N_1 \cap N_2$. Then we notice that the M_i's form a disjoint partition of N, because $\rho_r(A) = 2$. If $q \in N_3$, then the elements of A_{*q} can be written as follows: $a_{jq} = 0$ for $j \in M_o$ and $a_{jq} = 1$ for $j \notin M_o$. If $q \in N_1 - N_3$, then the elements of A_{*q} can be written as follows: $a_{jq} = 1$ for $j \in M_1 \cup M_3$ and $a_{jq} = 0$ for $j \notin M_1 \cup M_3$. If $q \in N_2 - N_3$, then the elements of A_{*q} can be written as $a_{jq} = 1$ for $j \in M_2 \cup M_3$ and $a_{jq} = 0$ for $j \notin M_2 \cup M_3$. This proves $\rho_c(A) \leq 2$. Thus clearly $\rho_c(A) \geq 1$ but $\rho_c(A) \neq 1$, because $\rho_r(A) \neq 1$. Hence $\rho_c(A) = 2$.

<u>Sufficiency:</u> Take the transpose of A.

(iii) The proof is somewhat long-winded. We restrict ourselves to an outlines of the proof. For more complete discussion see [1]. Whenever the indices i, j, and k appear in the same expression we assume that they are all distinct. Let $B_r(A) = \{v_1, v_2, v_3\}$. Let $N_i = \{h \in N: e_h \otimes v_i = e_h\}$, $i = 1, 2, 3$. We shall list all possible relations between the N_i's, $i = 1, 2, 3$. The proof will be consist of four cases and some subcases.

<u>Case (1):</u> $N_{\sigma(1)} \not\subseteq (N_{\sigma(2)} \cup N_{\sigma(3)})$ for every $\sigma \in P_3$, where P_3 denotes the set of all permutation on 3 symbols $\Rightarrow \rho_c(A) = 3$.

<u>Case (2):</u> $N_i \subseteq (N_j \cup N_k)$, $N_j \not\subseteq (N_i \cup N_k)$, and $N_k \not\subseteq (N_i \cup N_j)$.

<u>Subcase (1):</u> $N_i \subseteq (N_j \cap N_k) \Rightarrow \rho_c(A) = 3$,

<u>Subcase (2):</u> $N_k \not\supseteq N_i \subseteq N_j \Rightarrow \rho_c(A) = 3$,

<u>Subcase (3):</u> $N_k \not\supseteq N_i \not\subseteq N_j \Rightarrow \rho_c(A) = 4$.

<u>Case (3):</u> $N_i \subseteq (N_j \cup N_k)$, $N_j \subseteq (N_i \cup N_k)$, $N_k \not\subseteq (N_i \cup N_j)$.

<u>Subcase (1):</u> $N_i \subset N_j$ and $N_j \subset N_k \Rightarrow \rho_c(A) = 3$,

<u>Subcase (2):</u> $N_i \not\subset N_j \not\subset N_i$ but $N_i \subseteq N_k \Rightarrow \rho_c(A) = 3$,

<u>Subcase (3):</u> $N_i \subset N_j$ and $N_j \not\subset N_k \Rightarrow \rho_c(A) = 3$,

<u>Subcase (4):</u> $N_i \not\subset N_j$ for every $i \neq j \Rightarrow \rho_c(A) = 4$.

<u>Case (4):</u> $N_i \subseteq (N_j \cup N_k)$, $N_j \subseteq (N_i \cup N_k)$, $N_k \subseteq (N_i \cup N_j)$ $\Rightarrow \rho_c(A) = 3$. This completes the proof.

<u>LEMMA 6</u>. <u>Let</u> $A \in M_{m,n}(R)$ <u>and</u> $A^\circ \in M_{m+1,n}(R)$ <u>such that</u> $A^\circ_{i*} = A_{i*}$ <u>for every</u> $1 \leq i \leq m$ <u>and</u> $A^\circ_{m+1*} \in R(A)$. <u>Then</u>

(i) (a) $A_{*q} = \sum_{j\in\Gamma} A_{*i_j} \iff$ (b) $A^\circ_{*q} = \sum_{j\in\Gamma} A_{*i_j}$, $\Gamma \subseteq N = \{1,2,\ldots,n\}$,

(ii) $\rho_c(A) = \rho_c(A^\circ)$.

Proof. Our proof of theorem will proceed in turn.

(i) <u>Sufficiency:</u> $A^\circ_{*q} = \sum_{j\in\Gamma} A^\circ_{*i_j} \Rightarrow a^\circ_{pq} = \sum_{j\in\Gamma} a_{pi_j}$ for $1 \leq p \leq m+1$

$\Rightarrow a_{pq} = \sum_{j\in\Gamma} a_{pj}$ for $1 \leq p \leq m$, because $a^\circ_{ij} = a_{ij}$ for $i \leq m$.

Hence $A_{*q} = \sum_{j\in\Gamma} A_{*i_j}$.

<u>Necessity:</u> From the preceding computation we see that

$$a^\circ_{pq} = \sum_{j\in\Gamma} a^\circ_{pj},\ 1 \leq p \leq m.$$

Hence we need only consider $a_{m+1,q}$. Since $A^\circ_{m+1*} \in R(A)$, it is clear that

(K) $\quad A^\circ_{m+1*} = \sum_{h\in\Psi} A_{h*}$, $\Psi \subseteq M = \{1, 2, \ldots, m\}$.

Suppose $a_{m+1,q} = 0$. Then by (K), $a_{hq} = 0$ for every $h \in \Psi$. From (a) we observe that $a_{hj} = 0$ for every $h \in \Psi$ and $j \in \Gamma$. Again, by (K), $a_{m+1,j} = 0$ for every $j \in \Gamma$, and thus

$$0 = a^\circ_{m+1,q} = \sum_{j\in\Gamma} a^\circ_{m+1,j} = 0.$$

Now suppose $a_{m+1,q} = 1$. Then by (K) there exists $h_\circ \in \Psi$ such that $a_{h_\circ q} = 1$ and by (a) there exists $j_\circ \in \Gamma$ such that $a_{h_\circ j_\circ} = 1$. Again, by (K), $a_{m+1,j_\circ} = 1$. Hence $a_{m+1,q} = \sum_{j\in\Gamma} a_{n+1,j}$.

(ii) Since the basis of C(A) and C(A°) correspond in 1-to-1 fashion,

$$\rho_c(A) = \rho_c(A^\circ).$$

This completes the proof.

<u>LEMMA 7.</u> <u>If</u> A, B ε $M_n(R)$ <u>and</u> R(A) = R(B), <u>then</u>

$$\rho_c(A) = \rho_c(B).$$

Proof. By permutation and induction.

We now proceed to give a characterization of the Green's relations that will be useful in later work. For simplicity we let $S = M_n(R)$. Two elements A, B of a semigroup S are said to be $\mathcal{L}$ ($\mathcal{R}$)-equivalent iff they generate the same principal left (right) ideal in S. We denote the relation $\mathcal{L} \cap \mathcal{R}$ by $\mathcal{H}$ and join $\mathcal{L} \vee \mathcal{R}$ of $\mathcal{L}$ and $\mathcal{R}$ by $\mathcal{D}$, that is, $\mathcal{D}$ is the intersection of all the equivalence relations on S that contain $\mathcal{L}$ and $\mathcal{R}$.
By L_A (R_A, H_A, D_A) we mean that the set of all elements of S which are $\mathcal{L}$ ($\mathcal{R}$, $\mathcal{H}$, $\mathcal{D}$)-equivalent to A.

<u>THEOREM 8.</u> <u>If</u> A, B ε S, <u>then</u> (i) $A \mathcal{L} B \Leftrightarrow R(A) = R(B)$, (ii) $A \mathcal{R} B \Leftrightarrow C(A) = C(B)$, (iii) $A \mathcal{D} B \Leftrightarrow \exists\, C \in S \ni R(A) = R(C)$ <u>and</u> $C(C) = C(B)$.

Proof. We prove each of these in turn.

(i) <u>Necessity:</u> $A \mathcal{L} B \Rightarrow SA \cup A = SB \cup B \Rightarrow \exists\, X, Y \in S \ni XA = B$ and $YB = A$ but $R(B) = R(XA) \subseteq R(A)$ and $R(A) = R(YB) \subseteq R(B)$.

<u>Sufficiency:</u> Assume R(A) = R(B). Then by Theorem 1, both row spaces R(A) and R(B) has unique basis. Let $C = \{c_1, c_2, \ldots, c_k\}$, $c_i \in V_n(R)$, i = 1, 2, ..., k, be the nonzero linearly independent rows of A and B in some order. If we set $G = [c_1, c_2, \ldots, c_k, 0, \ldots, 0]^t_{nxn}$, then there exist $X_1, X_2, Y_1, Y_2 \in S$ such that $G = X_1A$, $A = X_2G$, $G = Y_1B$, and $B = Y_2G$. It follows that $A \mathcal{L} G$ and $G \mathcal{L} B$. Hence $A \mathcal{L} B$. (ii) Dual proof holds for the column space.
(iii) The proof follows from definition of $\mathcal{D}$ and (i) and (ii).
This proves the theorem.

<u>THEOREM 9.</u> <u>If</u> A <u>and</u> B <u>are</u> <u>elements</u> <u>of</u> <u>a</u> <u>semigroup</u> S, <u>then</u>

$A \mathcal{D} B \Rightarrow \rho_r(A) = \rho_r(B)$ and $\rho_c(A) = \rho_c(B)$.

Proof. Follows from Lemmas 6 and 7 and Theorem 8.

<u>LEMMA 10.</u> <u>If</u> A <u>and</u> B <u>are elements of a semigroup</u> S, <u>then</u> $A \mathcal{D} B \Leftrightarrow \exists\, X_1, X_2, Y_1, Y_2 \in S \ni$ (i) $Y_1X_1A = A$, (ii) $X_1Y_1B = B$, (iii) $AX_2Y_2 = A$, (iv) $BY_2X_2 = B$, (v) $X_1AX_2 = B$, (vi) (i)-(v) $\Rightarrow Y_1BY_2 = A$.

Proof. <u>Necessity:</u> $A \mathcal{D} B \Rightarrow \exists\, C \in S \ni A \mathcal{L} C$ and $C \mathcal{R} B \Leftrightarrow \exists\, X, Y, Z, W \in S \ni XA = C$, $YC = A$, $CZ = B$, and $BW = C$. Let $X_1 = X$, $X_2 = Z$, $Y_1 = Y$, and $Y_2 = W$. Then it is easy to see that (i)-(v) holds, because (i) $Y_1X_1A = YXA = YC = A$, (ii) $X_1Y_1B = XYB = XYCZ = X(YC)Z = (XA)Z = CZ = B$, (iii) as (i), (iv) as (ii), and (v) $X_1AX_2 = XAZ = CZ = B$.

<u>Sufficiency:</u> If (i)-(v), then $Y_1BY_2 = Y_1X_1AX_2Y_2 = AX_2Y_2 = A$. Furthermore, we observe that if (i)-(iv) and (vi) then (v). Let $C = X_1A$. Then $C = BY_2$. Hence $A \mathcal{L} C$ and $C \mathcal{R} B$. By (iii) and (v), $BY_2 = (X_1AX_2)Y_2 = X_1(AX_2Y_2) = X_1A$. By (i), $Y_1X_1A = Y_1C = A$. Since $C = X_1A$, $A \mathcal{L} C$. By (i) and (iv), $B = BY_2X_2 = CX_2$. By Theorem 8 (i), $C \mathcal{R} B$. This completes the proof.

<u>THEOREM 11.</u> $\mathcal{D}$ <u>on</u> S <u>is an equivalence relation.</u>

Proof. Let $A, B, C \in S$. (i) $A \mathcal{D} A$. Obvious. (ii) $A \mathcal{D} B \Rightarrow B \mathcal{D} A$. Let $X_1' = Y_1$, $Y_1' = X_1$, $X_2' = Y_2$, and $Y_2' = X_2$. Then use Lemma 10. (iii) $A \mathcal{D} B$ and $B \mathcal{D} C \Rightarrow A \mathcal{D} C$.
Let $G, F, J, K, X, Y, Z, W \in S$. Then $YXA = A$, $XYB = B$, $AZW = A$, $BWZ = B$, $XAZ = B$, $YBW = A$, $GFB = B$, $BJK = B$, and $FBJ = C$.
Let $X_1 = FX$, $X_2 = ZJ$, $Y_1 = YG$, and $Y_2 = KW$. Then

$$Y_1X_1A = Y_1X_1AZW = YGFXAZW = Y(GFB)W = YBW = A,$$

$$AX_2Y_2 = YXAX_2Y_2 = YXAZJKW = Y(BJK)W = YBW = A,$$

$$X_1AX_2 = FXAZJ = FBJ = C.$$

Hence $A \mathcal{D} C$. This completes the proof.

We define $D_r^n = \{A \in S:\ \rho(A) = r\}$.

LEMMA 12. If $A \in S$ has rank 1 iff there exist nonzero vectors

$u \in V^n(R)$ and $v \in V_n(R)$ such that $A = uv$.

Proof. Since $\rho_r(A) = 1$, the cardinality of $R(A)$ is 1 and thus every nonzero row of A is equal; let this row be denoted by v. Then $A_{i*} = u_i v$ ($u_i \in R$, $i = 1, 2, \ldots, n$). Now setting $u = (u_1, u_2, \ldots, u_n)^t$. Then $A = uv$. Reverse the argument to prove the converse. This completes the proof.

<u>LEMMA 13.</u> <u>If</u> $u, x \in V^n(R)$ <u>and</u> $v, y \in V_n(R)$, <u>then</u> $(uv)(xy) = u(vx)y$.

PROOF. Let $u = (u_1, u_2, \ldots, u_n)^t$, $x = (x_1, x_2, \ldots, x_n)^t$, $v = (v_1, v_2, \ldots, v_n)$, and $y = (y_1, y_2, \ldots, y_n)$. Then $uv = (u_i v_j)_{nxn}$ and $xy = (x_i y_j)_{nxn}$. Thus we obtain

$$(uv)(xy) = (\sum_{j=1}^{n} (u_i v_j x_j y_k))_{nxn} = (\sum_{j=1}^{n} v_j x_j)_{1x1}(u_i y_k)_{nxn},$$

$$u(vx)y = (vx)_{1x1}(uy)_{nxn} = (\sum_{j=1}^{n} v_j x_j)_{1x1}(u_i y_k)_{nxn}.$$

This proves the lemma.

<u>THEOREM 14.</u> D_1^n <u>is a</u> $\mathcal{D}$<u>-class.</u>

Proof. Let $A, B \in D_1^n$. Then by Lemma 12, $A = uv$ and $B = xy$ where $u, x \in V^n(R)$ and $v, y \in V_n(R)$. If we set $C = xv$ then we have $R(A) = R(C)$ and $C(B) = C(C)$. By Theorem 8, $A \mathcal{L} C$ and $C \mathcal{R} B$. Hence $A \mathcal{D} B$. This completes the proof.

<u>THEOREM 15.</u> $|D_1^n| = (2^n - 1)^2$.

Proof. Without loss of generality, we assume that $|V_n(R)| = 2^n$, because we have two different choices, i.e., 0 or 1, for each position of each row vector. Hence we obtain $|V_n(R) \setminus \{0\}| = 2^n - 1$. But $V_n(R) \setminus \{0\} = \{R(A): A \in D_1^n\}$. Similarly, we have $|V^n(R) \setminus \{0\}| = 2^n - 1$, but $V^n(R) \setminus \{0\} = \{C(A): A \in D_1^n\}$. But $D_1^n = (V^n(R) \setminus \{0\}) \times (V_n(R) \setminus \{0\})$. Hence $|D_1^n| = (2^n - 1)^2$. This proves the theorem.

Let $A \in D_2^n$. Then (i) the row space $R(A)$ is said to be of type I if whenever $B_r(A) = \{v_1, v_2\}$ then $v_1 \neq v_1 + v_2 \neq v_2$. Similarly, the column space $C(A)$ is said to be of type I if whenever $B_c(A) = \{u_1, u_2\}$ then $u_1 \neq u_1 + u_2 \neq u_2$. (ii) the row space $R(A)$ is said to be of type II if whenever $B_r(A) = \{v_1, v_2\}$ then $v_1 + v_2 = v_1$ or v_2. Similarly, the column space $C(A)$ is said to be of type II if whenever $B_c(A) = \{u_1, u_2\}$ then $u_1 + u_2 = u_1$ or u_2. It is easy to see that (i) $R(A)$ is of type I $\Rightarrow |R(A)| = 4$, (ii) $R(A)$ is of type II $\Rightarrow$ $|R(A)| = 3$.

LEMMA 16. *If* $A \in D_2^n$, *then* (i) $R(A)$ *is of type* I $\Rightarrow C(A)$ *is of type* I, (ii) $R(A)$ *is of type* II $\Rightarrow C(A)$ *is of type* II.

Proof. Let $B_r(A) = \{v_1, v_2\}$ such that $v_1 \neq v_1 + v_2 \neq v_2$.
Let $M_0 = \{h \in N: A_{h*} = 0\}$, $M_1 = \{h \in N: A_{h*} = v_1\}$,
$M_2 = \{h \in N: A_{h*} = v_2\}$, and $M_3 = \{h \in N: A_{h*} = v_1 + v_2\}$.
Let $N_1 = \{h \in N: e_h \otimes v_1 = e_h\}$, $N_2 = \{h \in N: e_h \otimes v_2 = e_h\}$, and $N_3 = N_1 \cap N_2$. Without loss of generality, we assume that there exist $i, j \in N$ such that $i \in M_1 \neq \phi$ and $j \in M_2 \neq \phi$ for every $i \neq j$, because $M_1 \cap M_2 = \phi$. Suppose $p \in N_1 - N_3$ and $q \in N_2 - N_3$. Then the elements of A_{*p} and A_{*q} can be written $a_{jp} = 1 \Leftrightarrow j \in M_1 \cup M_3$ and $a_{jq} = 1 \Leftrightarrow j \in M_2 \cup M_3$. Together with information we already have obtained, $A_{*p}, A_{*q} \in C(A)$. In particular,

$$a_{ip} = 1 \text{ and } a_{jp} = 0 \text{ for every } i \neq j,$$
$$a_{iq} = 0 \text{ and } a_{jq} = 1 \text{ for every } i \neq j,$$
$$a_{i(p+q)} = 1 = a_{j(p+q)} \text{ for every } i \neq j,$$

where $a_{ip}, a_{jp} \in A_{*p}$, $a_{iq}, a_{jq} \in A_{*q}$, and $a_{i(p+q)}, a_{j(p+q)} \in A_{*p} + A_{*q}$ Hence $A_{*p} \neq A_{*p} + A_{*q} \neq A_{*q}$. Since $\rho_r(A) = 2$, $\rho_c(A) = 2 \Rightarrow$ $|C(A)| \leq 4$. But $0, A_{*p}, A_{*q}, A_{*p} + A_{*q} \in C(A)$. Therefore $|C(A)| = 4$ and the column space $C(A)$ is generated by A_{*p} and A_{*q}. We conclude that $C(A)$ is of type I. The proof of (ii) is similar to that of (i). This proves the lemma.

COROLLARY 17. *If* $A \in D_2^n$, *then* (i) $R(A)$ *is of type* I $\Leftrightarrow C(A)$ *is of type* I, (ii) $R(A)$ *is of type* II $\Leftrightarrow C(A)$ *is of type* II.

Proof. (i) Let $C(A)$ be of type I. Then $R(A^t)$ is of type I, because $C(A) = R(A^t)$. If $C(A^t)$ is of type I, then $R(A)$ is of type I; because $C(A^t) = R(A)$. (ii) Follows from (i). This proves the corollary.

The elementary row transformation $T_r(A)$ on a matrix $A \in S$ is the interchanging of any two rows of A. The definition of elementary column transformation $T_c(A)$ and row transformation $T_r(A)$ are dual. Elementary row and column transformation will be referred to simply as elementary transformation $T(A)$.

<u>LEMMA 18.</u> <u>If</u> $A \in S$, <u>then</u> (i) <u>a</u> $\mathcal{L}$<u>-class of</u> S <u>is invariant under</u> $T_r(A)$, (ii) <u>a</u> $\mathcal{R}$<u>-class of</u> S <u>is invariant under</u> $T_c(A)$, (iii) a $\mathcal{D}$-class of S is invariant under $T(A)$.

Proof. The proof is an immediate consequence of definition.

<u>COROLLARY 19.</u> <u>If</u> $A \in D_2^n$, (i) <u>a type of a row space is invariant under</u> $T(A)$, (ii) <u>a type of column space is invariant under</u> $T(A)$.

Proof. (i) By simple computation, $R(A) = R(T_r(A))$. By Corollary 17, $C(T_r(A))$ and $R(T_r(A))$ are the same type, which in turn implies that $C(T_r(A))$ and $C(A)$ are the same type. (ii) A similar proof holds for the column space. This completes the proof.

For simplicity we let (i) $D_{21}^n = \left\{A \in D_2^n : R(A) \text{ is of type I}\right\}$, (ii) $D_{22}^n = \left\{A \in D_2^n : R(A) \text{ is of type II}\right\}$. Let

$$I_r^n = \left[\begin{array}{c|c} I_r & 0 \\ \hline 0 & 0 \end{array}\right]_{n\times n},$$ where I_r denotes the r x r identity matrix. Let

$$T_r^n = \left[\begin{array}{c|c} T_r & 0 \\ \hline 0 & 0 \end{array}\right]_{n\times n},$$ where T_r denotes the r x r upper triangular matrix, i.e., $t_{ij} = 1$ iff $i \geq j$, $t_{ij} \in T_r$.

<u>THEOREM 20.</u> (i) $A \in D_{21}^n \iff A \,\mathcal{D}\, I_2^n$, (ii) $A \in D_{22}^n \iff A \,\mathcal{D}\, T_2^n$.

Proof. We prove each of these in turn.

(i) <u>Necessity:</u> Let $B_r(A) = \left\{v_1, v_2\right\}$ such that $v_1 \neq v_1 + v_2 \neq v_2$. If we set $v_1 = (a_1, a_2, \ldots, a_n)$ and $v_2 = (b_1, b_2, \ldots, b_n)$ then there exist $i_o, j_o \in N$ such that $a_{i_o} = 1$ and $a_{j_o} = 0$ for every $i_o \neq j_o$ and $b_{i_o} = 0$ and $b_{j_o} = 1$ for every $i_o \neq j_o$.

For brevity we let $A' = T_r(A)$ and $A'' = T_c(A')$. Then by Lemma 18, $A \mathcal{D} A'$ and $A' \mathcal{D} A'' \Rightarrow A \mathcal{D} A''$. Let $G = [A''_{*1}\ A''_{*2} \mid 0]_{nxn}$. Then $\rho_r(G) = 2$ and $R(G) = R(I_2^n) \Rightarrow G \mathcal{L} I_2^n \Rightarrow G \mathcal{D} I_2^n$. We observe that $C(G) \subseteq C(A'')$ but $|C(G)| = 4$, because $R(G)$ is of type I, which in turn implies that $C(G)$ is of type I. Also, note that $|C(A'')| = 4$, because $A, A', A'' \in D_{21}^n$. Hence $C(G) = C(A'') \Rightarrow G \mathcal{R} A'' \Rightarrow G \mathcal{D} A''$. Hence $A \mathcal{D} I_2^n$.

<u>Sufficiency:</u> $A \mathcal{D} I_2^n \Rightarrow \exists\ B \in S \ni A \mathcal{L} B$ and $B \mathcal{R} I_2^n \Rightarrow C(B) = C(I_2^n) \Rightarrow C(B)$ is of type I. Hence $R(B)$ is of type I. Moreover, $R(A) = R(B)$. We immediately conclude that $A \in D_{21}^n$.

(ii) The proof is similar to that of (i).

<u>THEOREM 21.</u> <u>D_2^n contains exactly two distinct $\mathcal{D}$-classes.</u>

Proof. We shall show that $I_2^n \not\mathcal{D} T_2^n$. By $I_2^n \not\mathcal{D} T_2^n$ we mean that I_2^n is not $\mathcal{D}$-equivalent to T_2^n. Assume $I_2^n \mathcal{D} T_2^n$. Then there exist $F \in S$ such that $I_2^n \mathcal{R} F$ and $F \mathcal{L} T_2^n$. Since $I_2^n \mathcal{R} F$, $C(I_2^n) = C(F)$. By Corollary 17 (i), $C(F)$ is of type I $\Leftrightarrow R(F)$ is of type I. Since $F \mathcal{L} T_2^n$, $R(F) = R(T_2^n)$. By Corollary 17 (ii), $R(T_2^n)$ is of type II $\Rightarrow R(F)$ is of type II. Contradiction. This proves the lemma.

<u>LEMMA 22.</u> <u>If</u> $A \in S$ <u>has rank</u> 2 <u>then there exist nonzero matrices</u> $U \in M_{n,2}(R)$ <u>and</u> $V \in M_{2,n}(R)$ <u>such that</u> $A = UV$. <u>Furthermore, we can always pick</u> $U \in M_{n,2}(R)$ <u>and</u> $V \in M_{2,n}(R)$ <u>such that</u> $C(U) = C(A)$ <u>and</u> $R(V) = R(A)$.

Proof. The proof will be in two parts.

<u>Case (i):</u> Let $A \in D_{21}^n$. Let $B_r(A) = \{v_1, v_2\}$ and $B_c(A) = \{u_1, u_2\}$. Then clearly $|R(A)| = 4 = |C(A)|$. Let

$$V = \begin{bmatrix} v_1 \\ v_2 \end{bmatrix}_{2xn}. \quad \text{Let } U = \begin{bmatrix} U_{1*} \\ U_{2*} \\ \vdots \\ U_{n*} \end{bmatrix}_{nx2}, \quad U_{i*} \in V_2(R).$$

In particular,

$$U_i = \begin{cases} [1\ 0] \text{ if } A_{i*} = v_1, \\ [0\ 1] \text{ if } A_{i*} = v_2, \end{cases}$$

Thus $U_{i*} = [1\ 1]$ if $A_{i*} = v_1 + v_2$. Hence $A = UV$. We may assume $\rho_c(U) \leq 2$, because $U \in M_{n,2}(R)$. Therefore $|C(U)| \leq 4$. But $C(A) \subseteq C(U) \supseteq C(UV)$, which implies that $4 = |C(A)| \leq |C(U)| \leq 4$, and thus $|C(U)| = 4$. We conclude that $C(U) = C(A)$.

Case (ii): The proof is similar to that of Case (i).

<u>COROLLARY 23</u>. <u>Let</u> $A \in D^n_{22}$. <u>Then there exist</u> $U \in M_{n,2}(R)$ <u>and</u> $V \in M_{2,n}(R)$ <u>such that</u> $A = UV$, $R(V) = R(A)$, <u>and</u> $C(A) \subsetneq C(U)$ (i.e., $C(U)$ <u>is of type</u> I). <u>By</u> $C(A) \subsetneq C(U)$ <u>we mean that every element of</u> $C(A)$ <u>is an element of</u> $C(U)$, <u>but</u> $C(A) \neq C(U)$.

Proof. In the proof of the Lemma 22 (ii), let $i \in N$ such that $U_{i*} = [1\ 1]$. Just change this U_{i*} to $[1\ 0]$. The rest follows trivially.

<u>REMARK 24</u>. <u>Let</u> $A = UV$ <u>such that</u> $U \in M_{n,2}(R)$ <u>and</u> $V \in M_{2,n}(R)$.

(i) <u>Consider the Corollary 23, there exist several different</u> U <u>satisfying the conditions in the hypothesis since we may pick any</u> $U_{i*} = [1\ 1]$.

(ii) <u>We can obtain another true statement by letting</u> $U = [u_1\ u_2]$ $(u_1, u_2 \in V^n(R))$ <u>and finding appropriate</u> V.

(iii) <u>In the proof of the Lemma 22 Case (i) once we pick</u> V, <u>then there exists a unique</u> U <u>such that</u> $A = UV$.

(iv) <u>In the proof of the Lemma 22 Case (ii) once we pick</u> V, <u>then there exists a unique</u> U <u>such that</u> $C(U)$ <u>is of type</u> II <u>and</u> $A = UV$.

<u>THEOREM 25</u>. (i) $|D^n_{21}| = 2(2^{2n-1} + 2^{n-1} - 3^n)^2$,

(ii) $|D^n_{22}| = (3^n - 2^{n+1} + 1)^2$.

Proof. We shall first prove part (ii).

<u>Case (ii)</u>: Let $A \in D^n_{22}$ such that $B_r(A) = \{v_1, v_2\}$. We assume that $v_1 + v_2 = v_2$. Then clearly $\rho(v_1) < \rho(v_2)$. If $\rho(v_1) = 1$, then there exist $\binom{n}{1}$ v_1's such that $\rho(v_1) = 1$. For each such v_1 we have $(2^{n-1} - 1)$ v_2's so that the total number of such v_2 is $\binom{n}{1}(2^{n-1} - 1)$. If $\rho(v_1) = 2$, then there exist $\binom{n}{2}$ v_1's such that $\rho(v_1) = 2$.

For each such v_1 we have $(2^{n-2} - 1)$ v_2's so that the total number of such v_2 is $\binom{n}{2}(2^{n-2} - 1)$. If $\rho(v_1) = k$, then there exist $\binom{n}{k}$ v_1's such that $\rho(v_1) = k$. For each such v_1 we have $(2^{n-k} - 1)$ v_2's so that the total number of v_1 is $\binom{n}{k}(2^{n-k} - 1)$. Let $L_2^n = L_{21}^n \cup L_{22}^n$ and $R_2^n = R_{21}^n \cup R_{22}^n$, where L_{2i}^n (R_{2i}^n), $i = 1, 2$, denotes the set of all $\mathcal{L}$ $(\mathcal{R})$-classes in the $\mathcal{D}$-classes D_{2i}^n, $i = 1, 2$, respectively.

Together with information we already have obtained,

$$|L_{22}^n| = \sum_{k=1}^{n} \binom{n}{k}(2^{n-k} - 1)$$

$$= \sum_{k=1}^{n} \binom{n}{k}2^{n-k} - \sum_{k=1}^{n} \binom{n}{k}$$

$$= 3^n - 2^{n+1} + 1.$$

Similarly, if we let $B_c(A) = \{u_1, u_2\}$ such that $u_1 + u_2 = u_1$ then

$$|R_{22}^n| = 3^n - 2^{n+1} + 1.$$

It follows that

$$|D_{22}^n| = |L_{22}^n||R_{22}^n|$$

$$= (3^n - 2^{n+1} + 1)^2.$$

<u>Case (i):</u> Without loss of generality we may assume that there are exactly $\binom{2^n - 1}{2}$ $\mathcal{L}$-classes in the $\mathcal{D}$-classes D_2^n. Hence

$$|L_{21}^n| = |L_2^n \setminus L_{22}^n|$$

$$= \binom{2^n - 1}{2} - (3^n - 2^{n+1} + 1)$$

$$= (2^n - 1)^2 - (3^n - 2^{n+1} + 1)$$

$$= 2^{2n-1} + 2^{n-1} - 3^n.$$

Clearly analogous result holds for R_{21}^n.

$$|R_{21}^n| = |R_2^n \setminus R_{22}^n| = 2^{2n-1} + 2^{n-1} - 3^n.$$

Putting together, we have $|D_{21}^n| = |L_{21}^n||R_{21}^n|$.

It is easy to see that if we set $A = UV$ where $U \varepsilon M_{n,2}(R)$ and $V \varepsilon M_{2,n}(R)$ then there exists $B \varepsilon D^n_{21}$ such that $B = U''V = UV'$ where $U' = T_r(U)$ and $U'' = T_c(U')$ and $R(A) = R(B)$ and $C(A) = C(B)$. Since $A, B \varepsilon D^n_{21}$ we must multiply 2 to the above expression, that is

$$|D^n_{21}| = 2|L^n_{21}||R^n_{21}|$$
$$= 2(2^{2n-1} + 2^{n-1} - 3^n)^2.$$

This completes the proof.

In the following section we shall study the $\mathcal{D}$-classes D^n_3. An efficient method is presented for determining the $\mathcal{D}$-classes of matrices of dimension n and common row and column rank 3. As we showed earlier, matrices of S need not have the same row and column rank. The argument used in the proof of Theorem 5 (iii) has an important application that simplifies the classfication of the row and column spaces. With Theorem 5 already in hand, it seems most natural to use same notations. By use of Theorem 5 (iii) Case (1), we are led to the following definition. Let $A \varepsilon D^n_3$. We say R(A) (C(A)) is of type I iff (i) $N_{\sigma(1)} \not\subset (N_{\sigma(2)} \cup N_{\sigma(3)})$ for every $\sigma \varepsilon P_3$, where P_3 denotes the set of all permutation on 3 symbols, (ii) $|R(A)| = 8 = |C(A)|$. By use of Theorem 5 (iii) Case (2) Subcase (1), we are let to the following definition. Let $A \varepsilon D^n_3$. We say R(A) (C(A)) is of type III iff (i) $N_i \subset (N_j \cup N_k)$, $N_j \not\subset (N_i \cup N_k)$, $N_k \not\subset (N_i \cup N_j)$, $N_i \subset (N_j \cap N_k)$, (ii) $|R(A)| = 5 = |C(A)|$. By use of Theorem 5 (iii) Case (2) Subcase (2), we have the following definition. Let $A \varepsilon D^n_3$. We say R(A) (C(A)) is of type V iff (i) $\overline{N}_1 \subset (\overline{N}_2 \cup \overline{N}_3)$, $\overline{N}_2 \not\supset \overline{N}_1 \subset \overline{N}_3$, $\overline{N}_2 \not\subset (\overline{N}_1 \cup \overline{N}_3)$, $\overline{N}_3 \not\subset (\overline{N}_1 \cup \overline{N}_2)$, where $\overline{N}_i = \{h \varepsilon N\colon e^t_h \otimes v_i = e^t_h\}$, $i = 1, 2, 3$, (ii) $|R(A)| = 6 = |C(A)|$. By use of Theorem 5 (iii) Case (3) Subcase (1), we are led to the following definition.
Let $A \varepsilon D^n_3$. We say R(A) (C(A)) is of type VII iff (i) $N_i \subset (N_j \cup N_k)$, $N_j \subset (N_i \cup N_k)$, $N_k \not\subset (N_i \cup N_j)$, $N_i \subsetneq N_j$, and $N_j \subsetneq N_k$, (ii) $|R(A)| = 4 = |C(A)|$. By use of Theorem 5 (iii) Case (3) Subcase (2), we are led

to the following definition. Let $A \in D_3^n$. We say $R(A)$ $(C(A))$ is of type IV iff (i) $N_i \subset (N_j \cup N_k)$, $N_j \subset (N_i \cup N_k)$, $N_k \not\subset (N_i \cup N_j)$, $N_i \not\subset N_j \not\subset N_i$ but $N_i \subset N_k$, (ii) $|R(A)| = 5 = |C(A)|$.

<u>REMARK 26. Reviewing Case (2) Subcase (1), we see that the row space</u> $R(A)$ <u>of type</u> III <u>iff the column space</u> $C(A)$ <u>of type</u> IV <u>and the row space</u> $R(A)$ <u>of type</u> IV <u>iff the column space</u> $C(A)$ <u>of type</u> III.

By use of Theorem 5 Case (3) Subcase (3), we have the following definition. Let $A \in D_3^n$. We say $R(A)$ $(C(A))$ is of type VI iff (i) $N_i \subset (N_j \cup N_k)$, $N_j \subset (N_i \cup N_k)$, $N_k \not\subset (N_i \cup N_j)$, $N_i \subsetneq N_j$, $N_j \not\subset N_k$, (ii) $|R(A)| = 5 = |C(A)|$. By use of Theorem 5 Case (4), we are thus led to the following definition. Let $A \in D_3^n$. We say $R(A)$ $(C(A))$ is of type II iff (i) $N_i \subset (N_j \cup N_k)$, $N_j \subset (N_i \cup N_k)$, $N_k \subset (N_i \cup N_j)$, (ii) $|R(A)| = 5 = |C(A)|$. We are now in the happy position where all the pieces have been constructed and all we have to do is to put them together. The following lemma is an immediate consequence of the definitions.

<u>LEMMA 27. If</u> $A \in D_3^n$, <u>then</u>

(1) $R(A)$ <u>is of type</u> I <=> $C(A)$ <u>is of type</u> I,

(2) $R(A)$ <u>is of type</u> II <=> $C(A)$ <u>is of type</u> II,

(3) $R(A)$ <u>is of type</u> III <=> $C(A)$ <u>is of type</u> IV,

(4) $R(A)$ <u>is of type</u> IV <=> $C(A)$ <u>is of type</u> III,

(5) $R(A)$ <u>is of type</u> V <=> $C(A)$ <u>is of type</u> V,

(6) $R(A)$ <u>is of type</u> VI <=> $C(A)$ <u>is of type</u> VI,

(7) $R(A)$ <u>is of type</u> VII <=> $C(A)$ <u>is of type</u> VII.

We note here that exactly one of the above is true for each $A \in D_3^n$.

<u>COROLLARY 28. If</u> $A, B \in S$, <u>then</u>

(i) $A \mathcal{L} B$ <u>and</u> $\rho_r(A) = 3 = \rho_c(A) \Rightarrow \rho_c(B) = 3$,

(ii) $A \mathcal{R} B$ <u>and</u> $\rho_r(A) = 3 = \rho_c(A) \Rightarrow \rho_r(B) = 3$.

Proof. Lemma 27 implies that if $A \mathcal{D} B$ and $\rho_r(A) = \rho_r(B) = 3 = \rho_c(A) = \rho_c(B)$ then $R(A)$ and $R(B)$ are the same type and $C(A)$ and $C(B)$ are the same type. Now corollary follows.

<u>REMARK 29</u>. <u>Each $\mathcal{D}$-class contains all the elements which have permissible combinations of the row and column type</u>.

<u>THEOREM 30</u>. <u>There are exactly seven distinct $\mathcal{D}$-classes with $\rho_r(A) = 3 = \rho_c(A)$ for all $A \in S$.</u>

Proof. The proof will be in two parts.

(i) $n = 3$. We use Lemma 27 and the appendix.

(ii) $n \geq 3$. Let $B_r(A) = \{v_1, v_2, v_3\}$ and $\bar{A} = [v_1, v_2, v_3, 0, \ldots, 0]^t$. Clearly $A \mathcal{D} \bar{A}$, because $R(A) = R(\bar{A})$. By Corollary 28, $\rho_c(\bar{A}) = 3$.

Let $B_c(\bar{A}) = \{u_1, u_2, u_3\}$ and $\hat{A} = \left[\begin{array}{c|c} [u_1\ u_2\ u_3]_{3\times 3} & 0 \\ \hline 0 & 0 \end{array}\right]_{n\times n}$.

Then $\bar{A} \mathcal{D} \hat{A}$. In particular, $R(\hat{A})$ is of same type as $R(A)$ and $C(\hat{A})$ is of the same type as $C(A)$. Let $\pi_1 = \{\mathcal{D}\text{-classes of } A \in D_3^n\}$ and $\pi_2 = \{\mathcal{D}\text{-classes of } A \in D_3^3\}$. Then we have a mapping

$$f: \pi_1 \longrightarrow \pi_2$$
$$\mathcal{D}_A \qquad \mathcal{D}_{\tilde{A}}$$

where $\tilde{a}_{ij} = \hat{a}_{ij}$ for every $1 \leq i, j \leq 3$. A mapping f is well-defined, because $A \mathcal{D} B \Rightarrow R(A), R(B)$ is of the same type and $C(A), C(B)$ is of the same type $\Rightarrow \hat{A} \mathcal{D} \hat{B}$, $R(\hat{A}), R(\hat{B})$ is of the same type and $C(\hat{A}), C(\hat{B})$ is of the same type $\Rightarrow R(\tilde{A}), R(\tilde{B})$ is of the same type and $C(\tilde{A}), C(\tilde{B})$ is of the same type. By Remark 29, $\tilde{A} \mathcal{D} \tilde{B}$. Therefore f is a bijection (with Corollary 28 need only that it is an injection). This proves the theorem.

For brevity we let

(1) $D_{31}^n = \{A \in D_3^n : R(A) \text{ is of type I}\}$,

(2) $D_{32}^n = \{A \in D_3^n : R(A) \text{ is of type II}\}$,

(3) $D_{33}^n = \{A \in D_3^n : R(A) \text{ is of type III}\}$,

(4) $D_{34}^n = \{A \in D_3^n : R(A) \text{ is of type IV}\}$,

(5) $D_{35}^n = \{A \in D_3^n : R(A) \text{ is of type V}\}$,

(6) $D_{36}^n = \{A \in D_3^n : R(A) \text{ is of type VI}\}$,

(7) $D_{37}^n = \{A \in D_3^n : R(A) \text{ is of type VII}\}$.

LEMMA 31. *If we have k* ($k \geq 3$) *objects* $a_1, a_2, \ldots, a_k$, *then there exist* $k^n - 3(k-1)^n + 3(k-2)^n - (k-3)^n$ *different permutations of the k objects taken n* ($n > 0$) *at a time where repetitions are permitted and which contains each of* a_1, a_2, a_3 *at least once.*

Proof. There exist k^n permutations of k objects n at a time with repetitions. There exist

(i) $(k-1)^n$ permutations of the objects $a_2, a_3, \ldots, a_k$ n at a time with repetition.

(ii) $(k-1)^n$ permutations of the objects $a_1, a_3, \ldots, a_k$ n at a time with repetition.

(iii) $(k-1)^n$ permutations of the objects $a_1, a_2, a_4, \ldots, a_k$ n at a time with repetition.

(i)-(iii) each include the cases where none of a_1, a_2, a_3 appear,
(i) and (ii) include the cases where a_1, a_2 do not appear,
(ii) and (iii) include the cases where a_2, a_3 do not appear,
(i) and (iii) include the cases where a_1, a_3 do not appear.

(iv) $(k-2)^n$ permutations of the objects $a_3, a_4, \ldots, a_k$ n at a time with repetition.

(v) $(k-2)^n$ permutations of the objects $a_2, a_4, \ldots, a_k$ n at a time with repetition.

(vi) $(k-2)^n$ permutations of the objects $a_1, a_4, \ldots, a_k$ n at a time with repetition.

(iv)-(vi) each include the cases where none of a_1, a_2, a_3 appear.

(vii) $(k-2)^n$ permutations of the objects $a_4, a_5, \ldots, a_k$ n at a time with repetition.

Since we are interested only in those permutations where each of a_1, a_2, a_3 appear, it follows from Remark 29 that there exist

$$k^n - 3(k-1)^n + 3(k-2)^n - (k-3)^n$$

such permutations. This proves the lemma.

THEOREM 32. Let L^n_{3i} (R^n_{3i}, H^n_{3i}) be the set of all $\mathcal{L}$ $(\mathcal{R}, \mathcal{H})$-classes of D^n_{3i}, $i = 1, 2, \ldots, 7$. Then

(1) $|L^n_{31}| = \frac{1}{3!}(8^n - 3\cdot 7^n + 3\cdot 6^n - 5^n) = |R^n_{31}|$,

$|H^n_{31}| = 3!$,

$|D^n_{31}| = \frac{1}{3!}(8^n - 3\cdot 7^n + 3\cdot 6^n - 5^n)^2$.

(2) $|L^n_{32}| = \frac{1}{3!}(5^n - 3\cdot 4^n + 3\cdot 3^n - 2^n) = |R^n_{32}|$,

$|H^n_{32}| = 3!$,

$|D^n_{32}| = \frac{1}{3!}(5^n - 3\cdot 4^n + 3\cdot 3^n - 2^n)^2$.

(3) $|L^n_{33}| = \frac{1}{2!}(5^n - 3\cdot 4^n + 3\cdot 3^n - 2^n) = |R^n_{33}|$,

$|H^n_{33}| = 2!$,

$|D^n_{33}| = \frac{1}{2!}(5^n - 3\cdot 4^n + 3\cdot 3^n - 2^n)^2$.

(4) $|L^n_{34}| = \frac{1}{2!}(5^n - 3\cdot 4^n + 3\cdot 3^n - 2^n) = |R^n_{34}|$,

$|H^n_{34}| = 2!$,

$|D^n_{34}| = \frac{1}{2!}(5^n - 3\cdot 4^n + 3\cdot 3^n - 2^n)^2$.

(5) $|L^n_{35}| = (6^n - 3\cdot 5^n + 3\cdot 4^n - 3^n) = |R^n_{35}|$,

$|H^n_{35}| = 1!$,

$|D^n_{35}| = (6^n - 3\cdot 5^n + 3\cdot 4^n - 3^n)^2$.

(6) $|L^n_{36}| = (5^n - 3\cdot 4^n + 3\cdot 3^n - 2^n) = |R^n_{36}|$,

$|H^n_{36}| = 1$,

$|D^n_{36}| = (5^n - 3\cdot 4^n + 3\cdot 3^n - 2^n)^2$.

(7) $|L^n_{37}| = (4^n - 3\cdot 3^n + 3\cdot 2^n - 1^n) = |R^n_{37}|$,

$|R^n_{37}| = 1!$,

$|D^n_{37}| = (4^n - 3\cdot 3^n + 3\cdot 2^n - 1^n)^2$.

Proof. We will use Theorem 30 and Lemma 31, and the egg-box structure of $\mathcal{D}$-classes. From Theorem 30, we know all row spaces (column spaces) of the same type belong to the same $\mathcal{D}$-class.

Thus by the egg-box structure of $\mathcal{D}$-class we need to find all the possible different row spaces (up to permutation) which have a given column space (similar for column spaces). We prove each of these in turn.

(1) $\mathcal{D}$-class D_{31}^n is generated by $A = I_3^n = \begin{bmatrix} I_3 & 0 \\ 0 & 0 \end{bmatrix}_{n \times n}$, where I_3 denotes the 3 x 3 identity matrix. We wish to find the number of different (up to permutation) matrix F such that $C(F) = C(I_3^n)$. The unique basis of $C(I_3^n)$ is

$$\left\{ \begin{bmatrix} 1 \\ 0 \\ 0 \\ \vdots \\ 0 \end{bmatrix}, \begin{bmatrix} 0 \\ 1 \\ 0 \\ \vdots \\ 0 \end{bmatrix}, \begin{bmatrix} 0 \\ 0 \\ 1 \\ \vdots \\ 0 \end{bmatrix} \right\}$$

and $|C(I_3^n)| = 8$. Using Lemma 31, there exist

$$(8^n - 3 \cdot 7^n + 3 \cdot 6^n - 5^n)$$

different acceptable permutation, i.e., matrix F such that $C(F) = C(I_3^n)$. If $I_3^n \mathcal{R} F$, then

$$F = \begin{bmatrix} [\ ?\ ]_{3 \times n} \\ 0 \end{bmatrix}_{n \times n}.$$

For each F such that $I_3^n \mathcal{R} F$ there exist at least $3!$ and for each F_σ ($\sigma \varepsilon P_3$, where P_3 denotes the set of all permutation on 3 symbols) such that $F \mathcal{H} F_\sigma$, i.e.,

$$F_\sigma = \begin{bmatrix} F_{\sigma(1)*} \\ F_{\sigma(2)*} \\ F_{\sigma(3)*} \\ 0 \end{bmatrix}_{n \times n} \tag{32.1}$$

($F_\sigma = F$ if σ is an identity permutation). But there can not exist more than $3!$ matrices having the same row space of rank 3 and all the row below the 3rd row equal to zero. Hence we get

$$|L_{31}^n| = \frac{1}{3!}(8^n - 3 \cdot 7^n + 3 \cdot 6^n - 5^n).$$

Clearly analogous result holds for $\mathcal{R}$. By [4, p. 49, Theorem 2.3], $|H_C| = 3!$ for every $C \varepsilon D_{31}^n$. Let C be such that $C \mathcal{H} \hat{C}$. Let $\{C_{p_1*}, C_{p_2*}, C_{p_3*}\}$ be any set of rows which form a basis for R(C).

Let $\gamma \in P_n$, where P_n denotes the set of all permutation on n symbols, $\gamma = (1, p_1)(2, p_2)(3, p_3)$ (i.e., permutation). Let $D = \gamma(C)$ and $\hat{D} = \gamma(\hat{C})$ where $\gamma(X)$ means the matrix obtain by letting X permute the rows of X. Clearly $D \mathcal{H} \hat{D}$. Let

$$K = I_3^n \cdot D = \begin{bmatrix} [\ ?\ ]_{3\times n} \\ \hdashline 0 \end{bmatrix}_{n\times n}.$$

Then $\hat{K} = I_3^n \cdot \hat{D}$. Further, we have $K \mathcal{L} D$, $K \mathcal{R} \hat{K} \Rightarrow D \mathcal{D} \hat{K} \Rightarrow \rho_r(\hat{K}) = 3$ which in turn implies that $C_{p_1*}, C_{p_2*}, C_{p_3*}$ spans $R(\hat{C})$ and in fact

$\left\{C_{p_1*}, C_{p_2*}, C_{p_3*}\right\} = \left\{\hat{C}_{p_1*}, \hat{C}_{p_2*}, \hat{C}_{p_3*}\right\}$, because $C \mathcal{H} \hat{C}$. The order is completely arbitrary by (32.1). Since p_1, p_2, p_3 were arbitrary it follows that there exist $\sigma \in P_3$ such that if $\hat{C}_{h*} = v_i$ (where $\{v_1, v_2, v_3\}$ spans $R(C)$) then $\hat{C}_{h*} = v_{\sigma(i)}$. If $C_{h*} = 0$, then $\hat{C}_{h*} = 0$, because $C(C) = C(\hat{C})$. If $C_{h*} = v_i + v_j$, then we claim that $\hat{C}_{h*} = v_{\sigma(i)} + v_{\sigma(j)}$ where σ is the same as before. Let p_1, p_2 be such that $C_{p_1*} = v_1$, $C_{p_2*} = v_2$. Let $\gamma \in P_n$ be such that $\gamma = (1, p_1)(2, p_2)(3, k)$. Let $D = I_3^n \cdot \gamma(C)$ and $\hat{D} = I_3^n \cdot \gamma(\hat{C})$. Then $D \mathcal{R} \hat{D}$ but $\rho_r(D) = 2$, which implies that $\rho_r(\hat{D}) = 2$ by Theorem 5. Hence $\hat{C}_{h*} = \hat{C}_{p_1*} + \hat{C}_{p_2*}$. By symmetry and elimination,

$$C_{h*} = v_1 + v_2 + v_3 \Rightarrow \hat{C}_{h*} = v_1 + v_2 + v_3.$$

Hence $|H_C| = |P_3| = |H_{31}^n| = 3!$. We immediately conclude that

$$|D_{31}^n| = |H_{31}^n||L_{31}^n| = 3!\left(\frac{1}{3!}(8^n - 3\cdot 7^n + 3\cdot 6^n - 5^n)\right)^2$$

$$= \frac{1}{3!}(8^n - 3\cdot 7^n + 3\cdot 6^n - 5^n)^2.$$

This completes the proof of (i). The proofs of (ii)-(vii) are similar to that of (i).

CONCLUSION. The above methods are applicable to the study of the distribution of elements in any $\mathcal{D}$-class.

APPENDIX

We shall mention an egg-box structure [4] of a $\mathcal{D}$-class D. An egg-box table of D is an arrangement of all $\mathcal{H}$-classes of D such a way that if $\mathcal{H}$, $\mathcal{L}$, $\mathcal{R}$ are respectively $\mathcal{H}$, $\mathcal{L}$, $\mathcal{R}$-class of D with $\mathcal{H} = \mathcal{L} \cap \mathcal{R}$. Imagine the elements of D arranged in a square (or rectangular) pattern, like an egg-box, the rows corresponding to the $\mathcal{R}$-classes and the columns to the $\mathcal{L}$-classes contained in D. Each cell of the egg-box corresponds to an $\mathcal{H}$-class contained in D, and the no cell is empty. We do not arrange the elements in the $\mathcal{H}$-classes in any particular way. As we shall see presently, the $\mathcal{H}$-classes contained in D all have the same cardinal number, thus the cells of the egg-box are so to speak, equally full of elements of S. As an example, we shall write out all the $\mathcal{D}$-classes of the following (i) D_1^2, and (ii) D_2^2. For simplicity of notation we shall denote $B_r(A)$ by B_r and $B_c(A)$ by B_c, respectively.

D_1^2

B_c \ B_r	1 1	1 0	0 1
1 1	1 1 1 1	1 0 1 0	0 1 0 1
1 0	1 1 0 0	1 0 0 0	0 1 0 0
0 1	0 0 1 1	0 0 1 0	0 0 0 1

D_{21}^2

B_c \ B_r	1 0 0 1
1 0 0 1	1 0 0 1
0 1 1 0	0 1 1 0

D_{22}^2

B_c \ B_r	0 1 1 1	1 0 1 1
0 1 1 1	0 1 1 1	1 0 1 1
1 0 1 1	1 1 0 1	1 1 1 0

ACKNOWLEDGEMENT. The results announced here are contained in author's doctoral dissertation written at George Washington University under the guidance of Professor Jin Bai Kim of West Virginia Univsity and Professor Irving Katz of George Washington University.

REFERENCES

1. Butler, K. K. H. On (0, 1)-matrix semigroups, Ph.D. Dissertation, George Washington University, D. C., 1970.

2. ________ On (0, 1)-matrix semigroups, Notices of the Amer. Math. Soc., Issue NO. 122 (1970) 637.

3. Butler, K. H. K. The number of idempotents in (0, 1)-matrix semigroups, submitted to J. of Linear Algebra and Its Applications.

4. Clifford, A. H. and Preston, G. B. The Algebraic Theory of Semigroups, Amer. Math. Soc. Survey, No. 7-1, Providence, 1961.

5. Kim, J. B. On structures of linear semigroups, to appear.

6. Ore, O. Theory of Graph, Coll. Pub. vOl. 38, Amer. Math. Soc., Providence, 1962.

7. Plemmons, R. J. and Montague, J. S. Maximal subgroups of the semigroup of relations, J. of Algebra, 13 (1969) 575-587.

8. Ryser, H. J. Combinatorial Mathematics, Carus Math. Monograph 14, Wiley, New York, 1963.

ON CONNECTED CUBIC GRAPHS AND TRIVALENT TREES

C. C. Cadogan*

University of Waterloo

Waterloo, Ontario, Canada

1. INTRODUCTION

This paper sets up correspondences between subsets of the set of connected cubic graphs on n (even) nodes on the one hand and between trivalent trees and a class of connected cubic graphs on the other. A formula for trivalent trees in terms of other graphs is stated. The reader is referred to [2] for basic terminology on graphs and to [3] for a comprehensive bibliography on trees.

2. PRELIMINARIES

Let S be the set of connected graphs on n (even) trivalent nodes, loops and multiple edges allowed. We refer to the configuration 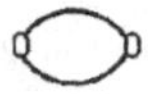 as a <u>pair</u> of edges.

Let the numbers of loops and pairs of edges in any graph $G \in S$ be β, γ respectively. Let

$$S_1 = \{G \in S: \beta = 0, \gamma \geq 1\}, \quad S_2 = \{G \in S: \beta \geq 1, \gamma = 0\},$$

$$S_3 = \{G \in S: \beta \geq 1, \gamma \geq 1\}, \quad S_4 = \{G \in S: \beta = \gamma = 0\}.$$

We denote the cardinality of set S' by $|S'|$.

In any graph $G \in S_1$, multiple edges occur only in pairs as follows:

(i) (ii)

---

*This paper was written while the author was an N.R.C. postdoctoral fellow at the University of Waterloo.

We denote by $C(n)$ the subset of S_1 in which circuits in the graphs occur in configurations (i) and (ii) only.

In any graph $G \in S_2$, loops occur as indicated in figures (iii) and (iv) only.

(iii)

(iv)

3. CARDINALITY OF SUBSETS OF S

<u>Theorem 1.</u> $|S_1| = |S_2|$.

<u>Proof.</u> We set up the correspondences (i)$\leftrightarrow$(iii) and (ii)$\leftrightarrow$(iv) between the configurations in section 2. This correspondence is one-to-one, hence the result.

<u>Theorem 2.</u> $|S_3| = |S| - 2|S_1| - |S_4|$.

<u>Proof.</u> $S_i (i = 1, 2, 3, 4)$ partition the set S .
By applying theorem 1 we obtain the result.

Methods for calculating $|S|$, $|S_1|$ and $|S_4|$ are outlined in [1]. See table 1 for some values of $|S_3|$ and figure 1 for graphs when $n = 6$.

TABLE 1

n	2	4	6	8
$\|S\|$	2	5	17	70
$\|S_1\|$	1	1	4	15
$\|S_4\|$	0	1	2	5
$\|S_3\|$	0	2	7	35

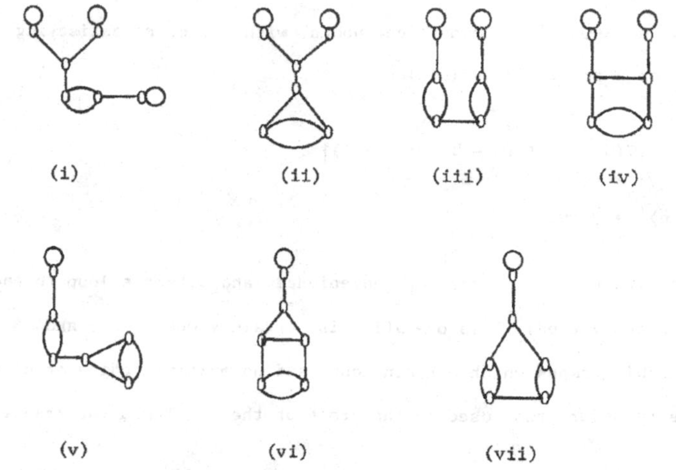

Figure 1

4. TRIVALENT TREES

The following two definitions are useful in this context.

Definition 1. A trivalent node in a trivalent tree is called exterior if two monovalent nodes are attached to it and interior otherwise.

Definition 2. An interior node in a trivalent tree is called reducible if there is a monovalent node attached to it and irreducible otherwise.

Let $T(n)$ denote the set of trivalent trees on n nodes $(n \geq 6)$. For $n = 2, 4$ we put $|T(n)| = 1$. Let α be the number of exterior nodes in any one of these trees. Then $\alpha \geq 2$.

Let $\alpha_i (i = 1, 3)$ be respectively the numbers of monovalent and trivalent nodes in any tree. By a simple calculation we obtain,

$$\alpha_1 = \frac{n}{2} + 1, \quad \alpha_3 = \frac{n}{2} - 1 . \tag{1}$$

Let $n = 2m$. We denote by $T'(p; q; r)$ the set of trees on p trivalent nodes, q bivalent nodes, and r monovalent nodes, with p, q, r satisfying the requirements for a tree. By definition,

$$(2) \qquad |T(n)| = |T'(m - 1; 0; m + 1)| .$$

<u>Theorem 3.</u> $|T(n)| = |C(n)|$.

<u>Proof.</u> We start with the set $T(n)$ for convenience, and attach a loop to each monovalent node in every tree. This operation is certainly one-to-one and creates a set of cubic graphs which contain loops and no multiple edges or circuits. We now employ the transformation used in the proof of theorem 1 and the result follows.

<u>Theorem 4.</u> $|T(n)| = \sum_{\alpha=2}^{v} |T'(\alpha - 2; m + 1 - 2\alpha; \alpha)|$, where

$v = \left[\frac{m+1}{2}\right]$, and $[x]$ denotes the greatest integer $\leq x$.

<u>Proof.</u> Suppose there are α exterior nodes in any tree on n nodes; then there are $m + 1 - 2\alpha$ reducible nodes, and $(m - 1 - \alpha) - (m + 1 - 2\alpha) = \alpha - 2$ irreducible nodes. Also, $m + 1 - 2\alpha \geq 0$ implies that maximum $\alpha = \left[\frac{m+1}{2}\right]$, since α is integral; thus $2 \leq \alpha \leq v$, where v is defined as above.

We now remove all monovalent nodes and incident edges and the theorem follows.

Let $N_\ell(a; b; c)$ and $N(a; b; c)$ represent respectively the numbers of connected linear graphs and of connected graphs on a trivalent nodes, b bivalent nodes and c monovalent nodes. The following theorem is a corollary of theorem 3 of [1] .

<u>Theorem 5.</u>

$$N_\ell(a; b; c) = \sum_{q=0}^{u} \left\{ (-1)^p X(\rho) N(a - q; b; (\sigma)) \right\} ,$$

where $$u = \begin{cases} a & \text{if } a \le b + 1 , \\ \left[\frac{a+b}{2}\right] & \text{if } a > b + 1 . \end{cases}$$

In this formula $(\rho) = (1^{j_1} 2^{j_2} \ldots q^{j_q})$ is a partition of the integer $q(\geq 1)$, $p = \sum_{i=1}^{q} j_i$, $X(\rho) = \frac{p!}{j_1!\, j_2! \ldots j_q!}$, and

$$(\sigma) = \begin{cases} \left(1^{j_1} 2^{j_2} \ldots c^{j_c+1} \ldots q^{j_q}\right) & \text{if } 1 \le c \le q , \\ \left(1^{j_1} 2^{j_2} \ldots q^{j_q} , c\right) & \text{if } c > q . \end{cases}$$

Evidently, $N_\ell(\alpha - 2; m + 1 - 2\alpha; \alpha) = |T'(\alpha - 2; m + 1 - 2\alpha; \alpha)|$, hence on combining theorems 4 and 5, we have,

Lemma.
$$|T(n)| = \sum_{\alpha=2}^{v} N_\ell\, (\alpha - 2; m + 1 - 2\alpha; \alpha) ,$$

where v is defined as in theorem 4.

REMARK

With $|T(2)| = |T(4)| = 1$, the formula of theorem 4 is useful in obtaining results for $|T(n)|$ when n is small. However, for large n the presence of the bivalent nodes does seem to complicate matters and although it appears from lemma immediately above that the problem is solved, the procedure by which the numbers $N(a; b; c)$ are obtained is somewhat involved. It would be useful, therefore, if further reductions on $|T'(\alpha - 2; m + 1 - 2\alpha; \alpha)|$ in terms of the numbers of smaller trees could be derived.

REFERENCES

(1) Cadogan, C.C. Graph-lattices and the enumeration of linear graphs, Proc. Conf. on Graph Theory, Baton Rouge, Louisiana, (To appear).

(2) Harary, F. Graph Theory, Addison-Wesley Pub. Co. (1969).

(3) Harary, F. Proof Techniques in Graph Theory, Academic Press, New York and London, (1969).

GRAPHICAL THEOREMS OF THE NORDHAUS-GADDUM CLASS

Gary Chartrand[1], Western Michigan University, Kalamazoo, MI 49001
John Mitchem[2], San Jose State College, San Jose, CA 95114

Dedicated to

Professor E.A. Nordhaus,
Michigan State University

In 1956 the following result by E.A. Nordhaus and J.W. Gaddum [13] appeared in the American Mathematical Monthly.

The Nordhaus-Gaddum Theorem: If G is a graph of order p, then the chromatic numbers $\chi(G)$ and $\chi(\overline{G})$ of G and its complement $\overline{G}$ satisfy the inequalities

(i) $2\sqrt{p} \le \chi(G) + \chi(\overline{G}) \le p + 1$

(ii) $p \le \chi(G) \cdot \chi(\overline{G}) \le \left(\frac{p+1}{2}\right)^2$.

Furthermore, these bounds are best possible for infinitely many values of p.

There is a variety of proofs of these inequalities; however, verification of the upper bound in (i) and of the lower bound in (ii) imply the truth of the remaining two inequalities. In order to see this, we note that (i) and (ii) may be written equivalently as

(i) $\sqrt{p} \le \frac{\chi(G) + \chi(\overline{G})}{2} \le \frac{p+1}{2}$

(ii) $\sqrt{p} \le \sqrt{\chi(G)\cdot\chi(\overline{G})} \le \frac{p+1}{2}$.

If we denote the arithmetic mean of two positive numbers a and b by A.M. (a,b) and their geometric mean by G.M.(a,b), then the upper bound in (i) states that

$$\text{A.M.}(\chi(G),\ \chi(\overline{G})) \le \text{A.M.}\ (1,p)$$

and the lower bound in (ii) gives

$$\text{G.M.}(1,p) \le \text{G.M.}(\chi(G),\ \chi(\overline{G})) .$$

Since the geometric mean of two positive numbers never exceeds their arithmetic mean, it follows that $\text{G.M.}(\chi(G), \chi(\overline{G})) \le \text{A.M.}(\chi(G), \chi(\overline{G}))$. Hence the inequalities (i) and (ii) of the Nordhaus-Gaddum Theorem may be expressed more compactly as

---

[1] Gary Chartrand is the first "academic son" of Professor Nordhaus.

[2] John Mitchem is the first "academic grandson" of Professor Nordhaus.

(iii) $G.M.(1,p) \le G.M.(\chi(G),\chi(\overline{G})) \le A.M.(\chi(G),\chi(\overline{G})) \le A.M.(1,p)$.

We now present a proof of the Nordhaus-Gaddum Theorem, beginning with the lower bound in (ii). In any $\chi(G)$ - coloring of a graph G, there exists a color class with at least $p/\chi(G)$ vertices; hence there exists a complete subgraph of $\overline{G}$ with at least $p/\chi(G)$ vertices. Thus $\chi(\overline{G}) \ge p/\chi(G)$, which gives the desired result.

For the upper bound in (i), we employ a method suggested by H.V. Kronk. By a theorem of Szekeres and Wilf [15], for any graph G,

$$\chi(G) \le 1 + \max_{G'<G} \delta(G') ,$$

where the maximum is taken over all induced subgraphs G' of G. (The symbol $\delta(G')$ denotes the minimum degree among the vertices of G' and $G' < G$ indicates that G' is an induced subgraph of G.)

Suppose $\max_{G'<G} \delta(G') = k$ for a given graph G of order p; hence every induced subgraph of G has minimum degree at most k. We now show that every induced subgraph of $\overline{G}$ has minimum degree at most $p - k - 1$. We note that every induced subgraph of $\overline{G}$ is the complement of an induced subgraph of G, i.e., every induced subgraph of $\overline{G}$ may be represented by $\overline{J}$, where J is an induced subgraph of G. Assume, to the contrary, that there exists an induced subgraph $\overline{H}$ of $\overline{G}$ such that $\delta(\overline{H}) \ge p-k$. Let h be the order of $\overline{H}$ (and, of course, H as well). Every vertex of H has degree at most $h - (p-k) - 1 = h - p+k - 1$ in H. Moreover, the vertices in H have degree at most $(h - p+k - 1) + (p-h) = k - 1$ in G. Since $\max_{G'<G} \delta(G') = k$, there exists an induced subgraph F of G such that $\delta(F) = k$. However, then, no vertex of H can be a vertex of F, i.e., F is an induced subgraph of $G - V(H)$. The subgraph F has at least $k+1$ vertices, implying that H has at most $p - k - 1$ vertices, contradicting the fact that $\delta(\overline{H}) \ge p - k$. We may therefore conclude that

$$\max_{\overline{G}'<\overline{G}} \delta(\overline{G}') = \max_{G'<G} \delta(\overline{G}') \le p-k-1 .$$

Hence,

$$\chi(G) \le 1 + \max_{G'<G} \delta(G') = 1 + k$$

and

$$\chi(\overline{G}) \le 1 + \max_{G'<G} \delta(\overline{G}') \le 1 + (p-k-1) = p-k ,$$

so that

$$\chi(G) + \chi(\overline{G}) \le (1+k) + (p-k) = p + 1 ,$$

thereby verifying the upper bound in (ii) and completing the proof of the Nordhaus-Gaddum Theorem.

For variables x and y, the inequalities $xy \ge p$ and $x + y \le p + 1$ (where p is a positive integer) determine the shaded region R shown in Figure 1.

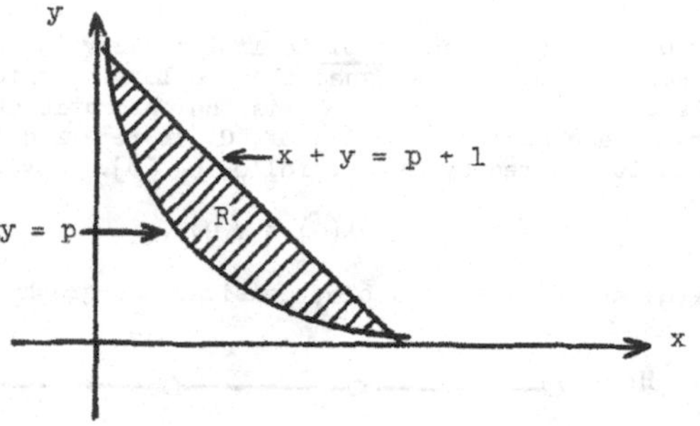

Figure 1

It therefore follows that if G is any graph of order p with $\chi(G) = x_0$ and $\chi(\overline{G}) = y_0$, then (x_0,y_0) is a point of the region R (including its boundary). Conversely, it has been shown independently by Finck [5] and Stewart [14] that if (x_0,y_0) is any lattice point which lies in the region R or on its boundary, then there exists a graph G of order p such that $\chi(G) = x_0$ and $\chi(\overline{G}) = y_0$. This shows the bounds given in (iii) to be best possible for all values of p. In addition, Finck characterized all graphs G such that equality holds in any of the four inequalities given in (i) and (ii).

We now arrive at the main object of this article, namely to present a survey of results which belong to the so-called Nordhaus-Gaddum class. We shall describe this class in some detail. For a given graph-theoretic parameter f and positive integer p, the problem is to determine upper and lower bounds (preferably sharp bounds) for

$$f(G) + f(\overline{G}) \quad \text{and} \quad f(G)\cdot f(\overline{G})$$

where G is a graph of order p. There are several variations of this problem. For example, one might consider distinct but related parameters f_1 and f_2 and develop bounds for

$$f_1(G) + f_2(\overline{G}) \quad \text{and} \quad f_1(G)\cdot f_2(\overline{G}) .$$

In addition, for a parameter f and graphs G and H related in some prescribed manner, the problem exists to investigate bounds for

$$f(G) + f(H) \quad \text{and} \quad f(G)\cdot f(H) .$$

In 1964 Dirac [4] considered a problem of this type. If G and H are edge-disjoint subgraphs of K_p which are critical (with respect to chromatic number), then it is a direct consequence of the Nordhaus-Gaddum Theorem that

$$\chi(G) + \chi(H) \le p + 1 .$$

Dirac generalized this result by showing that if G and H are chromatic-critical subgraphs of K_p having $n(\ge 0)$ edges in common, then

(iv) $$\chi(G) + \chi(H) \le p + \tfrac{1}{2} + \sqrt{2n + \tfrac{1}{4}} .$$

Dirac further characterized all chromatic-critical graphs G and H with the aforementioned properties such that equality holds in (iv).

A coloring of G is called complete if for every two distinct colors, there exist adjacent vertices of G assigned these colors. It is well known that the minimum complete coloring of a graph G is the chromatic number of G. The maximum number of colors in a complete coloring of G is referred to as the achromatic number of G and is denoted by $\psi(G)$, following [8]. Obviously,

$$\chi(G) \le \psi(G) .$$

To show that $\chi(G) < \psi(G)$ is possible, consider the graph H of Figure 2 colored as indicated.

H:

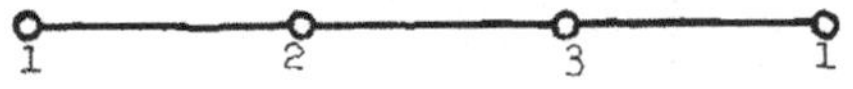

Figure 2

This complete 3-coloring illustrates the fact that $\psi(H) = 3$. Of course, $\chi(H) = 2$.

In [9] Hedetniemi conjectured that the upper bound in (i) can be improved to $\chi(G) + \psi(\overline{G}) \le p+1$ for any graph G of order p. Harary and Hedetniemi [8] (see also Geller and Hedetniemi [6]) then showed the conjecture to be true. This conjecture was also verified independently by Gupta [7] who then proceeded to improve this result. A pseudocomplete coloring of a graph G is not a coloring of G but rather is an assignment of colors to the vertices of G with the property that for every two distinct colors there exist adjacent vertices assigned these colors. We note that in a pseudocomplete coloring, adjacent vertices need not be assigned distinct colors. The pseudoachromatic number $\psi_s(G)$ of G is the maximum number of colors in a pseudocomplete coloring of G. Hence,

$$\psi(G) \le \psi_s(G) .$$

It was proved by Gupta that for any graph G of order p,

$$\chi(G) + \psi_s \le p+1 .$$

In addition, Gupta proved that for any graph G of order p,

$$\psi(G) + \psi(\overline{G}) \le \left\{\tfrac{4}{3}p\right\} ,$$

$$\psi(G) + \psi_s(\overline{G}) \le \left\{\tfrac{4}{3}p\right\} ,$$

$$\psi_s(G) + \psi_s(\overline{G}) \le \left\{\tfrac{4}{3}p\right\} ,$$

where all bounds are best possible for infinitely many values of p.

We also remark that a theorem of the Nordhaus-Gaddum class exists in the case of the edge-chromatic number $\chi_1(G)$; in particular, Alavi and Behzad[3] [1] and Vizing [17] verified independently that for any graph G of order p,

[3]Mehdi Behzad of the National University of Iran is the second "academic son" of Professor Nordhaus.

$$2\left[\frac{p+1}{2}\right] - 1 \le \chi_1(G) + \chi_1(\overline{G}) \le p + 2\left[\frac{p-2}{2}\right]$$

$$0 \le \chi_1(G) \cdot \chi_1(\overline{G}) \le (p-1)(2\left[\frac{p}{2}\right] - 1) .$$

It might also be noted that each of the aforementioned proofs makes use of the inequality proved by Vizing [16] :

$$\chi_1(G) \le 1 + \Delta(G) ,$$

where $\Delta(G)$ denotes the maximum degree among the vertices of G. The fact that no theorem of the Nordhaus-Gaddum class has been established in the case of the total chromatic number $\chi_2(G)$ may very well be due to the fact that no sharp upper bound for $\chi_2(G)$ has been found in terms of $\Delta(G)$. It has been conjectured by Behzad [3], however, that

$$\chi_2(G) \le 2 + \Delta(G) .$$

We now discuss Nordhaus-Gaddum problems with respect to non-chromatic parameters. The <u>vertex-arboricity</u> $a(G)$ of a graph G is the fewest number of subsets in a partition of the vertex set of G such that each subset induces an acyclic subgraph. Of course, $a(G) \le \chi(G)$ for any graph G. Indeed, it follows that $a(G) \le \chi(G) \le a(G)/2$. With the aid of this observation, Mitchem [12] proved the following inequalities for any graph G of order p:

$$\text{(v)} \qquad \sqrt{p} \le a(G) + a(\overline{G}) \le \frac{p+3}{2}$$

$$\text{(vi)} \qquad \frac{p}{4} \le a(G) \cdot a(\overline{G}) \le \left(\frac{p+3}{4}\right)^2 .$$

As with the Nordhaus-Gaddum Theorem, it is possible to show that the truth of the upper bound in (v) and lower bound in (vi) implies the remaining two inequalities. Mitchem further showed that given positive integers x_o, y_o, and p such that

$$x_o + y_o \le \frac{p+3}{2} \quad \text{and} \quad \frac{p}{4} \le x_o y_o ,$$

there exists a graph G of order p for which $a(G) = x_o$ and $a(\overline{G}) = y_o$.

In [10] Lick and White[4] introduced the concept of "k-degenerate graphs". A graph G is called <u>k-degenerate</u> if $\delta(H) \le k$ for every induced subgraph H of G. The parameter <u>$\rho_k(G)$ of a graph</u> G is then defined as the minimum number of subsets in a partition of the vertex set of G such that each subset induces a k-degenerate graph. The 0-degenerate graphs are the empty graphs (graphs with no edges) while the 1-degenerate graphs are precisely the acyclic graphs (forests); hence $\rho_o(G) = \chi(G)$ and $\rho_1(G) = a(G)$, i.e., the parameters $\rho_k(G)$ serve to generalize the chromatic number and vertex-arboricity.

Lick and White [11] then generalized the Nordhaus-Gaddum Theorem and the aforementioned result of Mitchem by proving for any nonnegative integer k and graph G

---

[4]Arthur White of Western Michigan University is the third "academic son" of Professor Nordhaus.

ff order p

$$\frac{\sqrt{p}}{k+1} \le G.M.(\rho_k(G),\rho_k(\overline{G})) \le A.M.(\rho_k(G),\rho_k(\overline{G})) \le \frac{p+2k+1}{2k+2} .$$

They have further shown that $\sqrt{p}/(k+1)$ is not a sharp bound, in general.

In [2] Alavi and Mitchem considered problems of the Nordhaus-Gaddum type for the parameters connectivity $\kappa(G)$ and edge-connectivity $\kappa_1(G)$. In particular, they showed for any graph G of order p that

$$1 \le \kappa_1(G) + \kappa_1(\overline{G}) \le p - 1 \quad \text{and}$$

$$0 \le \kappa_1(G) \cdot \kappa_1(\overline{G}) \le M(p),$$

where $M(p) = \left[\frac{(p-1)^2}{4}\right]$ if $p \not\equiv 3 \pmod 4$ and $M(p) = \frac{p^2-2p-3}{4}$ if $p \equiv 3 \pmod 4$. Moreover, all bounds are best possible. The same bounds hold for connectivity, but the upper bound for $\kappa(G)\cdot\kappa(\overline{G})$ has not been proved to be sharp although it is believed to be so.

We have presented a number of results which belong to the Nordhaus-Gaddum class. There are many open questions in this area, however; indeed there is probably one for nearly every graphical parameter not mentioned here. Thus the number of future theorems of the Nordhaus-Gaddum class appears to be unlimited.

REFERENCES

1. Y. Alavi and M. Behzad, "Complementary Graphs and Edge Chromatic Numbers", J. SIAM Appl. Math. (to appear).
2. Y. Alavi and J. Mitchem, "Connectivity and Line-Connectivity of Complementary Graphs", (this volume).
3. M. Behzad, Graphs and Their Chromatic Numbers, Doctoral thesis, Michigan State University, 1965.
4. G.A. Dirac, "Graph Union and Chromatic Number", J. London Math. Soc., 39 (1964), 451-454.
5. H.J. Finck, "On the Chromatic Numbers of a Graph and its Complement", Theory of Graphs (P. Erdos and G. Katona, eds.) Academic Press (1968), 99-113.
6. D. Geller and S. Hedetniemi, "A Proof Technique in Graph Theory", Proof Techniques in Graph Theory (F. Harary, ed.) Academic Press (1969), 49-59.
7. R.P. Gupta, "Bounds on the Chromatic and Achromatic Numbers of Complementary Graphs", Recent Progress in Combinatorics (W.T. Tutte, ed.) Academic Press (1969), 229-235.
8. F. Harary and S. Hedetniemi, "The Achromatic Number of a Graph", J. Combinatorial Theory, 8 (1970), 154-161.
9. S. Hedetniemi, "Homomorphisms of Graphs and Automata", University of Michigan, Tech. Rept., pp. 24-25. Ann Arbor, Michigan (1966).

10. D.R. Lick and A.T. White, "k-Degenerate Graphs", Canad. J. Math. (to appear).

11. D.R. Lick and A.T. White, "Point Partition Numbers of Graphs and Their Complements", (in preparation).

12. J. Mitchem, _On Extremal Partitions of Graphs_, Doctoral Thesis, Western Michigan University, 1970.

13. E.A. Nordhaus and J.W. Gaddum "On Complementary Graphs", Amer. Math. Monthly, 63 (1956), 175-177.

14. B.M. Stewart, "On a Theorem of Nordhaus and Gaddum", J. Combinatorial Theory, 6 (1969), 217-218.

15. G. Szekeres and H.S. Wilf, "An Inequality for the Chromatic Number of a Graph", J. Combinatorial Theory, 4 (1968), 1-3.

16. V.G. Vizing, "On an Estimate of the Chromatic Class of a p-Graph", (Russian), Diskret. Analiz. 3 (1964), 25-30.

17. V.G. Vizing, "The Chromatic Class of a Multigraph", Cybernetics, 1 (1965), No. 3. 32-41.

ISOMETRIC GRAPHS

Gary Chartrand[1], Western Michigan University, Kalamazoo, MI 49001
M. James Stewart, Lansing Community College, Lansing, MI 48910

Although there are several procedures available for associating a metric space with a connected graph G, the most common one is obtained by defining the distance $d(u,v)$ between two vertices u and v of G to be zero if $u = v$ and as the length of any shortest u-v path of G for $u \neq v$. This concept of distance in a connected graph is encountered in many aspects of the theory of graphs. For example, even the fundamental concept of "graph isomorphism" can be described via distance. The usual definition of isomorphism is as follows: Two graphs G_1 and G_2 are isomorphic if there exists a one-to-one correspondence

$$\phi: V(G_1) \to V(G_2)$$

such that u and v are adjacent vertices in G_1 if and only if ϕu and ϕv are adjacent vertices in G_2. An equivalent definition may now be given which employs the aforementioned metric d. Two connected graphs G_1 and G_2 are isomorphic if there exists a one-to-one correspondence

$$\phi: V(G_1) \to V(G_2)$$

such that $d(u,v) = d(\phi u, \phi v)$ for all $u, v \in V(G_1)$. One may then define isomorphism between two arbitrary graphs G_1 and G_2 by either considering their components or defining $d(w_1,w_2) = \infty$ if no w_1-w_2 path exists in the graph under discussion.

If, instead of considering the vertices of G_1 in total, we consider the vertices of G_1 individually, then the preceding definition of isomorphism suggests the following concept: A connected graph G_2 is said to be isometric from a connected graph G_1 if for each vertex v of G_1 there exists a one-to-one correspondence

$$\phi_v: V(G_1) \to V(G_2)$$

such that

$$d(v,u) = d(\phi_v v, \phi_v u)$$

for all $u \in V(G_1)$. For example, the graph G_2 of Figure 1 is isometric from the graph G_1. This can be seen by making the following definitions

---

[1] Research supported in part by National Science Foundation Grant GP-9435.

$$\phi_u u = c \qquad \phi_v v = b \qquad \phi_w w = c \qquad \phi_x x = b$$

$$\phi_v x = d \qquad\qquad \phi_x v = d ,$$

under the added restrictions that each of the functions ϕ_u, ϕ_v, ϕ_w, and ϕ_x are one-to-one mappings onto $V(G_2)$.

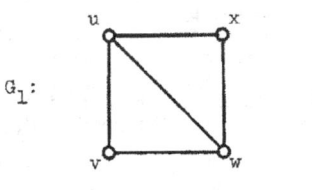

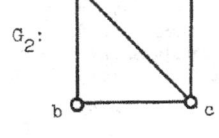

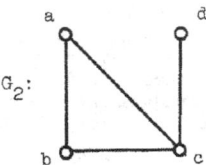

Figure 1. a graph G_2 isometric from a graph G_1

It should be noted that if G_2 is isometric from G_1 then for each such mapping ϕ_v described in the definition we have

$$\deg v = \deg \phi_v v.$$

With this observation it is now evident that the graph G_1 of Figure 1 is not isometric from G_2 since there is no mapping ϕ_d with the desired properties.

Should G_1 and G_2 be graphs which are isometric from each other, then we say G_1 and G_2 are isometric graphs. While isomorphic graphs are certainly isometric, the converse is not true in general; for instance, the graphs G_1 and G_2 of Figure 2 are isometric though not isomorphic.

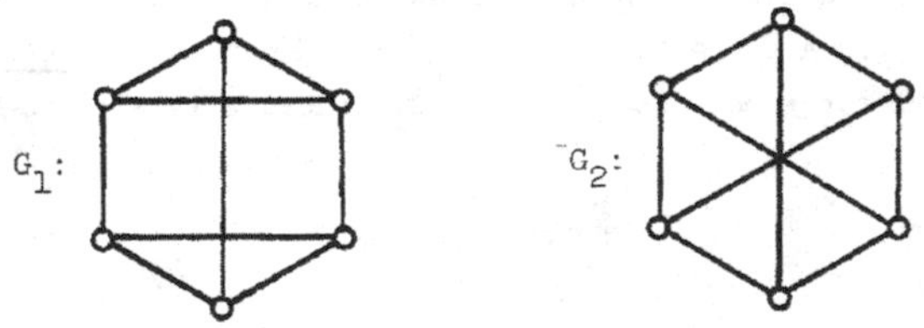

Figure 2. isometric, non-isomorphic graphs

The graphs of Figure 2 actually serve to illustrate the following result.

Theorem 1. If G_1 and G_2 are n-regular graphs of order p, where $n \geq (p-1)/2$, then G_1 and G_2 are isometric.

Proof. Let $u \in V(G_1)$ and $v \in V(G_2)$, and define $\phi_u u = v$. Further, if u_i, $1 \leq i \leq n$, are the vertices adjacent to u and v_i are the vertices adjacent to v, then let $\phi_u u_i = v_i$ for $1 \leq i \leq n$. Now extend ϕ_u so that it is a one-to-one mapping from $V(G_1)$ onto $V(G_2)$. The fact that $n \geq (p-1)/2$ implies that the distance between any two vertices of G_j, $j = 1, 2$, does not exceed 2. Thus

$$d(u,w) = d(v, \phi_u w)$$

for all $w \in V(G_1)$ which gives the desired result.

Although there exist a great many connected graphs G_1 for which there exists no graph G_2 (different from G_1) such that G_2 is isometric from G_1, it does follow that for each $v \in V(G_1)$ there is a graph G_v and a one-to-one correspondence ϕ_v: $V(G_1) \to V(G_v)$ such that

$$d(v,u) = d(\phi_v v, \phi_v u)$$

for all $u \in V(G_1)$. Indeed, it has been shown by Ore [1, p.103] that each such G_v may be chosen as a spanning tree of G_1. Since the proof of this result is of importance, we include it here.

Theorem 2. (Ore) For every connected graph G and vertex v of G, there exists a spanning tree T_v of G such that

$$d_G(v,u) = d_{T_v}(v,u)$$

for all $u \in V(G)$.

Proof. Denote by $d_i(v)$ those vertices of G at a distance i from v, $i = 0, 1, 2, \ldots$. Since G is connected, for $u \neq v$, it follows that $u \in d_j(v)$ for some $j \neq 0$. Hence each such vertex u is adjacent to at least one vertex in $d_{j-1}(v)$ and possibly also vertices in $d_j(v)$ and $d_{j+1}(v)$. Delete all edges, except one, of the type uw, $w \in d_{j-1}(v)$, together with every edge of the type uw, $w \in d_j(v)$. In the graph obtained by removing these edges, the distance between v and u is unaffected. This procedure is now employed for all $u \neq v$ resulting in a graph T_v. In order to see that T_v is a tree, suppose T_v contains a cycle C, and let w_1 be a vertex of C whose distance from v in T_v is maximum. Assume $w_1 \in d_k(v)$; then if w_2 and w_3 are adjacent to w_1 on C, $w_i \in d_{k-1}(v) \cup d_k(v)$, $i = 2, 3$. If $w_2 \in d_k(v)$ or $w_3 \in d_k(v)$, we have a contradiction. Thus $w_2 \in d_{k-1}(v)$ and $w_3 \in d_{k-1}(v)$ which, again, contradicts the way T_v was constructed. This completes the proof.

A spanning tree T_v of a graph G satisfying Theorem 2 is called an isometric tree at v. If there is only one such tree T_v (up to isomorphism) for a given vertex v of G, then v has a unique isometric tree. If G has the same unique isometric tree at each of its vertices, then G is said to have a unique isometric tree. For example, the graph G_1 of Figure 1 has a unique isometric tree at each of its vertices but does not have a unique isometric tree. On the other hand, the regular complete bipartite graph $K(3,3)$ does not have a unique isometric tree at any of its vertices while $K(2,2)$ has a unique isometric tree. Another graph with a unique isometric tree is the famed Petersen graph (see Figure 3).

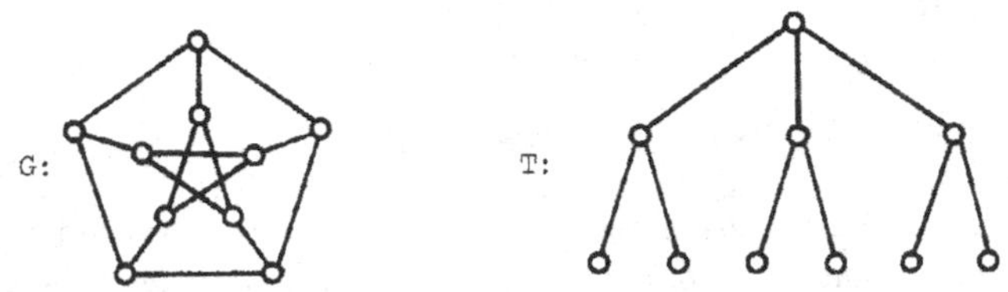

Figure 3. the Peterson graph G and its unique isometric tree T

We now consider graphs having unique isometric trees. Obviously, each tree possesses a unique isometric tree; therefore, we henceforth consider only cyclic graphs (connected graphs with cycles).

It is well known that the radius and diameter of a connected graph G, denoted $\text{rad}\ G$ and $\text{diam}\ G$ respectively, satisfy the inequalities

$$\text{rad}\ G \leq \text{diam}\ G \leq 2\ \text{rad}\ G .$$

Moreover, there are infinitely many non-isomorphic graphs G for which $\text{rad}\ G = \text{diam}\ G$ and infinitely many for which $\text{diam}\ G = 2\ \text{rad}\ G$. In connection with the latter class of graphs, we present the following result.

Theorem 3. Let G be a connected graph with $\text{diam}\ G = 2\ \text{rad}\ G$ such that G has a unique isometric tree. Then the endpoints of every diametrical path of G have degree 1.

Proof. Let u and v be the endpoints of a diametrical path P, i.e., P is a shortest u-v path and the length of P has the value $\text{diam}\ G$. Also, let w be a central vertex of G.

Suppose u, say, has degree exceeding 1, and consider the unique isometric tree T_u at u. Referring to the proof of Theorem 2, we see that T_u may be chosen so that it contains the path P. Necessarily, P contains only one vertex adjacent to u, but since u has degree at least 2, there exists a vertex u_1 adjacent to u in T_u such that u_1 does not lie on P. Hence the $u_1 - v$ path in T_u has length $1 + \text{diam}\ G$.

We next consider the tree T_w. Let w_1 and w_2 be distinct vertices of G different from w. Since w belongs to the center of G, the $w_1 - w$ path P_1 and $w_2 - w$ path P_2 in G each has length at most $2\ \text{rad}\ G = \text{diam}\ G$. Since w_1 and w_2 were selected arbitrarily, it follows that no path in T_w has length exceeding $\text{diam}\ G$. This shows that $T_w \neq T_u$ which contradicts the hypothesis that G has a unique isometric tree and proves the result.

Another necessary condition for a graph to possess a unique isometric tree is given in the next theorem. We denote the maximum degree among the vertices of a graph G by $\Delta(G)$.

Theorem 4. If G is a cyclic graph having a unique isometric tree, then G has at least two vertices of degree $\Delta(G)$.

Proof. Suppose v is the unique vertex of G such that $\deg v = \Delta(G)$. We now consider two cases.

Case 1. Assume v *belongs to cycles of* G. Let C be a cycle of minimum length n containing v. Then C contains one or two vertices (depending on whether n is even or odd) whose distance from v in G is $\left[\frac{n}{2}\right]$. Let u be such a vertex, and consider T_u. Again referring to the proof of Theorem 2, we have $v \in d\left[\frac{n}{2}\right](u)$. If n is even, v is adjacent with two vertices on C which belong to $d\left[\frac{n-2}{2}\right](u)$, while if n is odd, then v is adjacent to a vertex of C which also belongs to $d\left[\frac{n}{2}\right](v)$. In either case, an edge incident with v may be deleted in constructing T_u. This implies that in T_u, no vertex has degree $\Delta(G)$. However, the vertex v itself has degree $\Delta(G)$ in T_v; thus $T_v \neq T_u$. This contradicts the fact that G has a unique isometric tree.

Case 2. Assume every edge incident with v *is a bridge.* In this case, it now follows that the isometric tree at each vertex necessarily contains exactly one vertex of degree $\Delta(G)$; hence we arrive at a contradiction in a different manner here. Let w be a vertex belonging to a cycle of G at minimum distance m from v. It follows immediately now that the degrees of the vertices in T_v at a distance m from v are precisely the same as the degrees of the vertices in G at a distance m from v (where in all graphs we label the vertex of degree $\Delta(G)$ by v).

We now proceed as in Case 1. Let C be a cycle of minimum length n containing w, and let w_1 be one of the (possibly two) vertices of C at a distance $\left[\frac{n}{2}\right]$ from w. In the construction of T_{w_1}, at least one edge incident with w is necessarily deleted, reducing the degree of a vertex at a distance m from v in T_{w_1}. However, in T_{w_1} no other vertex at a distance m from v has its degree altered. In any case, the degrees of the vertices at a distance m from v in T_{w_1} are not the same as the corresponding numbers in T_v. Hence $T_v \neq T_{w_1}$, which again produces a contradiction and completes the proof.

We conclude with the following.

Conjecture. If a graph G has a unique isometric tree, then G has at most one cyclic block.

REFERENCE

1. O. Ore, *Theory of Graphs*, Amer. Math. Soc. Colloq. Publ. Vol. 38, Providence (1962).

THE FREQUENCY PARTITION OF A GRAPH

Phyllis Zweig Chinn
Towson State College, Baltimore, MD 21204

A partition of a nonnegative integer n is a sequence of nonnegative integers $n_1,\dots,n_r$ where $n_1 \geq n_2 \geq \dots \geq n_r$ and $n = n_1 + n_2 + \dots + n_r$. Such a partition is said to be of length r. To indicate which integers are repeated in the sequence, the partition $n_1,n_1,\dots,n_1,n_2,\dots,n_2,\dots,n_k,\dots,n_k$ where n_j occurs i_j times, and $n_1 > \dots > n_k$ will be written

$$\left[n_1^{i_1},\dots,n_j^{i_j},\dots,n_k^{i_k}\right].$$

The degree sequence of a (p,q)-graph, after a possible reordering, forms a partition of 2q. This partition is of length p, and is usually called the partition of a graph. Only some partitions of an integer 2q are partitions of graphs, and those which are, are usually called graphical.

Associated with the partition $\left[n_1^{i_1},\dots,n_k^{i_k}\right]$ is the sequence $i_1,\dots,i_k$. Properly ordered, this is a partition of the length of the original partition. The new partition may be called the frequency partition of the original partition. This process, applied to a graphical partition, yields a partition of p called the frequency partition of the graph. This partition may be obtained directly from the graph by recording the frequencies with which the various degrees in the degree set of the graph G are assumed.

Theorem 1. A graph and its complement both have the same frequency partition.

Proof. Assume that G has frequency partition $i_1,\dots,i_k$. Because G has exactly i_j points of degree d_j, its complement has exactly i_j points of degree $p-1-d_j$, and thus has the same frequency partition.

Since every graph must have at least two points of some degree, it is clear that there is no p-graph with frequency partition $[1^p]$.

<u>Theorem</u> 2. Given any partition of an integer $p \geq 2$, other than 1^p there is at least one connected p-graph having this partition as its frequency partition.

<u>Proof</u>. If $p = 2$, the only connected graph of order 2 has two points of the same degree (namely 1).

Assume the theorem holds for any partition of all nonnegative integers not greater than p, and consider a partition of $p + 1$ given by $n_1, n_2, \ldots, n_k$ where $n_1 > 1$ and $k \geq 2$. The integers $n_1, \ldots, n_{k-1}$ form a partition of $p+1-n$ with $n > 1$ and $p+1-n_k \leq p$. By the induction hypothesis, there is a connected graph with this as frequency partition. Add n_k isolated points to this graph. The new graph has the proper frequency partition but it is not connected. Its complement has the same frequency partition and is connected.

If $k = 1$, the complete (p+1)-graph has frequency partition $(p=1)^1$ and this completes the proof.

A GRAPH WITH p POINTS AND ENOUGH DISTINCT (p-2)-ORDER SUBGRAPHS IS RECONSTRUCTIBLE

Phyllis Zweig Chinn
Towson State College, Baltimore, MD 21204

Finite graphs with no loops or multiple edges will be considered. A _p-graph_ has p points; a _(p,g)-graph_ has p points and q lines.

Given the collection $G_i = G-v_i$ of induced (p-1)-subgraphs of a p-graph G. Consider the $\binom{p}{p-1}$ subgraphs induced by removing any point from any of the G_i.

Theorem. If a graph has the property that each induced (p-2)-subgraph of, say, G_1 occurs exactly once as an induced (p-2)-subgraph of any G_i, $i \neq 1$, then the graph G is reconstructible.

Since the collection G_i corresponds to a graph G, each induced (p-2)-subgraph of G will appear as an induced subgraph of two of the G_i. Thus, the hypothesis of the theorem corresponds to a condition that the (p-2)-subgraphs of G induced by removing v_1 and a second point are unique in type among all the (p-2)-subgraphs of G.

The theorem may be proven as follows.

Proof. Remove a point from G_1 and find the subgraph, say G_k, which contains the induced subgraph isomorphic to this (p-2)-subgraph. The point chosen must correspond to v_k in G. In this way G_1 can be uniquely labelled. Since the degree of each point in G can be determined from the collection G_i, those points of G_1 which are adjacent to v_1 in G must be exactly those points whose degree in G_1 is one less than in G.

A graph and its complement clearly satisfy the conditions of the theorem simultaneously.

No p-graphs for $p \leq 6$ satisfy the conditions of the theorem. For p = 7, several examples can be found, where a unique point functions as v_1. Figure 1 shows one of these graphs.

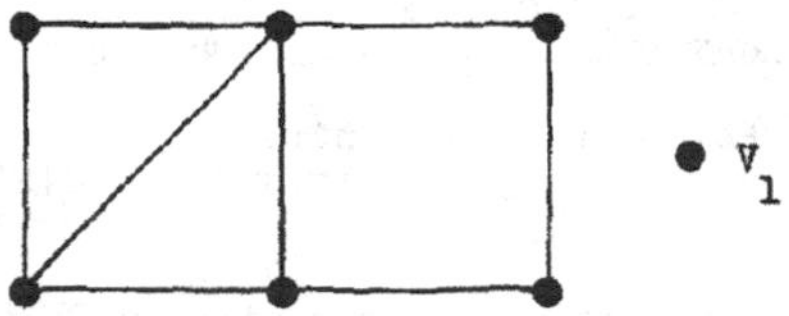

Figure 1

While the reconstruction problem has been solved for trees, it may be of some interest to note that no tree of order ≤ 10 satisfies the conditions of this theorem, and the 18 order tree of Figure 2 is the smallest tree the author has found thus far which does satisfy the conditions.

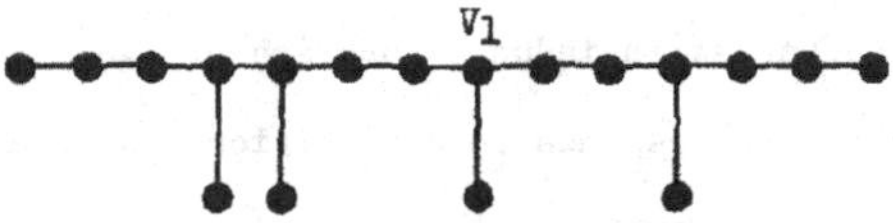

Figure 2

Regular graphs are also not of interest in terms of reconstruction, but the graph of Figure 3 gives another example of the prevalence of graphs satisfying the hypothesis.

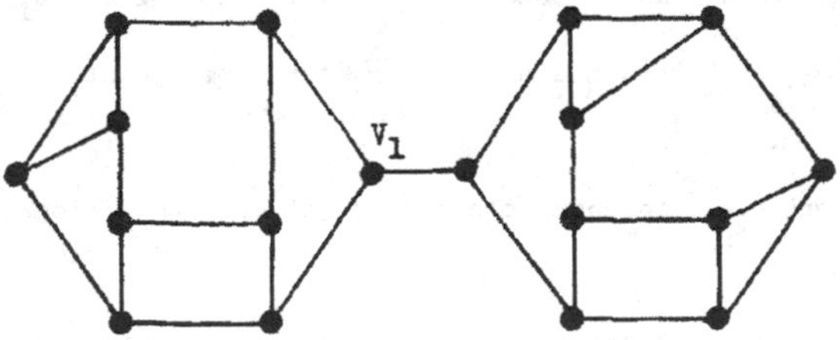

Figure 3

A necessary condition for graphs to satisfy the hypothesis of the theorem is that it have trivial automorphism group. Leo Moser, in an informal conversation, conjectured that "most" graphs have a trivial automorphism group. Thus there is a possibility that a large percentage of high order graphs are reconstructible by this theorem.

A stronger condition which might be easier to check for a particular collection of graphs G_i is that all the induced (p-2)-subgraphs of G are distinct (i.e. are induced in pairs among the G_i). Figure 4 shows an example of such a graph. It may be minimal in the sense that if a (p,g)-graph has $\binom{p}{2}$ distinct types of induced (p-2)-subgraphs, then $p < 13$ or $q < 17$. The author has shown that no p-graph for $p \leq 7$ has this property. It is easy to show that no tree has all induced (p-2)-subgraphs distinct.

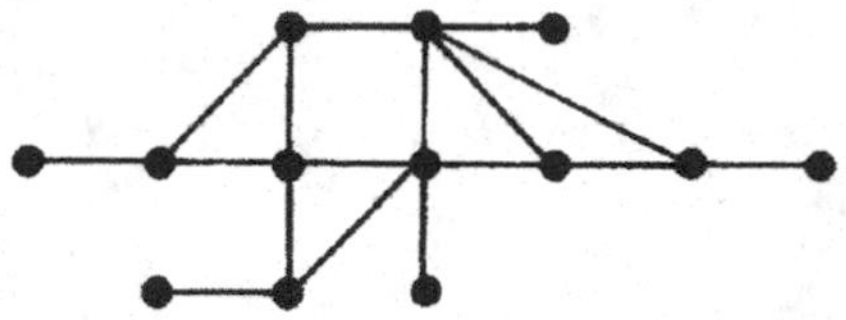

Figure 4

Acknowledgement. Thanks are due to Paul J. Kelly for his valuable comments.

ON A RESULT OF GOETHALS AND SEIDEL

Jane W. Di Paola
New York University
University Heights, Bronx, N.Y. 10453

Introduction

To recognize whether a balanced incomplete block design with appropriate parameters may be embedded as a residual design in a symmetric balanced incomplete block design is in general a very complicated problem. We show in this paper that a recent result of Goethals and Seidel [3] when combined with a theorem of Shrikhande [7] has as a corollary: An affine design with two points on a line is embeddable. While our exposition is straight-forward, this paper is not entirely self-contained. Some knowledge of graph theory and of block designs is assumed.

Notation

We use v, b, r, k, λ as the parameters of a balanced incomplete block design, hereinafter to be called simply a design. The parameter v is the number of elements, b the number of blocks, r the number of times each element appears, k the number of elements in a block, and λ the number of appearances of each pair of elements together in the same block. (See [2] or [5]). A design is said to be quasisymmetric if for any pair of distinct blocks b_i, b_j we have $|b_i \cap b_j| = \mu$ or η, with $\mu > \eta \geq 0$, and to avoid trivialities, we shall assume that in a quasisymmetric design there exist at least one pair b_s, b_t with $|b_s \cap b_t| = \mu$ and at least one pair b_p, b_q with $|b_p \cap b_q| = \eta$.

We define the (block) graph of a quasisymmetric design by taking the blocks as vertices and defining distinct blocks b_i, b_j to be adjacent if $|b_i \cap b_j| = \mu$ and non-adjacent if $|b_i \cap b_j| = \eta$.

We consider b_1 non-adjacent to b_1. Let K_m denote the complete graph on m vertices. Let $K_m(n)$ denote the graph which consists of n copies of K_m. Denote by $H_m(n)$ the complement of $K_m(n)$.

<u>A Result of Goethals and Seidel</u>

Goethals and Seidel [3] have studied designs having graph $H_2(n)$ and have shown that the only quasisymmetric designs having graph $H_2(n)$ have as parameters

$$\begin{aligned} v &= n + 1 \\ b &= 2n \\ r &= n \\ k &= \tfrac{1}{2}(n+1) \\ \lambda &= \tfrac{1}{2}(n-1) \\ \mu &= \tfrac{1}{4}(n+1) \\ \eta &= 0 \end{aligned} \tag{1}$$

Since μ is an integer it follows that $n \equiv 3 \pmod 4$. It is shown by Goethals and Seidel [3] and independently by Dembowski [2] that such designs exist if and only if there exists a normalized Hadamard matrix of order n+1.

<u>Embeddability of Affine Designs</u>

The preceding results are related to the problem of the embeddability of affine designs. By definition, a <u>parallelism</u> of an arbitrary (balanced incomplete block) design D is an equivalence relation, $/\!/$, among the blocks of D satisfying the following parallelism condition:

(2) Let p be any point of the design and B any block. Then there exists on p a unique block C such that $B /\!/ C$.

We note that if $p \in B$, then B = C.

An <u>affine design</u> is a design with a parallelism satisfying the condition: there exists an integer $\mu > 0$ such that, if, for any pair of blocks B, C, B is not parallel to C, then $|B \cap C| = \mu$. (For these and related definitions see Dembowski [2]).

It is clear that an affine design is quasisymmetric with $\eta = 0$. Conversely, it is clear from the definitions that a quasisymmetric design with graph $H_m(n)$ and $\eta = 0$ is an affine design. The crucial point here is that any quasisymmetric design with graph $H_2(n)$ has $\eta = 0$. This is shown in the algebraic part of the Goethals and Seidel work [3] and also appears in the work of Stanton and Kalbfleisch [8] which we shall discuss later in this paper.

It is known that in a design with parallelism every parallel class of blocks contains equally many blocks. Denote the number of blocks in each parallel class by m. This implies that an affine design has graph $H_m(n)$.

The parameters of an affine design are expressible in terms of m and μ as

$$\begin{aligned} v &= \mu m^2 \\ b &= m(\mu m^2-1)(m-1)^{-1} \\ r &= (\mu m^2-1)(m-1)^{-1} \\ k &= \mu m \\ \lambda &= (\mu m-1)(m-1)^{-1} \end{aligned} \tag{3}$$

This relation was found by Bose [1]. We note that the relation of Hadamard matrices to affine designs gives a restriction on λ when $m = 2$, that is, we must have λ odd for affine designs with $m = 2$.

The parameter m has additional significance. A set L of points of a design is said to be a <u>line</u> if, for some two distinct points, x and y, L consists of all points incident with every block containing the 2-subset (x,y). In an affine design each line contains m points. An affine design with $\lambda = 1$, and $\mu = 1$, is embeddable. For other cases Shrikhande [7] has shown the following:

<u>Theorem 1</u>. (Shrikhande). Let A be an affine design with $\lambda > 1$ and parameters as in (3). Then A is embeddable if and only if there exists a symmetric balanced incomplete block design D_o with parameters

$$\begin{aligned} v_0 &= b_0 = (\mu m^2 - 1)(m-1)^{-1} \\ r_0 &= k_0 = (\mu m - 1)(m-1)^{-1} \qquad (4) \\ \lambda_0 &= (\mu - 1)(m-1)^{-1} \end{aligned}$$

<u>Affine Designs with m = 2</u>

It is well-known that a symmetric (balanced incomplete block) design can be obtained from a normalized Hadamard matrix. (Cf. Hall [5]). Designs so constructed are called <u>Hadamard designs</u>. For affine designs with m = 2, the design D_0 required by Shrikhande in Theorem 1 is the design derived from the Hadamard matrix promised by Goethals and Seidel in showing that affine designs with m = 2 exist if and only if an appropriate Hadamard matrix exists. We combine these results in the following theorem for which a constructive proof is given.

<u>Theorem 2</u>. (Goethals, Seidel, Shrikhande). An affine design with m = 2 is embeddable as a residual design in an appropriate symmetric design.

<u>Proof</u>. Let A be an affine design with m = 2. Then, as noted above, A is quasisymmetric, has parameters given in (1) and has graph $H_2(n)$. The symmetric design D in which A will be shown to be embeddable has parameters $v, k, \lambda = 2n+1, n, \frac{1}{2}(n-1)$.

Denote the set of blocks of A by

$$B' = \{b_1', b_2', \ldots, b_{2n}'\} .$$

Consider the complement of the graph of A. Since A has graph $H_2(n)$ the complement is the "ladder graph" on 2n vertices. We obtain the complement of $H_2(n)$ from the set of blocks B' as follows. Let b_i' be adjacent to b_j' if $|b_i' \cap b_j'| = 0$. Since m = 2 each b_i' is disjoint from one and only one b_j'. Thus we have n copies of the complete graph on 2 vertices. Without loss of generality we list the blocks of B' so that b_r', b_s' are adjacent (disjoint) if and only if $s-r = n$. The graph is shown in Figure 1.

Figure 1.

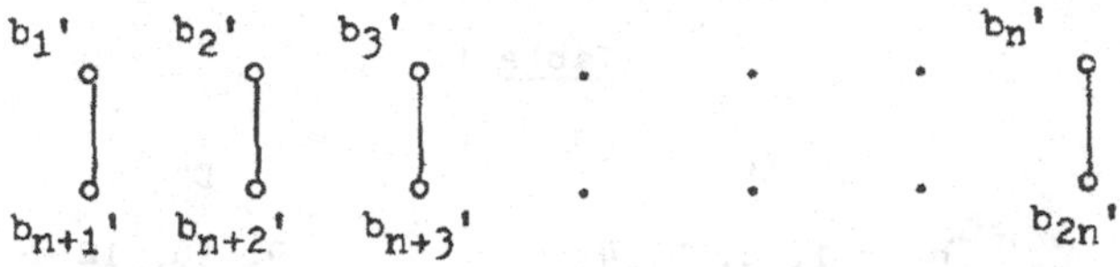

The existence of A insures the existence of a normalized Hadamard matrix of order n+1 which in turn yields a Hadamard design D* which is a symmetric design with parameters $v^*, k^*, \lambda^* = n, \frac{1}{2}(n-1), \frac{1}{4}(n-3)$. We use two copies of the design D* as the derived design D" of the required symmetric design D. Designate the set of blocks of D* by

$$B^* = \{b_1^*, b_2^*, \ldots, b_n^*\}.$$

Let V, B denote the set of points and the set of blocks respectively of the design D. Let V', B' denote these sets for A and V*, B* those for D*. Then the required symmetric design D has

$$V = V' \cup V^*$$

$$B = (v_1^*, v_2^*, \ldots, v_n^*), (b_i' \cup b_i^*), (b_{i+1}' \cup b_i^*) \quad \text{for } i = 1, \ldots, n.$$

An Illustrative Example

We show that design No. 13 of Hall [5] with $v, b, r, k, \lambda = 8, 14, 7, 4, 3$ is embeddable. If the blocks are listed in the order given by Hall (See D' in Table 1) the appropriate graph which is the complement of $H_2(7)$ is shown in Figure 2. The appropriate Hadamard design is obtained from D' by deleting one element and determining the structure of the $(k-1)$-element subsets of the blocks originally containing the deleted element. In this D* is isomorphic to PG(2,2). D" consists of two copies of D*.

Figure 2.

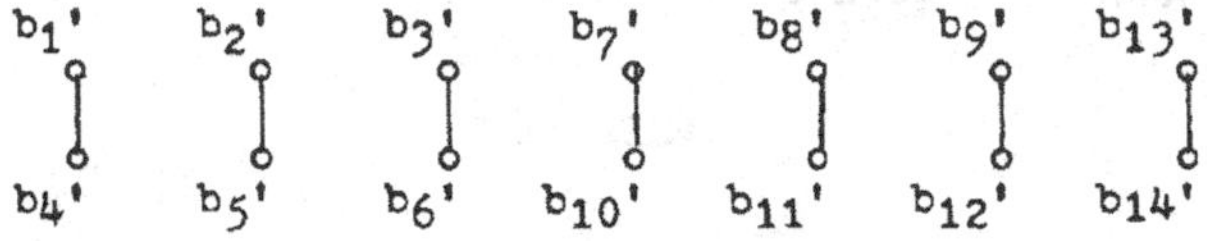

Table 1.

	D'	D"
b_1 =	1, 2, 3, 4,	9, 10, 12
b_2 =	1, 2, 7, 8,	10, 11, 13
b_3 =	1, 3, 6, 8,	11, 12, 14
b_4 =	5, 6, 7, 8,	9, 10, 12
b_5 =	3, 4, 5, 6,	10, 11, 13
b_6 =	2, 4, 5, 7,	11, 12, 14
b_7 =	1, 4, 6, 7,	12, 13, 15
b_8 =	1, 2, 5, 6,	13, 14, 9
b_9 =	1, 3, 5, 7,	14, 15, 10
b_{10}=	2, 3, 5, 8,	12, 13, 15
b_{11}=	3, 4, 7, 8,	13, 14, 9
b_{12}=	2, 4, 6, 8,	14, 15, 10
b_{13}=	1, 4, 5, 8,	15, 9, 11
b_{14}=	2, 3, 6, 7,	15, 9, 11

b_{15} = 9, 10, 11, 12, 13, 14, 15

Reading across the rows of Table 1 gives the required symmetric design in which Hall's No. 13 is now embedded. The given affine design (No. 13) appears as the residual design D' and D" is the derived design and b_{15} the block added to complete the design D.

Quasisymmetric Designs

Following Shrikhande and Goethals and Seidel we have defined quasisymmetric designs as those having (for distinct blocks) only two intersection numbers μ and η and have specified $\mu > \eta$. The use of μ as the adjacency number for the block graph of the design is arbitrary. Had we chosen η to express adjacency we would have obtained the complementary graph. This arbitrary choice is avoided by Stanton and Kalbfleisch [8] who call a design s-quasisymmetric if for each block b_i there are s blocks of the form b_j having the property that $|b_i \cap b_j| = x$ and $b-(s+1)$ blocks having the property that $|b_i \cap b_j| = y$. It is shown by Stanton and Kalbfleisch that the designs which are 1-quasisymmetric are of two types:

(1) Those which consist of two copies of a symmetric design with parameters v, k, λ.

(2) The affine designs with $m = 2$ which we have already discussed.

In the terms of Goethals and Seidel, type (1) consists of designs having as graph the ladder graph on 2n vertices. This is the complement of $H_2(n)$. These designs have $\mu = k$ and $\eta = \lambda$.

We have mentioned the known result that if a design has a parallelism the number of blocks in each parallel class is a constant. An extension of this result appears (differently phrased) in the work of Goethals and Seidel [3]. We state this extension as follows:

<u>Theorem 3</u>. (Goethals and Seidel) A quasisymmetric design is s-quasisymmetric for some s.

<u>Proof</u>. From the definitions we know that a design is s-quasisymmetric if and only if its block graph is regular. The theorem, therefore, follows as a corollary from the Goethals and Seidel [3] proof that the graph of a quasisymmetric design is strongly regular (and hence regular).

We note that affine designs are $(m-1)$-quasisymmetric.

Remarks

Our role here has been largely expository of the pioneering work of Goethals and Seidel, of Shrikhande, and of Stanton and Kalbfleisch. However, the embeddability of affine designs with $m = 2$, as shown in Theorem 2, seems to have escaped notice in the literature. Shrikhande [7] has shown that $R = k + 2$ in a set of parameters with $\lambda = 2$ is sufficient for embedding and general necessary and sufficient conditions for embedding have been devised by Hall and Connor [5] but these are difficult to apply. For the case of quasisymmetric designs there are no counter examples to the conjecture that a quasisymmetric design with appropriate parameters is embeddable.

We have seen that affine designs with $m = 2$ form a class of embeddable designs whose parameters can be expressed in terms of n where $n \equiv 3 \pmod 4$. Table 2 presents these parameters for small n. Any quasisymmetric design of this class with $v = 2^d$ is isomorphic to a design formed by the points and hyperplanes of an affine space of dimension d. This is easily shown by means of a characterization of affine spaces due to Kantor [6]. Moreover, for $n \equiv 1 \pmod 4$ parameters of the form of (1) (with different intersection numbers) are also of the form of parameters of residual designs. Table 3 presents these parameters for small n. We call these the Shrikhande family of parameters since the only work done on designs of this family is due to Shrikhande who showed the embeddability of the design having $n = 5$.

Not much is known about designs with the Shrikhande parameters. For $n = 5$, a quasisymmetric design exists and is embeddable since $\lambda = 2$. This design is 3-quasisymmetric and has for its graph the Peterson graph. We present this design in Table 4 since Hall's table of designs [5] has an error in the symmetric design (No.5). It is actually unknown whether there are other quasisymmetric designs in this family. The known residual designs in this family with $n > 5$ are not quasisymmetric. This leaves us with a large area of unsolved problems concerning both existence and embeddability of quasisymmetric designs with parameters of the form of (1).

Table 2.

Affine parameters with $m = 2$

n	v	b	r	k	λ	μ	η
3	4	6	3	2	1	1	0
7	8	14	7	4	3	2	0
11	12	22	11	6	5	3	0
15	16	30	15	8	7	4	0
19	20	38	19	10	9	5	0
23	24	46	23	14	11	6	0
.	.	.	.	.	.	.	0

Table 3.

The Shrikhande Parameters

n	v	b	r	k	λ	μ	η
5	6	10	5	3	2	2	1
9	10	18	9	5	4	?	?
13	14	26	13	7	6	?	?
17	18	34	17	9	8	?	?
21	22	42	21	11	10	?	?
.	.	.	.	.	.	.	.

Finally, we note that a next step to the Goethals and Seidel results would be to consider quasisymmetric designs having graph $H_m(m)$ with $m > 2$. An affine design A has graph $H_m(n)$ and if an appropriate symmetric design with $v'' = n$ exists we could take m copies of it as the derived design and show the embeddability of A as in Theorem 2. This procedure recovers for us Shrikhande's Theorem 1 (above) since solving for n in (3) using $b = mn$ yields $v'' = n = (\mu m^2-1)(m-1)^{-1}$ which is v_o of (4).

Table 4

The Shrikhande Design for $n = 5$

	D'	D''
b_1	= 2, 6, 10	4, 5
b_2	= 6, 7, 0	3, 5
b_3	= 6, 7, 8	4, 1
b_4	= 7, 8, 2	5, 9
b_5	= 6, 8, 10	9, 3
b_6	= 7, 0, 10	9, 4
b_7	= 0, 8, 10	1, 5
b_8	= 0, 2, 6	9, 1
b_9	= 2, 7, 10	1, 3
b_{10}	= 0, 2, 8	3, 4

$b_{11} = 1, 3, 4, 5, 9$

References

1. Bose, R. C., A note on the resolvability of balanced incomplete block designs. Sankhya 6 (1942) 105-110.

2. Dembowski, P., Finite Geometries, Springer-Verlag, New York, 1968.

3. Goethals, J.M. and Seidel, J.J., Strongly regular graphs derived from combinatorial designs, Canadian J. Math. (to appear).

4. Hall, M., Jr., and Connor, W.S., An embedding theorem for balanced incomplete block designs, Canadian J. Math. 6 (1954) 35-41.

5. Hall, M., Jr., Combinatorial Theory, Blaisdell, Waltham, 1967.

6. Kantor, W. M., Characterizations of finite projective and affine spaces, Canadian J. Math. 21 (1969) 61-75.

7. Shrikhande, S. S., On the nonexistence of affine resolvable balanced incomplete block designs, Sankhya, 11 (1951) 185-186.

8. ______________, On the dual of some balanced incomplete block designs, Biometrics, 8 (1952) 66-72.

9. Stanton, R.G., and Kalbfleisch, J.G., Quasi-symmetric balanced incomplete block designs, J. Combinatorial Theory, 4 (1968) 391-396.

GENERALIZED LOWER BOUNDS FOR THE ABSOLUTE VALUES OF THE COEFFICIENTS OF CHROMATIC POLYNOMIALS

Bernard Eisenberg, Mathematics Department
Kingsborough Community College,
City University of New York, Brooklyn, NY 11235

PRELIMINARY CONSIDERATIONS AND DEFINITIONS

A graph G is defined as a non-empty set of objects, N, called nodes or points and a set E called edges or lines, consisting of unordered pairs of distinct elements in N. It is assumed that G has no multiple edges. Let $|N| = n$ and $|E| = e$ be the cardinalities of these sets. Geometrically, one thinks of the graph, G, as a set of points with zero or more lines joining some pairs of points.

Let Λ be a set of colors and let $|\Lambda| = \lambda$ be the cardinality of Λ. Any assignment of colors from Λ to the elements of N is called a coloring of the graph G. If the coloring is restricted so that adjacent nodes must be colored differently, then the coloring is said to be proper. A coloring which is not proper is defined to be improper. If a proper coloring of G is to be considered distinct from another proper coloring of G obtained from the first by a permutation of the colors, then this is referred to as a proper coloring with color difference. Otherwise it is called a proper coloring with color indifference.

Examples: Let $N = \{1,2,3,4\}$ and let $E = \{(1,2), (1,3), (2,4)\}$ be the nodes and edges of the graph G. (refer to figure 1)

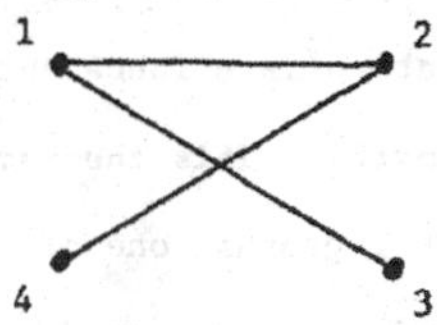

Graph G

Figure 1

Let $\Lambda = \{R,W,B,G,Y\}$ be the set of colors standing for Red, White, Blue, Green, and Yellow. Then the assignment of colors from Λ to N given by $C_1 = \{1 \to R, 2 \to R, 3 \to W, 4 \to W\}$ is an improper coloring, but the assignment, $C_2 = \{1 \to R, 2 \to W, 3 \to W, 4 \to R\}$ is a proper coloring. The coloring $C_3=\{1\to W, 2\to R, 3\to R, 4\to W\}$ is proper and is also a permutation of C_2. If C_2 is to be counted as distinct from C_3 then color difference is being considered. Otherwise C_2 and C_3 are color indifferent. From this point on, on, when the term "coloring" is used it will mean a proper coloring with color difference accounted for. The number of colorings of G in figure 1 is $5\times4\times4\times4=5\times4^3$ since node 3 may have any of 5 colors, node 1 may then have any of 4 colors, node 2 any of 4 colors and node 4 any of 4 colors. In general, if G is a tree on n nodes, as illustrated above, and $|\Lambda|=\lambda$ colors are available to color the nodes of G, then by reasoning as in the example, the number of colorings $M_G(\lambda)$ of G is given by

$$M_G(\lambda)=\lambda(\lambda-1)^{n-1}=\sum_{r=1}^{n}(-1)^{n-r}\binom{n-1}{r-1}\lambda^r$$

or

$$M_G(\lambda)=\sum_{r=1}^{n}(-1)^{n-r}a_r\lambda^r \text{ where } a_r=\binom{n-1}{r-1}$$

The polynomial, $M_G(\lambda)$, is defined as the chromatic polynomial of the graph, G, and represents the number of ways to color G with λ or less colors. It is known that G is a tree on n-nodes iff $M_G(\lambda)=\lambda(\lambda-1)^{n-1}$. This requires $a_r=\binom{n-1}{r-1}$ for all r such that $1\le r\le n$. It will be shown that this requirement is excessive. In fact, G is a tree iff $a_1=1$. Also G is a tree iff $a_r=\binom{n-1}{r-1}$ for <u>any</u> <u>one</u> value of r and $a_1\neq0$. The latter condition is equivalent to connectedness. It has also been shown (see [1]) that for all connected graphs on n nodes, $a_r\geq\binom{n-1}{r-1}$, the value of a_r for a tree. This bound will be improved when one considers the number of edges in the graph. It will be shown that $a_r\geq(e-n+2)\cdot\binom{n-2}{r-1}+\binom{n-2}{r-2}$ and that as a consequence some graphs may be characterized by the value of a_1.

In general, one can associate a chromatic polynomial, $M_G(\lambda)=\sum_{r=1}^{n}(-1)^{n-r}a_r\lambda^r$ with each graph, G. This polynomial may be obtained by the methods described in [1]. An important tool in this computation is a fundamental theorem for chromatic polynomials (abbreviated c.p. henceforth). This theorem asserts that the c.p. of any graph is the sum of the c.p.'s of 2 graphs, one of which is the given graph with an edge added to two non-adjacent nodes and the second of which is the graph obtained from the given graph by superimposing these 2 non-adjacent nodes. For example, if G_1 = , G_2 =  and G_3 = 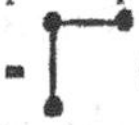 then  = 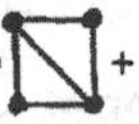 +  =  +  where the equality refers to the equality of c.p.'s of the graphs in the equation. (The graphic device for writing sums of c.p.'s is due to Zykov and is utilized in [1].) This follows from the fact that if v_1 and v_2 are non-adjacent in G, then the number of colorings of G is equal to the sum of the number of colorings of G when v_1 and v_2 are colored differently and the number of colorings of G when v_1 and v_2 are colored the same. Another form of the fundamental theorem is given by the equality  = 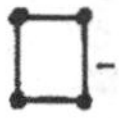 - , obtained from the previous equality by transposition. This form of the theorem is used in the main result which follows.

LOWER BOUNDS FOR THE CHROMATIC COEFFICIENTS

Theorem 1: If G is a connected graph on n-nodes and e-edges and $M_G(\lambda) = \sum_{r=1}^{n} (-1)^{n-r} a_r \lambda^r$,

then $$a_r \geq (e-n+2)\binom{n-2}{r-1} + \binom{n-2}{r-2} \quad . \qquad (1)$$

Proof: If G is a tree, $a_r = \binom{n-1}{r-1}$ so that (1) is verified for $e=n-1$.

Assume the theorem to be true for all graphs on n-nodes and e-edges where $(n-1) \leq e \leq \bar{e}$.

Consider a connected graph, G, on n-nodes and $(\bar{e}+1)$ edges. Then by the fundamental theorem, $M_G(\lambda) = M_{G'}(\lambda) - M_{G''}(\lambda)$, where the number of nodes and edges in G, G', and G'' are given in table 1 and

TABLE 1

Graph	Nodes	Edges
G	n	$\bar{e}+1$
G'	n	$\bar{e}$
G''	n-1	$\leq \bar{e}$

where $M_G(\lambda) = \sum_{r=1}^{n} (-1)^{n-r} a_r \lambda^r$, $M_{G'}(\lambda) = \sum_{r=1}^{n} (-1)^{n-r} a'_r \lambda^r$, and $M_{G''}(\lambda) = \sum_{r=1}^{n-1} (-1)^{n-1-r} a''_r \lambda^r$. Then $a_r = a'_r + a''_r$.

By the inductive assumption,

$$a'_r \geq (\bar{e}-n+2)\binom{n-2}{r-1} + \binom{n-2}{r-2}. \qquad (2)$$

Since G'' is a connected graph on (n-1)-nodes, we have (see [1]) $a''_r \geq \binom{(n-1)-1}{r-1} = \binom{n-2}{r-1}$ (3)

Therefore $a_r = a'_r + a''_r \geq (\bar{e}-n+2)\binom{n-2}{r-1} + \binom{n-2}{r-2} + \binom{n-2}{r-1}$ so that

$$a_r \geq [(\bar{e}+1)-n+2]\binom{n-2}{r-1} + \binom{n-2}{r-2}$$

The inductive proof is therefore complete.

Corollary 1: If G is connected on n-nodes and e-edges,

$$a_1 \geq (e-n+2) \quad (4)$$

$$a_2 \geq (e-n+2)(n-2)+1 \quad (5)$$

$$a_3 \geq (e-n+2)\,\frac{(n-2)(n-3)}{2} + (n-2) \quad \text{etc.} \quad (6)$$

Corollary 2: G is a tree iff $a_1=1$

Proof: If G is a tree, $a_1=\binom{n-1}{1-1}=1$

Conversely, if $a_1=1$, then G is connected.

By the previous corollary, $1 \geq e-n+2$.

Since G is connected, $e \geq n-1$ so that $e-n+2 \geq 1$.

Combining the last 3 inequalities yields $e-n+2=1$ or $e=n-1$ which together with G being connected, implies G is a tree.

Corallary 3: G is a tree iff $a_r=\binom{n-1}{r-1}$ for any single value of r and $a_1 \neq 0$.

Proof: If G is a tree, $a_r=\binom{n-1}{r-1}$ and $a_1=1 \neq 0$.

Conversely, if $a_r=\binom{n-1}{r-1}$ and $a_1 \neq 0$, then G is connected and by theorem 1,

$\binom{n-1}{r-1} \geq (e-n+2)\binom{n-2}{r-1}+\binom{n-2}{r-2}$. Therefore

$\binom{n-2}{r-1}+\binom{n-2}{r-2} \geq (e-n+2)\binom{n-2}{r-1}+\binom{n-2}{r-2}$. This implies

$1 \geq e-n+2$. Since G is connected $e-n+2 \geq 1$. Hence

$1=e-n+2$ or $e=n-1$ which implies G is a tree.

In reference 1, the author shows that G is a tree iff

$$M_G(\lambda)=\lambda(\lambda-1)^{n-1}=\sum_{r=1}^{n} (-1)^{n-r}\binom{n-1}{r-1}\lambda^r.$$ This requires $a_r=\binom{n-1}{r-1}$ for all r. Corollary 3 above requires $a_r=\binom{n-1}{r-1}$ for but one value of r and that G be connected. When $r=1$, connectedness is not required in the hypothesis since $a_1 \neq 0$ implies connectedness.

As a consequence of corollary 1, it is possible to characterize graphs whose values of a_1 are 1,2,3 or 4. The case $a_1=1$ has already been disposed of in corollary 2. In order to take care of the other cases, the chromatic polynomial and the value of a_1 for a graph which has precisely one circuit will be computed. Such a graph may be considered a generalization of a polygon since it is a polygon with one or more trees adjoined to the nodes. As such, the term pseudo-polygon will be used for this type of graph.

Definition: A pseudo-polygon is a connected graph having precisely 1 circuit as a sub-graph.

A pseudo-triangle is a pseudo-polygon whose 1 circuit has 3 edges.

A pseudo-p-gon is a pseudo-polygon whose 1 circuit has p edges.

Theorem 2: A graph, G, on n-nodes and e-edges is a pseudo-polygon iff e=n.

Proof: If G is pseudo-polygon on n-nodes, then the removal of 1 edge from the circuit leaves a tree. The n-1 edges of the tree together with the removed edge yields a total of (n-1)+1=n edges. Therefore e=n.

Conversely, if G is connected and e=n, then G has a spanning tree plus an additional edge. The spanning tree contains no circuits but provides a path between any two nodes. The additional edge joining v_1 to v_2 completes 1 circuit. If 2 or more circuits were completed by the edge joining v_1 to v_2, then a circuit had already existed prior to the addition of the edge joining v_1 to v_2 for there must have been 2 or more distinct paths from v_1 to v_2. But this is impossible since the spanning tree has no circuits. Therefore if G is connected and e=n, G contains precisely 1 circuit.

Theorem 3: If G is a pseudo-p-gon on n-nodes then $M_G(\lambda)=\left[(\lambda-1)^p+(-1)^p(\lambda-1)\right](\lambda-1)^{n-p}$ (1)

Proof : By the fundamental theorem, $M_G(\lambda)=M_{G'}(\lambda)-M_{G''}(\lambda)$ where G'=G minus a polygonal edge and G" is a pseudo-(p-1)-gon having (e-1)-edges and (n-1)-nodes. A pictorial representation of the corresponding c.p.'s is given in figure 2.

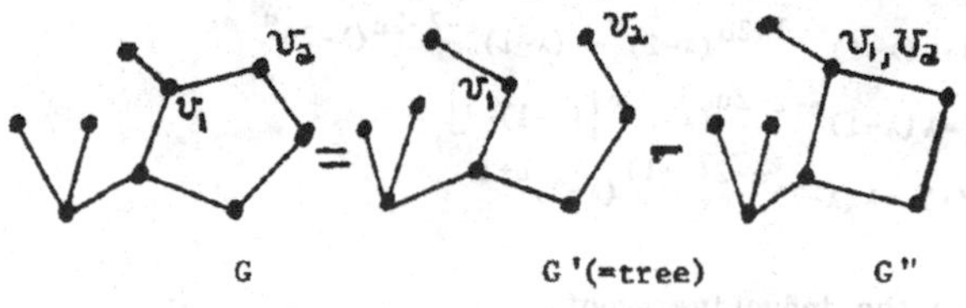

FIGURE 2

By applying this same procedure to G" and to the lower order pseudo-polygons which occur as a result, the c.p. for G is shown to be, $M_G(\lambda) = \lambda(\lambda-1)^{n-1} - \lambda(\lambda-1)^{n-2} + \lambda(\lambda-1)^{n-3} - \lambda(\lambda-1)^{n-4} + \text{-----}$

$$+ \text{---} (-1)^{j-1}\lambda(\lambda-1)^{n-j} + \text{----} + (-1)^2\lambda(\lambda-1)^{n-(p-2)} + (-1)^1\lambda(\lambda-1)^{n-(p-1)}$$

$$= \lambda(\lambda-1)^{n-1}\left[\frac{[-(\lambda-1)]^{-(p-1)}-1}{[-(\lambda-1)]^{-1}-1}\right] = (\lambda-1)^{n-p}\left[(\lambda-1)^p + (-1)^p(\lambda-1)\right]$$

Corollary 1 : If G is a pseudo-p-gon on n-nodes, $a_1 = p-1$. (a_1 is independent of n.) This is obtained by extracting the coefficient of λ' in (1) above.

Corollary 2: If k is any integer $\geqslant 2$, then there exists at least one connected graph for which $a_1 = k$. Such a graph is provided by any pseudo - (k+1)-gon.

Definition: A near tree is a connected graph whose largest circuit has 3 edges. The graph may have more than one such circuit. When there is precisely one circuit of 3 edges in a near tree, it is a pseudo 3-gon.

Theorem 4: If G is a near tree having e-edges and t-triangles (or 3-gons), then $M_G(\lambda) = \lambda(\lambda-1)^{e-2t}(\lambda-2)^t$ (1)

Proof: If G is a tree on e-edges (t=o), $M_G(\lambda) = \lambda(\lambda-1)^e = \lambda(\lambda-1)^{e-2(0)}(\lambda-2)^0$. This confirms the theorem when t=0. Assume (1) is valid for $t \leqslant u$. Let G be a near tree having e-edges and (u+1) triangles. Then $M_G(\lambda) = M_{G'}(\lambda) - M_{G''}(\lambda)$ where G' is obtained from G by deletion of a triangular edge and G" is obtained from G' by bringing together the 2 nodes of the deleted edge. Then the number of edges in G, G', and G" is e, (e-1), and (e-2) respectively and the corresponding number of triangles is (u+1), u, and u. Since G' and G" are near trees on u triangles, the inductive assumption may be applied

so that $M_G(\lambda) = \lambda(\lambda-1)^{e-1-2u}(\lambda-2)^u - \lambda(\lambda-1)^{e-2-2u}(\lambda-2)^u$

$$= \lambda(\lambda-1)^{e-2-2u}(\lambda-2)^u\left[(\lambda-1)-1\right]$$

Therefore $M_G(\lambda) = \lambda(\lambda-1)^{e-2(u+1)}(\lambda-2)^{u+1}$

This completes the inductive proof.

Corollary 1: If G is a near tree, $a_r = \sum_{i=0}^{r-1}\binom{e-2t}{r-i-1}\binom{t}{i}2^{t-i}$

This is obtained by extracting the coefficient of

λ^r in (1) and taking its absolute value.

Corollary 2: If G is a near tree having t-triangles, then $a_1 = 2^t$.

Theorem 5: If $a_1 = 2$, G is a pseudo-triangle.

If $a_1 = 3$, G is a pseudo-quadrilateral (4-gon)

If $a_1 = 4$, G is a pseudo-pentagon (5-gon) or G has precisely 2 triangles.

Proof: If $a_1=2$, then $2=a_1 \geq e-n+2 \geq (n-1)-n+2=1$.

Therefore $e-n+2=1$ or 2. $e-n+2=1$ implies $e=n-1$ which implies that G is a tree and that $a_1=1$.

This is impossible since $a_1=2$. If $e-n+2=2$, $e=n$.

Therefore G is a pseudo-polygon by theorem 2.

If p is the number of edges in this polygon, then by the corollary to theorem 3, $a_1 = p-1=2$ or $p=3$. Therefore G is a pseudo triangle.

If $a_1=3$, then $3=a_1 \geqq (e-n+2)$. Therefore $e-n+2=1$, 2, or 3.

The case $e-n+2=1$ leads to a contradiction as before. $e-n+2=2$ implies $e=n$ so that G may be a pseudo p-gon for which $a_1=p-1=3$ or $p=4$.

Therefore G may be a pseudo-quadrilateral. If $e-n+2=3$, $e=n+1=(n-1)+2$.

Therefore G is the union of a spanning tree and two additional edges which produce exactly 2 or 3 circuits (figure 3 a and b) of

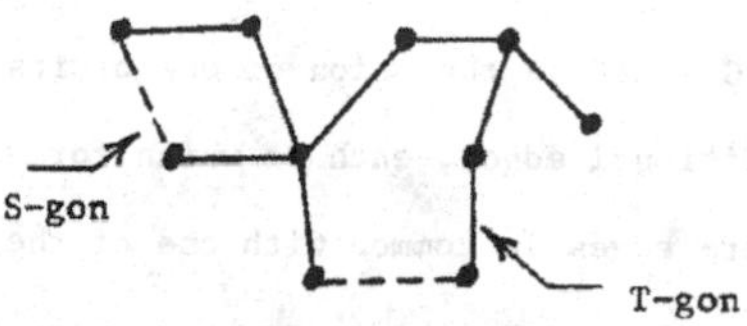

Figure 3 a - 2 circuits

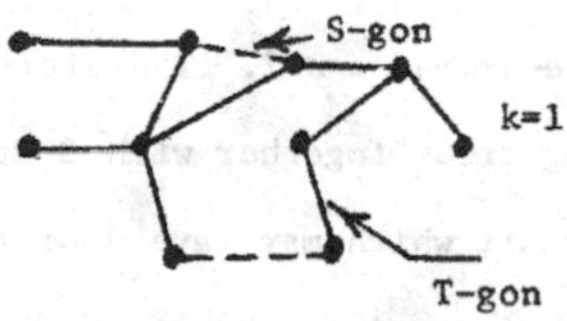

Figure 3 b - 3 circuits

lengths $\geqq$ 3. If there are 2 circuits, they have a node in common at most. The least value of a_1 occurs when the circuits are smallest in length. Then $a_1 \geqq 2^2 = 4$ because the smallest circuits are triangles and for a graph with precisely two triangles, $a_1=2^2=4$. This contradicts $a_1=3$ and so is impossible here. If there are three circuits (as in figure 3b), then they consist of an S-gon and a T-gon having k edges in common. Therefore

$a_1 = k(S+T-2k-1) + (S-k-1)\cdot(T-k-1)$* .

Here $S-k > 1$, $T-k > 1$ and $1 \leq k \leq \min(S-2, T-2)$. Also, $(S+T-2k-1) \geq 3$. $a_1 \geq 1\times3+1\times1=4$, (this occurs when $k=1$ & $S=T=3$), which contradicts the fact that $a_1=3$ here. Therefore if $a_1=3$, the only possiblity is that G is a pseudo - 4 - gon.

If $a_1=4$, then since $4=a_1 \geq e-n+2 \geq 1$, $e-n+2=1,2,3$, or 4. As before, the case $e-n+2=1$ leads to a contradiction. $e-n+2=2$ implies $e=n$ so that G may be a pseudo-p-gon for which $a_1=p-1=4$ or $p=5$. Therefore G may be a pseudo pentagon. If $e-n+2=3$, $e=n+1=(n-1)+2$. As before, G consists of a graph with precisely 2 or 3 circuits. If G has precisely 2 circuits, $a_1=(S-1)(T-1)$* where S&T are the number of edges in the circuits respectively and $S \geq 3$, $T \geq 3$. Therefore $(S-1)(T-1)=4$ which implies that $S=T=3$. Therefore G may be a near tree having precisely two triangles and no other circuits. If G has precisely three circuits (figure 3b), then as before $a_1=k(S+T-2k-1)+(S-k-1)(T-k-1)$* and $a_1 \geq 4$. Then since $S-k-1 \geq 1$, $S \geq 1+1+1=3$ and similarly $T \geq 3$. Also for $k=1$ and $a_1=4$, $S+T-2k-1 \leq 3$. This implies $S+T \leq 6$. Combining this with $S \geq 3$ and $T \geq 3$ gives $S=3$ and $T=3$. Therefore G is a graph having two triangles with a common edge and no other circuits.

If $e-n+2=4$, $e=n+2$. Therefore G would be the union of one of its spanning trees together with 3 additional edges, each of which forms a circuit which may have 0 or more edges in common with one of the other circuits. As a result of this commonness, there may be as many as 7 circuits. If there are precisely 3 circuits, then they may have a vertex in common but nothing more. In this case $a_1=(R-1)(S-1)(T-1)$* where R,S,&T are the number of sides in each of the circuits respectively.

---

* Based on unpublished results.

Therefore $4=(R-1)(S-1)(T-1)$. But $R\geq 3$, $S\geq 3$ & $T\geq 3$, for otherwise there would be multiple paths, which are excluded. Therefore $4 \geq 2x2x2=8$. This contradiction precludes that C has 3 circuits with a common vertex. The next possibility is that G has 4 circuits- three of which, taken in pairs have at least 1 edge in common, while the fourth circuit may have a node in common with the other three. In this case, (figure 4), $a_1=[k(R+S-2k-1)+(R-k-1)(S-k-1)](T-1)^*$

In this expression, $k\geq 1$, $R\geq 3$, $S\geq 3$, $T\geq 3$, $R-k-1\geq 1$, $S-k-1\geq 1$, and $R+S-2k-1=(R-k-1)+(S-k-1)+1\geq 3$, otherwise multiple edges appear.

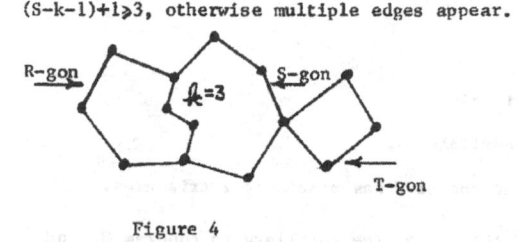

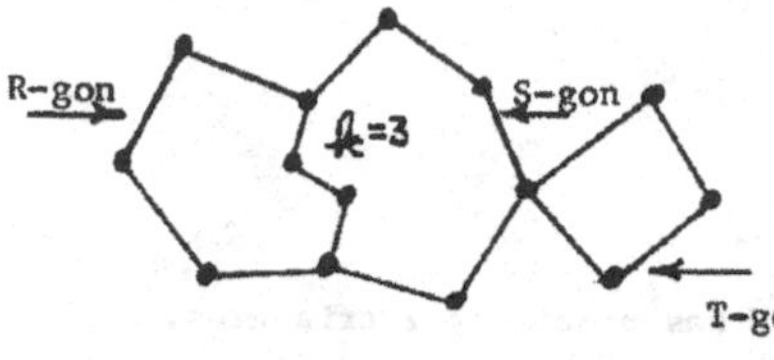

Figure 4

Example of 4 circuits

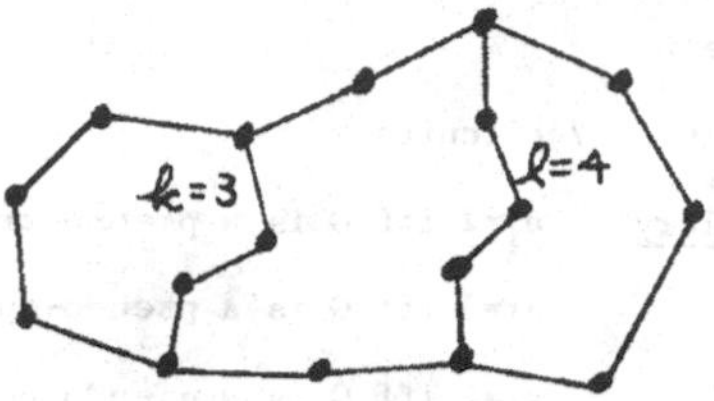

Figure 5

Example of 6 circuits

Therefore $a_1 \geq [1\cdot 3 + 1\cdot 1](2)=8$. But $a_1=4$ here. This contradiction precludes the possibility of figure 4. The next possibility is that G has six circuits as in figure 5. For this case $a_1\geq 2(kl)^*$, where k and l are the number of common edges that two of the circuits have with a third. Since $k\geq 1$, and $l\geq 1$, $a_1\geq 2kl=2$. But then the graph which will lead to the smallest value of a_1 is that given in figure 6.

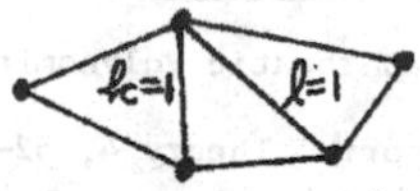

Figure 6

For the graph in figure 6, $M_G(\lambda)= \lambda(\lambda-1)(\lambda-2)^3$ so that $a_1=8\neq 4$ as required here. If k=2 and l=1 or if k=1 and l=2, the graph with the least value for a_1 is given in figure 7.

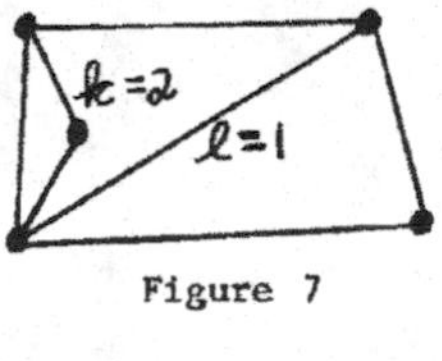

Figure 7

Since this graph is isomorphic to that in figure 6, the value of $a_1=8$ for this graph. Since $a_1\neq 4$ for the graph in figure 7, this situation is not possible here. If $k=l=2$, $a_1\geq 2(k)(l)=8\neq 4$. Any larger values of k and l lead to graphs whose $a_1> 4$. Therefore G can't have 6 circuits.

The only remaining possibility is that G contains 7 circuits as in figure 8 so that any 2 of the circuits have at least 1 edge in common. Then $a_1 \geq 2(jk+jl+kl)^* = 2(1+1+1)=6$ and $a_1 \neq 4$ here. In summary, if $a_1=4$, G is either a pseudo-pentagon or a connected graph having precisely 2 triangles. The two triangles may have a point, a line, or nothing in common.

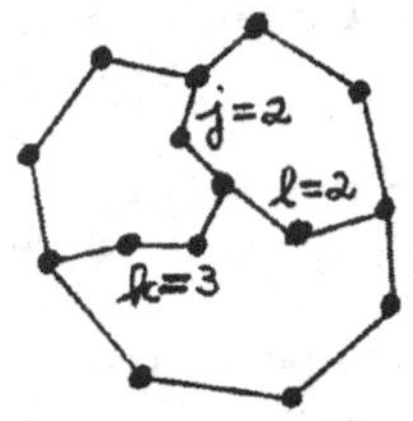

Figure 8

Example of 7 circuits

Corollary: $a_1=2$ iff G is a pseudo-triangle.

$a_1=3$ iff G is a pseudo-quadrilateral.

$a_1=4$ iff G is a pseudo pentagon or G has precisely 2 triangles.

This corollary combines the results of theorem 5, the corollary to theorem 3, and corollary 2 of theorem 4. It also requires a knowledge of the fact that if G contains as its only circuits precisely 2 triangles with a common edge, then $a_1=4$.

References

1. R. Read An Introduction to Chromatic Polynomials Journal of Combinatorial Theory 4, 52-71 (1968)

2. F. Harary Graph Theory, Addison Wesley, Mass. 1969

* Based on unpublished results

THE LEAST NUMBER OF EDGES FOR GRAPHS HAVING AUTOMORPHISM GROUP OF ORDER THREE

by

Roberto Frucht

Universidad Tecnica Federico Santa Maria
Valparaiso, Chile

Allan Gewirtz

Brooklyn College, CUNY
Brooklyn, N.Y., 11210, U.S.A.

Louis V. Quintas

Pace College
New York, N.Y., 10038, U.S.A.

§1. Introduction. By a graph we mean a finite undirected graph (as defined in Harary [6, p. 7]) without loops and without multiple edges. The automorphism group of a graph consists of those permutations of the vertex set of the graph which preserve adjacency relations (cf. Harary [6, p. 161]). Let $e(G,n)$ denote the least integer for which there exists a graph having $e(G,n)$ edges, n vertices, and automorphism group isomorphic to G. The value of $e(G,n)$ is known for the cases where $G = id$, the identity group (cf. Theorem 1) and $G = S_m$, the symmetric group of degree m (cf. Theorem 2). Analogous theorems for the case where the graphs are required to be connected can be found in Quintas [10] and Gewirtz and Quintas [5] and for regular graphs having identity group in Gewirtz, Hill, and Quintas [3]. A survey of extremum problems concerning graphs and their groups is given in Gewirtz, Hill, and Quintas [4].

In this paper the value of $e(C_3,n)$ is determined for all

admissible n, where C_3 is the group of order three (cf. Theorem 3). Since the greatest number $E(G,n)$ of edges realizable by a graph having n vertices and group G is obtained by considering the complementary graph of a minimum edge graph with the same number of vertices and group, our result yields

$$E(C_3,n) = n(n-1)/2 - e(C_3,n).$$

The result of this paper is to be viewed in the context of our ultimate objective, which is to determine $e(C_m,n)$, where C_m is the cyclic group of order m.

By an <u>asymmetric</u> graph we mean a graph whose automorphism group is isomorphic to the identity group.

Let a_i denote the number of asymmetric trees having i vertices and for each n (n = 8, 9, ...) let N and w be defined as follows:

$$\sum_{i=1}^{N} ia_i \leq n < \sum_{i=1}^{N+1} ia_i$$

$$n = \sum_{i=1}^{N} ia_i + w(N+1) + r \quad (0 \leq w < a_{N+1};\ 0 \leq r < N+1).$$

<u>Remark</u>. The values of N and w depend on n but for convenience we have not included this in the notation. We also note that the values of a_i for all i have been determined by Harary and Prins (cf. [8]).

<u>Theorem 1</u>. If id is the identity group, then $e(id,n)$ is not defined for n = 2, 3, 4, and 5, and

$$e(id,n) = \begin{cases} 0 & n = 1 \\ 6 & n = 6, 7 \\ n - \sum_{i=1}^{N} a_i - w & n = 8, 9, \ldots . \end{cases}$$

<u>Proof</u>. Quintas (cf. [10, Theorem 1]).

<u>Theorem 2</u>. If S_m is the symmetric group of degree m ($m \geq 2$), then $e(S_m,n)$ is not defined for $n < m$,

$$e(S_2,n) = n - 2 \qquad n = 2, 3, \ldots, 8,$$

$$\text{if } m \geq 3, \text{ then } e(S_m,n) = \begin{cases} 0 & n = m \\ n - 1 & n = m + 1, m + 3 \\ n - 2 & n = m + 2, m + 4, m + 5 \\ 6 & n = m + 6, \end{cases}$$

and if $m \geq 2$, then

$$e(S_m,n) = e(id,n - m + 1) \qquad n = m + 7, m + 8, \ldots .$$

<u>Proof</u>. Quintas (cf. [11, Theorem 2]).

<u>Theorem 3</u>. If C_3 is the group of order three, then $e(C_3,n)$ is not defined for $n < 9$, and

$$e(C_3,n) = \begin{cases} 15 & n = 9, 10, \ldots, 14 \\ 16 & n = 15 \\ 17 & n = 16 \\ 18 & n = 17, 18, 19 \\ 21 & n = 20, 21, 22 \\ 23 & n = 23 \\ 18 + e(id,n - 18) & n = 24, 25, \ldots . \end{cases}$$

<u>Proof</u>. (Cf. §2).

§2. <u>Proof of Theorem 3.</u> If $n < 9$, then there does not exist a graph having n vertices and automorphism group isomorphic to C_3. This is one case of a 1963 theorem of Meriwether, which gives the least number of vertices that a graph can have and have group C_m, the cyclic group of order m (cf. [9] and for the statement of Meriwether's Theorem see Gewirtz, Hill, and Quintas [4]).

In 1966 Harary and Palmer established

$$e(C_3, 9) = 15$$

(cf. [7]). The graph shown in Fig. 2.1 is the minimal edge C_3-graph having 9 vertices displayed by Harary and Palmer in [7].

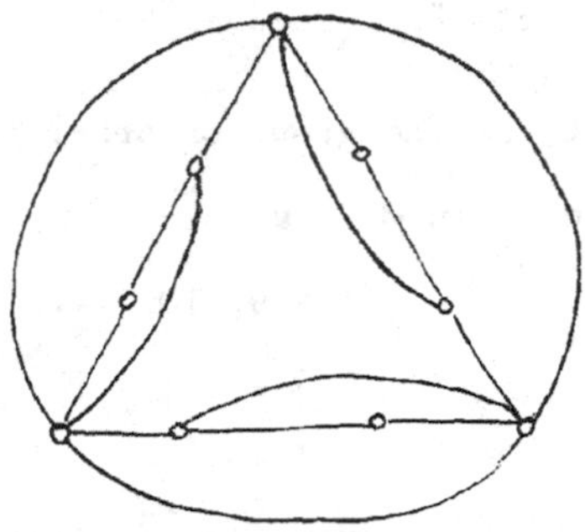

Fig. 2.1

Since the least number of vertices that a C_3-graph can have is 9 and asymmetric graphs having 2, 3, 4, or 5 vertices do not exist, it follows that for n = 10, 11, 12, 13, and 14 the possible C_3-graphs must either be connected or consist of two components one of which is a singleton. Thus, we have

$$e(C_3, n) = \min\left\{e_c(C_3, n-1), e_c(C_3, n)\right\} \quad n = 10, 11, 12, 13, \text{ and } 14 \qquad (2.1)$$

where $e_c(C_3, n)$ denotes the least integer for which there exists a connected C_3-graph having n vertices. In a paper to appear in

Scientia the values of $e_c(C_3,n)$ are determined for all n (cf. Frucht, Gewirtz, and Quintas [2]). These values are stated here in the following theorem.

Theorem 4. $e_c(C_3,n)$ is not defined if $n < 9$, and

$$e_c(C_3,n) = \begin{cases} 15 & n = 9,\ 10 \\ 16 & n = 11 \\ 15 & n = 12 \\ 17 & n = 15 \\ n + 2 & n = 3k + 1 \text{ or } 3k + 2 \quad k = 4,\ 5,\ \ldots \\ n & n = 3k \quad k = 6,\ 7,\ \ldots \end{cases}.$$

Proof. Frucht, Gewirtz, and Quintas (cf. [2]).

Using (2.1) and Theorem 4 we have

$$e(C_3,n) = 15 \quad n = 10,\ 11,\ 12,\ 13,\ 14.$$

For $15 \leqq n \leqq 23$ we have that C_3-graphs must be of the following types:

i) connected on n vertices,

ii) a connected C_3-graph on $n - 1$ vertices and a singleton,

or

iii) a connected C_3-graph on $n - p$ vertices $(9 \leqq n - p \leqq 17)$ and an asymmetric graph on p vertices $(p \geqq 6)$.

For these graphs the least number of edges is as follows:

i) $e_c(C_3,n)$,

ii) $e_c(C_3,n-1)$,

and

iii) $e_c(C_3,n-p) + e(id,p)$.

Thus, for $15 \leqq n \leqq 23$, $9 \leqq n - p \leqq 17$, and $p \geqq 6$ we have

$$e(C_3,n) = \min\left\{e_c(C_3,n), e_c(C_3,n-1), e_c(C_3,n-p) + e(id,p)\right\} \qquad (2.2)$$

Using (2.2), Theorem 4, and Theorem 1 (from which we get $e(id,n) = 6$ for $n = 6, 7, 8$ and $e(id,n) = n - 2$ for $n = 9, 10, \ldots, 14$) we obtain

$$e(C_3,n) = \begin{cases} 16 & n = 15 \\ 17 & n = 16 \\ 18 & n = 17, 18, 19 \\ 21 & n = 20, 21, 22 \\ 23 & n = 23 \end{cases}$$

For the remaining cases, $n \geqq 24$, we prove the following lemma.

<u>Lemma</u>. $e(C_3,n) = 18 + e(id, n - 18) \qquad n \geqq 24$.

<u>Proof</u>. Let K be a C_3-graph on n vertices. Then K must be of the form $K = M + A$, where M is a connected C_3-graph on $\alpha_0(M)$ vertices and A is an asymmetric graph on $n - \alpha_0(M)$ vertices. By Meriwether's Theorem $\alpha_0(M) \geqq 9$ and the number of edges of K is $\alpha_1(K) = \alpha_1(M) + \alpha_1(A)$. Thus,

$$\alpha_1(K) \geqq e_c(C_3, \alpha_0(M)) + e(id, n - \alpha_0(M)) \qquad (2.3)$$

We now consider the following cases:

1) $\alpha_0(M) = 9$

2) $\alpha_0(M) = 10$

3) $\alpha_0(M) = 11$

4) $\alpha_0(M) = 12$

5) $\alpha_0(M) = 15$

6) $\alpha_0(M) = 3k + 1$ or $3k + 2 \qquad k = 4, 5, \ldots$

7) $\alpha_0(M) = 3k \qquad k = 6, 7, \ldots$.

<u>Remark</u>. The following inequalities are easy to establish by examining the structure of the minimum edge forests given in Quintas [10, Theorem 1]. For example (2.5) and (2.6) were noted in Quintas [11, Lemma 5].

$$e(id,p) + s \leqq e(id,p+1+s) \quad s = 0, 1, 2, \ldots, 7 \text{ and } p \geqq 7 \tag{2.4}$$

$$e(id,p) \leqq e(id,p+s) \text{ whenever both sides of the inequality are defined} \tag{2.5}$$

$$e(id,p+s) \leqq e(id,p) + s \text{ whenever both sides of the inequality are defined except for the case } (p = 1, s = 5) \tag{2.6}$$

<u>Case 1</u>. $\alpha_o(M) = 9$

$$\begin{aligned} \alpha_1(K) &\geqq e_c(C_3,9) + e(id,n-9) && \text{by (2.3)} \\ &\geqq 15 + e(id,n-13) + 3 && \text{by Theorem 4 and (2.4)} \\ &\geqq 18 + e(id,n-18) && \text{by (2.5)} \end{aligned}$$

<u>Case 2</u>. $\alpha_o(M) = 10$

$$\begin{aligned} \alpha_1(K) &\geqq e_c(C_3,10) + e(id,n-10) && \text{by (2.3)} \\ &\geqq 15 + e(id,n-18) + 7 && \text{by Theorem 4 and (2.4)} \\ &> 18 + e(id,n-18) \end{aligned}$$

<u>Case 3</u>. $\alpha_o(M) = 11$

$$\begin{aligned} \alpha_1(K) &\geqq e_c(C_3,11) + e(id,n-11) && \text{by (2.3)} \\ &\geqq 16 + e(id,n-18) + 6 && \text{by Theorem 4 and (2.4)} \\ &> 18 + e(id,n-18) \end{aligned}$$

<u>Case 4</u>. $\alpha_o(M) = 12$

$$\begin{aligned} \alpha_1(K) &\geqq e_c(C_3,12) + e(id,n-12) && \text{by (2.3)} \\ &\geqq 15 + e(id,n-18) + 5 && \text{by Theorem 4 and (2.4)} \\ &> 18 + e(id,n-18) \end{aligned}$$

<u>Case 5.</u> $\alpha_0(M) = 15$

$$\begin{aligned}
\alpha_1(K) &\geq e_c(C_3, 15) + e(id, n-15) && \text{by (2.3)}\\
&\geq 17 + e(id, n-18) + 2 && \text{by Theorem 4 and (2.4)}\\
&> 18 + e(id, n-18)
\end{aligned}$$

<u>Case 6.</u> $\alpha_0(M) = 3k + 1$ or $3k + 2 \quad k = 4, 5, \ldots$

$$\begin{aligned}
\alpha_1(K) &\geq e_c(C_3, \alpha_0(M)) + e(id, n-\alpha_0(M)) && \text{by (2.3)}\\
&= \alpha_0(M) + 2 + e(id, n-\alpha_0(M)) && \text{by Theorem 4}\\
&= 13 + t + 2 + e(id, n-13-t) && 13 + t = \alpha_0(M) \not\equiv 0 \bmod 3\\
&\geq 15 + t + e(id, n-18-t) + 4 && \text{by (2.4)}\\
&\geq 19 + e(id, n-18) && \text{by (2.6)}\\
&> 18 + e(id, n-18)
\end{aligned}$$

<u>Case 7.</u> $\alpha_0(M) = 3k \quad k = 6, 7, \ldots$

$$\begin{aligned}
\alpha_1(K) &\geq e_c(C_3, \alpha_0(M)) + e(id, n-\alpha_0(M)) && \text{by (2.3)}\\
&= \alpha_0(M) + e(id, n-\alpha_0(M)) && \text{by Theorem 4}\\
&= 18 + 3t + e(id, n-18-3t) && 18 + 3t = \alpha_0(M)\\
&\geq 18 + e(id, n-18) && \text{by (2.6)}
\end{aligned}$$

Thus, we have shown that if K is a C_3-graph having at least 24 vertices, then $\alpha_1(K) \geq 18 + e(id, n-18)$. To show that for each $n \geq 24$ this bound is realized, consider the graph consisting of the connected C_3-graph on 18 vertices having one cycle (see Fig. 2.2, this graph was displayed by Frucht in [1, Fig. 4, p. 246]) and a minimal edge asymmetric graph on n - 18 vertices. This completes the proof of the lemma and consequently completes the proof of Theorem 3.

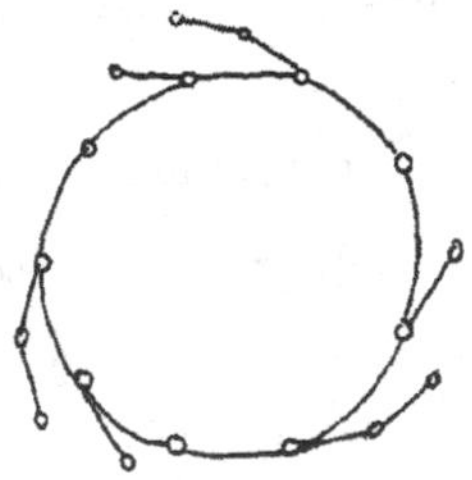

Fig. 2.2

References

1. Frucht, R., Herstellung von Graphen mit vorgegebener abstrakter Gruppe, Compositio Math. 6 (1938), 239-250.

2. Frucht, R., Gewirtz, A., and Quintas, L. V., El número mínimo de líneas para grafos conexos con grupo de automorfismos de orden 3, Scientia, (in press).

3. Gewirtz, A., Hill, A., and Quintas, L. V., El número mínimo de puntos para grafos regulares y asimétricos, Scientia 138 (1969), 103-111.

4. Gewirtz, A., Hill, A., and Quintas, L. V., Extremum problems concerning graphs and their groups, Proceedings of the Calgary International Conference on Combinatorial Structures and their Applications, Gordan and Breach, New York, 1970, 103-109.

5. Gewirtz, A. and Quintas, L. V., Connected extremal edge graphs having symmetric automorphism group, Recent Progress in Combinatorics (W. T. Tutte, ed.) Academic Press, New York, 1969, 223-227.

6. Harary, F., Graph Theory, Addison-Wesley, Reading, 1969.

7. Harary, F. and Palmer, E. M., The smallest graph whose group is cyclic, Czech. Math. J. 16 (1966), 70-71.

8. Harary, F. and Prins, G., The number of homeomorphically irreducible trees, and other species, Acta Math. 101 (1959), 141-162.

9. Meriwether, R. L., Smallest graphs with a given cyclic group (1963) unpublished, but see Math Reviews 33 (1967) #2563.

10. Quintas, L. V., Extrema concerning asymmetric graphs, J. Combinatorial Theory 3 (1967), 57-82.

11. Quintas, L. V., The least number of edges for graphs having symmetric automorphism group, J. Combinatorial Theory 5 (1968), 115-125.

CONNECTIVITY IN DIGRAPHS *

Dennis Geller and Frank Harary
The University of Michigan, Ann Arbor, MI 48104

Connectivity in graphs has been extensively investigated, and in fact an entire book has been written on the subject by Tutte [13]. On the other hand, connectivity in digraphs has until recently been almost completely neglected, with the exception of a modest beginning in the book [9, Chapter 9]. More recent results are contained in [1], [4], [5], [8], [10], [11], and [12]. In this article, we begin with an expository review of connectivity concepts and results concerning both graphs and digraphs. We then compare seven classical equivalent conditions for a graph to be 2-connected with corresponding statements, not all equivalent, for digraphs. The primordial inequalities relating connectivity, line-connectivity, and the minimum degree of a graph are extended to derive three corresponding inequalities for directed graphs, one for each of the connectedness categories.

CONNECTIVITY

The degree of connectedness of a graph G is accurately described in terms of just two invariants. The <u>point-connectivity</u> κ is the minimum number of points whose removal disconnects G or results in the trivial graph. The <u>line-connectivity</u> λ is the minimum number of lines whose removal disconnects G.

For digraphs the situation is more complicated because of the various kinds of connectedness which occur. A digraph D is <u>strong</u> if every two points are mutually reachable; it is <u>unilateral</u> if every point can either reach or be reached from every other point; and it is <u>weak</u> if the underlying graph GD is connected. Digraphs which are not even weak are called <u>disconnected</u>.

Each digraph D is in exactly one of the four connectedness categories C_i, i = 0, 1, 2, 3, defined as follows:

<u>Category</u>	<u>Consists of all</u>
C_0	disconnected graphs
C_1	weak but not unilateral
C_2	unilateral but not strong
C_3	strong.

* Research has been supported in part by Grants GN-566 and GJ-519 from the National Science Foundation, and by Grant GM-12236 from the National Institutes of Health.

The connectivity invariants of digraphs are every bit as complicated as the above categories might suggest. The <u>ij connectivity</u> κ_{ij} of a digraph D is the minimum number of points whose removal changes D from a C_i digraph to one in C_j. If D is in C_i but no induced subdigraph of D is in C_j, we take κ_{ij} as undefined, unless $j = 0$ in which case $\kappa_{ij} = p - 1$. Thus for D in C_i, $i > 0$, $\kappa_{i0} = \kappa GD$.

The invariants λ_{ij} are defined similarly but only for $i > j$. The κ or λ invariants for a particular digraph D can be combined to yield a single number describing just how firmly D is attached to its connectedness category. For example, we define the <u>strong connectivity</u> of a strong digraph D as

$$\kappa_3(D) = \min \{\kappa_{30}, \kappa_{31}, \kappa_{32}\}.$$

Thus $\kappa_3(D)$ is the minimum number of points whose removal renders D non-strong or trivial. Similar definitions of κ_0, κ_1, κ_2 and $\lambda_0 - \lambda_3$ are as expected. We say D is <u>2-strong</u> if D is strong and $\kappa_3 \geq 2$. This corresponds to a graph G being 2-connected, defined by $\kappa \geq 2$.

In the next section, we present some results concerning strong digraphs and κ_3 which are analogous to simple results on 2-connected graphs. In the final section we give bounds on the κ_i and λ_i, which imitate the well known graphical inequality, $\kappa \leq \lambda \leq \delta$.

STRONG DIGRAPHS

Graphs which are 2-connected have been characterized in many different ways [7, p. 27].

<u>Theorem 1</u>. The following statements are equivalent for any graph G:

(0) G is 2-connected.

(1) Every two points lie on a common cycle.

(2) Every point and line lie on a common cycle.

(3) Every two lines lie on a common cycle.

(4) For every three distinct points there is a path joining any two which contains the third.

(5) For every three distinct points there is a path joining any two which avoids the third.

(6) For every two points and one line there is a path joining the points which contains the line.

These statements can be interpreted for digraphs but then they are not all equivalent. Here is a list of the directed analogues, with a name for each one which has been used in the literature [9, Ch. 9]. Following the practice of Professor Nash-Williams, who calls them dicircuits, we refer to directed cycles as dicycles.

(0) D is 2-strong.

(1) Every two points lie on a common dicycle (<u>cyclically connected</u>).

(2) Every point and arc lie on a common dicycle.

(3) Every two lines lie on a common dicycle.

(4) For every three distinct points u, v, w, there is a u-v dipath containing w (flexible).

(5) For every three distinct points u, v, w, there is a u-v dipath avoiding w (point invulnerable).

(6) For every two points u, v, and one line x, there is a u-v dipath containing x.

(7) For every two points u, v, and one line x, there is a u-v dipath avoiding x (line invulnerable).

Note that (7) does not correspond to a condition on the graphical list since the corresponding undirected condition does not imply that a graph is 2-connected.

A theorem relating (0), (1), (4), (5), and (7) is given in [9, p. 256].

Theorem 2. A flexible digraph is both cyclically connected and point invulnerable. A cyclically connected digraph is 2-strong. A point invulnerable digraph is line invulnerable and 2-strong.

These implication are indicated in Figure 1.

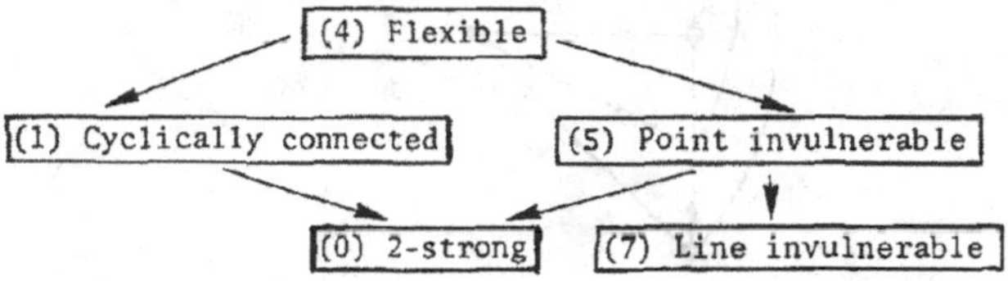

FIGURE 1

The conditions for D to be 2-strong are our principal concern and the following theorem tells just which conditions are necessary and which are sufficient for statement (0).

Theorem 3. For a strong digraph D, only conditions (4), (5), or (6) imply that D is 2-strong. On the other hand, $\kappa_3 \geq 2$ implies only condition (5).

Proof. Theorem 2 tells us that flexible or point invulnerable digraphs are 2-strong. Since condition (6) obviously implies that D is flexible, (4), (5), and (6) all imply $\kappa_3 \geq 2$. Furthermore, it is clear that every 2-strong D satisfies (5).

To show that these are the only necessary or sufficient conditions, we need merely present examples. We avoid using dicycles for this purpose because they might be questioned on grounds of over-simplicity.

Condition (1) is neither necessary nor sufficient for $\kappa_3 \geq 2$ as shown in Figure 2, where the two points of D_1 marked u_2 are identified.

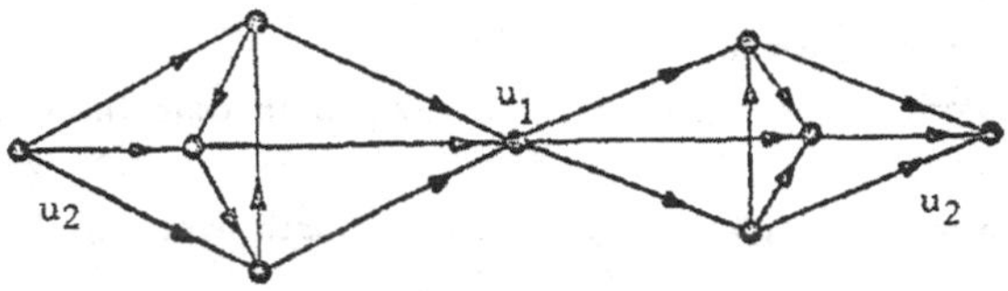

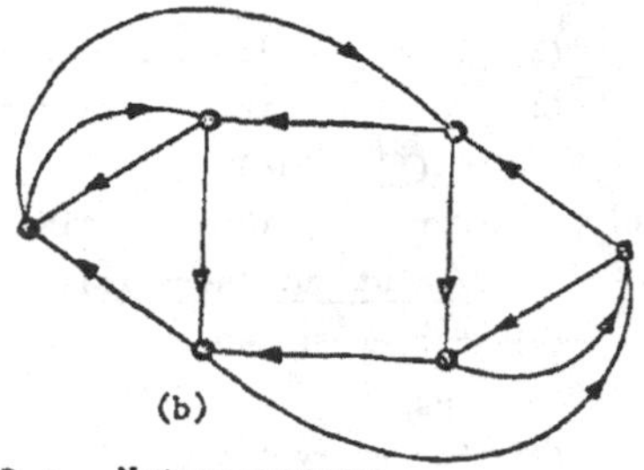

D_1: Not sufficient D_2: Not necessary

Figure 2. (0) and (1) are unrelated.

Condition (2) is neither necessary not sufficient for $\kappa_3 \geq 2$. It is not sufficient by D_1 above, while D_3 of Figure 3 shows it is not necessary.

D_3:

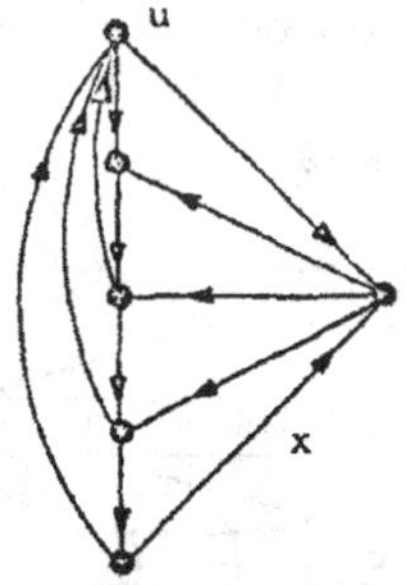

Figure 3. Condition (2) is not necessary for (0).

Condition (3) is neither necessary nor sufficient if every two arcs lie on a dicycle, D itself must be a dicycle, so $\kappa_3 = 1$. Finally, conditions (4) and (6) are not necessary by D_4 of Figure 4 and D_3 of Figure 3.

D_4:

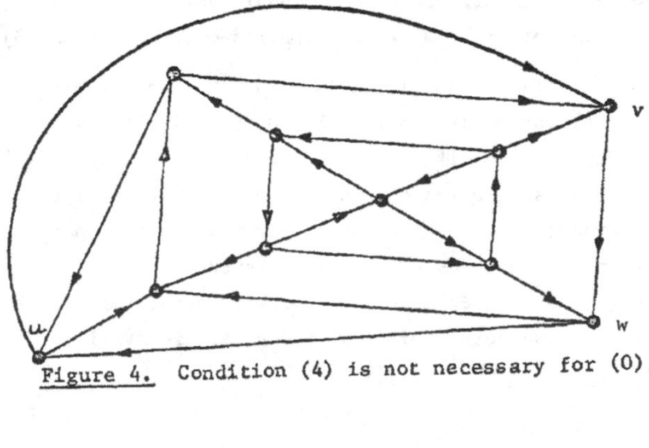

Figure 4. Condition (4) is not necessary for (0).

The existence of only a single property (point invulnerability) equivalent to $\kappa_3 \geq 2$ is certainly surprising, considering the abundance of equivalent formulations in the undirected case. There are other implications among the conditions, some of which are pointed out in [8, Ch. 9]. In particular, it is shown there that condition (7), line invulnerability, is equivalent to $\lambda_3 \geq 2$.

CONNECTIVITY INVARIANTS

It was first observed by Whitney [14] that connectivity, line-connectivity, and minimum degree of a graph are related by an inequality.

Theorem 4. For any connected graph G,

$$0 < \kappa(G) \leq \lambda(G) \leq \delta(G).$$

Chartrand and Harary [3] constructed a family of graphs which shows that these inequalities can always be realized.

Theorem 5. For all integers a, b, c such that $0 < a \leq b \leq c$, there exists a connected graph G with $\kappa(G) = a$, $\lambda(G) = b$, and $\delta(G) = c$.

In the same paper it was pointed out that if δ is large enough, then κ must also be large.

Theorem 6. If G has p points and $\delta \geq (p-2+n)/2$ where $1 \leq n \leq p-1$, then $\kappa(G) \geq n$.

Chartrand [2] showed that if δ is large enough, the second inequality of Theorem 4 becomes an equality.

Theorem 7. If G has p points and $\delta \geq [p/2]$, then $\lambda = \delta$.

A result expressing the maximum possible connectivity for graphs with a given number of points and lines is given in [6].

Theorem 8. Among all graphs with p points and q lines, the maximum connectivity is 0 when $q < p-1$ and is $[2q/p]$ when $q \geq p-1$.

Each of these results for graphs has a rather natural directed analogue. In fact, there are distinct analogues of Theorem 4 for weak, unilateral, and strong digraphs.

The indegree id(v), outdegree od(v), and total degree td(v) of a point v are famililar concepts: let δ_{id}, δ_{od}, and δ_1 be the respective minima of these functions over all the points of a digraph D, and set $\delta_3 = \min(\delta_{id}, \delta_{od})$. Let $\delta_{id}^{(2)}$ be min $\{id(v) + \delta_{id}(D-v) | v \in D\}$; define $\delta_{od}^{(2)}$ similarly and then set $\delta_2 = \min(\delta_{id}^{(2)}, \delta_{od}^{(2)})$. Note that each of $\delta_{id}^{(2)}$ and $\delta_{od}^{(2)}$ can be realized by starting with a point v of minimum in- or outdegree. With this notation, we can state the analogue of Theorem 4.

Theorem 9. For any digraph D for which the parameters are defined,

a. $\kappa_1 \leq \lambda_1 \leq \delta_1$

b. $\kappa_2 \leq \lambda_2 \leq \delta_2$

c. $\kappa_3 \leq \lambda_3 \leq \delta_3$.

Proof. The first result actually concerns GD and so is just Theorem 4. For (c), let v be a point in D with $od(v) = \delta_{od}$ and remove the outbundle of v from D to get D'. Then if $u \varepsilon D$ there will be no v - u path in D', which therefore cannot be strong; thus $\lambda_3 \leq \delta_{od}$ and so by duality, $\lambda_3 \leq \delta_3$. Now, note that if D has $\lambda_{3,2} = 1$ then $\lambda_3 = \kappa_3 = 1$. Otherwise, choose $\lambda_3 - 1$ arcs of D whose removal leaves a digraph D' with $\lambda_{3,2} = 1$, and let uv be an arc such that $D' - uv \varepsilon C_2$. Remove one endpoint of each of the other $\lambda_3 - 1$ arcs from D, leaving both u and v if possible. The resulting digraph, if still strong, has $\lambda_{3,2} = 1$, and hence $\kappa_3 \leq \lambda_3$.

The proof for part (b) is similar. We begin by choosing u and v such that $id(u) = \delta_{id}$ and $id_{D-u}(v) = \delta_{id}(D-u)$. Remove the inbundles of both u and v from D and the points will be weakly separated in the resulting digraph. Application of duality then yields $\lambda_2 \leq \delta_2$. That $\kappa_2 \leq \lambda_2$ follows as above.

Having demonstrated these inequalities we naturally next show that the three parameters are otherwise independent. Samples of these constructions are given in Figure 5-7.

Theorem 10. For any set of numbers satisfying a, b, or c of Theorem 8, there exists a digraph possessing those numbers as the given parameters.

Proof. Construction 1. To construct digraphs with $\kappa_1 = k$, $\lambda_1 = \ell$, $\delta_1 = d$, take two tournaments T_{d+1} and T'_{d+1} with point sets $\{u_1,\ldots,u_{d+1}\}$ and $\{v_1,\ldots,v_{d+1}\}$ respectively. To the digraph $T_{d+1} \cup T'_{d+1}$ add ℓ arcs from points $u_1,\ldots,u_k$ to points $v_1,\ldots,v_\ell$, as evenly distributed among the u_i as possible. Then unless $\kappa = d$ we have $\delta_1 = d$, $\lambda_1 = \ell$, and $\kappa_1 = k$. If $\kappa = d$ we instead take T_{d+1} and add a new point w with arcs $\{wu_i \mid i = 1,\ldots,d\}$. Note that either of these graphs can be made strong or unilateral by appropriate reorientation.

Construction 2. We are given $\kappa_2 = k \leq \lambda_2 = \ell \leq \delta_2 = d = \min\left(\delta_{id}^{(2)}, \delta_{od}^{(2)}\right)$; without loss of generality take $j = \delta_{od} \leq \delta_{id} \leq d$. We need concern ourselves with six different sets of points: two copies of DK_{2d}, the symmetric digraph with underlying graph K_{2d}, one copy of DK_d, and two points w_1 and w_2. There will be an arc between w_1 and each point of $DK_{2d}^{(1)}$: d-j of the arcs will be directed from w_1. There will be arcs from DK_k to $DK_{2d}^{(1)}$, distributed as evenly as possible among the points of DK_k. We then add all arcs from $DK_{2d}^{(2)}$ to DK_k, again distributed as evenly as possible. There is an arc between w_2 and each point of $DK_{2d}^{(2)}$, k of these directed away from w_2. There are all arcs from DK_d to each of w_1 and w_2, and all arcs in both directions between DK_d and DK_k.

Construction 3. Begin with two copies of DK_d, $d = \delta_3$, and $k = \kappa_3$ additional points w_i. Add all arcs from $DK_d^{(1)}$ to the w_1 and from the w_1 to $DK_d^{(2)}$, and $\ell = \lambda_3$ arcs from $DK_d^{(2)}$ to $DK_d^{(1)}$, distributed as evenly as possible among the points.

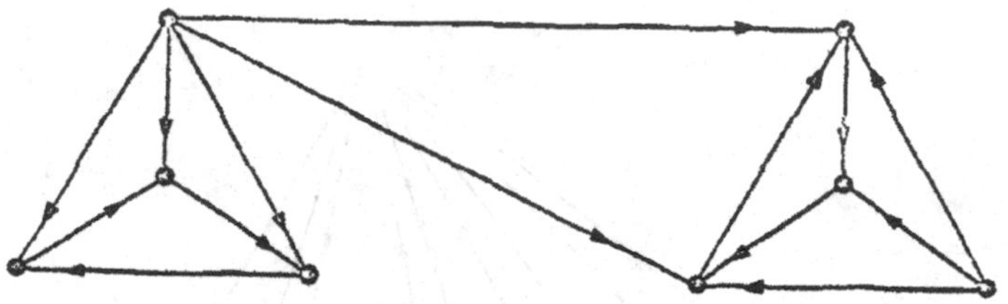

$\kappa_1 = 1,\ \lambda_1 = 2,\ \delta_1 = 3$

Figure 5.

We now proceed to develop some additional bounds on the connectivity of κ_3 corresponding to Theorems 6, 7, and 8. Recall that the condensation D^* of a digraph D has for points the strong components S_i of D, with $S_iS_j \in D^*$ if and only if there are points $u \in S_i, v \in S_j$ such that $uv \in D$.

Theorem 11. For any strong digraph D with p points, if $\delta_3 \geq \frac{p+n-2}{2}$ then $\kappa_3 \geq n$.

Proof. Suppose that $\delta_3 \geq \frac{p+n-2}{2}$ but there is a set S with $|S| = n - 1$ such that D-S is not strong. Let $D_1, D_2, \ldots, D_t$ be the strong components of D-S and without loss of generality choose D_1 such that $|D_1| \geq (p-n+1)/2$. Now, any point $v \in D_1$ can be adjacent both to and from all the points of S. However, if D_i is any other strong component of D-S then all lines between D_1 and D_i are either all into D_i or all out of D_i. So, since the condensation $(D-S)^*$ has at least one transmitter and one receiver, we also choose D_1 so that $v \in D_1$ implies that in D, $\delta_3(v) \leq \frac{p-n-1}{2} + n-1 = \frac{p+n-3}{2}$, a contradiction.

Theorem 12. If D is a strong digraph with $\delta_3 \geq (p-1)/2$, then $\lambda_3 = \delta_3$.

Proof. We know that $\lambda_3 \leq \delta_3$. Suppose $\lambda_3 < \delta_3$. Then we can find an $A \leq V(D)$ such that there are exactly λ_3 arcs from A to V-A. Suppose that the λ_3 arcs are incident from $m(m \leq \lambda_3)$ points of A. Since A has at least m points there are at least $m\delta_3 - \lambda_3$ arcs between points of A. But $m\delta_3-\lambda_3 > m\delta_3-\delta_3 > m(m-1)$ and thus there must be points in both A and V-A adjacent only to (respectively from) points in their respective subset. Then each has at least $\delta_3 + 1$ points. So $|V| \geq 2\delta_3 + 2 > 2\delta_3 + 1 \geq p$, a contradiction.

Theorem 13. If D is a strong digraph with p points and q arcs, then $\kappa_3 \leq [q/p]$.

Proof. If $[q/p] = r$ then if D has a point v_1 with min $(id(v_1), od(v_1)) > r$ there is

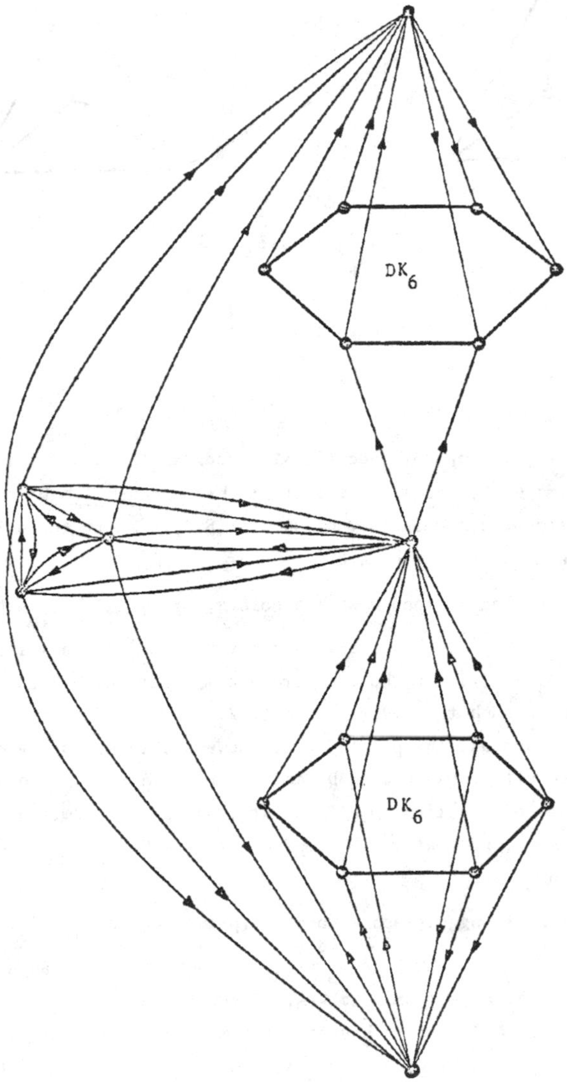

$\kappa_2 = 1,\ \lambda_2 = 2,\ \delta_2 = 3$

Figure 6

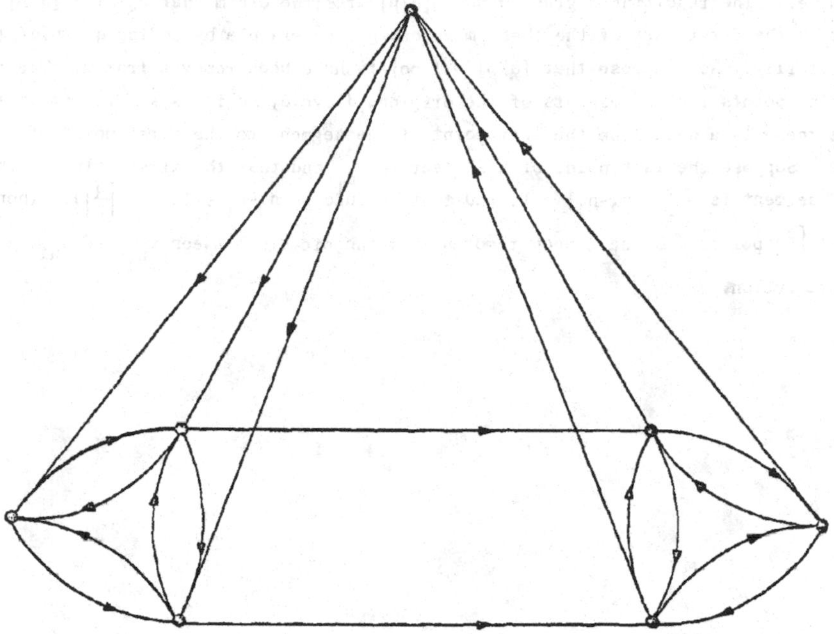

$\kappa_3 = 1,\ \lambda_3 = 2,\ \delta_3 = 3$

Figure 7.

also a point v_2 with min $(idv_2, odv_2) \leq r$. Otherwise all points v have idv = odv = r and, in either case $\kappa_3 \leq r$ by Theorem 8.

To show that equality is possible for a given q and p, we begin with a dicycle with p points having arcs $\{v_i v_{i+1}\}$, and then add arc $v_i v_j$ whenever $j-i \equiv m \pmod p$ for $2 \leq m \leq [q/p]$. (Of course if $[q/p] = 1$ then the dicycle is itself the desired example.) The resulting digraph D has $p[q/p]$ arcs; we claim that $\kappa_3(D) = [q/p]$ so that by the first part of the theorem we can get an example by adding $q - p[q/p]$ arcs arbitrarily. Now suppose that $[q/p] - 1$ points have been removed from D. The remaining points lie on segments of the original dicycle, so it is sufficient to show that there is a walk from the last point of one segment to the first point of the next. Suppose the last point of a segment is v_{n_1} and that the first point of the next segment is v_{n_2}, $|n_2-n_1| > 1$, and that (modulo p) $n_2-n_1 \in \{2,\ldots,\left[\frac{q}{p}\right]\}$. Then at least $\left[\frac{q}{p}\right]$ points must have been removed from the dicycle between v_{n_1} and v_{n_2}, a contradiction.

REFERENCES

(1). W. G. Brown and F. Harary, Extremal digraphs. Combinatorial Mathematics and its Applications (P. Erdős, Ed.) Budapest, to appear.

(2). G. Chartrand, A graph-theoretic approach to a communications problem. J. SIAM Appl. Math. 14 (1966) 778-781.

(3). G. Chartrand and F. Harary, Graphs with prescribed connectivities. Theory of Graphs (P. Erdős and G. Katona, Eds.) Akadémiai Kiadó, Budapest, 1968, 61-63.

(4). D. P. Geller, Minimally strong digraphs, Proc. Edinburgh Math. Soc., 17, Part 1 (1970), to appear.

(5). R. Halin, A theorem on n-connected graphs, J. Combinatorial Theory 7 (1969) 150-154.

(6). F. Harary, The maximum connectivity of a graph. Proc. Nat. Acad. Sci. USA 1142-1146.

(7). F. Harary, Graph Theory. Addison-Wesley, Reading, Mass., 1969.

(8). F. Harary, Variations on a theorem by Menger, SIAM Studies in Applied Math. Vol. 4, to appear.

(9). F. Harary, R. Z. Norman, and D. Cartwright, Structural Models: an introduction to the theory of directed graphs. Wiley, New York, 1965.

(10). B. Manvel, P. Stockmeyer, and D. J. A. Welsh, On removing a point of a digraph, Studia Sci. Math. Hung., to appear.

(11). J. W. Moon, Topics on Tournaments, Holt, Rinehart and Winston, New York, 1968.

(12). C. St. J. A. Nash-Williams, Well-balanced orientations of finite graphs and unobtrusive odd-vertex-pairing, Recent Progress in Combinatorics (W. T. Tutte, Ed.) Academic Press, New York, 1969, pp. 133-149.

(13). W. T. Tutte, The Connectivity of Graphs, Toronto Univ. Press, Toronto, 1967.

(14). H. Whitney, Congruent graphs and the connectivity of graphs. Amer. J. Math. 54 (1932) 150-168.

ANTITONE FIXED POINT MAPS

A. Gewirtz, Brooklyn College, C.U.N.Y., Brooklyn, NY 11210
Louis V. Quintas, Pace College, New York, NY 10038

A Lattice is a poset where each pair of elements has a glb, lub.

A complete Lattice is a lattice such that every subset has a glb and a lub.

We only talk about complete lattices.

The purpose of this paper is to determine under what conditions antitone maps of complete lattices have fixed points. Birkhoff has shown [1] that every isotone function from a complete lattice to itself has a fixed point. This is not the case for antitone functions.

<u>Definition</u> 1. Let L be a complete lattice. Then $L \xrightarrow{f} L_n$ is called <u>antitone</u> iff $a \le b \longrightarrow f(a) \ge f(b)$ for all $a, b \in L$.

<u>Proposition</u> 1. Let $a_1 > \ldots, > a_{2n-1}$, $n \ge 1$ be $2n - 1$ points which constitute a $\le$ lattice L. Then $\exists$ one and only one 1 - 1 antitone map f and a_n is the one and only one fixed point of f.

Proof:

Define f by $f(a_i) = a_{2n-i}$ $i = 1, \ldots, 2n - 1$.

<u>Proposition</u> 2. Let $a_1 > \ldots > a_{2n}$ $n \ge 1$ be a $\le$ lattice L. Then $\exists$ one and only one 1 - 1 antitone map f of L and f has no fixed points.

Proof:

Define f as in proposition 1. f is clearly 1 - 1 antitone with no fixed points.

<u>Definition</u> 2. A lattice $a_1 > \ldots > a_n$ is called a <u>chain</u>.(cl)

If $n = 2m - 1$ an <u>odd chain</u> (ocl)

If $n = 2m$ an <u>even chain</u> (ecl)

We write $\{a_i\}$ for either type chain.

If $a_1 > \ldots > a_n$ is a cl we call n the **length** of the chain.

Note: Let $a^1 = \{a^1_i\}$ and $a^2 = \{a^2_j\}$ be two cl's such that $a^1, a^2 \subset L$, L a complete lattice. Since L is complete we know $a^1_1 = a^2_1$ and $a^1_n = a^2_m$ where n is the length of a^1 and m is the length of a^2. It may well happen that a^1 and a^2 intersect at intermediate points. We provide a notation that will show at each point of a cl which other cl's intersect it.

We replace $a^1_1 > \ldots > a^1_n$ by $a^1_{(1,0)} > \ldots > a^1_{(n,o)}$ and $a^2_1 > \ldots > a^2_m$ by $a^2_{(o,1)} > \ldots > a^2_{(o,m)}$

Suppose, using old notation $a^1_1 = a^2_1$, $a^1_3 = a^2_7$, $a^1_n = a^2_m$ where $n > 4$ $m > 8$.

Then a^1 and a^2 would be written as

a) $a^1_{(1,1)} > a^1_{(2,0)} > a^1_{(3,7)} > a^1_{(4,0)} > \ldots > a^1_{(n-1,0)} > a^1_{(n,m)}$

b) $a^2_{(1,1)} > a^2_{(0,2)} > \ldots > a^2_{(0,6)} > a^2_{(3,7)} > a^2_{(0,8)} > \ldots > a^2_{(0,m-1)} > a^2_{(n,m)}$

Observe that either chain tells you

1) its' length.

2) The length of the intersecting chain.

3) The number of points of intersection.

4) At which points the chain is intersected.

5) Which points of the intersecting chain are at the intersection.

In an obvious manner this notation can be extended to chains $a^1, a^2, \ldots, a^n$; Forinstance

a) $a^1_{(3,\, 0,\, 0,\, 0,\, 6,\, 2,\, 0,\, 0,0)}$ means

1) There are 9 chains in L, $a^1, a^2, \ldots, a^9$.

2) We are looking at the 3'd point of a^1.

3) The 6'th point of a^5 is intersecting a^1 at this point.

4) The 2'nd point of a^6 is intersecting a^1 at this point.

5) No other chain intersects a^1 at the 3'd point.

b) $a^7_{(4, 2, 0, 0, 0, 0, 11, 0, 6)}$

1) This is the 11'th point of a^7.

2) The 4'th point of a^1 is intersecting a^7 at this point.

3) The 2'nd point of a^2 is intersecting a^7 at this point.

4) The 6'th point of a^9 is intersecting a^7 at this point.

5) No other chain intersects a^7 at the 11'th point.

c) Clearly $a^i_{(1, 1, \dots, 1)}$ is always a point of a^i, $i = 1, \dots, 9$ since L is complete.

d) If $l_1, l_2, \dots, l_9$ are the lengths of $a^1, a^2, \dots a^9$ respectively then $a^i_{(l_1, l_2, \dots, l_9)}$ is always a point of a^i.

e) $a^i_{(S_1, S_2, S_3, 1, S_5, S_6, S_7, S_8, S_9)}$ implies $S_j = 1$ for $j = 1, 2, 3, 5, 6, 7, 8, 9$ for all $i = 1, \dots, 9$.

and

$a^i_{(X_1, X_2, l_3, X_4, X_5, X_6, X_7, X_8, X_9)}$ implies $X_j = l_j$

$j = 1, 2, 4, 5, 6, 7, 8, 9$ for all $i = 1, \dots, 9$ where l_j is the length of a^j.

f) $a^1_{(1, 1, 1)}$, $a^1_{(2, 2, 0)}$, $a^1_{(3, 3, 0)}$, $a^1_{(4, 0, 3)}$ gives the following information.

1) There are 3 chains.

2) a^1 and a^2 have their first 3 points in common and separate at the 4'th point.

3) a^3 meets a^1 and a^2 at the 1'st point. a^3 has no other contact with a^2 up to their 3'd points but its' 3'd point meets the 4'th point of a^1.

4) $a^1_{(4, 0, 3)}$ is not (by e) the end of the chain.

<u>Proposition 3.</u> Let L be a lattice given by

1) $a^1_{(1,\dots,1)}$ belongs to L

2) $a^1_{(l_1,\dots, l_n)}$ belongs to L where $n \geq 2$, for all $i, j = 1,\dots,n$

$$l_i = l_j = 2m_i - 1 \quad m_i \geq 2$$

3) For all $i = 1,\dots, n$ if $1 < X_i < l_i$ then

$a^i_{(0,\dots, 0, X_i, 0,\dots,0)}$ belongs to L, X_i in the i'th position

4) For each a^i if $1 < X_i < Y_i < l_i$ then

$$a^1_{(1,\dots,1)} > a^i_{(0,\dots,X_i,\dots, 0)} > a^i_{(0,\dots,Y_i,\dots,0)} > a^1_{(l_1,\dots,l_n)}.$$

(Observe $a^1_{(1,\dots, 1)} \equiv a^i_{(1,\dots, 1)}$ $i = 1,\dots n$).

Then there are 1 - 1 antitone maps with and without fixed points according to the following plan

Let $i = 0,\dots, n$ be the number of fixed points. Then there are $\binom{n}{i} R_{n - i}$ 1 - 1 antitone maps with i fixed points and every fixed point is of the form $a^i_{(0,\dots, m_i,\dots, 0)}$

where $R_0 = 1$, $R_1 = 0$ and for $m \geq 2$

$$R_m = m! \left(1 + \sum_{k=1}^{m} (-1)^k \cdot \frac{1}{k!}\right)$$

Remarks:

1) $a^1_{(1,\dots, 1)} \equiv a^i_{(1,\dots, 1)}$ For all $i = 1,\dots, n$

2) $a^1_{(l_1,\dots, l_n)} \equiv a^i_{(l_1,\dots, l_n)}$ For all $i = 1,\dots, n$

3) Therefore $a^1_{(1,\dots, 1)} > a^i_{(0,\dots, X_i,\dots, 0)} > a^1_{(l_1,\dots, l_n)}$

makes sense for all $i = 1,\dots, n$.

Proof:

We first show that there is a 1 - 1 antitone map f with

n fixed points and that the fixed points are of the type specified. Suppose

1) $f\left(a^1_{(1,\dots,1)}\right) \neq a^1_{(1_1,\dots,1_n)}$

Then either

2) $f\left(a^1_{(1,\dots,1)}\right) = a^1_{(1,\dots,1)}$ or

3) $f\left(a^1_{(1,\dots,1)}\right) = a^i_{(0,\dots,X_i,\dots,0)}$ for some i

But 2) is obviously impossible since

4) For all X in a^i, $i = 1,\dots,n$ if $X \neq a^1_{(1,\dots,1)}$

then $a^1_{(1,\dots,1)} > X$

5) And therefore, $f\, a^1_{(1,\dots,1)} \geq f(X)$ C!

Suppose 3) holds. Then since $a^1_{(1,\dots,1)}$ is related to each X in L then $f\, a^1_{(1,\dots,1)} = a^i_{(0,\dots,X_i,\dots,0)}$ must be related to f(X) for all X in L.

6) Therefore f(X) is in a^i for all X in L.

But there are nl_1 points of L and only l_1 points in a^i and therefore such a map could not be 1 - 1.

7) There $f\left(a^1_{(1,\dots,1)}\right) = a^1_{(1_1,\dots,1_n)}$

Similarly

8) $f\left(a^1_{(1_1,\dots,1_n)}\right) = a^1_{(1,\dots,1)}$

We now define f on the remaining points of L by

9) $f\left(a^i_{(0,\dots,X_i,\dots,0)}\right) = a^i_{(0,\dots,2m_i - X_i,\dots,0)}$

For all $i = 1,\dots,n$ $\quad 1 < X_i < 1_i$

For each chain a^i this map is the same as the map defined in proposition 1 and therefore

10) For all $i = 1,\dots, n$ $f\left(a^i_{(0,\dots,\, m_i,\dots,\, 0)}\right) = a^i_{(0,\dots,\, m_i,\dots,\, 0)}$

By proposition 1 this is the only 1 - 1 antitone map (fixed point) which sends elements of a cl a^i into itself.

We observe that if X,Y belong to a^i and f(X) is in a^j $i \neq j$ then f(Y) belongs to a^j in order to preserve relations and if $X < Y$ then $f(X) > f(Y)$ to preserve antitone property. This means that if there are 1 - 1 antitone maps f such that X in a^i implies f(X) in a^j then for each such i,j there can be no fixed points. Therefore there is one and only one 1 - 1 antitone map with n fixed points.

We now demonstrate the existence of another 1 - 1 antitone map f'. Let

11) $a^1_{(1,\dots,\, 1)} \longleftrightarrow a^1_{(l_1,\dots,\, l_n)}$

12) $a^1_{(X,\, 0,\dots,\, 0)} \longleftrightarrow a^2_{(0,\, 2m_2 - X,\, 0,\dots,\, 0)}$

$1 < X < l_1 = l_2 = 2m_2 - 1$

13) $f'\left(a^i_{(0,\dots,\, X_i,\dots,\, 0)}\right) = a^i_{(0,\dots,\, 2m_i - X_i,\dots,\, 0)}$

For all $i = 3,\dots, n$

f' defined by 11) 12) 13) is clearly 1 - 1 antitone.

12) By above discussion gives no fixed points.

13) is the map defined in 9) and gives n - 2 fixed points.

It is clear that a generalization of this map can be defined. First decide how many fixed points are required so that the required amount can be defined by 13) (Clearly all fixed points are of the hypothesized type) and then map the remaining cl's into each other as in 12).

Therefore, for a given p $0 \leq p < n$ where p is the number of required fixed points the only remaining question is how many 1 - 1 antitone maps with p fixed points can be defined?

The answer is given by the following observations.

a) The number of fixed points p is equal to the number of chains which map into themselves. Since there are n chains these can be picked in $\binom{n}{p}$ ways.

b) There are $(n - p) = m$cl's remaining. These mcl's must be mapped into each other without any cl going into itself. We denote by R_m the number of ways this can be done for a fixed m.

Clearly if $(n - p) = 1$ $R_1 = 0$

If $m = (n - p) \geq 2$ then

$$R_m = m! \left(1 + \sum_{k=1}^{m} (-1)^k \frac{1}{k!}\right) \quad [2]$$

Therefore $m < n - 1$ fixed points can be gotten with $\binom{n}{n-m} R_m$ different maps.

<u>Proposition</u> 4. Let L be a lattice of the same type as in Proposition 3 except that if $i \neq j$ this implies $l_i \neq l_j$ For all $i,j = 1,\ldots, n$. Then there is one and only one 1 - 1 antitone map f and f has n fixed points given by $a^i_{(0,\ldots,\, m_i,\ldots,\, 0)}$ For all $i = 1,\ldots, n$.

Proof: That such an f exists is proven in proposition 3. Also, by proposition 3, if there is another 1 - 1 antitone map g then g must map some a^i into an a^j $i \neq j$.

By hypothesis $l_i \neq l_j$. Suppose $l_i < l_j$. Then there are more points in a_j then a_i and therefore g cannot be 1 - 1.

<u>Corollary</u> 1: L be a lattice of the same type as in proposition 4 except that all chains are ecl's. Then there is one and only one 1 - 1 antitone map f and f has no fixed points.

Proof: Proposition 2 and Proposition 4.

Theorem 1: Let L be a lattice given by

1) $a^1_{(1,\dots,1)}$ is in L

2) $a^1_{(l_1,\dots,l_n)}$ is in L $\quad n \geq 2$

3) $l_1 \neq l_2 \neq, \dots, \neq l_{n_o}$ For all $i \leq n_o$ $\; l_i = 2m_i - 1$

$l_{n_o+1} = l_{n_o+2} =, \dots, = l_{n_o+n_1} = 2m_{n_o+n_1} - 1$

.

.

.

.

.

.

$l_{\sum_{i=0}^{r-1} n_i + 1} =, \dots, = l_{\sum_{i=0}^{r} n_i} = 2m_{\sum_{i=0}^{r} n_i} - 1$

a) $m_i \geq 2$

b) $M_{\Sigma n_i} \geq 2$

c) $\sum_{i=0}^{r} n_i = n$

d) $n_1 \leq n_2 \leq \dots \leq n_r$

e) $l_{\sum_{i=0}^{g} n_i} \neq l_{\sum_{i=0}^{h} n_i}$ if $g \neq h$

f) $l_i \neq \sum_{j=1}^{h} n_j$ for $i = 1,\dots, n_o$ and $h = 2,\dots, r$

4) For all i $\; 1 < X_i < l_i$ implies $a^i_{(0,\dots, X_i,\dots, 0)}$ is in L

5) For each a^i if $1 < X_i < Y_i < l_i$ then

$$a^1_{(1,\dots,1)} > a^i_{(0,\dots,X_i,\dots,0)} > a^i_{(0,\dots,Y_i,\dots,0)} > a^1_{(l_1,\dots,l_n)}$$

Then the following are true.

a) There is one and only one 1 - 1 antitone map with n fixed points. The fixed points are given by

i) $a^i_{(0,\dots, m_i,\dots, 0)}$ For all $i = 1,\dots, n_o$

ii) $a^i_{(0,\dots, m_{\Sigma n_j},\dots, 0)}$ For all $i = n_o + 1 \dots \sum_{j=1}^{r} n_j$

b) Every 1 - 1 antitone map has at least n_o fixed points.

c) There is no 1 - 1 antitone map with n - 1 fixed points.

d) There are at least

i) $R_{\sum_{k=2}^{r} n_k}$ 1 - 1 antitone maps with $\sum_{k=0}^{1} n_k$ fixed points.

ii) $R_{\sum_{k=3}^{r} n_k}$ 1 - 1 antitone maps with $\sum_{k=0}^{2} n_k$ fixed points.

.

.

.

j) $R_{\sum_{k=j+1}^{r} n_k}$ 1 - 1 antitone maps with $\sum_{k=0}^{j} n_k$ fixed points.

.

.

.

r-1) $R_{\sum_{k=r}^{r} n_k}$ 1 - 1 antitone maps with $\sum_{k=0}^{r-1} n_k$ fixed points.

Proof:

a) Proposition 3, Proposition 4

b) Proposition 4

c) Proposition 3

d) We prove for d j)

Define f as

i) $a^1_{(1,\dots,1)} \longleftrightarrow a^1_{(1_1,\dots,1_n)}$

ii) For $i \leq \sum_{k=0}^{j} n_k$

$$f\left(a^i_{(0,\dots,X_i,\dots,0)}\right) = a^i_{(0,\dots,2m_i-X_i,\dots,0)}$$

iii) Permute all other chains as in proposition 3 so that no chain maps into itself.

<u>Corollary</u> 1. Let L be a lattice of the type given in theorem 1 with s ecl's adjoined. That is we have a new lattice L' whose top is $a^1_{(1,\dots,1)}$ and whose bottom is $a^1_{(1_1,\dots,1_n,\dots,1_{n+s})}$ where

$l_{n+1} = l_{n+2} = \dots = l_{n+s} = 2t$. Then the conclusions of theorem 1 are true except that the number of maps must be multiplied by s!

Corollary 2.

In Corollary 1 if

a) $l_{n+1} = \dots = l_{n+s_1}$

b) $l_{n+s_1+1} = , \dots , = l_{n+s_1+s_2}$

.

.

.

$l_{n+\sum_{k=1}^{h-1} s_k + 1} = , \dots , = l_{n+\sum_{k=1}^{h} s_k}$ where $\sum_{k=1}^{h} s_k = s$

Then the number of 1 - 1 antitone fixed point maps of theorem 1 must be multiplied by $(s_1!)(s_2!)\dots(s_n!)$.

In particular note that if $l_{n+i} \neq l_{n+j}$ For all i,j $i \neq j$ then the conclusions of theorem 1 hold exactly as stated in the the theorem.

Definition 4. A lattice of the type hypothesized in propositions 3, 4 and theorem 1 is called a finite black box (fbb).

We also call the midpoint of an ocl a finite black box if the ocl is not part of a finite black box. We call $a^1_{(1,\dots,1)}$ the top of the finite black box and designate it T.

We call $a^1_{(l_1,\dots,l_n)}$ the bottom of the finite black box and designate it B.

Definition 5. The map f on a fbb such that $f\left(a^i_{(0,\dots,X_i,\dots,0)}\right) = a^i_{(0,\dots,2m_i-X_i,\dots,0)}$ where $l_i = 2m_i$ or $l_i = 2m_i-1$ will be called symmetric on the chain a^i.

Definition 6. If $f\left(a^i_{(0,\dots,X_i,\dots,0)}\right) = a^j_{(0,\dots,2m_j-X_i,\dots,0)}$ $i \neq j$ $l_i = l_j$ Then f will be called symmetric from the chain a^i

to the chain a^j.

Definition 7. For all i if f is either symmetric on a^i or symmetric from a^i to a^j and $a^1_{(1,\dots,1)} \longleftrightarrow a^1_{(1_1,\dots,1_n)}$ Then f will be called _symmetric on the fbb_.

Definition 8. For all $n \geq 3$ let p_n be a set of chains a^{i_n} of length n where the chains intersect only at the top and the bottom. i_n is contained in Λ_n where Λ_n is an index set of cardinality less than or equal to the cardinality of the reals. Let L^∞ be the lattice of all such cl's given by

1) $a^1_{(1,\dots,1,\dots)}$ belongs to L^∞
2) $a^1_{(3,\dots,3,\dots,4,\dots,4,\dots,n,\dots,n,\dots)}$ belongs to L^∞
3) If a^{i_n} belongs to p_n then $1 < X_{i_n} < n$ implies $a^{i_n}_{(0,\dots,X_{i_n},\dots,0,\dots)}$ belongs to L^∞
4) If $1 < X_{i_n} < Y_{i_n} < n$ then

$$a^1_{(1,\dots,1,\dots)} > a^{i_n}_{(0,\dots,X_{i_n},\dots,0,\dots)} >$$

$$a^{i_n}_{(0,\dots,Y_{i_n},\dots,0,\dots)} > a^1_{(3,\dots,3,4,\dots,4,n,\dots,n,\dots)}$$

We call L^∞ an _infinite black box_ (ibb).

We call $a^1_{(1,\dots,1,\dots)}$ the _top_ of the ibb and designate it T.

We call $a^1_{(3,\dots,3,4,\dots,4,n,\dots,n,\dots)}$ the _bottom_ of the ibb and designate it B.

Definition 9. The map f on an ibb such that

$$f\left(a^{i_n}_{(0,\dots,X_{i_n},\dots,0,\dots)}\right) = a^{i_n}_{(0,\dots,2m_{i_n}-X_{i_n},\dots,0,\dots)}$$

will be called _symmetric on_ a^{i_n}.

Definition 10. The map f on an ibb such that if $i_n \neq j_n$

$$f\left(a^{i_n}_{(0,\dots,\, X_{i_n},\dots,\, 0,\dots)}\right) = a^{j_n}_{(0,\dots,\, 2m_{j_n} - X_{i_n},\dots,\, 0,\dots)}$$ will

be called <u>symmetric</u> <u>from</u> <u>a^{i_n}</u> <u>to</u> <u>a^{j_n}</u>.

<u>Definition</u> 11. For all p_n if for all i_n f is either symmetric on a^{i_n} or symmetric from a^{i_n} to a^{j_n} $i_n \neq j_n$ and $T \leftrightarrow B$ then f will be called <u>symmetric</u> on L^∞.

<u>Definition</u> 12. f will be called <u>symmetric</u> <u>on</u> <u>a</u> <u>bb</u> if f is symmetric on a fbb or is symmetric on an ibb.

<u>Proposition</u> 5. Let L^∞ be an ibb and $n \geq 3$ be an odd integer. Then any f symmetric on L^∞ and symmetric on some a^{i_n} contained in p_n is $1 - 1$ antitone with fixed point $a^{i_n}_{(0,\dots,\, m_{i_n},\dots,\, 0,\dots)}$ where $n = 2m_{i_n} - 1$.

Proof: Theorem 1.

<u>Proposition</u> 6. If f is symmetric on L^∞ and for all p_n f is not symmetric on a^{i_n} for all i_n then f is $1 - 1$ antitone with no fixed points.

Proof: Theorem 1.

<u>Corollary</u> 1. Let f be symmetric on L^∞, $n \geq 3$ be an odd integer $n = 2m - 1$ and $p_n = \{a^{1_n}\}$ Then f is $1 - 1$ antitone with fixed point $a^{1_n}_{(0,\dots,\, m,\dots,\, 0,\dots)}$.

<u>Proposition</u> 7. If f is not symmetric on L^∞ then f is not $1 - 1$ antitone.

Notation: We write $a^{i}_{(0,\dots,\, X_i,\dots,\, 0,\dots)}$, $a^{1}_{(1,\dots,\, 1,\dots)}$ and $a^{1}_{(1_1,\dots,\, 1_n,\dots)}$ as points of a bb even if the bb is a fbb unless we care to distinguish the fbb.

<u>Proposition</u> 8. Let L be a bb with more than 1 chain and let $t_1 >,\dots,> t_r = T$, $B = b_1 > \dots > b_m$ be cl's where $r,m \geq 1$. Then

a) $\{t_i\} \cup L \cup \{b_i\} = C$ is a lattice.

b) $f_* : C \longrightarrow C$ is 1 - 1 antitone iff $f_* | L = f : L \longrightarrow L$ is 1 - 1 antitone and $t_k \longleftrightarrow b_{m-k}$ where $m = r$, $k = 1,\dots, r$

If f_* is 1 - 1 antitone then f_* and f have the same fixed points points.

c) $m \neq r$ implies C has no 1 - 1 antitone maps.

Proof: a) We rename the points of C as follows.

1) $t_k = C^i_{(k,\dots,k,\dots)}$ i in Λ, $k = 1,\dots, r$

2) For all a^i in L $a^i_{(0,\dots,X_i,\dots,0,\dots)} = C^i_{(0,\dots,X_i+(r-1),\dots,0,\dots)}$

3) $b_j = C^i_{(l_1+(r-1)+(j-1),\dots,l_n+(r-1)+(j-1),\dots)}$, i in Λ, $j = 1,\dots, m$ where l_i is the length of a^i.

Clearly C is a lattice

b) suppose $m = r$ and f is symmetric on L (therefore f is 1 - 1 antitone on L)

1) Define f_* on C so that $f_* = f$ on L and $C^i_{(k,\dots,k,\dots)} \longleftrightarrow C^i_{(l_1+(r-1)+(m-k),\dots,l_n+(r-1)+(m-k),\dots)}$, $k = 1,\dots, r$

f_* is clearly 1 - 1 antitone with fixed points of f.

2) A map g such that $g\left(C^i_{(k,\dots,k,\dots)}\right) = C^i_{(0,\dots,X_i,\dots,0,\dots)}$ C^i of length $l_i + 2m = l_i + m + r$ is impossible since $C^i_{(k,\dots,k,\dots)}$ is related to each point of C. Therefore, for all X in C, g(X) would have to be in C^i which means g cannot be 1 - 1.

Similarly $g\left(C^i_{(l_1+(r-1)+(j-1),\dots,l_n+(r-1)+(j-1),\dots)}\right) \neq$

$C^{i}_{(0,\dots,X_i,\dots,0,\dots)}$ $j = 1,\dots, m,$ i in Λ.

3) Also it is impossible for $g\left(C^{i}_{(0,\dots,X_i,\dots,0,\dots)}\right) = C^{i}_{(k,\dots,k,\dots)}$

since $C^{i}_{(0,\dots,X_i,\dots,0,\dots)}$ is only related to the $1_i + 2m$ points of C^i and $C^{i}_{(k,\dots,k,\dots)}$ is related to all points of C.

Similarly $g\left(C^{i}_{(0,\dots,X_i,\dots,0)}\right) \neq C^{i}_{(1_1+(r-1)+(j-1),\dots,1_n+(r-1)+(j-1),\dots)}$,

i in Λ, $j = 1,\dots, m$.

4) Therefore any 1 - 1 antitone map f_* must send bb points to bb points (and therefore by proposition 7 must be symmetric on L). f_* on points outside the bb must clearly be defined as in b) or else the antitone porperty would be lost (see propositions 1, 2)

c) By b_4) if $m \neq r$ the 1 - 1 property would be lost.

Proposition 9. Let R' be the lattice ordered by the real line. Then The map $f: R' \longrightarrow R'$ given by

$$f(X) = -X,\ f(0) = 0 \text{ is } 1 - 1 \text{ antitone with fixed point } 0.$$

Proposition 10. Let L be a bb with more than 1 chain and $\{t_i\}$, i in $[0,1)$ and $\{b_j\}$ j in $(-1,0]$ be chains ordered by $t_{i_1} > t_{i_2}$ iff $i_1 > i_2$ and $b_{j_1} > b_{j_2}$ iff $j_1 > j_2$.

Let $L \cap \{t_i\} = t_0 = a^{1}_{(1,\dots,1,\dots)}$

$L \cap \{b_j\} = b_0 = a^{1}_{(1_1,\dots,1_n,\dots)}$

Then

a) $L \cup \{t_i\} \cup \{b_j\} = C$ is a lattice

b) $f_*: C \longrightarrow C$ is 1 - 1 antitone if

$$f_* \text{ is given by} \begin{cases} f \text{ on } L \text{ where } f \text{ is } 1-1 \text{ antitone} \\ t_i \longleftrightarrow \underline{b}_i \quad i \text{ containing } [0,-1] \end{cases}$$

(See note below Corollary 1)

c) The fixed points of f_* are those of f.

Proof:

a) Clear

b) and c) Define f_* on C so that

1) $f_* = f$ on L and

$f_*(t_i) = b_{-i}$, $0 \leq i < 1$

$f_*(b_j) = a_{-j}$, $-1 < j \leq 0$

This is obviously 1 - 1 antitone with the fixed points of f.

Corollary 1:

If the index set of $\{t_i\}$ and $\{b_j\}$ is Z then the conclusions of proposition 10 still hold.

Note: As long as $\{t_i\}$ and $\{b_j\}$ map to each other in some 1 - 1 fashion f_* would still be 1 - 1 antitone with the same fixed points as f provided we insist $f_* \mid L = f$.

What happens when we lift the restrictions just mentioned is the subject of proposition 17.

There are no difficulties in any finite case, because in the finite case if a is not related to b then f(a) cannot be related to f(b) whereas in the infinite case there is no such restriction.

Definition 13. A lattice of the type hypothesized in Proposition 8 is called a finite normal black box.

A lattice of the type hypothesized in Proposition 10 or Corollary 1 of Proposition 10 is called an infinite normal black box.

Proposition 12: C as defined in Proposition 8, Proposition 10 or Proposition 10, Corollary 1 has 1 - 1 antitone maps (with or without fixed points.) iff the cardinality of $\{t_i\}$ is equal to the cardinality of $\{b_j\}$. If the cardinalities are equal the 1 - 1 antitone maps are given by the conclusions of Proposition 8, Proposition 10 and Corollary 1 to Proposition 10.

Note: In the note following Corollary 1 of Proposition 10 we discussed the "large" number of maps that could occur if the $\{t_i\}$ and $\{b_j\}$ are infinite. These maps would tend to obscure the next sequence of propositions by requiring the hypothesis to be extremely complicated in order to consider all the maps. We therefore will restrict ourselves to finite nbb's. Note that the bb's themselves may or may not be finite. We return to infinte nnb's in Proposition 17 to show the only condition under which a bb does not determine the antitone properties of a map.

<u>Proposition</u> 13: Let C be a finite nbb and let $\{d_j\}$ be a cl intersecting C on the top at 2 points t_i and t_m, $t_i < t_m$. Let f be a 1 - 1 antitone map of C. (with or without fixed points). Let $\{e_k\}$ be a cl. Then

f_*: $C \cup \{d_j\} \cup \{e_k\} \longrightarrow C \cup \{d_j\} \cup \{e_k\}$ is 1 - 1 antitone with the same fixed points as f iff

1) $f_* \big| C = f$
2) The number of elements in $\{d_j\}$ is equal to the number of elements in $\{e_k\}$ (finite)
3) $\{e_k\}$intersects the bottom of C in 2 points b_1 and b_h $b_1 < b_h$ such that $b_1 \overset{f}{\longleftrightarrow} t_m$, $b_h \overset{f}{\longleftrightarrow} t_i$
4) f_* sends $\{e_k\}$ and$\{d_j\}$ into each other in the sense of Proposition 3 or the 1st Corollary to theorem 1.

Proof: Trivial

<u>Proposition</u> 14: Let C be a finite nbb and let $\{d_j\}$ be a finite cl intersecting C on the top at one point t_h and in the bb at one point $a^i_{(0,\ldots,X_i,\ldots,0)}$. Let f be a 1 - 1 antitone map of C (with or without fixed points). Let $\{e_k\}$ be a cl. Then f_*: $C \cup \{d_j\} \cup \{e_k\}$ $\longrightarrow$ $C \cup \{d_j\} \cup \{e_k\}$ is 1 - 1 antitone withe same fixed points as

f iff

1) $f_* | C = f$

2) $\{d_j\}$ and $\{e_k\}$ have the same number of elements.

3) $\{e_k\}$ intersects the bottom of C at a point b_p and the bb at a point $a^j_{(0,\ldots,\, 2m_j - X_i,\ldots,\, 0,\ldots)}$ such that $b_p \overset{f}{\longleftrightarrow} t_h$ and $a^j_{(0,\ldots,\, 2m_j - X_i,\ldots,\, 0,\ldots)} \overset{f}{\longleftrightarrow} a^i_{(0,\ldots,\, X_i,\ldots,\, 0,\ldots)}$

4) f_* sends $\{e_k\}$ and $\{d_j\}$ into each other in the sense of Proposition 3 or the first Corollary to theorem 1.

Proof: trivial. <u>Note</u> $a^i_{(0,\ldots,\, m_i,\ldots,0,\ldots)}$ is a fixed point iff $i = j$ and $l_i = 2m_i - 1$

<u>Proposition</u> 15: Let C be a finite nbb and let $\{d_j\}$ be a finite cl intersecting the top of C at t_p and the bottom of C at b_h. Then

1) $C \cup \{d_j\}$ has a 1 - 1 antitone map $\bar{f}$ iff $b_h \overset{f}{\longleftrightarrow} t_p$ where f is 1 - 1 antitone on C.

2) If there is $\bar{f}$ as in 1) then the fixed points of $\bar{f}$ are the fixed points of f and if the cardinality of $\{d_j\}$ is equal to $2r - 1$ then the midpoint of $\{d_j\}$ is also fixed.

<u>Corollary</u> 1. Let C be a finite nbb and let $\{H_j\}$ be a finite set of finite nbb's or an infinite set of finite nbb's such that for each H_i the greatest point of H_i intersects the top of C at t_i and the least point of H_i intersects the bottom of C at b_i where any 1 - 1 antitone map f of C is such that $b_i \overset{f}{\longleftrightarrow} t_i$.

Let f_i be any 1 - 1 antitone map of H_i.

Then $f_*: C \cup \{H_j\} \longrightarrow C \cup \{H_j\} = C'$ which is defined to be f on C and f_i on H_i for all i, is 1 - 1 antitone with fixed points the fixed points of f plus the fixed points of f_i for all i.

<u>Definition</u> 14. C' of Corollary 1, Proposition 15 is called a <u>generalized finite nbb</u>.

<u>Proposition</u> 16: Let C' be a generalized finite nbb and let g be

a 1 - 1 antitone map, with or without fixed points, defined on C'. Let $e_1, e_2, \ldots, e_n$ be n cl's which have the following properties. (each e_j of finite length)

1) For all j the greatest point of e_j intersects the top of C or the top of one of the H_i.
2) The least point of e_j intersects the top of C or the top of one of the H_i or one of the chains in one of the bb's.
3) All intermediate points of e_j may intersect a point of the top of C or a point of the top of one of the H_i or may not intersect any other chain or may intersect some e_k, $k \neq j$
4) Corresponding to each e_j there is one and only one d_j with the property that for each point t_1 at which e_j intersects the top of C there is a corresponding point of d_j which intersects the bottom of C at b_q the point which is given by $b_q \overset{g}{\longleftrightarrow} t_1$.
5) If e_j intersects a bb at $a^i_{(0,\ldots,\, X_i,\ldots,\, 0,\ldots)}$

   Then g must be such that

   $$a^j_{(0,\ldots\, 2m_j - X_i,\ldots,\, 0,\ldots)} \overset{g}{\longleftrightarrow} a^i_{(0,\ldots,\, X_i,\ldots,\, 0,\ldots)}$$

   and the corresponding d_j intersects the bb at a

   $$a^j_{(0,\ldots,\, 2m_j - X_i,\ldots,\, 0,\ldots)}.$$
6) If e_j intersects e_k, $j \neq k$ then the corresponding d_j intersects the corresponding d_k
7) If a point of e_j does not intersect any other chain then the corresponding point of d_j does not intersect any other chain.

Note: When we say points of e_j and d_j correspond we mean in the sense that the points of $\{t_i\}$ and $\{b_j\}$ correspond in the definition of the finite normal black box so that we can "naturally" define the antitone map on e_j and d_j.

Then g_* which is g extended to the e_j and d_j in the sense just described is 1 - 1 antitone with the same fixed points as g.

Proof: By construction of the hypotheses.

Proposition 17: (Destruction Proposition)

Let C be an infinite normal finite black box. Let the finite black box be such that every chain has odd length and if l_i is the length of a^i, l_j the length of a^j, $i \neq j$ then $l_i \neq l_j$. $i, j = 1, \dots, n$. Let f be the one and only one 1 - 1 antitone map on the bb (by Proposition 4 such a map exists) where $a^i_{(0,\dots, m_i, \dots, 0)}$ $i = 1, \dots, n$, $l_i = 2m_i - 1$ are the n fixed points.

Then $\exists$ a 1 - 1 antitone map g: $C \to C$ such that g has no fixed points.

(i.e. If the top and bottom are infinite the bb can be destroyed)

Proof: Without loss of generality we consider the bb to have 2 chains, a^1 of length n, a^2 of length m and $\{t_i\}$ and $\{b_j\}$ to be countably infinite.

We let $\{t_i\} \cap bb = a^1_{(1, 1)} = t_o$ and $j > k$ implies $t_j > t_k$

$\{b_q\} \cap bb = a^1_{(n, m)} = b_o$ and $r > s$ implies $b_r > b_s$

We define g as follows

1) For some integer j, $g(t_j) = a^2_{(0, \frac{m-1}{2})}$

2) $g(t_{j-i}) = t_{j+1}$, For all i in Z such that $0 < i \leq j$

3) $g\left(a^1_{(n-k, 0)}\right) = t_{2j + (n-k)}$ For all k in Z such $1 \leq k \leq n-2$

4) $g\left(a^2_{(0, m-\ell)}\right) = t_{2t + (n-1) + m-\ell)}$ For all ℓ in Z such that $1 \leq \ell \leq m-2$

5) $g\left(a^1_{(n, m)}\right) = t_{2j + (n-1) + m}$

6) $g(b_q) = t_{2j + (n-1) + m - q}$ For all q in Z such that $q < 0$

7) $g\left(t_{j+i}\right) = b_{j+i}$ For all i in Z such that $i \geq 1$

Example: Let S_n be a set with n elements. Then there are 2^n subsets of S_n including S_n itself and the null set. We form $\tilde{S}_n$ the lattice of subsets of S_n by defining $A \leq B$ iff $A \subseteq B$ and $A < B$ iff $A \subset B$. We have $2^n = \binom{n}{0} + \ldots + \binom{n}{r} + \ldots + \binom{n}{n}$ where $\binom{n}{r}$ is the number of distince subsets containing r elements.

We observe that the mapping which sends each set into its' complement is 1 - 1 antitone with no fixed points for any $\tilde{S}_n$ $n \geq 1$. We now investigate $\tilde{S}_n$ for all $n \geq 1$ for other 1 - 1 antitone maps.

Proposition 18: $\tilde{S}_1$ has no other 1 - 1 antitone maps and therefore no 1 - 1 maps with fixed points.

Proposition 19: $\tilde{S}_2$ has one other 1 - 1 antitone map f and this map has 2 fixed points.

Proof: If a_1, a_2 are the elements of S_2.

Define $\{a_1, a_2\} \overset{f}{\longleftrightarrow} \phi$

$f(\{a_1\}) = \{a_1\}$

$f(\{a_2\}) = \{a_2\}$

Proposition 20: $\tilde{S}_n$, $n = 2m - 1$, $m > 1$ has no 1 - 1 antitone maps with fixed points.

Proof: If n is odd every chain has even length and therefore there are no fixed points.

References

1. G. Birkhoff, Lattice Theory A.M.S. 1961

2. H. Ryser, Combinatorial Mathematics M.A.A. 1963

ON SMALL GRAPHS WITH FORCED MONOCHROMATIC TRIANGLES

by

R. L. Graham
Bell Telephone Laboratories, Incorporated
Murray Hill, New Jersey

and

J. H. Spencer
The Rand Corporation
Santa Monica, California

Let us denote by $S(k,\ell;\ r)$ the following statement:

There exists a graph G which does not contain a complete subgraph on ℓ vertices but which has the property that any r-coloring of the edges of G must contain a monochromatic complete subgraph on k vertices.

It is immediate from Ramsey's Theorem (cf. [5]) that for any fixed k and r, $S(k,\ell;\ r)$ is true for ℓ sufficiently large. In partiular, it follows that $S(3,7;\ 2)$ holds by taking G to be K_6, the complete graph on 6 vertices. Recently, Erdös and Hajnal [1] asked whether $S(3,6;\ 2)$ holds. This was first answered affirmatively by J. H. van Lint (unpublished) who gave as an example of a graph which establishes $S(3,6;\ 2)$, the <u>complement</u> of the graph shown in Fig. 1.

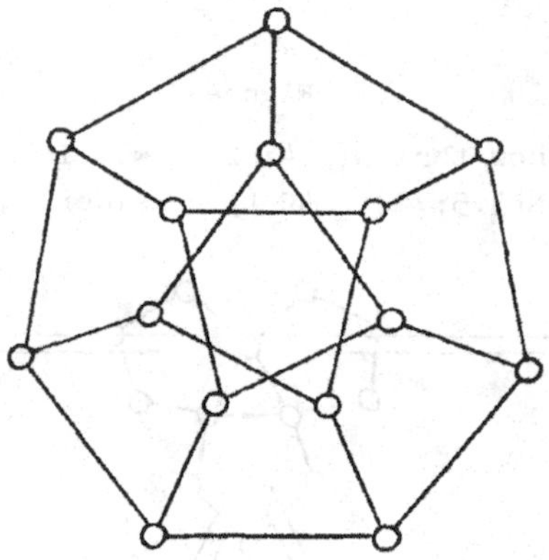

Figure 1

Soon thereafter, L. Pósa (unpublished) proved the existence of a graph G for which $S(3,5;\ 2)$ holds, basing his work on some previous existence proofs of Erdös.

The final step in this direction was achieved by the late J. H. Folkman [2] who established $S(3,4;\ 2)$ by the explicit construction of an appropriate (very large) graph G. More generally, Folkman also established $S(k,k+1;\ 2)$ in [2] for all $k \geq 3$. Furthermore, Folkman asserted in 1968 that he had a proof of $S(3,4;\ 3)$ and a very complicated proof of $S(3,4;\ 4)$ but no notes on these ideas have as of yet been discovered. It was conjectured by Folkman and independently by Erdős and Hajnal that $S(k,k+1;\ r)$ holds for all k and r.

Erdős has pointed out that it would be of interest to determine the least number $N(k,\ell;\ r)$ of vertices a graph may have which can be used to establish $S(k,\ell;\ r)$. It was shown by one of the authors in [3] that $N(3,6;\ 2) = 8$. The unique graph G which achieves this bound is the complement of the 8 vertex graph shown in Fig. 2. Thus, G has 8 vertices and 23 edges.

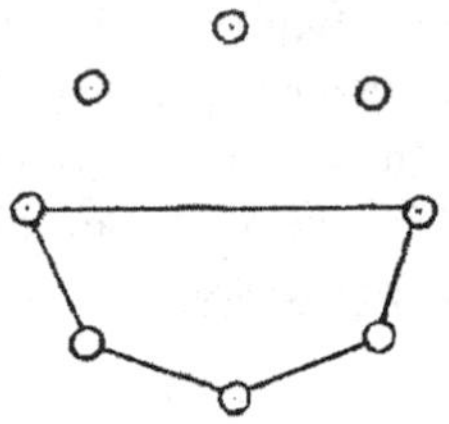

Figure 2

The results of [2] show that $N(3,4;\ 2) < \infty$. In a recent paper, Schäuble [6] proves $N(3,5;\ 2) \leq 42$ by considering the graph shown in Fig. 3.

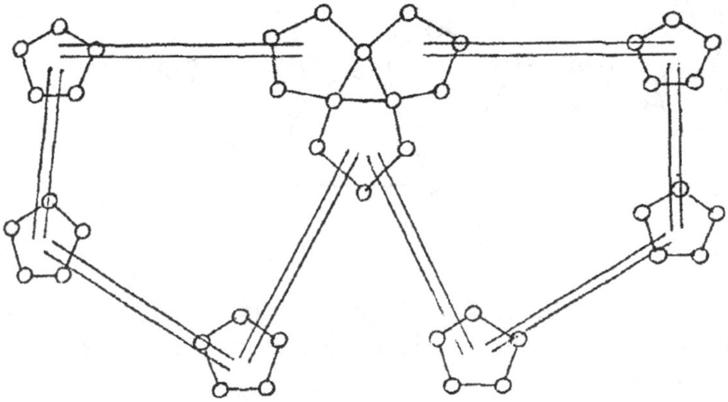

Figure 3

Here, we use the notation

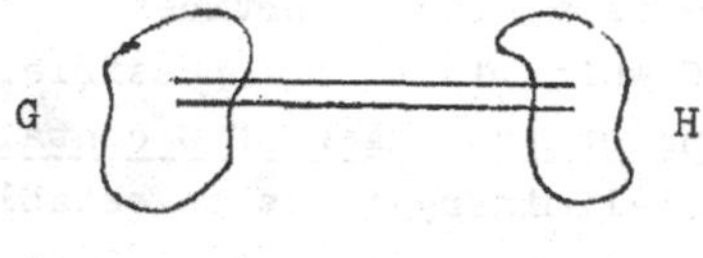

to indicate that all vertices of G are connected to all vertices of H.

In this note we prove the following result:

Theorem: $N(3,5;\ 2) \leq 23$.

Proof: Consider the graph G given in Fig. 4.

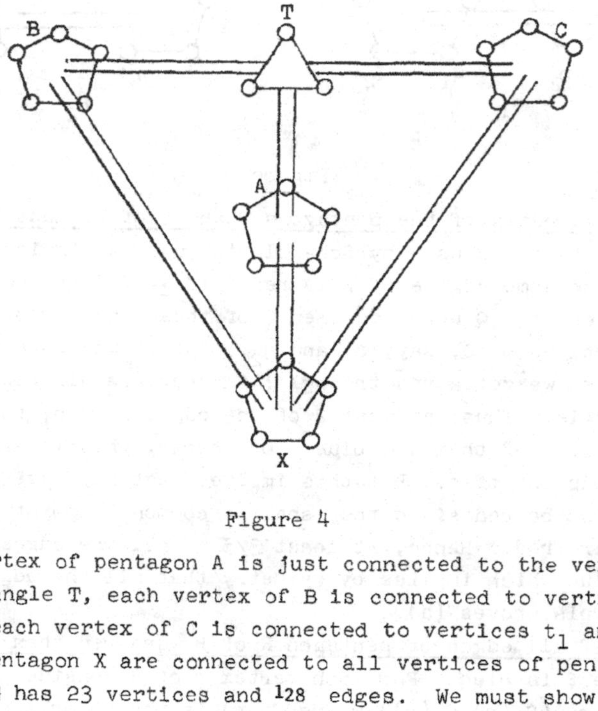

Figure 4

In G, each vertex of pentagon A is just connected to the vertices t_2 and t_3 of triangle T, each vertex of B is connected to vertices t_1 and t_2 of T, and each vertex of C is connected to vertices t_1 and t_3. All vertices of pentagon X are connected to all vertices of pentagons A, B, C. Thus, G has 23 vertices and 128 edges. We must show that G can be used to establish $S(3,5;\ 2)$.

(i) $K_5 \not\subseteq G$. Consider the possible locations of the vertices of a hypothetical subgraph K_5. We cannot have ≥ 3 vertices of this K_5 in one pentagon A, B, C or X since they all contain no triangles. Also, since there are no edges between pentagons A, B and C, no vertex of the K_5 can be in X. If the K_5 had ≥ 3 vertices not in T, at least two of the pentagons A, B, C would have to contain a vertex of the K_5

which is impossible since these pentagons have no interconnecting edges. The only possibility left is if all 3 vertices of T were also vertices of the K_5. The remaining 2 vertices of the K_5 must then belong to one of A, B, C which is also impossible.

(ii) Any 2-coloring of the edges of G contains a monochromatic triangle. We need two preliminary facts to establish (ii). We refer to Fig. 5 for the graphs under consideration. Assume the graphs H_1 and H_2 have been 2-colored so that no monochromatic triangles have been formed.

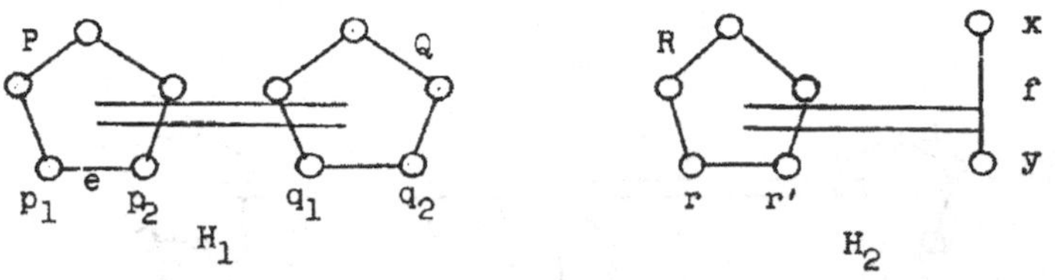

Figure 5

(a) All edges of the pentagons P and Q of H_1 must be the same color. This fact was used by Schauble in [6]. We indicate a short proof. Assume some edge e of P is red. If ≥ 3 of the edges from some endpoint p_1 of e to Q were red then 2 of these edges must go to adjacent vertices of Q, say, q_1 and q_2. But if any edge between p_2, q_1, q_2 is red then we get a red triangle; if they are all blue then we get a blue triangle. Thus, at most 2 of the edges from p_1 to Q can be red, i.e., at least 3 of them are blue. Of course, this is also true for the other endpoint of e. But this implies that any edge of P adjacent to e must also be red since they share a common endpoint. Hence, all edges of P are red. Hence, at least 3/5 of all the edges between P and Q must be blue which implies by symmetry that all the edges of Q are also red. This proves (a).

(b) If all edges of pentagon R of H_2 are red then the edge f is red. Assume f is blue. For each vertex r of R consider the ordered pair of colors $(C_x(r), C_y(r))$ where $C_x(r)$ is the color assigned to the edge from r to x, with $C_y(r)$ defined similarly. We certainly cannot have $(C_x(r), C_y(r))$ = (blue,blue) since this forms a blue triangle r,x,y Also $(C_x(r), C_y(r))$ = (red,red) is impossible because any red edge between r',x,y forms a red triangle and if these edges are all blue then a blue triangle is formed. Hence, we must have $(C_x(r), C_y(r))$ = (red,blue) or (blue,red). However, we cannot have $(C_x(r), C_y(r)) = (C_x(r'), C_y(r'))$ because the red component, say, $C_x(r) = C_x(r')$ = red, would form a red triangle r, r', x. Hence adjacent vertices in H_2 must have distinct pairs $(C_x(r), C_y(r))$. This is <u>impossible</u> however because H_2 is an odd cycle. This proves (b).

The proof of (ii) is now immediate. Assume without loss of generality that some edge of pentagon X in G is red. Hence by (a), all edges of A, B and C are also red. Finally, by (b), all edges of triangle T are red. This proves the Theorem.

It might be conjectured that $N(3,5;\ 2) = 23$ although admittedly there is not too much evidence for such an assertion. It seems very difficult to establish any nontrivial lower bounds on the $N(k,\ell;\ r)$. S. Lin [4] has recently shown $N(3,5;\ 2) \geq 10$. However, it is not known even if $N(3,5;\ 2) \geq 11$.

REFERENCES

1. Erdös, P. and Hajnal, A., "Research Problem 2-5", Journal of Combinatorial Theory, Vol. 2, 1967, p. 104.
2. Folkman, J. H., "Graphs with Monochromatic Complete Subgraphs in Every Edge Coloring", Rand Corporation Memorandum RM-53-58-PR, October 1967 (to appear in SIAM Journal of Applied Mathematics).
3. Graham, R. L., "On Edgewise 2-Colored Graphs with Monochromatic Triangles and Containing No Complete Hexagon", Journal of Combinatorial Theory, Vol. 4, No. 3, April 1968, p. 300.
4. Lin, S., "On Ramsey Numbers and $\vec{K}_r$-Coloring of Graphs", to appear.
5. Ryser, H. J., "Combinatorial Mathematics, Wiley, New York, 1963.
6. Schäuble, M., "Zu einem Kantenfärbungsproblem", Wiss. Z. TH Ilmenau 15, Vol. 2, 1969, pp. 55-58.

LATEST RESULTS ON CROSSING NUMBERS

Richard K. Guy

The University of Calgary, Calgary 44, Alberta, Canada

0. INTRODUCTION

A *graph* G is a non-empty set of objects, V, called *vertices*, *points* or *nodes*, together with a subset E of the set of (unordered) pairs of those objects, $E \subset [V]^2$, called *edges*, *lines* or *arcs*. A *drawing* D of a graph G is a mapping of G into a surface, which in this paper will be the Euclidean plane or (equivalently except when considering rectilinear crossing number) the (surface of the) sphere S^2 in Euclidean 3-space, in which the set V is mapped into $|V|$ distinct points, called *nodes*, and the set E is mapped into $|E|$ open (Jordan) *arcs*, each of which is closed by its two defining nodes. No (interior) point of an arc is a node, and we will also assume that no point of the surface belongs to more than two arcs. P. Erdös has suggested the study of the corresponding problems when this last condition is relaxed.

In particular we are concerned with *good drawings* in which

(a) no arc intersects itself (because the arcs are Jordan arcs),

(b) any pair of arcs have at most one point in common, and such a point is a *crossing* (we suppose that the tangents to the arcs at the point are distinct),

(c) two arcs sharing a common node have no (other) point in common,

and with *optimal drawings* which, among all possible drawings of G, exhibit the minimum number of crossings. It can be verified that an optimal drawing is a good drawing. This least number of crossings is called the *crossing number* of G and is denoted by $\nu(G)$. If the surface is the plane, and the arcs are straight line segments, then the corresponding concept is called the *rectilinear crossing number* and denoted by $\bar{\nu}(G)$.

There are few graphs for which the crossing number is known and hardly any infinite families. It is known that $\nu = 2$ for the Petersen graph [13, p. 124] and that $\nu = 1$ for the Möbius ladders [10]. The most studied infinite families are the complete graphs, K_n, the complete bipartite graphs, $K_{m,n}$ and the (1-skeleton of the)

n-dimensional cube Q_n. Only very partial results are known in each case. The present paper lists recent results; proofs are either outlined or omitted, some of them being in papers submitted to the *Journal of Combinatorial Theory* and the *Canadian Journal of Mathematics*.

1. THE n-DIMENSIONAL CUBE

The graph Q_n has for vertices the 2^n points of the n-dimensional vector space over $GF(2)$; a pair of vertices comprise an edge if and only if their vectors differ in exactly one component. Recently R.B. Eggleton and the author have shown [3] that

$$\nu(Q_n) \leq \frac{5}{32}\,4^n - \left[\frac{n^2+1}{2}\right]2^{n-2} \tag{1}$$

by a construction which replaces each node in a best (in the sense of having the least known number of crossings) drawing of Q_n with n odd, by a set of 2 or 4 nodes as in Figures 1 and 2, suitable care being taken with the routing of the n pairs and quartets (shown as sets of 2 and 4 parallels in Figures 1 and 2) of arcs which replace the n arcs emanating from the original node in Q_n.

They have also shown that there is equality in (1) when $n = 4$, and have obtained all non-isomorphic optimal drawings of Q_4. Here and elsewhere in this paper, a weak definition of isomorphism is used in order to avoid a further increase in the already large numbers of non-isomorphic drawings that exist for all but the first few values of n. Tutte [24] has introduced a crossing function by labelling the nodes of a drawing with the integers $1,2,\ldots,|V|$. This induces an orientation on each arc, say from the node with the smaller label to that with the larger. As one goes from i to j ($i < j$) along an arc, one crosses another arc (k,l), $(k < l)$, h times from left to right and h' from right to left. The function $\lambda(ij,kl) = -\lambda(kl,ij) = -\lambda(ji,kl)$ is defined by $h-h'$. Since we are considering only good drawings λ takes only the values $0, \pm 1$. We say that two drawings of the graph G are *isomorphic* if and only if there is a labelling of the nodes of each drawing, so that the crossing function $\lambda(ij,kl)$ takes the same values for each drawing. The usual definition of isomorphism would distinguish between drawings in which the same crossings on an arc were made in different orders; we do not do this. Some non-isomorphic optimal drawings of Q_4 are shown in Figures 3 to 14.

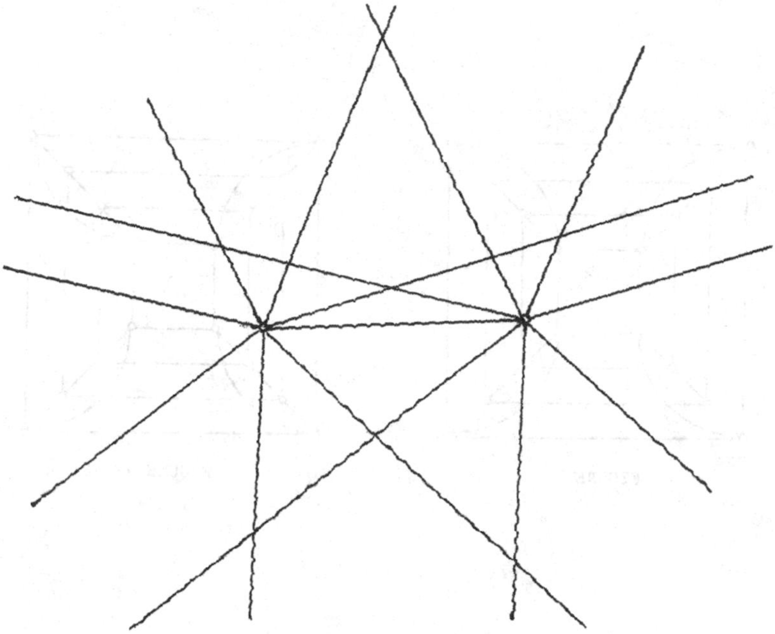

FIGURE 1.

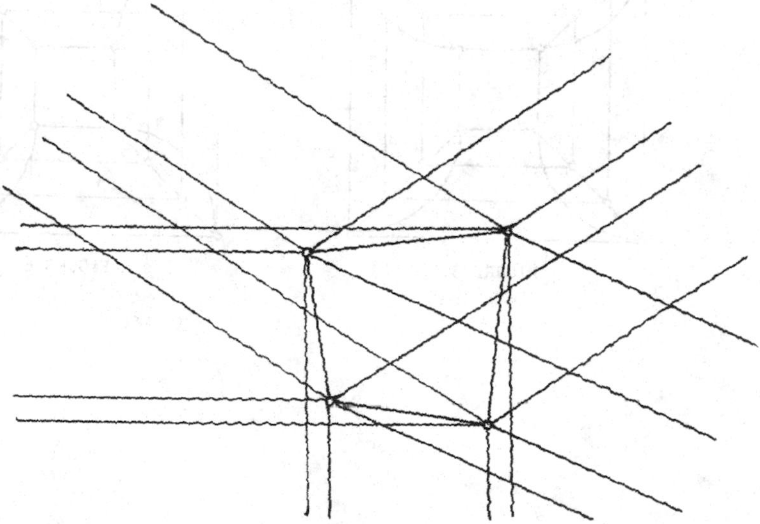

FIGURE 2.

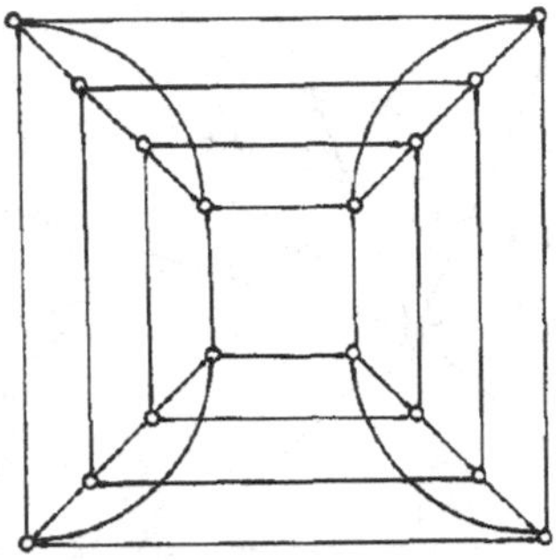

FIGURE 3.

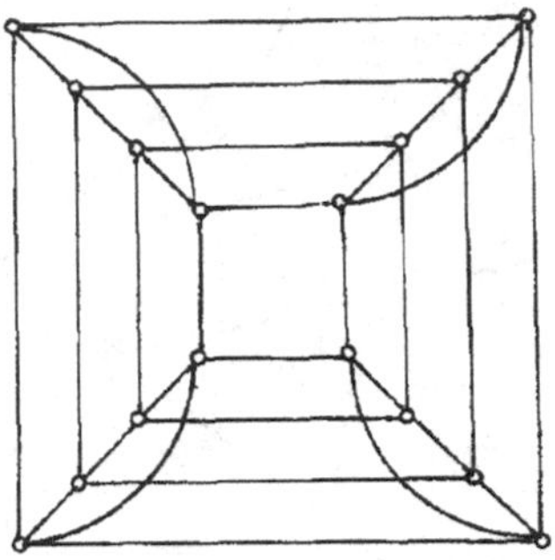

FIGURE 4.

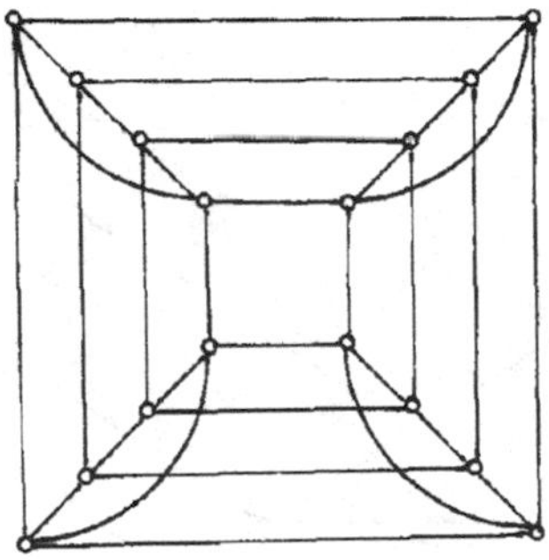

FIGURE 5.

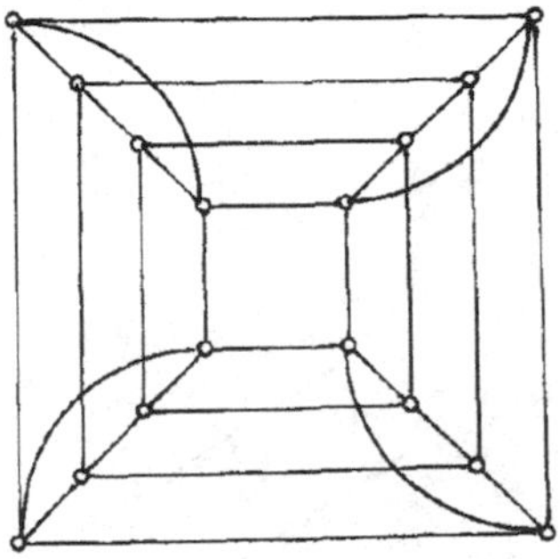

FIGURE 6.

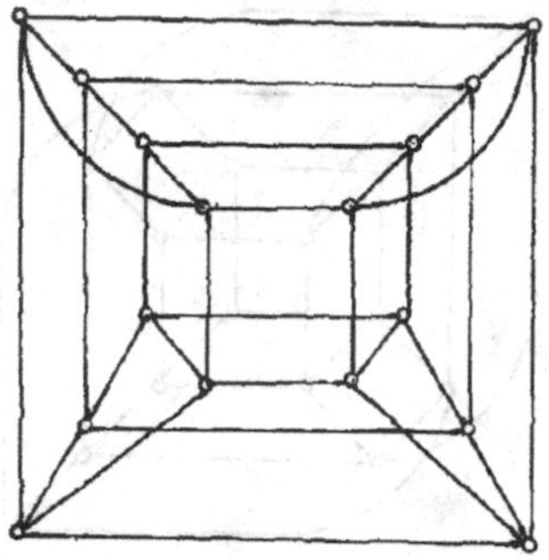

FIGURE 7.

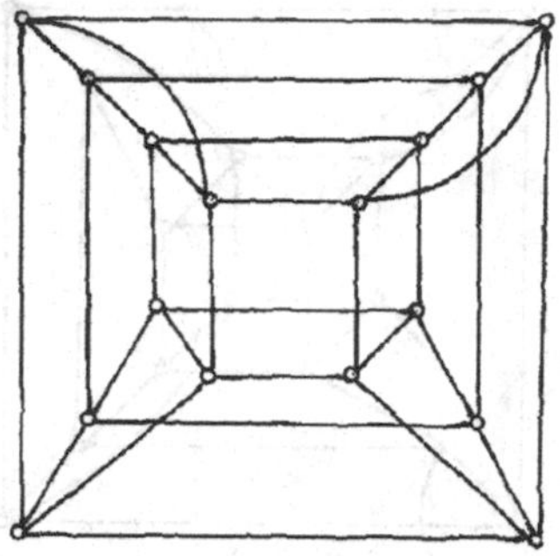

FIGURE 8.

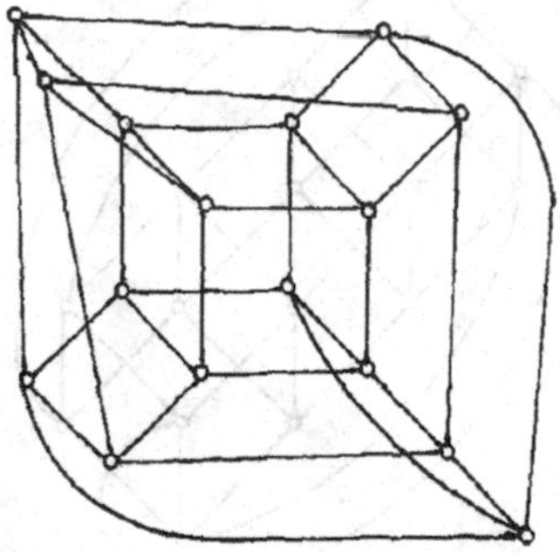

FIGURE 9.

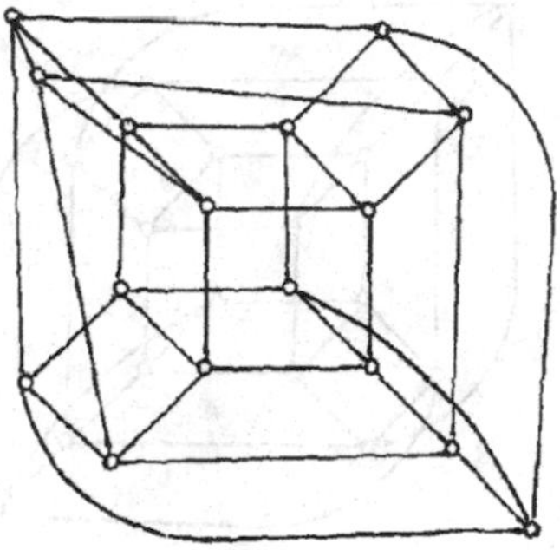

FIGURE 10.

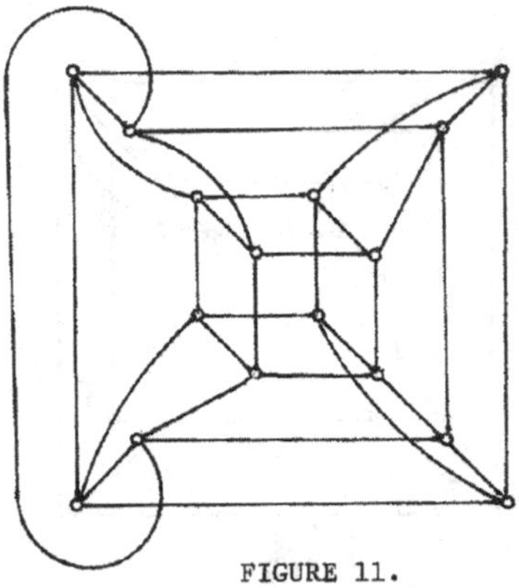

FIGURE 11.

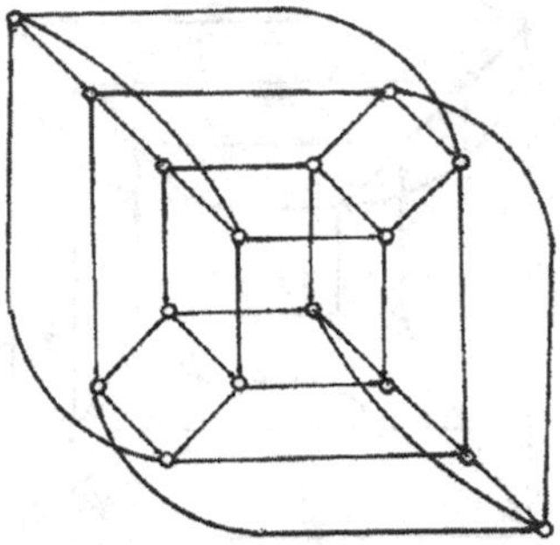

FIGURE 12.

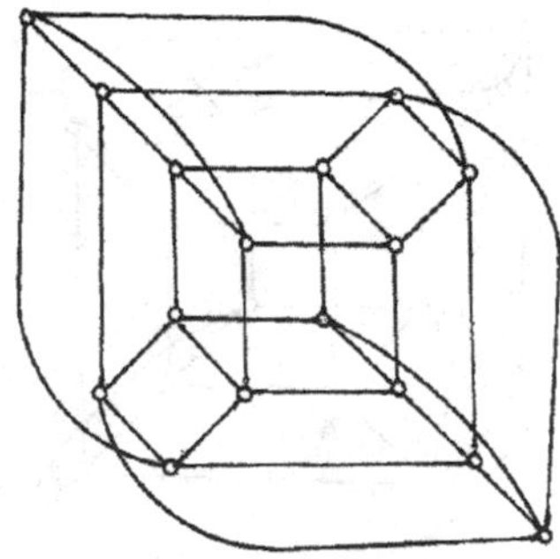

FIGURE 13.

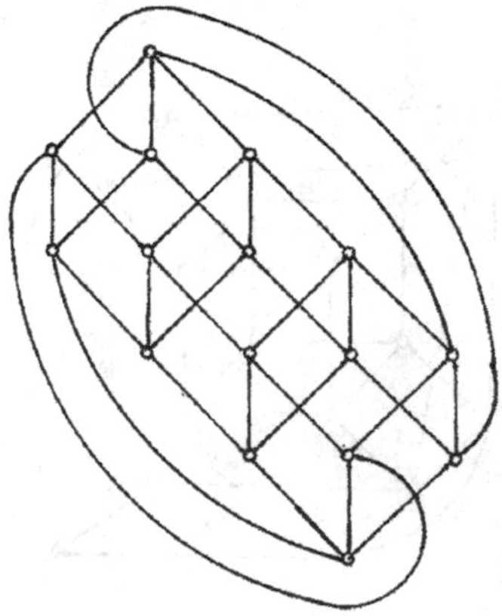

FIGURE 14.

Eggleton and Guy conjecture that equality holds in (1) for all $n \geq 3$; it seems that formula (22) in [7] is false.

2. THE COMPLETE BIPARTITE GRAPH

The graph $K_{m,n}$ has $m+n$ vertices, and mn edges, each of the m vertices forming an edge with each of the n. For some ten years it was thought that Zarankiewicz [26] and Urbanik [25] had settled the problem of determining $\nu(K_{m,n})$, but it turned out otherwise [8,17]. Zarankiewicz gave a construction to show that

(2) $$\nu(K_{m,n}) \leq [\tfrac{1}{2}m][\tfrac{1}{2}(m-1)][\tfrac{1}{2}n][\tfrac{1}{2}(n-1)]$$

and showed that equality holds for $m = 3$. Indeed, since his construction uses only straight line segments, we may replace ν by $\bar{\nu}$ in (2). A counting argument shows that if equality holds in (2) when m is odd, then it also holds with $m+1$ in place of m. Recently, Kleitman [18] has shown equality in (2) for $m = 5$ and hence for $m = 6$.

Kleitman's result enables us [9] to deduce

(3) $$\nu(K_{m,n}) \geq m(m-1)n(n-2)/20, \quad (n \geq m \geq 5).$$

3. THE COMPLETE GRAPH

The graph K_n consists of n vertices, every pair of which form an edge. There is even more mythology surrounding this problem than the ones already considered. The main conjecture [1,4,6,7,14,17,21,22,23] is that equality holds in the formula

(4) $$\nu(K_n) \leq \tfrac{1}{4}[\tfrac{1}{2}n][\tfrac{1}{2}(n-1)][\tfrac{1}{2}(n-2)][\tfrac{1}{2}(n-3)].$$

This bound has been established by two independent constructions given by Blažek and Koman [1]; see also [7,14] and many of the other papers already quoted.

It seems that equality has been established in (4) only for $n \leq 10$, and even this result is hard to find in the literature; [23] purports to give such a proof, but it contains errors. We give here an outline of such a proof, and its extension to $n \leq 12$, which R.B. Eggleton and the author hope to produce [4].

We may show that $\nu(K_5) = 1$ by an appeal to Kuratowski's theorem [19], or by use of Euler's formula, or by a parity argument, which yields the following theorem:

THEOREM 1 (Eggleton). If n is odd, $\nu(K_n) \equiv \binom{n}{4}$, modulo 2. I.e. $\nu(K_n)$ is even if $n \equiv 1$ or 3, modulo 8, and odd if $n \equiv 5$ or 7, modulo 8. A counting argument similar to that mentioned in section 2 gives the following theorem.

THEOREM 2 (Guy [6]). Equality in (4) for n odd, implies equality with $n+1$ in place of n.

Hence it follows that $\nu(K_6) = 3$. Moreover the optimal drawing of K_6 is unique within isomorphism. This may be seen by defining the *responsibility*, r_i, of a node in a drawing as the total number of crossings on arcs terminating at that node. Then in an optimal drawing $\Sigma r_i = 4\nu$, since each crossing is in the responsibility of 4 nodes. For K_6, no node has $r_i = 3$, since its deletion would lead to a crossing-free drawing of K_5, which is impossible. Hence, for K_6, $r_i = 2$ for each node, and deletion of any node from an optimal drawing of K_6 leads to an optimal drawing of K_5. Addition of a node to such a drawing leads to an optimal drawing of K_6 in essentially only one way.

By counting and Theorem 1, $\nu(K_7) = 7$ or 9, the total responsibility being 28 or 36, so some $r_i = 4$ or 6 respectively and deletion of this node leads to an optimal drawing of K_6. Insertion of a node in such a drawing in every possible way shows that $\nu(K_7) = 9$, and that there are just five non-isomorphic optimal drawings of K_7. These are shown in Figures 15 to 19. It follows from Theorem 2 that $\nu(K_8) = 18$. The total responsibility in an optimal drawing of K_8 is 72, so every $r_i = 9$, and deletion of any node leaves an optimal drawing of K_7. By adding a node to Figures 15 to 19 we find that there are only three non-isomorphic optimal drawings of K_8. These are shown in Figures 20 to 22.

As before, $\nu(K_9) = 34$ or 36 and the total responsibility is 136 or 144, showing that there is a node with $r_i \geq 16$, deletion of which leaves a drawing of K_8 with 18 or 20 crossings. No node may be added to Figures 20 to 22 without forming at least 18 additional crossings, so $\nu(K_9) = 36$. It follows from Theorem 2 that $\nu(K_{10}) = 60$.

A census of optimal drawings of K_9 has yielded 181 non-isomorphic drawings in which at least one vertex has responsibility 18, but there are several drawings with no responsibility greater than 17 and at least one in which each node has responsibility 16. This is the drawing (Figure 23) found by Harary and Hill [14] which demonstrates that $\bar{\nu}(K_9) = \nu(K_9) = 36$. They also noted that $\bar{\nu}(K_n) = \nu(K_n)$ for $n \leq 7$. Optimal drawings of K_7 using only straight line segments can be made using Figures 15 and 18 as bases; these are the only such drawings within isomorphism. It can be

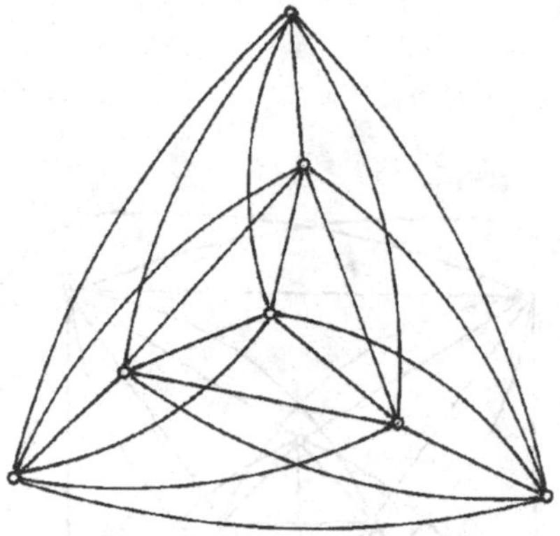

FIGURE 15.

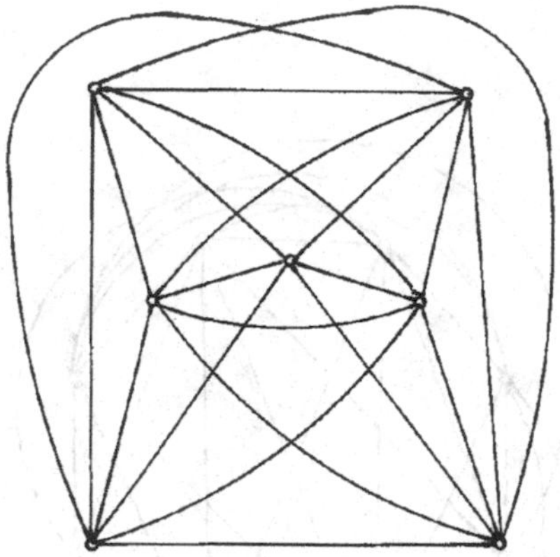

FIGURE 16.

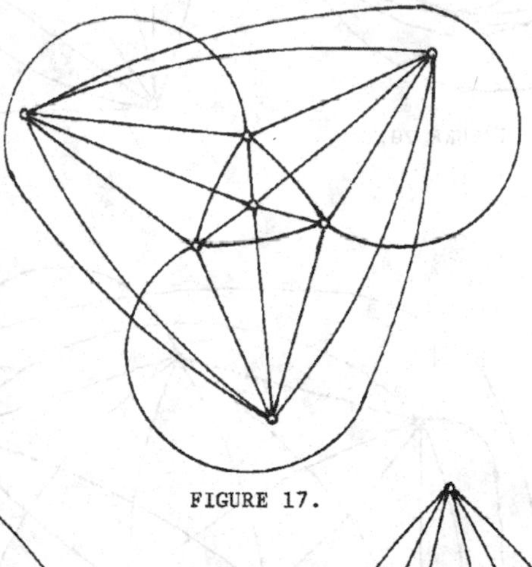

FIGURE 17.

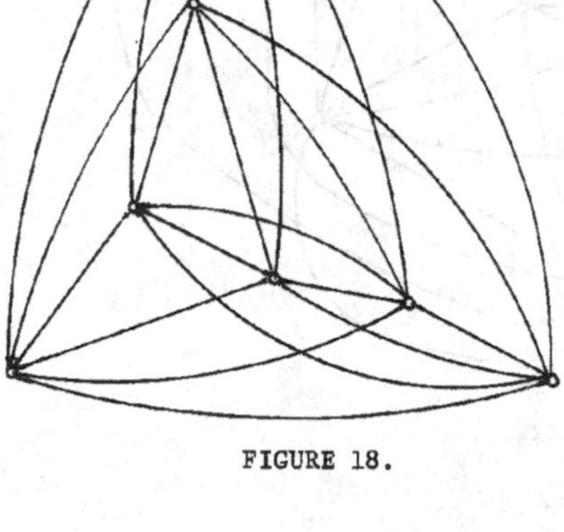

FIGURE 18.

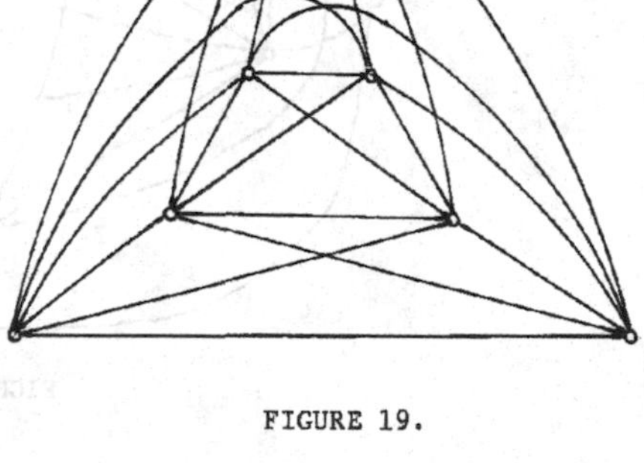

FIGURE 19.

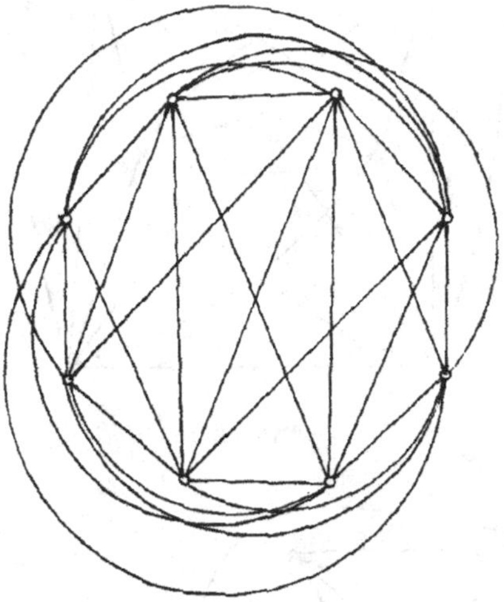

FIGURE 20.

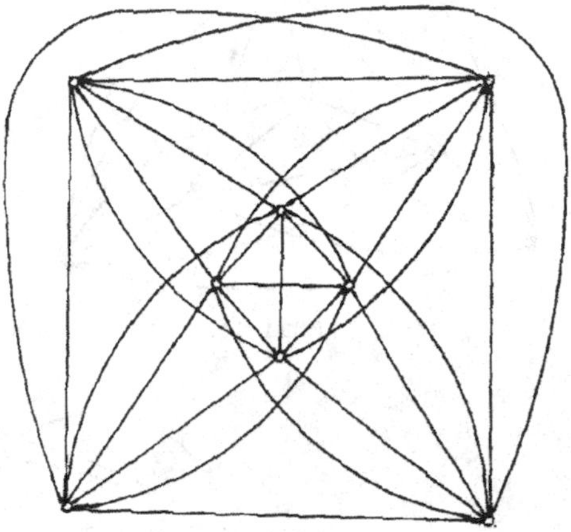

FIGURE 21.

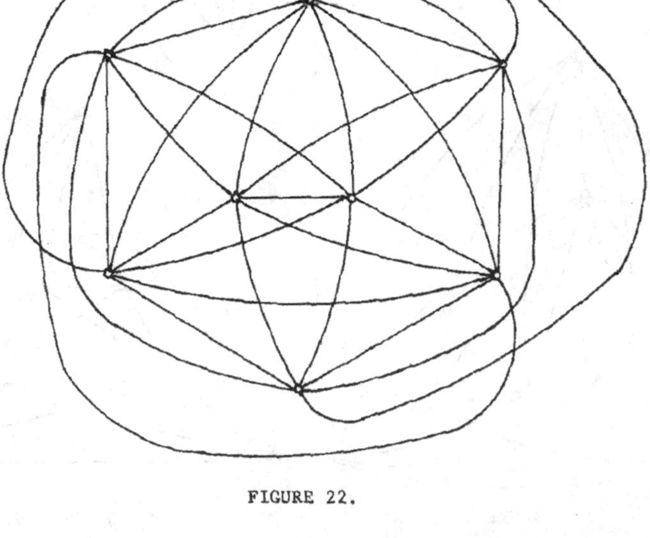

FIGURE 22.

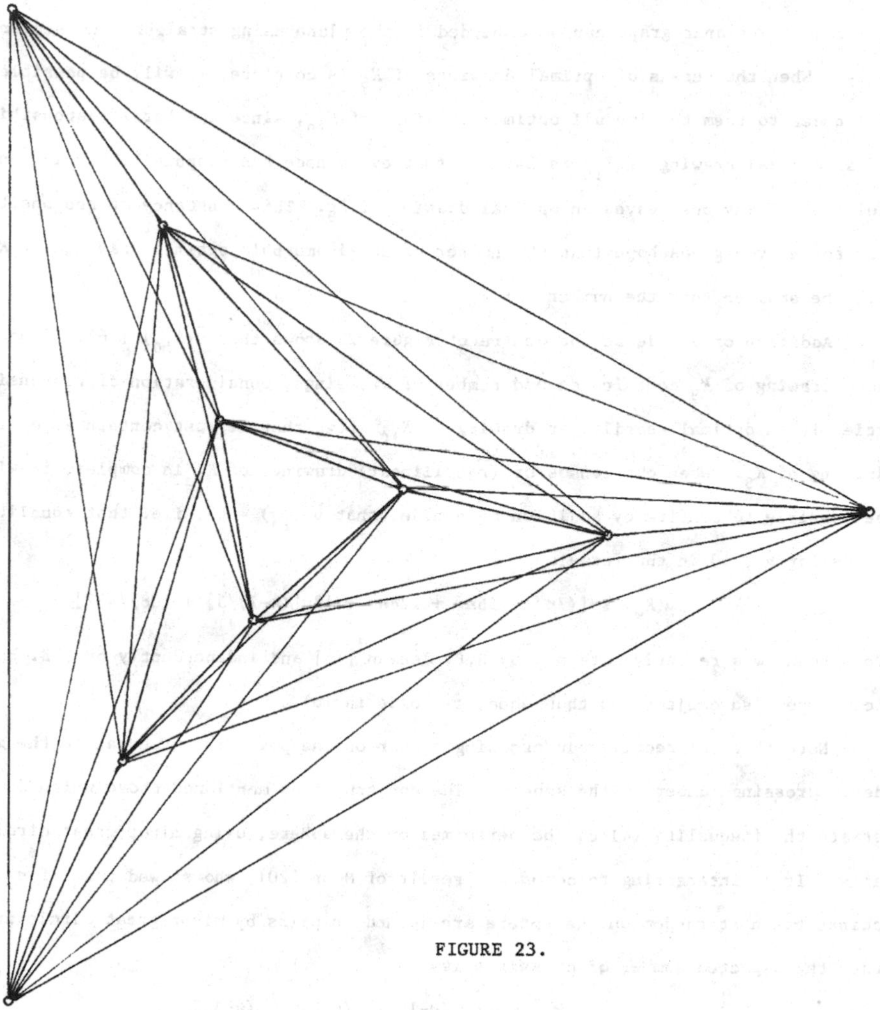

FIGURE 23.

shown that no drawing can be made which is isomorphic to any of Figures 20 to 22, so this confirms a conjecture of Harary and Hill that $\bar{\nu}(K_8) = 19 > \nu(K_8)$; they exhibited a rectilinear drawing of K_8 with 19 crossings. This contrasts with a theorem of Fáry [5] that any planar graph can be embedded in the plane using straight line segments only. When the census of optimal drawings of K_9 is complete, it will be possible to add nodes to them to give all optimal drawings of K_{10}, since the total responsibility of an optimal drawing of K_{10} is 240, so that every node has responsibility 24, and deletion of any one leaves an optimal drawing of K_9. This constancy of responsibility for n even gives hope that the number of non-isomorphic optimal drawings of K_{10} will be smaller than the number for K_9.

Addition of a node at the centre of Figure 23 shows that $\bar{\nu}(K_{10}) \leq 63$. Since no good drawing of K_9 contains an odd number of crossings, consideration of responsibilities in an optimal rectilinear drawing of K_{10} shows that it must contain an optimal drawing of K_9. When our census of (rectilinear) drawings of K_9 is complete it will be possible to confirm by addition of a node, that $\bar{\nu}(K_{10}) = 63$, i.e. that equality holds for $n \leq 10$ in the formula

$$\bar{\nu}(K_n) \leq [(7n^4 - 56n^3 + 128n^2 + 48n[(n-7)/3] + 108)/432] \qquad (5)$$

This bound was recently obtained by H.F. Jensen [16] and independently by R.B. Eggleton. One also conjectures that equality holds in (5).

Note that the rectilinear crossing number on the plane is not equal to the geodesic crossing number on the sphere. The constructions mentioned above which demonstrate the inequality (4) can be performed on the sphere, using minor great circle arcs. It is interesting to compare a result of Moon [20], who showed that if n points taken at random on the sphere are joined in pairs by minor great circle arcs, then the expected number of crossings is

$$\tfrac{1}{4} \cdot \tfrac{1}{2}n \cdot \tfrac{1}{2}(n-1) \cdot \tfrac{1}{2}(n-2) \cdot \tfrac{1}{2}(n-3).$$

As he points out, if airlines plan their routes economically they will also virtually minimize the risk of collision!

4. CONCLUSION

Crossing number problems have also been investigated on the torus [11,12] and the crossing number of the multipartite graph has been examined by Blazek and Koman [2], but nowhere is any complete result established. Although the elementary methods exhibited here seem to be reaching the limit of their effectiveness, there may be hope of combining the ideas of Kleitman [18] with those of Tutte [24] to obtain further results.

I am indebted to R.B. Eggleton for stimulating discussions, and permission to reproduce some of his results.

REFERENCES

1. Blažek, J. and Koman, M., A minimal problem concerning complete plane graphs, in Fiedler, M. (ed.), *Theory of graphs and its applications*. Proc. Symp. Smolenice, 1963, Publ. House, Č.A.S., Prague, 1964, 113-117; M.R. 30(1965), #4249.
2. Blažek, J. and Koman, M., On an extremal problem concerning graphs, *Comm. Math. Univ. Carolinae*, 8(1967), 49-52; M.R. 35(1968), #1506.
3. Eggleton, R.B. and Guy, R.K., The crossing number of the n-cube, *Amer. Math. Soc. Notices*, 17(1970), 757. A fuller version will be submitted to *Canad. J. Math.*
4. Eggleton, R.B. and Guy, R.K., The crossing number of the complete graph. To be submitted to *J. Combinatorial Theory*.
5. Fáry, I., On straight line representation of planar graphs, *Acta Sci. Math. (Szeged)*, 11(1948), 229-233; M.R. 10(1949), 136.
6. Guy, R.K., A combinatorial problem, *Nabla (Bull. Malayan Math. Soc.)*, 7(1960), 68-72.
7. Guy, R.K., The crossing number of the complete graph. *Univ. of Calgary Math. Res. Paper* #8, 1967.
8. Guy, R.K., The decline and fall of Zarankiewicz's theorem, in Harary, F. (ed.), *Proof techniques in graph theory*, Academic Press, 1969, 63-69.
9. Guy, R.K., Sequences associated with a problem of Turán and other problems, *Proc. Balatonfüred Combinatorics Conf.*, 1969, Bolyai Janos Matematikai Tarsultat, Budapest, 1970.
10. Guy, R.K. and Harary, F., On the Möbius ladders, *Canad. Math. Bull.*, 10(1967), 493-496; also Mat. Lapok, 18(1967), 59-60; M.R. 37(1969), #98, 2627.

11. Guy, R.K. and Jenkyns, T.A., The toroidal crossing number of $K_{m,n}$, *J. Combinatorial Theory*, 6(1969), 235-250; M.R. 38(1969), #5660.

12. Guy, R.K., Jenkyns, T.A. and Schaer, J., The toroidal crossing number of the complete graph, *J. Combinatorial Theory*, 4(1968), 376-390; M.R. 36(1968), #3682.

13. Harary, F., *Graph Theory*, Addison-Wesley, Reading, Mass., 1969.

14. Harary, F. and Hill, A., On the number of crossings in a complete graph, *Proc. Edinburgh Math. Soc.* (2), 13(1962-3), 333-338; M.R. 29(1965), #602.

15. Holroyd, E.M., A lower bound for the crossing number of a complete graph, *Road Research Lab. (Britain)*, Technical Note TN303, May, 1968.

16. Jensen, H.F., An upper bound for the rectilinear crossing number of the complete graph, *J. Combinatorial Theory*, 9(1970),

17. Kainen, P.C., On a problem of Erdös, *J. Combinatorial Theory*, 5(1968), 374-377; M.R. 38(1969), #72.

18. Kleitman, D.J., The crossing number of $K_{5,n}$, *Univ. of Calgary Math. Res. Report* #65, Nov. 1968. To appear in *J. Combinatorial Theory*, 9(1970).

19. Kuratowski, K., Sur la problème des courbes gauches en topologie, *Fund. Math.*, 15(1930), 271-283.

20. Moon, J.W., On the distribution of crossings in random complete graphs, *J. Soc. Indust. App. Math.*, 13(1965), 506-510; M.R. 31(1966), #3357.

21. Saaty, T.L., The minimum number of intersections in complete graphs, *Proc. Nat. Acad. Sci., U.S.A.*, 52(1964), 688-690; M.R. 29(1965), #4045.

22. Saaty, T.L., Two theorems on the minimum number of intersections for complete graphs, *J. Combinatorial Theory*, 2(1967), 571-584; M.R. 35(1968), #2796.

23. Saaty, T.L., Symmetry and the crossing number for complete graphs, *J. Res. Nat. Bureau Standards*, 73B(1969), 177-186.

24. Tutte, W.T., Towards a theory of crossing numbers, *J. Combinatorial Theory*, 8(1970), 45-53.

25. Urbaník, K., Solution du problème posé par P. Turán, *Colloq. Math.*, 3(1955), 200-201.

26. Zarankiewicz, K., On a problem of P. Turán concerning graphs, *Fund. Math.*, 41(1954), 137-145; M.R. 16(1955), 156.

GRAPHS OF (0,1)-MATRICES*

Stephen T. Hedetniemi

Department of Computer Science
The University of Iowa, Iowa City, IA 52240

Let A be a (0,1)-matrix. The <u>graph of the matrix</u> A is the graph $G(A)$ whose points $V(G(A))$ correspond 1 - 1 with the set of 1's in A and two points are adjacent in $G(A)$ if and only if the corresponding 1's lie in the same row or same column of A. Special subclasses of the class of graphs of (0,1)-matrices have received a great deal of attention in the literature by Hoffman [9] and Ray-Chaudhuri [10], Moon [13], Shrikhande [16], Bose and Laskar [4], Aigner [1], Dowling [5], and others. They have considered, for example, the line-graphs of projective planes π, which can be imagined to be the graphs of the incidence matrices of π, and they have considered the <u>triangular</u> graphs, the points of which correspond 1 - 1 with the set of unordered pairs of n symbols, where two pairs are adjacent if and only if they have a symbol in common. Characterizations of these special classes of graphs have been obtained either in terms of eigenvalues of their adjacency matrices, or in terms of the numbers and degrees of the points, the number of points adjacent to two adjacent points, and the distances between points.

In this paper we obtain several characterizations of the entire class of graphs of (0,1)-matrices, including a forbidden, or excluded, subgraph characterization, and in the process establish connections between these graphs and line graphs, clique graphs, bipartite graphs, comparability graphs, unimodular graphs and perfect graphs. Finally, as a by-product of these characterizations we obtain the result that if the clique graph $K(G)$ of a graph G is bipartite, then the chromatic number $\chi(G)$ of G equals $\omega(G)$, the number of points in the largest clique of G.

An alternate definition of a graph of a (0,1)-matrix is the following: a graph G is the graph of a (0,1)-matrix A if the points of G can be represented by pairs of integers (i,j) in such a way that two points (i,j) and (k,ℓ) are adjacent in G if and only if either $i = k$ or $j = \ell$. The 1's of the matrix A then correspond precisely with the pairs (i,j) which represent points of G. In Figure 1 we present a (0,1)-matrix A and its graph $G(A)$.

---

*This research was supported in part by Grant number N00014-68-A-0500, Office of Naval Research.

FIGURE 1

$$A = \begin{bmatrix} 1 & 1 & 0 & 1 & 0 \\ 1 & 1 & 1 & 0 & 0 \\ 1 & 0 & 1 & 0 & 1 \\ 0 & 0 & 0 & 1 & 0 \end{bmatrix} \qquad G(A) =$$

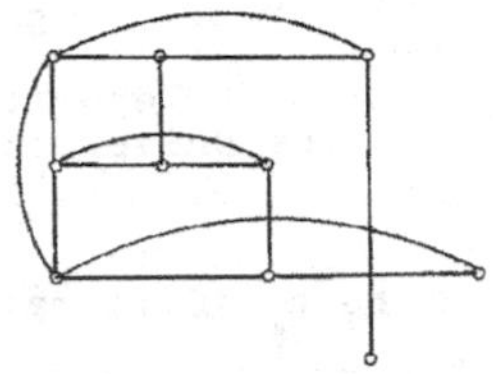

It is well known that with every (0,1)-matrix one can naturally associate another graph, which is always bipartite (cf. Ore [14]). Given a (0,1)-matrix A of size m by n, let us denote by H(A) the bipartite graph H whose points $V(H(A)) = \{u_i \mid i=1,\cdots,m\} \cup \{v_j \mid j=1,\cdots,n\}$ correspond 1 - 1 with the set of m rows together with the set of n columns of A. Point u_i is adjacent to point v_j in H(A) if and only if $a_{ij} = 1$ in A. Figure 2 illustrates H(A).

FIGURE 2

$$A = \begin{bmatrix} 0 & 0 & 0 & 1 \\ 0 & 0 & 1 & 1 \\ 0 & 1 & 0 & 1 \\ 1 & 1 & 1 & 0 \end{bmatrix} \qquad H(A)$$

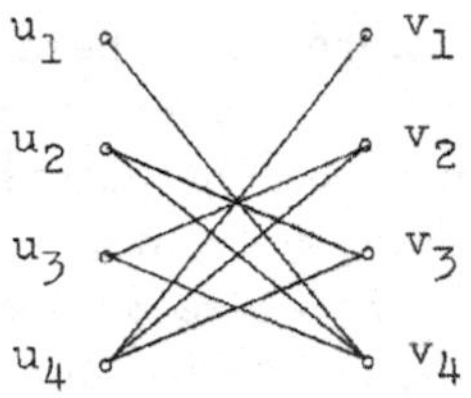

The line graph L(G) of a graph G = (V,E) has a set of points V(L(G)) which corresponds 1 - 1 with the set of lines E(G) of G, and two points of L(G) are adjacent if and only if the corresponding lines have a point in common. For a thorough discussion of line graphs, see Harary [8]. The following characterization is immediate.

Theorem 1. Given a (0,1)-matrix A,

$$G(A) = L(H(A)) .$$

Proof: If the points of H(A) are $V(H(A)) = \{u_i \mid i=1,\cdots,m\} \cup \{v_j \mid j=1,\cdots,n\}$, then the points of L(H(A)) are $V(L(H(A))) = \{(u_i,v_j) \mid a_{ij} = 1\}$. Thus V(L(H(A))) and V(G(A)) naturally correspond 1 - 1. By definition of the linc graph of a graph, two lines are adjacent if and only if they have a point in common. Thus, in

$L(H(A))$, (u_i,v_j) is adjacent to (u_k,v_ℓ) if and only if either $u_i = u_k$ or $v_j = v_\ell$; but that is to say if and only if a_{ij} and $a_{k\ell}$ lie in either the same row or the same column.

Thus we see from Theorem 1 that graphs of (0,1)-matrices are line graphs of the bipartite graphs naturally associated with (0,1)-matrices. But since it also follows that to any bipartite graph G we can associate a (0,1)-matrix A such that $H(A) = G$, we observe that the class of graphs of (0,1)-matrices is precisely the class of line graphs of all bipartite graphs.

We turn now to a second characterization of the graphs of (0,1)-matrices, which actually is not surprising when you consider the matrix involved. Given a graph G, a K-partition of $V(G)$ is a partition $\pi = \{V_1,V_2,\cdots,V_m\}$ such that for every i, $i = 1,\cdots,m$, the subgraph $\langle V_i\rangle$ induced by V_i is a complete graph; the K-partition π contains the line $u_iu_j \in E(G)$ if and only if for some k, $u_i,u_j \in V_k$. We will say that two partitions $\pi_1 = \{V_1,V_2,\cdots,V_m\}$ and $\pi_2 = \{W_1,W_2,\cdots,W_n\}$ are orthogonal if and only if $\pi_1 \cdot \pi_2 = 1$, i.e., for every i,j, $|V_i \cap W_j| \le 1$.

Theorem 2. A graph G is the graph of a (0,1)-matrix A if and only if there exist two orthogonal K-partitions of $V(G)$ containing all the lines of G.

Proof: Let $\pi_1 = \{V_1,V_2,\cdots,V_m\}$ and $\pi_2 = \{W_1,W_2,\cdots,W_n\}$ be two orthogonal K-partitions of $V(G)$ containing all the lines of G. We define an $m \times n$ matrix A as follows: $a_{ij} = 1$ if and only if $V_i \cap W_j \ne \emptyset$; observe that $|V_i \cap W_j| \le 1$. We must show that $G \cong G(A)$. We define the isomorphism $\phi : V(G) \to V(G(A))$ by $\phi(u) = a_{ij}$ if and only if $V_i \cap W_j = \{u\}$. The mapping ϕ is clearly 1-1 and onto. We must show that it preserves adjacency. Let $uv \in E(G)$ and let $\phi(u) = a_{i_1,i_2}$, $\phi(v) = a_{j_1,j_2}$. Then by the assumptions on the K-partitions, either there exists an i for which $u,v \in V_i$, or there exists a j for which $u,v \in W_j$, but not both. This implies that either

$$\phi(u) = a_{i,i_2}, \quad \phi(v) = a_{i,j_2}$$

or that $\phi(u) = a_{i_1 j}$, $\phi(v) = a_{j_1 j}$. In either case $\phi(u)$ is adjacent to $\phi(v)$, by the definition of $G(A)$.

Now assume $\phi(u)$ is adjacent to $\phi(v)$. This means that either $\phi(u) = a_{i,i_2}$, $\phi(v) = a_{i,j_2}$, or that $\phi(u) = a_{i_1 j}$, $\phi(v) = a_{j_1 j}$. This means in turn that either $u,v \in V_i$ or $u,v \in W_j$, or in either case, since $\langle V_i \rangle$ and $\langle W_j \rangle$ are complete subgraphs, that u is adjacent to v.

The reverse implication, given a (0,1)-matrix A to construct two orthogonal K-partitions, is obvious.

We illustrate this characterization with Figure 3.

FIGURE 3

H: 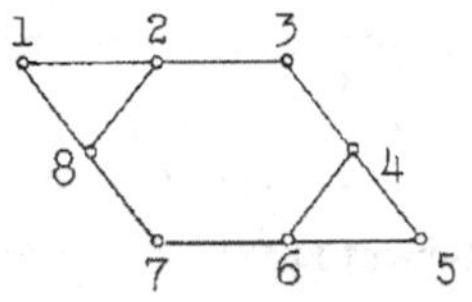

$\pi_1 = \{\overline{128}, \overline{34}, \overline{76}, \overline{5}\}$

$\pi_2 = \{\overline{456}, \overline{23}, \overline{78}, \overline{1}\}$

$$\begin{array}{c} \\ 128 \\ 34 \\ 76 \\ 5 \end{array} \begin{array}{c} \begin{array}{cccc} 456 & 23 & 78 & 1 \end{array} \\ \begin{bmatrix} 0 & 1 & 1 & 1 \\ 1 & 1 & 0 & 0 \\ 1 & 0 & 1 & 0 \\ 1 & 0 & 0 & 0 \end{bmatrix} \end{array}$$

Another characterization of graphs of (0,1)-matrices can be given which somewhat resembles Theorem 2, and as such is still reasonably close to the definition. We need the following definition. A clique of a graph G is a maximal complete subgraph of G. The clique graph K(G) of G has a set of points which corresponds 1 - 1 with the cliques of G, and two points of K(G) are adjacent if and only if the corresponding cliques have a point in common.

Theorem 3. A graph G is the graph of a (0,1)-matrix if and only if the clique graph K(G) of G is bipartite and for any two cliques C_i, C_j of G, $|C_i \cap C_j| \leq 1$.

Proof. The graph G(A) of a (0,1)-matrix A clearly has the stated properties required of its cliques.

Conversely, suppose K(G) is bipartite and for any two cliques $|C_i \cap C_j| \leq 1$. We construct a (0,1)-matrix as follows: Let the cliques of G be arranged into two groups $C_1, C_2, \cdots, C_m$, $D_1, D_2, \cdots, D_n$, such that for any i,j, $C_i \cap C_j = \emptyset = D_i \cap D_j$.

We first construct an $m \times n$ matrix B such that $b_{ij} = 1$ if and only if $|C_i \cap D_j| = 1$; if $C_i \cap D_j = \{u\}$, then b_{ij} will be the unique representation in our (0,1)-matrix A of the point u. Note that it is not possible for the point u to appear in any other intersection of two cliques, since then u would belong to at least three cliques, thereby creating a triangle in K(G) which is assumed to be bipartite.

Having constructed the matrix B we now construct the desired matrix A by adding on to B additional columns and rows, essentially one for each point of G not accounted for in B. That is, if $|C_i| = m$ and there appear in B only $n < m$ 1's in the ith row, we add on $m - n$ new columns, each containing a 1 in the ith row and zeroes everywhere else. Similarly, we add on new rows, with a single 1 in the jth column, for every D_j for which the jth column of B contains fewer 1's than the number of points in D_j.

The points in G which correspond to the 1's in the rows and columns which are added on to B are points which belong to only one clique of G.

Inspection of the resultant matrix A reveals that $G(A) \cong G$. We omit the tedious details.

The preceding two characterizations of graphs of (0,1)-matrices are very similar to Krausz's characterization of line graphs [12] and to Robert's and Spencer's characterization of clique graphs [15]. All of these characterizations, however, leave something to be desired.

A much better characterization of line graphs was obtained by Beineke [2] and Robertson (unpublished) who displayed exactly those 9 graphs which cannot appear as subgraphs of a line graph.

We next make good use of Theorem 3 to obtain a corresponding forbidden subgraph characterization of graphs of (0,1)-matrices. Let C_n denote a cycle of length n, $K_n - x$, the complete graph on n points, minus any one line x, and $K_{m,n}$ the complete bipartite graph on two sets of m and n points.

Lemma 4. If G is the graph of a (0,1)-matrix A, then G contains no induced subgraph isomorphic to either

(i) $K_{1,3}$,

(ii) $K_4 - x$, or

(iii) C_{2n+1}, for any $n \geq 2$.

Proof: (i) Since $K_{1,3}$ contains one point of degree 3, if $K_{1,3}$ is to be an induced subgraph of $G(A)$ then A must contain at least three 1's on some row or column. But these three 1's will induce a triangle, by definition of $G(A)$.

(ii) If $G(A)$ is to contain $K_4 - x$ as an induced subgraph, then A must contain three 1's on some row or column to account for one of the two triangles in $K_4 - x$. The remaining point of $K_4 - x$ must be adjacent to two points of this triangle, but any other 1 in A can only be adjacent to either 1 or 3 of these 1's.

(iii) Suppose $G(A) \cong C_{2n+1}$, $n \geq 2$. By Theorem 1, we know that

$L(H(A)) = G(A) \cong C_{2n+1}$. But $H(A)$ is by definition bipartite, while the only graph H such that $L(H) = C_{2n+1}$ is $H = C_{2n+1}$, which is not bipartite; a contradiction.

We now see that the graphs in Lemma 4 are sufficient to characterize graphs of $(0,1)$-matrices.

Theorem 4. A graph G is the graph of a $(0,1)$-matrix A if and only if G contains no induced subgraph isomorphic to either

(i) $K_{1,3}$,

(ii) $K_4 - x$, or

(iii) C_{2n+1}, for $n \geq 2$.

Proof: Lemma 4 establishes the implication from left to right. Suppose then that G is not the graph of any $(0,1)$-matrix. Consider the cliques of G.

We first show that if for any two cliques C_i, C_j of G, $|C_i \cap C_j| \geq 2$ then G contains an induced subgraph isomorphic to $K_4 - x$. Let $u_i, u_j \in C_i \cap C_j$. Since $C_i \not\subset C_j$ there must exist a point $u_1 \in C_i$ which is not adjacent to some point $u_2 \in C_j$. But then $\langle\{u_1, u_i, u_j, u_2\}\rangle \cong K_4 - x$.

Thus, let us assume that for any two cliques C_i, C_j of G, $|C_i \cap C_j| \leq 1$.

Now consider $K(G)$. If $K(G)$ is bipartite, and $|C_i \cap C_j| \leq 1$, then by Theorem 3 we know that G is the graph of a $(0,1)$-matrix; a contradiction. So then, let us further assume that $K(G)$ is not bipartite.

Suppose, first, that $K(G)$ contains a triangle. Let C_1, C_2, C_3 be three distinct cliques forming a triangle in $K(G)$. Since for any i,j we are assuming $|C_i \cap C_j| \leq 1$, we can let $C_1 \cap C_2 = \{u_{12}\}$, $C_2 \cap C_3 = \{u_{23}\}$, and $C_3 \cap C_1 = \{u_{31}\}$.

Case 1: Assume u_{12}, u_{23}, and u_{31} are all distinct. Clearly these three points induce a triangle in G. Then let C^1 be the unique clique of G containing these three points. Now, however, $|C_1 \cap C^1| \geq 2$, so we must have that $C_1 = C^1$. But also $|C_2 \cap C^1| \geq 2$, so we must have $C_1 = C^1 = C_2$; a contradiction.

Case 2: Assume two of these three points are equal, say $u_{12} = u_{23}$. But then, $u_{12} = u_{23} \in C_1$, C_2, and C_3. Let u_1 be a point of C_1 not adjacent to some point $u_2 \in C_2$, we have a subgraph then like that in Figure 4.

FIGURE 4

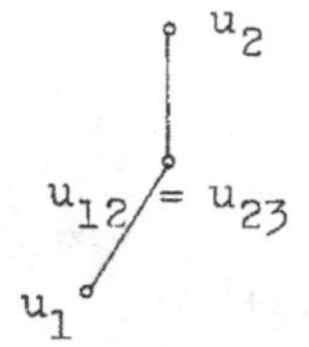

Next, let u_3 be a point of C_3 which is not a point of C_2, u_3 is of course adjacent to u_{23} and $u_3 \neq u_1$, else $|C_1 \cap C_3| \geq 2$.

Subcase 1: u_3 is not adjacent to u_2. Then if u_3 is also not adjacent to u_1 we have found a $K_{1,3}$ in G. Thus, let us assume u_3 is adjacent to u_1 but not u_2. But now let C^1 be the clique containing u_1, u_{23}, and u_3. Clearly, $|C^1 \cap C_1| \geq 2$ and $|C^1 \cap C_3| \geq 2$. So we must have $C^1 = C_1$ and $C^1 = C_3$; a contradiction.

Subcase 2: u_3 is adjacent to u_2. Then let C^1 be the clique of G containing u_2, u_{23}, and u_3. As above, $|C^1 \cap C_2| \geq 2$, $|C^1 \cap C_3| \geq 2$; which leads to a contradiction.

Finally then let us assume that K(G) is not bipartite and does not contain a triangle. Let P denote an odd cycle $C_1, C_2, \cdots, C_{2n}, C_{2n+1}, C_1$ of (distinct) cliques in K(G), for $n \geq 2$. From this odd cycle P we can construct an odd cycle P^1 in G, $u_{1,2}, u_{2,3}, \cdots, u_{2n,2n+1}, u_{2+1,1}$, where for any i, $C_i \cap C_{i+1} = \{u_{i,i+1}\}$. Note that all of the points $u_{i,i+1}$ are distinct; since if $u_{i,i+1} = u_{j,j+1}$ then at least three cliques, say C_i, C_{i+1}, C_j (and C_{j+1}) will have a point in common and hence form a triangle in K(G), which contradicts our assumptions.

If this odd cycle P^1 in G has no chords then according to the theorem we are done. So let us assume there exists a chord $u_{i,i+1}u_{j,j+1}$ in P^1. We have two cases to consider, whether the chord defines a triangle in P^1 or not. Figure 5 illustrates the case where $u_{i,i+1}u_{j,j+1}$ defines a triangle in P^1; the corresponding cliques are shown to the right.

Consider in this case the clique C^1 containing $u_{i,i+1}$, $u_{i+1,j+1}$, and $u_{j,j+1}$. Clearly, $|C^1 \cap C_{i+1}| \geq 2$, implying that $C^1 = C_{i+1}$. But also, $|C^1 \cap C_{j+1}| \geq 2$, implying that $C^1 = C_{j+1}$. This contradicts our assumption that $C_{i+1} \neq C_{j+1}$.

FIGURE 5

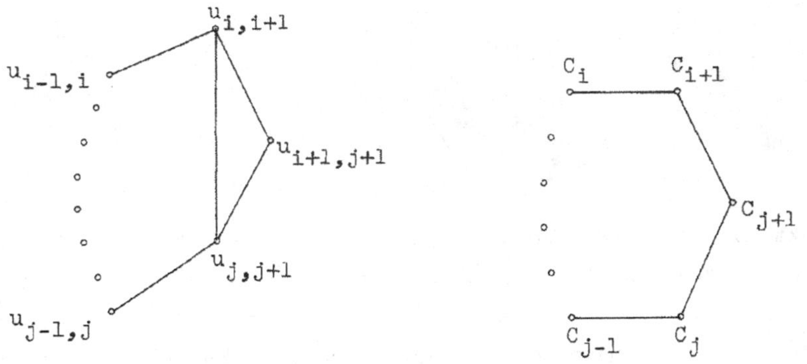

Finally, we consider the case where $u_{i,i+1}u_{j,j+1}$ does not determine a triangle in P^1; Figure 6 illustrates this case.

FIGURE 6

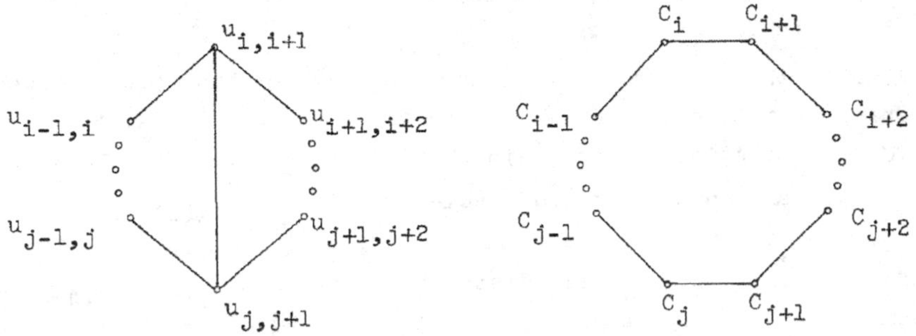

Consider first whether or not the set $u_{i-1,i}$, $u_{j,j+1}$, $u_{i+1,i+2}$ is an independent set. If it is, then G contains $K_{1,3}$ and we are done. Suppose then that $u_{i-1,i}$ is adjacent to $u_{i+1,i+2}$. This line would be a chord of P^1 which defines a triangle and this case has already been considered above.

We conclude then that either $u_{i-1,i}$ or $u_{i+1,i+2}$, but not both, is adjacent to $u_{j,j+1}$. If both were adjacent, we would have a $K_4 - x$.

By exactly the same reasoning, in considering the set $u_{j-1,j}$, $u_{i,i+1}$, $u_{j+1,j+2}$, we can conclude that $u_{j-1,j}$ is not adjacent to $u_{j+1,j+2}$, and that either $u_{j-1,j}$ or $u_{j+1,j+2}$, but not both, is adjacent to $u_{i,i+1}$.

From this we can conclude that either G contains a $K_4 - x$, containing $u_{i,i+1}$, $u_{j,j+1}$, one of $u_{i-1,i}$ and $u_{i+1,i+2}$, and one of $u_{j-1,j}$ and $u_{j+1,j+2}$, or that these four points actually form a K_4. In this last case, let C^1 denote the clique of G containing this K_4. We observe that either $|C^1 \cap C_i| \geq 2$ or $|C^1 \cap C_{i+1}| \geq 2$, i.e., either $C^1 = C_i$ or $C^1 = C_{i+1}$. But we also observe that either $|C^1 \cap C_j| \geq 2$ or $|C^1 \cap C_{j+1}| \geq 2$, i.e., either $C^1 = C_j$ or $C^1 = C_{j+1}$. But since we have here assumed that all of C_i, C_{i+1}, C_j, C_{j+1} are distinct, we have a contradiction.

Having established several characterizations of graphs of (0,1)-matrices we now proceed to show how intimately these graphs are related to other well-known classes of graphs. The first connection arises from a classical theorem of König-Egerváry on (0,1)-matrices (cf. Hall [7]).

Theorem 5. (König-Egerváry) Let A be a (0,1)-matrix of size $m \times n$. The minimum number of lines that cover all the 1's in A is equal to the maximum number of 1's in A with no two of the 1's on a line.

A little reflection shows that this theorem is in fact a striking theorem about graphs of (0,1)-matrices; for the minimum number of lines which cover all the 1's in A is equal to $\theta(G(A))$, the minimum number of complete subgraphs which contain all the points of $G(A)$, and the maximum number of 1's in A with no two of the 1's on a line is equal to $\beta_o(G(A))$, the point independence number of $G(A)$, i.e., the largest number of points in a set no two of which are adjacent. Thus we can immediately restate the König-Egerváry Theorem as follows:

Corollary 5. Let A be a (0,1)-matrix. Then for the graph of A, $G(A)$, $\beta_o(G(A)) = \theta(G(A))$.

A second classical theorem of König [11] tells us even more about these graphs. Recall that $\chi(G)$ and $\Delta(G)$ denote, respectively, the chromatic number and maximum degree of any point of G.

Theorem 6. (König) If G is a bipartite graph, then

$$\chi(L(G)) = \Delta(G).$$

Corollary 6. If A is a (0,1)-matrix, then the chromatic number $\chi(G(A))$ of $G(A)$ equals the maximum number of 1's in any row or column of A.

Proof: This follows from Theorem 1 and the observation that $\Delta(H(A))$ equals the maximum number of 1's on any row or column of A, i.e.,

$$\chi(L(H(A))) = \chi(G(A)) = \Delta(H(A)).$$

What this corollary, together with the Konig-Egervary Theorem, tells us is that the class of graphs of $(0,1)$-matrices is "perfect."

According to Berge [3], a graph G is perfect if $\theta(G) = \beta_0(G)$, $\omega(G) = \chi(G)$, and these two equalities hold for every induced subgraph of G as well. One only has to observe that $\omega(G(A))$, the number of points in the largest complete subgraph of $G(A)$, equals the maximum number of 1's in any row or column of A.

The fact that graphs of $(0,1)$-matrices are perfect is not actually a new result since in [3], Berge first observes that any line graph of a bipartite graph is unimodular, and then proves that unimodular graphs are perfect. It is a pleasant surprise, however, that two classical results of König also provide us with this result.

Yet another relationship involving graphs of $(0,1)$-matrices is obtained from condition (iii) of Theorem 4. Since Gilmore and Hoffman [6] have shown that a graph is a comparability graph if and only if each odd cycle has at least one triangular chord, we can conclude that graphs of $(0,1)$-matrices are also comparability graphs.

It was Theorem 3, together with the realization that graphs of $(0,1)$-matrices are perfect, that led to the next result, for which we need the following lemma.

Lemma 7. Let G be a graph and let G^1 be obtained from G by deleting any point u from G, i.e., $G^1 = G - u$; then $K(G^1)$ is isomorphic to a subgraph of $K(G)$.

Proof: Let the cliques of G (and the points of $K(G)$) be labeled $C_1, C_2, \cdots, C_m, C_{m+1}, \cdots, C_n$, where point u belongs to cliques $C_1, C_2, \cdots, C_m$. Note first that since point u does not belong to C_j, for $j > m$, C_j is also a clique of $G^1 = G - u$.

Next, let $C_i^1 = C_i - u$, for $i = 1, 2, \cdots, m$, and let $D_i^1 = C_i^1$ if C_i^1 is a clique of G^1. We claim now that the cliques of G^1 correspond 1 - 1 with the set of cliques $\{D_i^1\} \cup \{C_{m+1}, \cdots, C_n\}$.

Let D^1 be a clique of G^1 which does not correspond to (is not equal to) one of the cliques $C_{m+1}, \cdots, C_n$. If D^1 is also a clique of G then it must correspond to one of the D_i^1. Suppose then that D^1 is not a clique of G. Then D^1 is a proper subset of at least one clique D of G. It follows that $D = D^1 \cup \{u\}$, for if there is a point $v \neq u$ in D which is not in D^1 then D^1 cannot be maximal in G^1 since v is also a point of $G^1 = G - u$. But this means that $D = D^1 \cup \{u\}$ is a clique of G, i.e., D^1 corresponds to one of the D_i^1.

Notice finally that in this correspondence of cliques of G^1 with those of G, either a clique D^1 of G^1 is the same as a clique of G, i.e., $D^1 \in \{C_{m+1}, \cdots, C_n\}$, or else $D^1 \cup \{u\}$ is a clique of G. Thus we assert that this correspondence is 1 - 1. It only remains to show that if two cliques are adjacent in $K(G^1)$ then the corresponding cliques are adjacent in $K(G)$. But this follows immediately from the definition of adjacency of two cliques; if two cliques of G^1 have a point in common, then clearly the corresponding cliques of G also have the same point in common.

Lemma 7 enables us to prove the desired result by induction. We are fascinated by the similarity of the following proof with that given by König for Theorem 6.

Theorem 7. For any graph G, if $K(G)$ is bipartite, then $\chi(G) = \omega(G)$.

Proof: We proceed by induction on the number of points p in G. For graphs having $p \leq 3$ points, the theorem can be verified easily by inspection. Assume then that for all graphs having $p - 1$ or fewer points, if $K(G)$ is bipartite, then $\chi(G) = \omega(G)$.

Let G have p points and let $K(G)$ be bipartite. Suppose first that G contains a point u which belongs to only one clique of G, let $G^1 = G - u$. Notice that if u belongs to only one clique, C_u, then $N(u) = \{v | uv \in E(G)\}$, the set of points adjacent to u, induces a complete subgraph of G. Now, by Lemma 7, $K(G^1)$ is isomorphic to a subgraph of $K(G)$ and since $K(G)$ is bipartite, $K(G^-)$ is also bipartite. Thus, by our induction hypothesis, $\chi(G^1) = \omega(G^1) = n$.

Color G^1 with n colors. Then two cases arise in attempting to extend this to a coloring of G by assigning a color to point u. If $|N(u)| = n$, then $\omega(G) = \omega(G^1) + 1$, and we can assign a new color to u, thereby coloring G with $n+1$ colors. This implies that $\chi(G) \leq n+1$, and since $\chi(G) \geq \omega(G) = n+1$, we conclude that $\chi(G) = \omega(G) = n+1$. If, on the other hand, $|N(u)| < n$, then we can assign one of the n colors to u which is not assigned to any point in $N(u)$. Thus $\chi(G) \leq n$, and since $\omega(G) = \omega(G^1) = n$, we conclude that $\chi(G) = \omega(G) = n$.

Suppose next that G contains a point u which belongs to only two cliques C_1 and C_2 of G; it is obvious that no point of G can belong to more than two cliques of G since then $K(G)$ would contain a triangle and hence not be bipartite.

Suppose we delete point u from G and let $G^1 = G - u$. Again, by Lemma 7, $K(G^1)$ is bipartite, and hence by our induction hypothesis, $\chi(G^1) = \omega(G^1) = n$. Let us color G^1 with n colors.

Now, we can also color G with n colors if fewer than n colors are used to color the points in $N(u)$. In this case, $\chi(G) \le n$, and since $\chi(G) = \omega(G) \ge \omega(G^1) = n$, we conclude that $\chi(G) = \omega(G) = n$.

Let us assume then that n colors are used to color the points in $N(u)$. Since point u belongs to two cliques C_1 and C_2, let $N(u) = N_1 \cup N_2$, where N_i denotes the points of C_i adjacent to u for $i = 1,2$. Note that possibly $N_1 \cap N_2 \ne \phi$.

Again we consider two cases. In the first case, either $|N_1| = n$ or $|N_2| = n$. In this case $\omega(G) = \omega(G^1) + 1$, and point u can be assigned a new color, thereby proving that $\chi(G) = n + 1 = \omega(G)$.

In the second case, $|N_1| < n$, $|N_2| < n$ and there exist two points, say $u_1 \in N_1$, $u_2 \in N_2$ such that u_1 and u_2 are colored differently, say with colors 1 and 2, respectively, no point of N_2 is colored 1, and no point of N_1 is colored 2.

Consider then the subgraph G_{12} induced in G_1 by the set of points colored 1 and 2, together with the point u. We must consider two cases. In the first case, point u is a cutpoint of G_{12}. Let G_1 and G_2 be the components of $G_{12} - u$ containing points u_1 and u_2, respectively. Then let us interchange in G_2 the colors 1 and 2 so that points originally colored 1 are now colored 2, and vice versa. In this fashion, point u_2 is now colored 1; note that since u is a cutpoint of G_{12}, u_2 is not adjacent to u_1. We can therefore color point u with the color 2, producing an n coloring of G. This implies that $\chi(G) \le n$ and since $\chi(G) \ge \omega(G) \ge \omega(G^1) = n$, we conclude that $\chi(G) = \omega(G) = n$.

In the second and last case, u is not a cutpoint of G_{12}. But in this case there must exist in G_{12}, and hence in G, a cycle containing points u_1, u and u_2; this cycle has odd length since the points on this cycle must alternate in colors 1 and 2 assigned to them. But this implies in turn that $K(G)$ contains an odd cycle, which is defined by the sequence of cliques containing successively the individual lines of this odd cycle. Note that none of these cliques can contain more than one of the lines in this cycle, since two points having the same color assigned to them cannot be adjacent. But since we have assumed that $K(G)$ is bipartite we reach a contradiction in this final case, thereby completing the proof.

We summarize the logical flow of this proof with the tree of Figure 7.

FIGURE 7

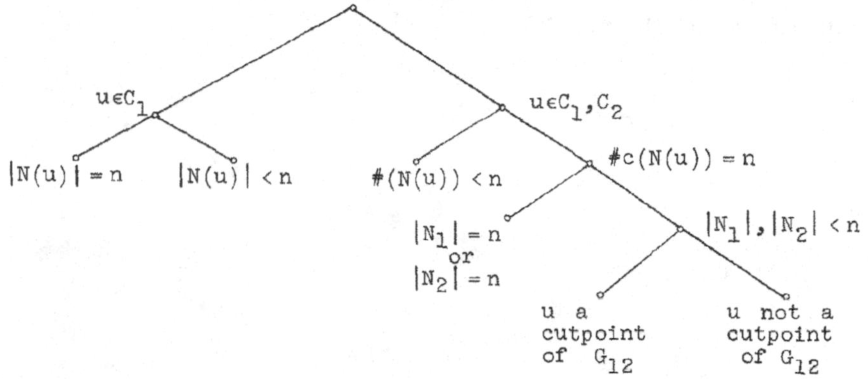

Theorem 7 naturally raises the question, when is the clique graph $K(G)$ of a graph G bipartite? Our last result provides an answer to this question; since the proof of this result very closely resembles that of Theorem 4, it is omitted.

Theorem 8. The clique graph $K(G)$ of a graph G is bipartite if and only if no point of G belongs to more than two cliques and G contains no induced subgraph isomorphic to an odd cycle of length $2n+1$, for any $n \geq 2$.

The conditions of Theorem 8 tell us that if $K(G)$ is bipartite then G is a comparability graph, again by the characterization of Gilmore and Hoffman [6]. Therefore, Theorem 8, Gilmore and Hoffman's result, and Berge's result that comparability graphs are perfect, provide another proof of Theorem 7.

It is interesting to note that a result very similar to Theorem 8 can easily be obtained for line graphs: the line graph $L(G)$ of a graph G is bipartite if and only if no point of G belongs to more than two lines (i.e., $\Delta \leq 2$) and G contains no (induced) subgraph isomorphic to an odd cycle.

Our final observation leads one to believe that a nice relationship exists between graphs for which $\chi(G) = \omega(G)$ and graphs having some kind of general bipartite property:

(i) if G is bipartite, then $\chi(G) = \omega(G)$,

(ii) if $L(G)$ is bipartite, then $\chi(G) = \omega(G)$, and

(iii) if $K(G)$ is bipartite, then $\chi(G) = \omega(G)$.

The author gratefully acknowledges H. J. Ryser, Alan Hoffman, and Claude Berge for their helpful comments, and Curt Cook for carefully examining all of the proofs.

REFERENCES

1. Aigner, M. The uniqueness of the cubic lattice graph. J. Combinatorial Theory 6, 282-297 (1969).
2. Beineke, L. Derived graphs and digraphs, Beiträge zur Graphentheorie (H. Sachs, H. Voss, and H. Walther, eds.). Teukner, Leipzig, 17-33 (1968).
3. Berge, C. Some classes of perfect graphs, Graph Theory and Theoretical Physics, edited by Frank Harary. Academic Press, London, 155-165 (1967).
4. Bose, R.C. and Laskar, R. A characterization of tetrahedral graphs. J. Combinatorial Theory 3, 366-385 (1967).
5. Dowling, T.A. A characterization of the T_m graph. J. Combinatorial Theory 6, 251-263 (1969).
6. Gilmore, P. and Hoffman, A.J. A characterization of comparability graphs and interval graphs. Canad. J. Math. 16, 539-548 (1964).
7. Hall, M., Jr. Combinatorial Theory. Blaisdell, Waltham, Mass., 1967.
8. Harary, F. Graph Theory. Addison-Wesley, 1969.
9. Hoffman, A.J. On the line graph of a projective plane. Proc. Amer. Math. Soc. 16, 297-302 (1965).
10. Hoffman, A.J. and Ray-Chaudhuri, D.K. On the line graph of a symmetric balanced incomplete block design. Trans. Amer. Math. Soc. 116, 238-252 (1965).
11. König, D. Theorie der endlichen and unendlichen graphen. Leipzig, 1936; reprinted Chelsea, New York, 1950.
12. Krauz, J. Démonstration nouvelle d'une théoreme de Whitney sur les réseaux. Mat. Fiz. Lapek 50, 75-89 (1943).
13. Moon, J.W. On the line-graph of the complete bigraph. Ann. Math. Stat. 34, 664-667 (1963).
14. Ore, O. Theory of Graphs. Amer. Math. Soc. Colloq. Publ. 38, Providence, 1962.
15. Roberts, F. and Spencer, J. A characterization of clique graphs. Memorandum RM-5933-PR, The Rand Corp., Santa Monica, Calif., February, 1969.
16. Shrikhande, S.S. On a characterization of the triangular association scheme. Ann. Math. Stat. 30, 39-67 (1959).

COMBINATORIAL ASPECTS OF GERSCHGORIN'S THEOREM

by

A. J. Hoffman*

1. Introduction

A well-known theorem of Gerschgorin asserts that, if $A=(a_{ij})$ is a complex matrix of order n, every eigenvalue λ of A lies in the union of the n disks:

$$(1.1) \qquad |a_{kk} - \lambda| \leq \sum_{j \neq k} |a_{kj}|, \qquad k = 1,\ldots,n .$$

There exist many generalizations of Gerschgorin's theorem (see [3]), and we shall be particularly concerned with generalizations in which the right hand sides of (1.1) are replaced by more general functions of the moduli of the off diagonal entries. Specifically, we shall speak of nonnegative functions f of n(n-1) nonnegative arguments, and by f (A) we shall mean the value of such a function when the arguments are the moduli of the off-diagonal entries of A. A set $\{f_1,\ldots,f_n\}$ of such functions will be called a G-generating family if, for every complex matrix A, every eigenvalue of A lies in the union of the n disks:

$$(1.2) \qquad |a_{kk} - \lambda| \leq f_k(A), \qquad k = 1,\ldots,n;$$

equivalently, $\{f_1,\ldots,f_n\}$ form a G-generating family if, for every complex matrix A,

$$(1.3) \qquad |a_{kk}| > f_k(A), \qquad k = 1,\ldots,n \quad \text{implies A nonsingular.}$$

This concept appears to have been first introduced by Nowosad in [4], and a systematic investigation initiated in [1]. In [2], the following combinatorial problem was raised:

Let us say that f depends on (i, j) if there exist two complex matrices A and B of order n such that $|a_{k\ell}| = |b_{k\ell}|$ if $(k,\ell) \neq (i,j)$, but $f(A) \neq f(B)$. Then define

$$D(f) = \{(i,j) \mid f \text{ depends on } (i,j)\}.$$

---

* This work was supported (in part) by the office of Naval Research under Contract NONR 3775(00). It contains portions of material presented in a lecture under this title given at the New York Graph Theory Conference, sponsored by St. John's University, in June, 1970.

What is the "pattern of dependencies" of a G-generating family $\{f_1, \ldots, f_n\}$? More formally, if each of $D_1, \ldots, D_n$ is a subset of $\{(i,j)\}_{i \neq j}$, what are necessary and sufficient conditions that there exist a G-generating family $\{f_1, \ldots, f_n\}$ such that $D(f_k) = D_k$, $k = 1, \ldots, n$? In [2], the following result was established:

<u>Theorem 1.</u> There exists a G-generating family $\{f_1, \ldots, f_n\}$ of functions each of which is homogeneous (of degree one) and bounded on bounded sets (e.g., continuous) such that $D(f_k) = D_k$, $k=1, \ldots, n$, if and only if, for every subset $S \subset \{1, \ldots, n\}$, with $|S| \geq 2$, every cyclic permutation σ of S, and every subset $T \subset S$,

$$|\{(i, \sigma i)\}_{i \in S} \cap \bigcup_{k \in T} D_k| \geq |T|. \tag{1.4}$$

(When (1.4) is satisfied, it is shown in [2] that the $\{f_k\}$ can be taken to be linear, so the requirement that the f's be linear, e.g., imposes no restriction on the $\{D_k\}$ in addition to (1.4)).

The purposes of this note are: (i) to recast the general problem in terms of the language of directed graphs; (ii) in that language, to find a more perspicuous and more easily testable restatement of (1.4); (iii) to consider the problem with other (or no) restrictions on the G-generating family.

We shall denote by D the complete directed graph, without loops, on n vertices. Thus, the (directed) edges of D consist of all $n(n-1)$ ordered pairs (i,j), $i, j = 1, \ldots, n$, $i \neq j$. We shall say $E \subset D$ if E has the same vertex set as D, and every edge of E is an edge of D. If $E \subset D$, $\bar{E}$ is the graph, with the same vertex set, whose edges are precisely all ordered pairs (i,j), $i \neq j$, which are not edges of E. If $E_1, E_2 \subset D$, $E_1 \cap E_2$ is the graph with the same vertex set, whose edges are precisely all edges in both E_1 and E_2. A path in $E \subset D$ is a sequence $\{i_1, \ldots, i_k\}$, $k \geq 2$, of distinct vertices such that (i_r, i_{r+1}) is an edge of E, $r=1, \ldots, k-1$, together with all these edges. The vertices i_1 and i_k are respectively initial and terminal vertices. A cycle in $E \subset D$ is a sequence $\{i_1, \ldots, i_k\}$, $k \geq 2$, of distinct vertices such that (i_r, i_{r+1}) is an edge of E, $r=1, \ldots, k-1$, (i_k, i_1) is an edge of E, together with all these edges.

We can now restate Theorem 1.

<u>Theorem 1'.</u> There exists a G-generating family $\{f_1, \ldots, f_n\}$ of functions each of which is homogeneous (of degree one) and bounded on bounded sets (e.g. continuous) such that $D(f_k) = D_k$, $k=1, \ldots, n$ if and only if

(1.4.1) for every $k=1, \ldots n$, $\bar{D}_k$ contains no cycle including k, and

(1.4.2) for every pair i, j, $i \neq j$, $\bar{D}_i \cap \bar{D}_j$ contains no path whose initial vertex is i

and whose terminal vertex is j.

Our other results are stated in the following theorems.

Theorem 2. There exists a G-generating family $\{f_1, \dots, f_n\}$ with $D(f_k) = D_k$, $k=1, \dots, n$, if and only if for every $i \neq j$,

(1.5) (i, j) is not an edge of $\bar{D}_i \cap \bar{D}_j$.

The statement of Theorem 3 is somewhat more complicated. Assume $D_i, \dots, D_n$ given, let C be any cycle in D, and let V be the set of vertices contained in C. Let $G(C)$ be the following (undirected) graph: the vertex set of $G(C)$ is V, and two vertices $i \neq j$ of V are adjacent if $D_i \cap D_j$ contains an edge of C.

We shall say $G(C)$ is balanced if, for each connected component K of $G(C)$, we have

$|V(K)| = |$ number of edges in $\bigcup_{i \in K} D_i$ which are also edges of $C|$.

Theorem 3. There exists a G-generating family $\{f_1, \dots, f_k\}$ of functions each of which is homogeneous (of degree one) with $D(f_k) = D_k$, $k=1, \dots, n$, if and only if, for every cycle C in G

(1.6) $G(C)$ is balanced.

We shall prove these theorems in reverse order.

2. Proof of Theorem 3.

We begin by first noting that it is not difficult to prove that (1.6) is equivalent to: for every subset $S \subset \{1, \dots k\}$, $|S| \geq 2$, every cyclic permutation σ of S, and every subset $T \subset S$, we have (1.4) or

$$\{(i, \sigma i)\}_{i \in S} \cap \bigcup_{k \in T} D_k \cap \bigcup_{k \in S-T} D_k \neq \emptyset \tag{2.1}$$

We first prove necessity. Assume $\{f_1, \dots, f_n\}$ G-generating, homogeneous, but there exists S, σ, T such that the $D(f_k)$ satisfy neither (1.4) nor (2.1) Let $|T| = t$, and let $i_1, \dots, i_r \in S$, be such that $r < t$, $i \in S - \{i_1, \dots, i_r\}$ implies $(i, \sigma i) \notin \bigcup_{k \in T} D(f_k)$. This follows from the negation of (1.4). Let $\mathcal{E} > 0$ be given, and define the off diagonal entries of a matrix $A(\mathcal{E})$ by

$$\begin{aligned} a_{i_j, \sigma i_j} &= -\mathcal{E} & j &= 1, \dots, r \\ a_{i, \sigma i} &= -1 & i &\in S - \{i_1, \dots, i_n\} \\ a_{k\ell} &= 0 & &\text{otherwise, } k \neq \ell. \end{aligned} \tag{2.2}$$

It follows from the homogeneity of the f_k that

(2.3) $f_k(A(\epsilon)) = \epsilon\, f_k(A(1))$, for all $k \in T$.

From the negation of (1.6), we have

(2.4) $\quad (A(\varepsilon)) = f_k(A(1))$ for all $k \in S - T$.

Hence, from (2.3) and (2.4)

(2.5) $\quad \prod_{k\in S} f_k(A(\epsilon)) = \epsilon^t \prod_{k\in S} f_k(A(1))$.

It follows from (2.5) that, since $r < t$, there exists a positive number ε such that

$$\varepsilon^r > \prod_{k\in S} f_k(A(\varepsilon)).$$

Hence, we can choose positive numbers $c_k = c_k(\varepsilon)$, $k \in S$ such that

(2.6) $\quad c_k > f_k(A(\varepsilon))$, $k \in S$,

(2.7) $\quad \varepsilon^r = \prod_{k\in S} c_k$.

Now define diagonal entries of $A(\varepsilon)$

(2.8) $\quad a_{i,i}(\varepsilon) = c_i \quad i \in S$

(2.9) $\quad a_{i,i}(\varepsilon) > f_i(A(\varepsilon))$, $i \notin S$.

From (2.6), (2.8), (2.9), $|a_{kk}| > f_k(A(\varepsilon))$ for all k. Since $\{f_1, \ldots, f_n\}$ is assumed a G-generating family, it follows from (1.3) that $A(\varepsilon)$ is nonsingular. But using (2.1), (2.7) and (2.8), $\det A = 0$. This contradiction establishes the necessity of at least one of (1.4) and (2.1).

To prove the sufficiency, we must exhibit a G-generating family $\{f_1, \ldots, f_n\}$ of homogeneous functions with each $D(f_k) = D_k$. To that end, we first define

$$D_k(A) = \{(i,j) \mid (i,j) \in D_k,\ a_{ij} \neq 0\};$$

$$f_k(A) = \begin{cases} 0 & \text{if } D_k(A) = \emptyset \\ n! \dfrac{\max\limits_{(i,j)\in D_k(A)} |a_{ij}|^n}{\min\limits_{(i,j)\in D_k(A)} |a_{ij}|^{n-1}} & \text{if } D_k(A) \neq \emptyset \end{cases} \qquad k = 1, \ldots, n.$$

Clearly, all we need prove is that the f's so defined form a G-generating family. Recalling (1.3) and the definition of a determinant as the sum of products, all we need show is

(2.10) $\quad$ for every $S \subset \{1, \ldots, k\}$, $|S| \geq 2$, every cyclic permutation σ of S, and every matrix $A = (a_{ij})$ such that $\prod_{i\in S} |a_{i,\sigma i}| \neq 0$,

$$\prod_{i\in S} |a_{i,\sigma i}| \leqq \prod_{k\in S} \left(\max_{(i,\sigma i)\in D_k} |a_{i,\sigma i}|^n \Big/ \min_{(i,\sigma i)\in D_k} |a_{i,\sigma i}|^{n-1} \right)$$

(Note that our assumptions that at least one of (1.4) and (2.1) holds guarantees that, for each $k \in S$, $\{i, \sigma i\}_{i\in S} \cap D_k \neq \phi$.

Let

(2.11) $b_1 \geqq b_2 \geqq \dots \geqq b_{|S|}$

be a rearrangement in descending order of the quantities $\{\log |a_{i,\sigma i}|_{i \in S}$. For simplicity of notation, assume $S = \{1, \dots, |S|\}$. Let $M = (m_{ij})$ be the (0, 1) matrix of order $|S|$ with $m_{ij} = 1$ $(i, j \in S)$ if and only if D_j contains $(Ti, \sigma Ti)$, where T is the permutation of S such that $b_i = \log |a_{Ti, \sigma Ti}|$. By taking the logarithm of both sides in (2.10), our problem reduces to proving that (2.11) implies

$$\sum_{i=1}^{|S|} b_i \leq n \sum_{k=1}^{|S|} b_{\alpha(k)} - (n-1) \sum_{k=1}^{|S|} b_{\beta(k)}, \tag{2.12}$$

where $\alpha(k)$ is the smallest i with that $m_{ik} = 1$,

$\beta(k)$ is the largest i such that $m_{ik} = 1$.

We think of (2.12) as an inequality which we must show is satisfied by every vector b satisfying (2.11).

Let u^j be the vector with $|S|$ components, of which the first j are 1, the remainder 0. Any b satisfying (2.11) can be expressed

$$b = \sum_{j=1}^{|S|} \lambda_j u^j, \qquad \lambda_j \geqq 0, \; j = 1, \dots, |S| - 1. \tag{2.13}$$

Since (2.12) holds as an equality for $j = |S|$, it follows from (2.13) that all we need prove is that (2.12) holds if $b = u^j$, $j < |S|$. We rewrite this case of (2.12) as

$$j \leq n \sum_{k=1}^{|S|} (u^j_{\alpha(k)} - u^j_{\beta(k)}) + \sum_{k=1}^{|S|} u^j_{\beta(k)} . \tag{2.14}$$

If, for some k_o, $\alpha(k_o) \leqq j$, $\beta(k_o) > j$, then the right hand side of (2.14) becomes

$$n + n \sum_{k \neq k_0} (u^j_{\alpha(k)} - u^j_{\beta(k)}) + \sum_k u^j_{\beta(k)} . \tag{2.15}$$

Since $u^j_{\alpha(k)} \geqq u^j_{\beta(k)}$ for all k, and $u^j_{\beta(k)} \geqq 0$ for all k, (2.19) is at least $n > j$, verifying (2.14).

Suppose such a k_o does not exist. This means that, for each $k \in S$, either $\alpha(k) > j$ or $\beta(k) \leqq j$. If we let $T = \{k \in S \mid \alpha(k) > j\}$, then (2.1) is violated.

It follows that (1.4) holds, which means

$$(2.16) \qquad |S| - j \geqq |T|.$$

But under these circumstances, the first term on the right-hand side of (2.14) vanishes and the second term is $|S| - |T|$. Thus, (2.16) proves (2.14).

3. Proof of Theorem 2

It is more convenient to restate (1.5) as

(3.1) for every $S \subset \{1, \ldots, n\}$, $|S| \geqq 2$ and every cyclic permutation σ of S,

$$\{(i, \sigma i)\}_{i \in S} \subset \bigcup_{k \in S} D_k.$$

We first show the necessity of (3.1). Assume (3.1) false, so that there exists $i_o \in S$ with

$$(3.\epsilon) \quad (i_o, \sigma i_o) \notin \bigcup_{k \in S} D(f_k).$$

Let B be a matrix in which $b_{i, \sigma i} = -1$ for all $i \in S$, $i \neq i_o$, all other off diagonal elements 0. Let $\{c_k\}_{k \in S}$ be positive numbers such that

$$(3.3) \qquad c_k > f_k(B) \text{ for all } k \in S.$$

Let $D=(d_{ij})$ be a matrix in which all off diagonal elements are 0 except

$$(3.4) \qquad d_{i, \sigma i} = -1 \text{ for } i \in S, \; i \neq i_o$$

$$d_{i_o, \sigma i_o} = -\prod_{k \in S} c_k.$$

Further, let

$$(3.5) \qquad d_{ii} = c_i \quad \text{for} \quad i \in S$$

$$(3.6) \qquad d_{ii} > f_i(D) \text{ for } i \notin S.$$

By comparing B with D, we see from (3.3), (3.5) and (3.6) that $|d_{kk}| > f_k(D)$ for all k. But (3.4) and (3.5) show that det D = 0, a contradiction.

To prove sufficiency of (3.1), define

$$f_k(A) = n! \; (1 + \sum_{\phi \neq T \subset D_k} \prod_{(i, j) \in T} |a_{ij}|), \; k = 1, \ldots, n.$$

It is clear that the $\{f_k\}$ so defined form a G-generating family with each $D_k = D(f_k)$. (Indeed, we have other choices of the f_k, and could have made them polynomials of degree at most two).

Note that the hypothesis that the f's be continuous, or even polynomials, would not change the condition (1.5) or (3.1).

4. Proof of Theorem 1. We must prove that (1.4) implies both (1.4.1) and (1.4.2) and conversely. Assume (1.4) holds. Suppose

(1.4.1) false for some k, then this violates (1.4) with $T=\{k\}$. Suppose (1.4.2) false for some $i \neq j$. Let

$$i = i_1, i_2, \dots i_k = j$$

be the sequence of vertices in a path contained in $\tilde{D}_i \cap \tilde{D}_j$. Then the cycle obtained by adjoining to the path the edge (j, i) violates (1.4) with $T = \{i, j\}$.

Conversely, assume (1.4.1) and (1.4.2). Suppose (1.4) is false for some cycle determined by an S and σ and some subset $T \subset S$. If $|T| = 1$, this violates (1.4.1). Assume $|T| = t > 1$, and let $i_1, i_2, \dots i_t$ be the vertices in T in the order in which they occur around the cycle. Let the number of edges in the subpath of the cycle from i_1 to i_2 be $a_1, \dots,$ from i_{t-1} to i_t be a_{t-1}, from i_t to i_1 be a_t. Since (1.4.2) holds, there are at most $a_k - 1$ edges of the cycle in $\tilde{D}_{i_k} \cap \tilde{D}_{i_{k+1}}$ (all k taken mod t) in the subpath from i_k to i_{k+1}. Consequently $\bigcap_{k=1}^{t} \tilde{D}_{i_k}$ contains at most $\sum_{i=1}^{t} (a_i - 1) = |S| - t$ edges. There $\bigcup_{k=1}^{t} D_{i_k}$ contains at least t edges of the cycle, contradicting the presumed falsity of (1.4).

We are grateful to Ellis Johnson, Peter Lax, Michael Rabin and Richard Varga for useful conversations about this material.

REFERENCES

[1] A. J. Hoffman, "Generalizations of Gerschgorin's Theorem: G-Generating Families", lecture notes, University of California at Santa Barbara, August, 1969, 46 pp.

[2] A. J. Hoffman and R. S. Varga, Patterns of Dependence in Generalizations of Gerschgorin's Theorem, to appear in SIAM Journal of Numerical Analysis.

[3] M. Marcus and H. Minc, "A Survey of Matrix Theory and Matrix Inequalities", Allyn and Bacon, Boston, 1964.

[4] P. Nowosad, "On the Functional (x^{-1}, Ax) and Some of Its Applications", An. Acad. Brasil Ci. 37 (1965), 163-165.

ON SUBDOMINANTLY BOUNDED GRAPHS - SUMMARY OF RESULTS

by LEONARD HOWES

THE BERNARD M. BARUCH COLLEGE
of The City University of New York

By a <u>graph</u> G, is meant a set of n-points, called the set of <u>vertices</u>, $V(G)$; and a set $E(G)$, of lines, or <u>edges</u>, joining some pairs of vertices, so that no pair of vertices is joined by more than one edge, and no edge joins a vertex to itself. When a pair of vertices in a graph are joined by an edge they are called <u>adjacent</u>. The <u>adjacency matrix</u> of a graph G, denoted $A(G)$, is a square $0 - 1$ matrix of order n whose rows and columns correspond to the vertices of G and for which $A_{ij} = 1$ if and only if vertex i and vertex j are adjacent. Thus, for each of the graphs considered, the associated adjacency matrix is symmetric with zeros on the diagonal. The <u>eigenvalues of a graph</u> are the eigenvalues of its adjacency matrix, and hence are real. For any graph G, the eigenvalues will be denoted $\lambda_1(G) \geq \lambda_2(G) \geq \ldots$ in descending order, and $\lambda^1(G) \leq \lambda^2(G) \leq \ldots$ in ascending order. The complement, $\bar{G}$, of the graph G is the graph described by $V(\bar{G}) = V(G)$, where two vertices in $\bar{G}$ are adjacent if and only if these two vertices are not adjacent in G. The <u>valence</u> of a vertex v in a graph G is the number of edges for which that vertex is an end-point. Two graphs G and H are said to be <u>"L away from each other"</u> if there exists graphs $\tilde{G}$, and $\tilde{H}$ such that $A(G) + A(\tilde{G}) = A(H) + A(\tilde{H})$ and every vertex of $\tilde{G}$ and $\tilde{H}$ has valence at most L. A <u>sub-graph</u> G' of G is the graph on the non-empty subset $V(G')$ of vertices of $V(G)$ where two vertices in G' are adjacent if and only if they were adjacent in the original graph G.

The following notation will be used:

$\bullet_\ell$ will be a complete graph on ℓ vertices, or clique, abbreviated K_ℓ, where every vertex is adjacent to every other vertex.

 will be a graph formed by K_ℓ and one more vertex adjacent to all the vertices of K_ℓ ; that is, $K_{\ell+1}$.

 will be the <u>independent set of</u> ℓ vertices, abbreviated $\bar{K}_\ell$, in which no two vertices are adjacent.

 will be a graph formed by two cliques on ℓ vertices, where every vertex in each clique is adjacent to all other vertices, that is $K_{2\ell}$.

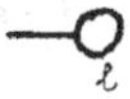 will be a graph formed by $\bar{K}_\ell$ and one more vertex adjacent to all the vertices of $\bar{K}_\ell$.

In short, a solid line joining graphs A and B forms a graph where every vertex in V(A) is adjacent to every vertex in V(B) .

A.J. Hoffman proved the following

<u>Theorem</u>: Let $\mathcal{G}$ be an infinite set of graphs, then the following statements about $\mathcal{G}$ are equivalent:

(1) There exists λ such that $\lambda(G) \geq \lambda$, $\forall\, G \in \mathcal{G}$. Where for any G , $\lambda(G)$ is the least eigenvalue of A(G) .

(2) There exists a positive integer ℓ such that no $G \in \mathcal{G}$ contains either 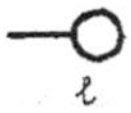 , or  , as a sub-graph.

(3) There exists a positive integer L , such that for each $G \in \mathcal{G}$ there exist graphs $\tilde{G}$ and H with the following being true:

(3a) $A(G) + A(\tilde{G}) = A(H)$.

(3b) Every vertex of $\tilde{G}$ has valence at most L ; and H contains a family of cliques $K^1, K^2, \ldots$ such that:

(3c) Each edge of H is in at least one K^i ,

(3d) Each vertex of H is in at most L of the cliques $K^1, K^2, \ldots$

(3e) $|V(K^i) \cap V(K^j)| \leqq L$, $i \neq j$.

The main result of the present investigation is the following

<u>Theorem 1</u>: Let $\mathcal{G}$ be an infinite set of graphs, then the following statements about $\mathcal{G}$ are equivalent:

(I) There exists a real number λ such that $\lambda_2(G) \leq \lambda$ for every $G \in \mathcal{G}$.

(II) There exists a positive integer ℓ such that for each $G \in \mathcal{G}$ the following graphs are not subgraphs of G .

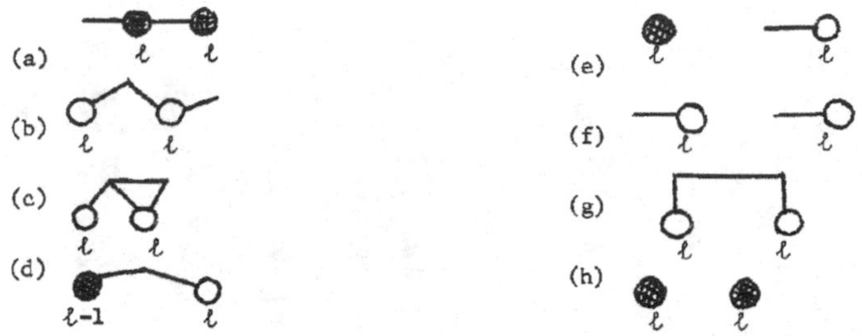

(III) There exists a positive integer L such that each $G \in \mathcal{G}$ is no more than L away from a graph H in which the following is true:

(IIIa) The set of vertices of H , $V(H)$, can be partitioned into two classes $V(H_1)$, $V(H_2)$ with associated subgraphs H_1 and H_2 where

(IIIb) H_1 contains a distinguished family of independent sets $\bar{K}^1, \bar{K}^2, \ldots$ such that every pair of non-adjacent vertices of H_1 is in at least one of the independent sets $\bar{K}^\ell$.

(IIIc) Every vertex of H_1 is in at most L of the independent sets $\bar{K}^1, \bar{K}^2, \bar{K}^3, \ldots$.

(IIId) $|V(\bar{K}^i) \cap V(\bar{K}^j)| \leq L$, $i \neq j$.

(IIIe) Every vertex in H_1 is adjacent to fewer than L vertices of H_2 .

(IIIf) For any $v \in V(H_2)$, there exist indices $\{i_1, i_2, \ldots, i_r\}$, $r < L$ such that

$$\{x \mid x \in V(H_1),\ v \text{ adjacent to } x\} = \bigcup_{j=1}^{r} V(\bar{K}^{i_j}) .$$

(IIIg) H_2 is an independent set.

A CLASS OF POINT PARTITION NUMBERS

Don R. Lick*
Western Michigan University, Kalamazoo, MI 49001

1. Introduction

One of the most studied parameters in graph theory is the chromatic number. Undoubtedly, its popularity as a research topic is due to its intimate relationship with the famous Four Color Problem. A coloring number for graphs closely related to the chromatic number is the point arboricity (see [2]). The similarity of many of the results for chromatic number and point arboricity suggests an extension of these parameters to a class of "coloring numbers" or "point partition numbers". We present some of the known results for these parameters as well as introduce a new bound given in terms of the maximum eigenvalue of the adjacency matrix. The point partition numbers of closed 2-manifolds are also discussed.

2. Definitions and preliminary results

The complete n-partite graph $K(p_1, p_2, \ldots, p_n)$ has its point set V partitioned into nonempty subsets V_i with $|V_i| = p_i$, $i = 1, 2, \ldots, n$, such that two points u and v are adjacent if and only if $u \in V_j$, $v \in V_h$, with $j \neq h$. If $p_i = 1$ for each i, then the graph is called the complete graph with n points. The complete graph with n points is denoted by K_n.

The smallest degree among the points of a graph G is called the minimum degree of G and is denoted by $\delta(G)$. The symbol $\Delta(G)$ denotes the maximum degree among the points of G. In [4] a graph G was defined to be k-degenerate, for k a nonnegative integer, if $\delta(H) \leq k$ for each induced subgraph H of G. The symbol Π_k will be used to denote the class of all k-degenerate graphs. It follows that the complete graph K_{p+2} with $p+2$ points is (p+1)-degenerate, but not p-degenerate. Hence the class Π_p is properly contained in the class Π_{p+1}. The graph $G = K(2,3)$ illustrated in Figure 1 is a 2-degenerate graph which is not 1-degenerate.

G:

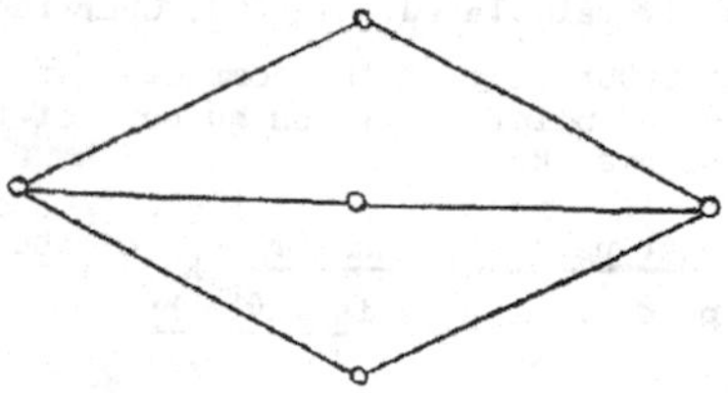

Figure 1.

* Work supported in part by grants from the National Science Foundation (GP-9435) and Western Michigan University (Faculty Research Fellowship).

A totally disconnected graph is one which has no lines. It is clear that a graph is totally disconnected if and only if it is 0-degenerate. Thus Π_0 is the class of all totally disconnected graphs. A forest (or acyclic graph) is a graph with no cycles, and these are precisely the 1-degenerate graphs.

It is easily seen that for any graph G there is a nonnegative integer k such that $G \in \Pi_k$. However, for a given graph G and a fixed nonnegative integer k, G need not be k-degenerate. This leads to the following definition. A point partition number $\rho_k(G)$ of the graph G, for a given nonnegative integer k, is the smallest number of sets which the point set $V(G)$ can be partitioned into so that each set induces a k-degenerate subgraph of G.

One readily sees that the parameters $\rho_0(G)$ and $\rho_1(G)$ are the chromatic number and the point arboricity of the graph G, respectively. The chromatic number of a graph has been extensively investigated, while the point arboricity has more recently become a topic of research. For values of $k \geq 2$, no special name has been given to the parameter $\rho_k(G)$, but these point partition numbers have been studied in [4], [5]. Some of these results are listed below. It might be pointed out that each of the parameters ρ_k, $k \geq 0$, may be thought of as a coloring number, since it gives the minimum number of colors in any coloring of the points of G so that each color class induces a k-degenerate subgraph of G.

A graph G is said to be n-critical with respect to ρ_k if $\rho_k(G) = n$, but $\rho_k(G-v) = n - 1$ for each point v of G. It is well-known that if G is n-critical with respect to ρ_0 (chromatic number), then the minimum degree satisfies the inequality $\delta(G) \geq n - 1$. Chartrand and Kronk [1] proved that if G is n-critical with respect to ρ_1 (point arboricity), then $\delta(G) \geq 2(n-1)$. In [4], these two inequalities were extended to include all nonnegative integers k.

THEOREM A. *If the graph G is n-critical with respect to ρ_k, where k is a nonnegative integer, then* $\delta(G) \geq (k+1)(n-1)$.

We now consider the point partition numbers in a bit more detail. As in the cases of chromatic number and point arboricity, for most graphs G and for small values of k, the numbers $\rho_k(G)$ are difficult to determine. However, for one large important class of graphs, the complete n-partite graphs, the point partition numbers $\rho_k(G)$ are easily calculated. In [2], Chartrand, Kronk, and Wall presented a formula for the point arboricity of the complete n-partite graph, and in [4], this formula was extended to the point partition numbers of the complete n-partite graphs, for all nonnegative integers k.

THEOREM B. *The point partition number ρ_k of the complete n-partite graph $K(p_1,p_2,\ldots,p_n)$, $p_1 \leq p_2 \leq \ldots \leq p_n$, is given by*

$$\rho_k(K(p_1,p_2,\ldots,p_n)) = n - \max\{j: \sum_{i=0}^{j} p_i \leq (n-j)k\},$$

where p_0 is defined to be zero.

As a corollary to Theorem B, we list the point partition numbers of the complete graphs. We use the notation $\{r\}$ to mean the least integer greater than or equal to the number r.

COROLLARY C. *For the complete graph* K_p *with* p *points and any nonnegative integer* k,

$$\rho_k(K_p) = \left\{\frac{p}{k+1}\right\} .$$

Since any graph G of order p may be thought of as a subgraph of the complete graph K_p, we have the following upper bound for the point partition numbers $\rho_k(G)$, $k \geq 0$.

COROLLARY D. *For any graph* G *with* p *points and for any nonnegative integer* k,

$$\rho_k(G) \leq \left\{\frac{p}{k+1}\right\} .$$

3. Bounds for the parameters ρ_k

The bounds for the point partition numbers $\rho_k(G)$, $k = 1, 2, \ldots,$ given in Corollary D may differ by a large amount from the actual values of the parameter $\rho_k(G)$. For example, the point partition number of the *n-star*, $K(1,n)$, is one for all $k \geq 1$, while the bound given in Corollary D is large for large values of n.

It is well-known that for any graph G, $\rho_0(G) \leq 1 + \Delta(G)$. Chartrand, Kronk, and Wall [2] gave an analogue of this result for point arboricity, namely,

$$\rho_1(G) \leq 1 + \left[\frac{\Delta(G)}{2}\right] .$$

In [4], it was proved that

$$\rho_k(G) \leq 1 + \left[\frac{\Delta(G)}{k+1}\right] , \tag{1}$$

for each nonnegative integer k. In many cases this upper bound is not particularly good, for example, $\rho_k(K(1,n)) = 1$, for each integer $k \geq 1$, while $\Delta(K(1,n)) = n$.

In light of the two upper bounds given above, it seems desirable to give an upper bound of the parameter $\rho_k(G)$, $k = 1, 2, \ldots,$ which is more global in nature, and less sensitive to the idiosyncrasies of a few uninfluential points. Such an upper bound is now given in terms of the largest eigenvalue of the adjacenty matrix of the graph.

For a graph G of order p, let $A = A[G]$ denote the $p \times p$ adjacency matrix of G. Let $\lambda = \lambda[A] = \lambda(G)$ be the largest eigenvalue of A. In [9], Wilf proved that $\rho_0(G) \leq 1 + \lambda(G)$, with equality if and only if G is a complete graph or an odd cycle. Mitchem [6] showed that

$$\rho_1(G) \leq 1 + \left[\frac{\lambda(G)}{2}\right] ,$$

and that this inequality is the best possible. These results are now extended to the parameters $\rho_k(G)$, $k = 2, 3, \ldots$.

THEOREM. *For any graph* G *and any nonnegative integer* k,

$$\rho_k(G) \le 1 + \left[\frac{\lambda(G)}{k+1}\right], \tag{2}$$

and this inequality is the best possible.

Before proving this result, we remark that it follows from the Perron-Frobenuis Theorem that $\lambda(G) \le \Delta(G)$, and so the inequality (2) never produces a larger bound for ρ_k than does (1).

Proof of the theorem. If $\rho_k(G) = 1$, then the inequality (2) holds since $\lambda(G) \ge 0$. Thus we assume that $\rho_k(G) = n \ge 2$. Let G_c be an n-critical subgraph of the graph G with respect to ρ_k . We consider the following three matrices: $A[G]$, the $p \times p$ adjacency matrix of G; $A[G_c]$, the $p' \times p'$ adjacency matrix of G_c; and A', the $p \times p$ matrix obtained from $A[G]$ by replacing by zeros those rows and columns that correspond to points of G which are deleted in forming G_c. Then

$$\lambda(G_c) = \lambda_{max}(A') \le \lambda(G) = \lambda ,$$

the first equality follows since the eigenvalues of A' are those of $A[G_c]$ plus an additional $p - p'$ zeros, and the inequality follows from the entry-by-entry domination of $A[G]$ over A'.

Since G_c is an n-critical graph with respect to ρ_k, it follows from Theorem A that $\delta(G_c) \ge (k+1)(n-1)$. From well-known results about matrices with nonnegative entries, it follows that

$$\lambda(G_c) \ge (k+1)(n-1) ,$$

and the result follows.

That this result is best possible follows from the fact that

$$\rho_k(K_p) = \left\{\frac{p}{k+1}\right\} = 1 + \left\{\frac{p-1}{k+1}\right\} ,$$

and

$$\lambda(K_p) = p - 1 .$$

4. Point partition numbers of closed 2-manifolds

The *point partition number* $\rho_k(M)$ *of a closed 2-manifold* M is the maximum point partition number $\rho_k(G)$ of all graphs G which can be imbedded in M. Then $\rho_0(M)$ and $\rho_1(M)$ are respectively the chromatic number and point arboricity of M.

In 1968, Ringel and Youngs [8] announced their solution to the long-standing Heawood Map-Coloring Conjecture: The chromatic number ρ_0 of the closed orientable 2-manifold of genus $\gamma(\gamma > 0)$ is $[(7+\sqrt{1+48\gamma})/2]$. This statement is now known as the Heawood Map-Coloring Theorem; for $\gamma = 0$, it continues to be called the Four Color Conjecture. In 1959, Ringel [7] showed that the chromatic number of the closed non-orientable 2-manifold of genus $\gamma(\gamma > 0)$ is $[(7+\sqrt{1+24\gamma})/2]$, if $\gamma \neq 2$, and for the Klein bottle $(\gamma = 2)$ the chromatic number is six.

In 1969, Kronk [3] showed that the point arboricity ρ_1 of the closed orientable 2-manifold of genus $\gamma(\gamma > 0)$ is $[(9+\sqrt{1+48\gamma})/4]$. The sphere has point-arboricity three, as shown by Chartrand and Kronk [1]. The similarity of these three formulas suggested the generalization discussed in this section.

Let $S_\gamma(\tilde{S}_\gamma)$ denote the closed orientable (non-orientable) 2-manifold of genus γ and let M_χ denote a closed 2-manifold of characteristic χ. For the case where $M_\chi = S_\gamma$, we have $\chi = 2 - 2\gamma$, while the case where $M_\chi = \tilde{S}_\gamma$ yields $\chi = 2 - \gamma$.

In [5], the following generalization of the three results mentioned above was proved.

THEOREM. <u>The point partition numbers for a closed 2-manifold</u> M_χ <u>are given by the formula</u>

$$\rho_k(M_\chi) = \left[\frac{(2k+7) + \sqrt{49-24\chi}}{(2k+2)}\right],$$

<u>except that</u>

(1) <u>in the orientable case</u>, $\rho_0(S_0) = 4$ <u>or</u> 5, $\rho_1(S_0) = 3$, $\rho_3 = (S_0) = \rho_4(S_0) = 2$; <u>and</u>

(2) <u>in the non-orientable case</u>, $\rho_0(\tilde{S}_2) = 6$, $\rho_1(\tilde{S}_2) = 3$ <u>or</u> 4, $\rho_2(\tilde{S}_2) = 2$ <u>or</u> 3.

<u>Outline of proof</u>: Let

$$f_k(t) = \frac{(2k+7) + \sqrt{49-24t}}{(2k+2)}.$$

With the aid of the above Theorem A and two consequences of the generalized Euler polyhedral formula, it is possible to show that $\rho_k(M_\chi) \leq [f_k(\chi)]$, unless $M_\chi = S_0$. Next, using the formulas established by Ringel [7] and Ringel and Youngs [8] for the non-orientable and orientable genus of the complete graph on n points respectively, it follows that $\rho_k(M_\chi) \geq [f_k(\chi)]$, if $M_\chi \neq \tilde{S}_2$. Finally, for the exceptional cases -- the sphere and the Klein bottle -- it is possible to resolve the issue in all but three situations: ρ_0 for S_0 and ρ_1 and ρ_2 for $\tilde{S}_2$. In each of these three cases, the ambiguity rests between two possible choices for the value of $\rho_k(M_\chi)$.

References

1. G. Chartrand and H. Kronk, "The point-arboricity of planar graphs", J. London Math. Soc., 44 (1969), 612-616.

2. G. Chartrand, H. Kronk, and C. Wall, "The point-arboricity of a graph", Israel J. Math., 6 (1968), 169-175.

3. H.V. Kronk, "An analogue to the Heawood map-coloring problem", J. London Math. Soc., 1 (Ser. 2), (1969), 750-752.

4. D.R. Lick and A.T. White, "k-Degenerate graphs", Canad. Math. J., to appear.

5. D.R. Lick and A.T. White, "The point partition numbers of closed 2-manifolds", submitted for publication.

6. J. Mitchem, On Extremal Partitions of Graphs, Thesis, Western Michigan University Kalamazoo, Michigan, 1970.

7. G. Ringel, Farbungsprobleme auf Flachen and Graphen, Deutscher Verlag, Berlin, 1959.

8. G. Ringel, and J.W.T. Youngs, "Solution of the Heawood map-coloring problem", Nat. Acad. Sci., 60 (1968), 438-445.

9. H.S. Wilf, "The eigenvalues of a graph and its chromatic number", J. London Math. Soc., 42 (1967), 330-332.

ON THE LENGTHS OF CYCLES IN PLANAR GRAPHS

Joseph Malkevitch, York College (CUNY), Flushing, NY 11365

A graph G with n vertices is called pancyclic if it contains cycles of length m, $3 \leq m \leq n$. (For standard graph - theoretic terminology, see Harary [5].)

In Bondy [1], one finds:

Conjecture 2. A planar hamiltonian graph in which every vertex has valence at least four is pancyclic.

The purpose of this note is to give some counterexamples to this conjecture, and to raise some further questions.

Remark. An earlier conjecture in Bondy [2], related to conjecture 2, was first disproved by R. Cori, et al. (Bondy [3]). The examples given below also disprove this earlier conjecture.

We begin the construction of a family of counterexamples to conjecture 2, with the graph in Figure 1, where the cycle composed of curved lines has length 3p ($p \geq 2$).

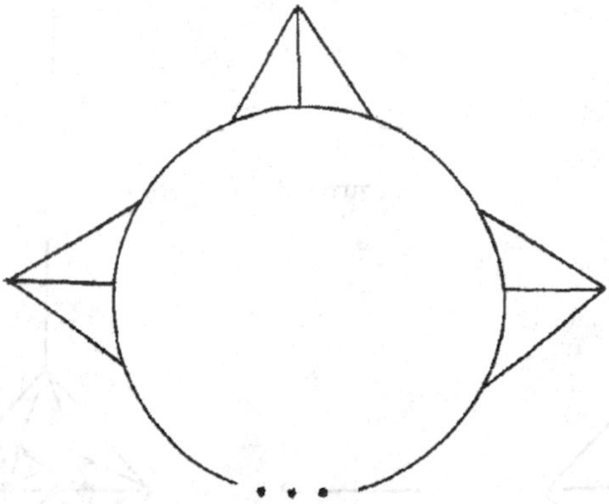

Figure 1

Each configuration consisting of two adjacent triangles in Figure 1, is treated as a "block box," into which we insert various fillings (see Malkevitch [6]), each filling yielding a different type of example. Figures 3 and 4 show how the configuration in Figure 2 can be filled to obtain a 4-valent and 5-valent configuration with 6 and 12 vertices respectively.

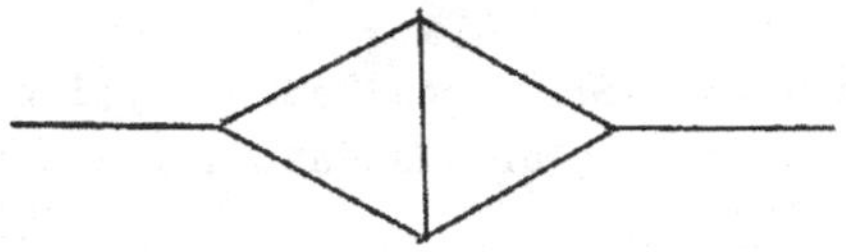

Figure 2

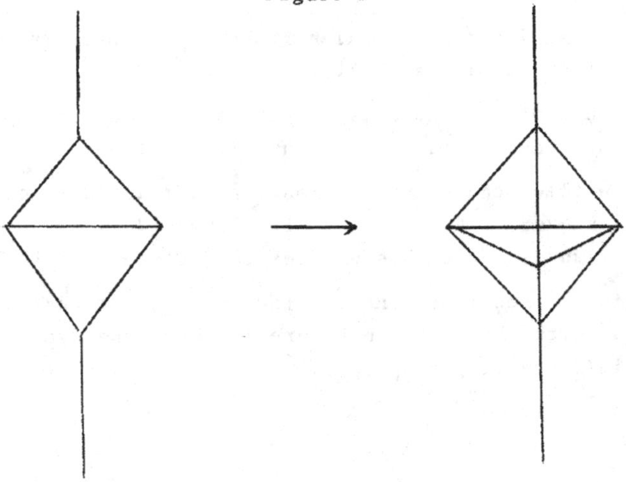

Figure 3

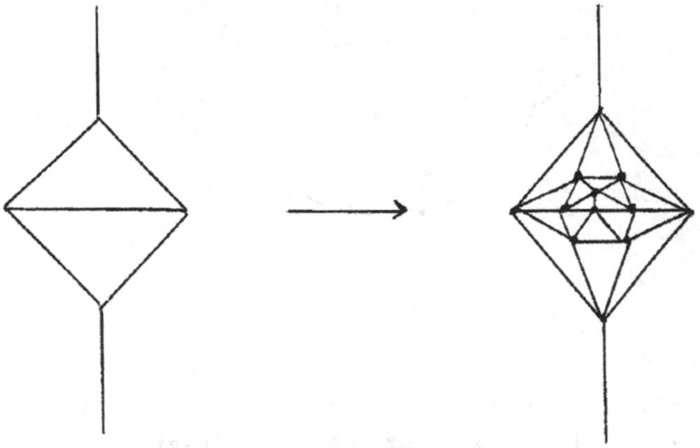

Figure 4

Note that the configurations in Figures 3 and 4 have cycles of all lengths (≥ 3) up to 6 and 12 respectively.

To obtain 4-valent and 5-valent graphs which are planar and hamiltonian but not pancyclic, insert the configurations of Figures 3 and 4 respectively, into every configuration of two adjacent triangles in the graph in Figure 1, with p chosen ≥ 3 and ≥ 5 respectively. (A typical 4-valent counterexample to conjecture 2 is shown in Figure 5.) It is easy to see that the graphs obtained are planar and hamiltonian, but they have no cycles of length s, $7 \leq s < 3p$ ($p \geq 3$) in the 4-valent case, and no cycles of length t, $13 \leq t < 3p$ ($p \geq 5$) in the 5-valent case. Note that the gap of missing cycles can be made arbitrarily large.

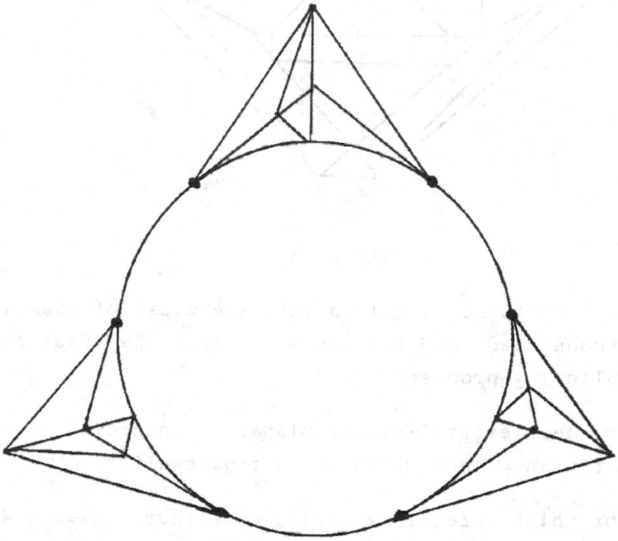

Figure 5

The family of examples constructed above are 2-connected but not 3-connected. Planar 3-connected graphs are of especial interest since by a theorem of E. Steinitz they are 3-polytopal. See Grünbaum [1]. Figure 6 shows a planar 4-connected graph (by a theorem of W. T. Tutte, a planar 4-connected graph has a hamiltonian circuit) which is not pancyclic since it has no cycle of length four. (This is in fact the graph of an Archimedean solid).

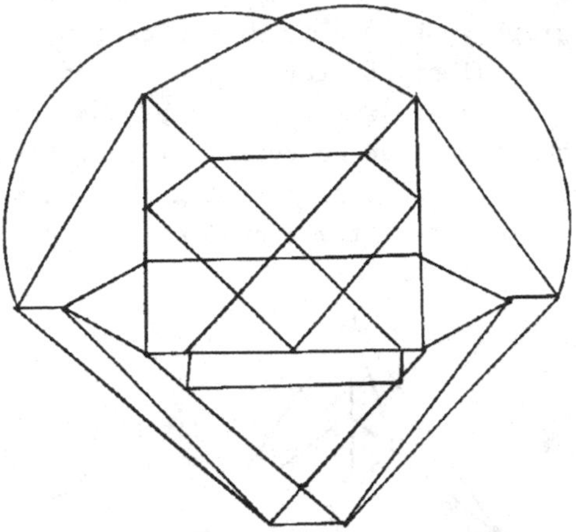

Figure 6

It is possible to construct an infinite class of examples which are planar, 4-connected, and have no 4-cycles. The ideas above suggest the following problems:

Problem 1 Do there exist 5-valent planar, 3-connected (4 or 5 connected?) hamiltonian graphs which are not pancyclic?

Problem 2 For which integers k ($k \geq 4$), does there exist a 4-valent (5-valent) planar, hamiltonian graph which has n ($>k$) vertices, and cycles of all lengths other than k? (The case when G is also 3-connected or more highly connected might also be considered.)

Problem 3 For what values of n ($n \geq 5$) does there exist some graph G_j with n vertices, such that G_j has cycles of length m, $3 \leq m$ ($\neq j$) $\leq n$, where j takes on values between 4 and $n-1$? (The G_j are assumed to be planar).

REFERENCES

1. Bondy, J. A., Pancyclic graphs, to appear.
2. Bondy, J. A., Cycles in graphs, Combinatorial Structures and Their Applications, Gordon and Breach, New York, 1970 15-18.
3. Bondy, J. A., Private Communication.
4. Grünbaum, B., Convex Polytopes, Wiley, New York, 1967.
5. Harary, F., Graph Theory, Addison-Wesley, Reading, 1969.
6. Malkevitch, J., 3-valent 3-polytopes with faces having fewer than 7 edges. Pro. N. Y. Int. Conf., to appear.

HAMILTONIAN ARCS AND CIRCUITS

C.St.J.A. Nash-Williams

University of Waterloo, Waterloo, Ontario, Canada

In this paper, all graphs will be understood to be without loops or multiple edges. A track in a graph G is a finite sequence

$$\xi_0, \lambda_1, \xi_1, \lambda_2, \xi_2, \lambda_3, \ldots, \lambda_n, \xi_n \qquad (1)$$

or an infinite sequence of one of the forms

$$\xi_0, \lambda_1, \xi_1, \lambda_2, \xi_2, \lambda_3, \ldots \qquad (2)$$

$$\ldots, \lambda_{-2}, \xi_{-2}, \lambda_{-1}, \xi_{-1}, \lambda_0, \xi_0, \lambda_1, \xi_1, \lambda_2, \ldots \qquad (3)$$

where the ξ_i are vertices of G and each λ_j is an edge joining ξ_{j-1} to ξ_j. Tracks of the form (2) are one-way infinite and those of the form (3) are two-way infinite. A track in which all terms are distinct is an arc. The track (1) is a circuit if $n \geq 3$, $\xi_0 = \xi_n$ and $\xi_1, \xi_2, \ldots, \xi_n$ are distinct. An arc or circuit in G is Hamiltonian if all vertices of G appear in it. The set of vertices of G will be denoted by $V(G)$ and its set of edges by $E(G)$. If $|V(G)| = \aleph_0$, G is denumerable.

This lecture will make some remarks about the following problems. (I) Which finite graphs have Hamiltonian circuits? (II) Which finite graphs have Hamiltonian arcs? (III) Which infinite graphs have one-way infinite Hamiltonian arcs? (IV) Which infinite graphs have two-way infinite Hamiltonian arcs? Just as finite graphs with Euler paths (or "Euler trails") and finite graphs with 1-factors (or "perfect matchings") have been satisfactorily characterised [6, Theorems 7.1 and 9.4], so it is tempting to wonder whether, if we were clever enough, we could in some reasonable way characterise those with Hamiltonian circuits. This problem has been a prominent theme in graph theory throughout the history of the subject, and of course Problems (II), (III) and (IV) suggest themselves naturally as close relatives. In this paper our attention will focus mainly on what can be inferred about these problems from information about the valencies of the vertices of the graphs concerned; and we shall also glance at some related questions involving directed graphs.

In fact, (I) and (II) are essentially the same problem. Suppose, first, that (I) had been solved and we wished to know whether a finite graph G had a Hamiltonian arc. This could be decided by forming a new graph H from G by the addition of a single new vertex and edges joining that vertex to all vertices of G : then G has a Hamiltonian arc if and only if $|V(G)| = 1$ or H has a Hamiltonian circuit. Thus a solution to problem (I) would solve (II) also. Now let us suppose that (II) was solved and we wished to know whether a certain finite graph G had a Hamiltonian circuit. If $V(G) \neq \emptyset$ select any vertex α of G and form a new graph K from G by adding three new vertices α', β, β', edges joining α' to all vertices which are adjacent to α in G, an edge joining α to β and an edge joining α' to β'. Then G has a Hamiltonian circuit if and only if $|V(G)| \geq 3$ and K has a Hamiltonian arc. Graph theory abounds in such constructions for transforming one problem into another.

The valency (or degree) of a vertex ξ will be denoted by $v(\xi)$. A graph is <u>regular</u> if all its vertices have the same valency and <u>irregular</u> otherwise. If $\xi_1, \ldots, \xi_n$ is a list of the vertices of a finite graph G, arranged in some order, the sequence $v(\xi_1), \ldots, v(\xi_n)$ will be said to be a <u>valency sequence</u> of G. A <u>rearrangement</u> of a sequence $a_1, \ldots, a_n$ is a sequence of the form $a_{\rho(1)}, \ldots, a_{\rho(n)}$ where ρ is a permutation of $\{1, \ldots, n\}$. Since any rearrangement of a valency sequence of G is also a valency sequence of G, an irregular graph will have more than one valency sequence: for example, the sequences 1, 1, 1, 2, 2, 4, 5 and 5, 1, 4, 2, 1, 2, 1 are both valency sequences of the graph in Figure 1. A sequence of numbers $a_1, \ldots, a_n$ will be said to be <u>greater than or equal to</u> a sequence of numbers $b_1, \ldots, b_n$ if $a_i \geq b_i$ for $i = 1, \ldots, n$. Later in this paper, we shall require some analogues of the foregoing concepts involving denumerable graphs, directed graphs and infinite sequences, and their definitions can be left to the reader. A sequence

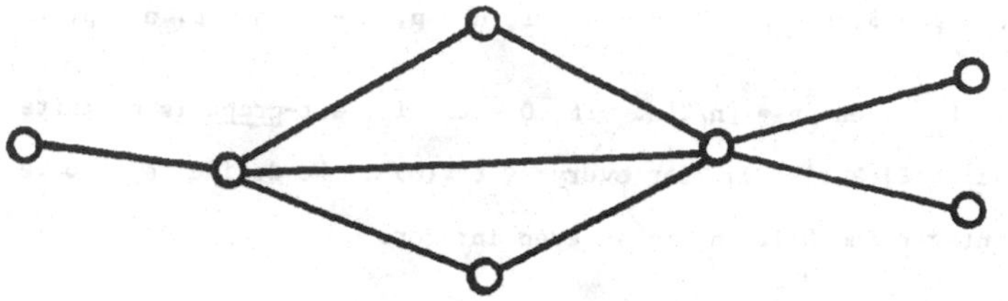

FIGURE 1

$a_1, \ldots, a_n$ is <u>graphic</u> if it is the valency sequence of some finite graph: graphic sequences can be characterized [6, Theorem 6.2] and indeed characterisation of graphic sequences is a special case of a more general problem solved by Tutte [24; 6, Theorem 9.5]. A graphic sequence $a_1, \ldots, a_n$ will be called (i) <u>forcibly Hamiltonian</u> if all graphs with valency sequence $a_1, \ldots, a_n$ have Hamiltonian circuits, (ii) <u>forcibly non-Hamiltonian</u> if no graphs with valency sequence $a_1, \ldots, a_n$ have Hamiltonian circuits. The greatest integer which is $\le x$ will be denoted by $[x]$. If n is an integer greater than or equal to 3, P_n will denote the sequence with n terms which is of the form

$$2, 3, 4, 5, \ldots, \frac{n}{2} - 2, \frac{n}{2} - 1, \frac{n}{2}, \frac{n}{2}, \frac{n}{2}, \frac{n}{2}, \ldots, \frac{n}{2}$$

if n is even and of the form

$$2, 3, 4, 5, \ldots, \frac{n-7}{2}, \frac{n-5}{2}, \frac{n-3}{2}, \frac{n-1}{2}, \frac{n-1}{2}, \frac{n+1}{2},$$

$$\frac{n+1}{2}, \frac{n+1}{2}, \frac{n+1}{2}, \ldots, \frac{n+1}{2}$$

if n is odd. (It should be understood that P_3 is the sequence 1, 2, 2 and P_4 is 2, 2, 2, 2 and P_5 is 2, 2, 3, 3, 3 .) More generally, if p, n are integers such that $2 \le p \le n/2$, then P_n^p denotes the following sequence with n terms:

$$2, 3, 4, 5, \ldots, p-2, p-1, p, p, p+1, p+2, p+3, \ldots$$

$$\ldots, n-p-3, n-p-2, n-p-1, n-p, n-p, n-p, n-p, \ldots, n-p$$

which is P_n when $p = [n/2]$. If $0 < t < 1$, a *t-graph* is a finite graph G such that $v(\xi) \geq t|V(G)|$ for every $\xi \in V(G)$. We define ε_n to be 1 if n is an odd integer and 0 if n is an even integer.

With these conventions, we can state some known results relating the existence of a Hamiltonian circuit in a graph to the valencies of its vertices. The first such theorem historically was probably Theorem 1, which is due to Dirac [3], and this was strengthened to Theorem 2 by Pósa [18; 12; 6, Theorem 7.3]. Pósa's theorem was in turn strengthened by Bondy [2], and a slight further strengthening of Bondy's theorem yields the result given below as Theorem 3. The proof of Theorem 3 requires only an easy adaptation of either of the known proofs [12, 18] of Theorem 2. For large graphs, Theorem 4, which is proved in [15], strengthens Dirac's theorem in a different direction.

THEOREM 1. *Every $\frac{1}{2}$-graph with at least three vertices has a Hamiltonian circuit.*

THEOREM 2. *If G is a finite graph and $|V(G)| = n \geq 3$ and some valency sequence of G is greater than or equal to P_n, then G has a Hamiltonian circuit.*

THEOREM 3. *If G is a finite graph and $|V(G)| = n$ and p is an integer and $2 \leq p \leq n/2$ and some valency sequence of G is greater than or equal to P_n^p, then G has a Hamiltonian circuit.*

THEOREM 4. *Every $\frac{1}{2}$-graph with n vertices has $[5(n + \varepsilon_n + 10)/224]$ edge-disjoint Hamiltonian circuits.*

Theorem 3 is equivalent to the statement that all sequences greater than or equal to P_n^p $(2 \leq p \leq n/2)$, and more generally all rearrangements of such sequences, are forcibly Hamiltonian. However, although in this way the theorem seems to catch a very large proportion of forcibly Hamiltonian sequences, it does not catch them all. For example, it can be shown that if n, p, r are positive integers and $p \le n/4$ then both the sequence

$$2p, 2p, \ldots, 2p, n - 2p - 1, n - 2p - 1, \ldots, n - 2p - 1,$$

with $2p$ terms equal to $2p$ and $n - 2p$ terms equal to $n - 2p - 1$, and the sequence

$$r + 1, r + 1, \ldots, r + 1, 3r, 3r, \ldots, 3r$$

with $2r$ terms equal to $r + 1$ and $2r$ terms equal to $3r$, are forcibly Hamiltonian. I suspect that it would be a feasible research problem, although one requiring much hard work, to characterize all forcibly Hamiltonian sequences, and possibly also all forcibly non-Hamiltonian sequences. However, since very many graphic sequences are in neither of these categories, this would still leave us far from solving Problem (I).

It is an open problem how far the number $[5(n + \varepsilon_n + 10)/224]$ in Theorem 4 can be improved, and a best possible result in this direction may have to await the discovery of more powerful proof techniques, since even the existing theorem requires a very long (although perhaps not exceptionally difficult) proof. At first, I had suspected that this number could be improved to $[(n + 1)/4]$ until a Hungarian undergraduate Mr. L. Babai produced a counter-example at the Balatonfüred Conference last year, and reflections suggested by his construction show that there are infinitely many values of n for which $[5(n + \varepsilon_n + 10)/224]$ certainly cannot be improved beyond $[(n + 4)/8]$: for details see [13]. However, since Mr. Babai's idea depends fairly heavily on using irregular graphs, one could still conjecture that every regular graph with n vertices of valency $k \ge n/2$ has $[k/2]$ edge-disjoint Hamiltonian circuits. For the very special case in which n is odd and $k = n - 1$, this has long been known to be true [1, pages 187-188].

Although Theorem 1 can fairly easily be seen to become false if $\frac{1}{2}$ is replaced by any smaller positive number, I had at one time thought that it might be feasible to characterize, say, the set of all $\frac{1}{3}$-graphs with Hamiltonian circuits, and then solve the same problem with $\frac{1}{3}$ replaced by some smaller number such as $\frac{1}{4}$, and so forth, thus taking Theorem 1 as a starting point for making further and further inroads into Problem (I). Subsequently I realized that, using a further trick for constructing one graph from another, one can prove that, for any fixed t such that $0 < t < \frac{1}{2}$, the problem of characterising t-graphs with Hamiltonian circuits is equivalent to Problem (I) itself. It therefore now seems unrealistic to hope for any complete solution to our problem for the class of $\frac{1}{3}$-graphs or $\frac{1}{4}$-graphs or any similar class of graphs. Nevertheless, I still feel that important insights might conceivably be achieved by studying such classes of graphs to find out <u>as much as we can</u> about which of them have Hamiltonian circuits and which have not. In fact, [15, Lemma 4], for which Dr. J.A. Bondy must share the credit, can be considered as a first step in this direction, although it was originally motivated as a step in the proof of Theorem 4.

Results very similar in kind to Theorems 1, 2 and 3 concerning the existence of Hamiltonian <u>arcs</u> in finite graphs can be deduced as corollaries to these three theorems using the connection between Problems (I) and (II) mentioned earlier. In [11], I proposed the problem of whether valency conditions on the lines of those in Theorem 2 might ensure the existence of one-way or two-way infinite Hamiltonian arcs in denumerable graphs of a suitable kind. The restriction "of a suitable kind" is easily seen to be necessary. To see this, let a finite subset X of $V(G)$ be called an r-<u>divisor</u> of a graph G (where r is a positive integer) if the graph $G - X$, obtained from G by removing the vertices in X and their incident edges, has at least r infinite components. Call G r-<u>divisible</u> if it has an r-divisor. Now let G be a denumerable graph with a two-way infinite Hamiltonian arc and let X be any finite subset of $V(G)$. If we suppose the given two-way infinite Hamiltonian arc of G to be the sequence (3) then there are integers m, n such that $m < n$ and

$X \subseteq \{\xi_m, \xi_{m+1}, \ldots, \xi_n\}$. Each of the one-way infinite arcs $\xi_{n+1}, \lambda_{n+2}, \xi_{n+2}, \lambda_{n+3}, \ldots$ and $\xi_{m-1}, \lambda_{m-1}, \xi_{m-2}, \lambda_{m-2}, \ldots$ is an arc in $G - X$ and so each of them must be contained in an infinite component of $G - X$. Since the vertices of any other component of $G - X$ must belong to the finite set $\{\xi_m, \xi_{m+1}, \ldots, \xi_n\}$, it follows that $G - X$ can have no further infinite components. Hence $G - X$ has at most 2 infinite components and consequently X cannot be a 3-divisor of G . This shows that a graph with a two-way infinite Hamiltonian arc cannot be 3-divisible, and similar considerations show that a graph with a one-way infinite Hamiltonian arc cannot be 2-divisible. Thus no hypotheses on the valencies of the vertices could ensure the existence of a one-way infinite Hamiltonian arc in a 2-divisible denumerable graph or a two-way infinite Hamiltonian arc in a 3-divisible denumerable graph.

It is easily seen that the graph of Figure 2 is 3-divisible, and, as this

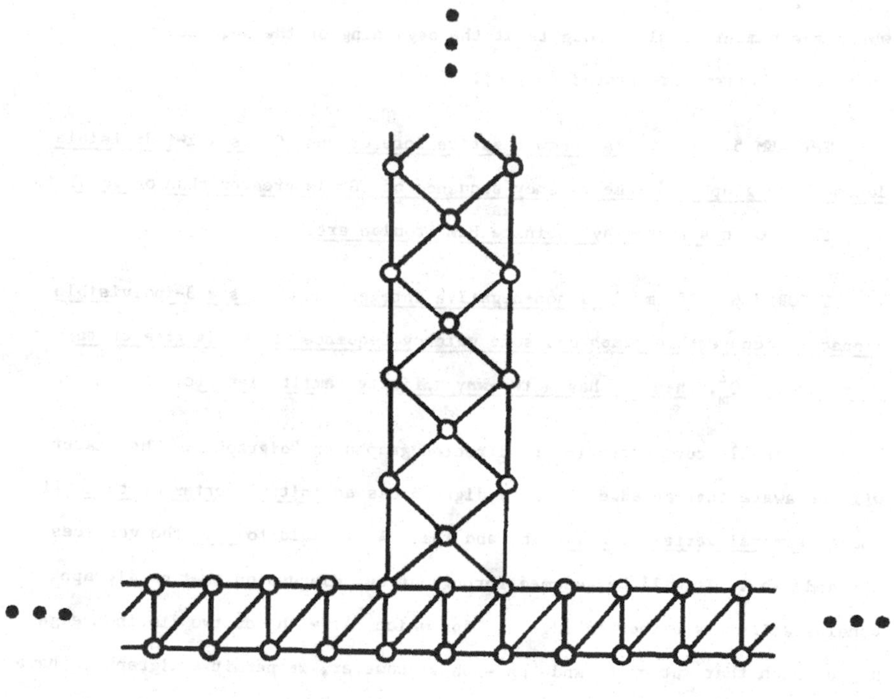

FIGURE 2

example suggests, one can intuitively picture an r-divisible graph as one which "has at least r infinite wings branching out of a finite centre". Thus the preceding paragraph indicates, intuitively, that if we wish to have Hamiltonian arcs in denumerable graphs we must restrict ourselves to graphs which "do not have too many infinite wings branching out of a finite centre". Restrictions of the kind which say, roughly speaking, that a graph "may not have too many infinite wings branching out of a finite centre" have already been found to be important in connection with the classical theorem of Erdős, Gallai and Vázsonyi characterising denumerable graphs with Euler paths [4; 17, Section 3.2] and in other work on problems of this general nature concerning denumerable graphs [9, 10, 11, 14, 20, 21, 22].

If m, q are non-negative integers, let Q_m^q denote the infinite sequence

$$\aleph_0, \aleph_0, \aleph_0, \ldots, \aleph_0, q, q+1, q+2, q+3, \ldots$$

where the number of aleph-noughts at the beginning of the sequence is m. The following theorems are proved in [16].

THEOREM 5. *If m is a non-negative integer and G is a 2-indivisible denumerable graph and some valency sequence of G is greater than or equal to Q_m^1, then G has a one-way infinite Hamiltonian arc.*

THEOREM 6. *If m is a non-negative integer and G is a 3-indivisible connected denumerable graph and some valency sequence of G is greater than or equal to Q_m^2, then G has a two-way infinite Hamiltonian arc.*

We finally turn attention to directed graphs or "digraphs". The reader will be aware that an edge λ of a digraph has an *initial vertex* or *tail* λt and a *terminal vertex* or *head* λh and that λ is said to *join* the vertices λt and λh. It will be assumed throughout our discussion that no digraph under consideration has either an edge λ for which $\lambda t = \lambda h$ or two distinct edges μ, υ such that $\mu t = \upsilon t$ and $\mu h = \upsilon h$. However, we permit a digraph to have distinct edges μ, υ such that $\mu t = \upsilon h$ and $\mu h = \upsilon t$. Only finite digraphs will be considered; I am not aware of any known results about infinite digraphs related to the theme of this lecture.

If ξ is a vertex of a digraph, then the <u>outvalency</u> $v_{out}(\xi)$ of ξ is the number of edges with tail ξ and the <u>invalency</u> $v_{in}(\xi)$ of ξ is the number of edges with head ξ. A <u>ditrack</u> in a finite digraph D is a sequence

$$\xi_0, \lambda_1, \xi_1, \lambda_2, \xi_2, \lambda_3, \ldots, \lambda_n, \xi_n \qquad (4)$$

where the ξ_i are vertices of D, the λ_j are edges of D and

$$\lambda_j t = \xi_{j-1}, \lambda_j h = \xi_j \qquad (j = 1, \ldots, n) .$$

A ditrack in which all terms are distinct in a <u>diarc</u>. The ditrack (4) is a <u>dicircuit</u> if $n \geq 2$, $\xi_0 = \xi_n$ and $\xi_1, \xi_2, \ldots, \xi_n$ are distinct. Various other items of notation and terminology used in connection with digraphs do not need definition since they are closely analogous to corresponding notation and terminology associated with graphs.

A digraph D is <u>symmetric</u> if, for every edge λ of D, there exists an edge μ of D such that $\lambda t = \mu h$ and $\lambda h = \mu t$. Thus two vertices of a symmetric digraph are necessarily joined either by two edges or by none. Any graph G can be "represented" in a fairly natural sense by a symmetric digraph D such that $V(D) = V(G)$ and, for each pair of distinct vertices ξ, η of G, ξ and η are joined by two edges of D if and only if they are joined by an edge of G: in these circumstances, we shall say that D is a <u>duplicate</u> of G.

The following analogues of Problems (I) and (II) suggest themselves. (I*) Which finite digraphs have Hamiltonian dicircuits? (II*) Which finite digraphs have Hamiltonian diarcs? By something not unlike our proof of the equivalence of Problems (I) and (II), it can be shown that (I*) and (II*) are equivalent problems. It is also obvious that a solution to (I*) would settle (I) because a finite graph G with three or more vertices has a Hamiltonian circuit if and only if a duplicate of G has a Hamiltonian dicircuit. Perhaps somewhat unexpectedly, one can show by a less obvious construction [8] that a solution to (I) would solve (I*); so that (I), (II), (I*) and (II*) can really all be regarded as different formulations of the same problem. This does not of course mean that partial results (such as Theorem 1) concerning (I) necessarily yield analogous partial results (such as Theorem 7 below) concerning (I*): on the contrary, it appears more appropriate in connection with results of this kind to think of graphs as equivalent to symmetric digraphs, so that results concerning Hamiltonian dicircuits in digraphs tend to be more general than corresponding results concerning Hamiltonian circuits in graphs, and consequently more difficult to prove.

As a first example of this, we mention the following counterpart of Theorem 1.

THEOREM 7. *If a finite digraph* D *has* $n\ (>0)$ *vertices and* $v_{out}(\xi) \geq n/2$ *and* $v_{in}(\xi) \geq n/2$ *for every* $\xi \in V(D)$, *then* D *has a Hamiltonian dicircuit.*

Although Theorem 7 can be proved by a fairly simple argument, this argument would probably be harder to think of (if one did not know it) than a proof of Theorem 1. In fact, Theorem 7 is a generalization of Theorem 1, since Theorem 1 is equivalent (using the above method of representing graphs by symmetric digraphs) to the special case of Theorem 7 in which D is symmetric.

A digraph D is *strongly connected* if, for every non-empty proper subset X of $V(D)$, there exists an edge λ of D such that $\lambda t \in X$ and $\lambda h \in V(D) \setminus X$. (Note that this implies that, for every non-empty proper subset Y of $V(D)$, there exists an edge μ of D such that $\mu h \in Y$ and $\mu t \in V(D)\setminus Y$, so that the definition is more symmetrical than it looks.) Since it can be trivially shown that any digraph satisfying the hypotheses of Theorem 7 is strongly connected, it follows that the following ingenious and attractive theorem of Ghouila-Houri [5] is a generalization of Theorem 7.

THEOREM 8. *If a finite strongly connected digraph* D *has* $n\ (>0)$ *vertices and* $v_{out}(\xi) + v_{in}(\xi) \geq n$ *for every* $\xi \in V(D)$, *then* D *has a Hamiltonian dicircuit*.

We remarked above that as corollaries to Theorems 1, 2 and 3 one can obtain similar theorems affirming the existence of Hamiltonian arcs in graphs satisfying certain conditions. A somewhat similar type of argument [5] enables us to deduce the following corollary from Theorem 8.

COROLLARY 8A. *If a finite digraph* D *has* n (> 0) *vertices and* $v_{out}(\xi) + v_{in}(\xi) \geq n - 1$ *for every* $\xi \in V(D)$, *then* D *has a Hamiltonian diarc*.

Letting $V(D) = \{\xi_1, \ldots, \xi_n\}$, the proof of Corollary 8A involves applying Theorem 8 to a digraph obtained from D by adding a new vertex ω and $2n$ new edges $\lambda_1, \ldots, \lambda_n, \mu_1, \ldots, \mu_n$ such that

$$\lambda_i t = \xi_i \ , \ \lambda_i h = \omega \ , \ \mu_i t = \omega \ , \ \mu_i h = \xi_i \quad (i = 1, \ldots, n) \ .$$

A well known special case of Corollary 8A is

COROLLARY 8B. *Every non-empty finite tournament has a Hamiltonian diarc*.

Corollary 8B can also be proved independently of Corollary 8A using an easy induction argument, or it can be viewed as a special consequence of the following theorem of Rédei [19; 7, Theorem 14], whose proof, although simplified somewhat by Szele [23], still entails some ingenuity.

THEOREM 9. *The number of Hamiltonian diarcs in any non-empty finite tournament is odd*.

By analogy with Theorem 4, one might conjecture that, if n is large, then a digraph satisfying the hypotheses of Theorem 7 might have a large number of edge-disjoint Hamiltonian dicircuits, but I have no idea how to prove or disprove this. Fairly recently, I worked through a considerable portion of what I think to be a potential proof that such a digraph has two edge-disjoint Hamiltonian dicircuits if $n \geq 5$, but even the proof of this rather weak conjecture involved analysing very many alternatives and threatened to reach a length of the order of 100 printed pages, with the result that I was hesitant to take the necessary time to complete it. Note that the truth of this last conjecture for symmetric digraphs follows immediately from Theorem 1, since a symmetric digraph with at least three vertices which satisfies the hypotheses of Theorem 7 is a duplicate of a graph which satisfies the hypotheses of Theorem 1 and which consequently has a Hamiltonian circuit c : to each edge in c there correspond two edges of the symmetric digraph, and the edges of the digraph corresponding to edges in c yield two edge-disjoint Hamiltonian dicircuits of the digraph.

Another unsolved problem which I personally find very intractable is whether Theorem 2 can be generalized to digraphs in something like the manner in which Theorem 7 generalizes Theorem 1. For example, one might conjecture the following generalization.

CONJECTURE. *If* D *is a finite digraph such that* $|V(D)| = n \geq 3$ *and some invalency sequence of* D *is greater than or equal to* P_n *and some outvalency sequence of* D *(not necessarily derived from the same ordering of the vertices) is greater than or equal to* P_n *, then* D *has a Hamiltonian dicircuit.*

Weaker versions of this conjecture could be obtained by requiring the invalency sequence and outvalency sequence to be derived from the same ordering of the vertices or even adding the additional hypothesis that $v_{in}(\xi) = v_{out}(\xi)$ for every $\xi \in V(D)$. Stronger versions could be obtained in something like the manner in which Theorem 3 strengthens Theorem 2 or the manner in which Theorem 8 strengthens Theorem 7, or by conjecturing that, if n is greater than or equal to about 5, then the hypotheses of the conjecture imply that D has two edge-disjoint Hamiltonian dicircuits (which follows from Theorem 2 when D is symmetric). All these variants are equally unproved and undisproved.

REFERENCES

[1] BERGE, C., The theory of graphs (1958), English translation by A. Doig (Methuen, London, 1962).

[2] BONDY, J.A., Properties of graphs with constraints on degrees, Studia Sci. Math. Hungar., to appear.

[3] DIRAC, G.A., Some theorems on abstract graphs, Proc. London Math. Soc. 2, 68-81 (1952).

[4] ERDOS, P., GRUNWALD, T. and VASZONYI, E., Uber Euler-Linien unendlicher Graphen, J. Math. and Phys. 17, 59-75 (1938).

[5] GHOUILA-HOURI, A., Une condition suffisante d'éxistence d'un circuit hamiltonien, C.R. Acad. Sci. Paris 251, 495-497 (1960).

[6] HARARY, F., Graph Theory (Addison-Wesley, Reading, Mass., 1969).

[7] MOON, J., Topics on tournaments (Holt, Reinhart and Winston, New York, 1968).

[8] NASH-WILLIAMS, C.St.J.A., Hamiltonian circuits in graphs and digraphs, The many facets of graph theory (Springer-Verlag, Berlin, Heidelberg and New York, 1969 [Lecture Notes in Mathematics, Number 110]), 237-243.

[9] NASH-WILLIAMS, C.St.J.A., Decomposition of graphs into two-way infinite paths, Canad. J. Math. 15, 479-485 (1963).

[10] NASH-WILLIAMS, C.St.J.A., Euler lines in infinite directed graphs, Canad. J. Math. 18, 692-714 (1966).

[11] NASH-WILLIAMS, C.St.J.A., Infinite graphs - a survey, J. Combinatorial Theory 3, 286-301 (1967).

[12] NASH-WILLIAMS, C.St.J.A., On Hamiltonian circuits in finite graphs, Proc. Amer. Math. Soc. 17, 466-467 (1966).

[13] NASH-WILLIAMS, C.St.J.A., Hamiltonian lines in graphs whose vertices have sufficiently large valencies, to appear in the Proceedings of the International Symposium on Combinatorics held at Balatonfüred, Hungary in August 1969.

[14] NASH-WILLIAMS, C.St.J.A., Euler lines in infinite directed graphs, in Theory of graphs, Proceedings of the Symposium at Tihany, Hungary in September 1966, ed. P. ERDÖS and G. KATONA (Hungarian Academy of Sciences, Budapest, 1968), 243-249.

[15] NASH-WILLIAMS, C.St.J.A., Edge-disjoint Hamiltonian circuits in graphs with vertices of large valency, to appear. (Preprints obtainable from author.)

[16] NASH-WILLIAMS, C.St.J.A., Hamiltonian lines in infinite graphs with few vertices of small valency, submitted to Aequationes Mathematicae.

[17] ORE, O., Theory of Graphs (American Mathematical Society, Providence, 1962 [American Mathematical Society Colloquium Publications Vol. XXXVIII]).

[18] POSA, L., A theorem concerning Hamiltonian lines, Magyar Tud. Akad. Mat. Kutató Int. Közl. 7, 225-226 (1962).

[19] REDEI, L., Ein kombinatorischer Satz, Acta Litt. Sci. Szeged 7, 39-43 (1934).

[20] ROTHSCHILD, B., The decomposition of graphs into a finite number of paths, Canad. J. Math. 17, 468-479 (1965).

[21] SEKANINA, M., On an ordering of the set of vertices of a connected graph, Spisy Přirod. Fak. Univ. Brno 412, 137-142 (1960).

[22] SEKANINA, M., On an ordering of the set of vertices of a graph, Časopis Pěst. Mat. 88, 265-282 (1963).

[23] SZELE, T., Kombinatorische Untersuchungen über gerichtete vollständige graphen, Publ. Math. Debrecen 13, 145-168 (1966).

[24] TUTTE, W.T., A short proof of the factor theorem for finite graphs, Canad. J. Math. 6, 347-352 (1954).

GRAPH-VALUED FUNCTIONS AND HAMILTONIAN GRAPHS

Wayne S. Petroelje, Western Michigan University, Kalamazoo, MI 49001
Curtiss E. Wall, Olivet College, Olivet, MI 49076

The total graph $T(G)$ of a graph G is that graph whose vertex set can be put in one-to-one correspondence with the set of vertices and edges of G such that two vertices of $T(G)$ are adjacent if and only if the corresponding elements of G are adjacent or incident. A second manner of describing the total graph can be given. In order to present this alternative description we define two additional concepts.

The square of a graph G, denoted G^2, is that graph having the same vertex set as G where two vertices are adjacent in G^2 if the distance between these two vertices in G is at most two. For a non-negative integer n, the nth subdivision graph $S_n(G)$ of a graph G is that graph obtained from G by replacing each edge uv by a u-v path of length $n + 1$. The graph $S_1(G)$ is referred to as the subdivision graph of G and is also denoted $S(G)$ while $S_0(G)$ is G itself. It is easily verified for every graph G that $[S(G)]^2 = T(G)$.

Behzad and Chartrand [1] have shown that if G is any non-trivial connected graph, then $T(T(G))$ is hamiltonian. From our earlier remarks this implies that $T([S(G)]^2) = T([S_1(G)]^2)$ is hamiltonian. This suggests a generalization. In order to prove this, we make use of the fact that the total graph of any graph having a spanning eulerian subgraph is hamiltonian (see [1]).

<u>Theorem</u>: For any non-negative integer n and nontrivial connected graph G, the graph $T([S_n(G)]^2)$ is hamiltonian.

<u>Proof</u>: That the result is true for $n = 1$ follows from the aforementioned theorem of Behzad and Chartrand. We now consider two cases.

<u>Case 1</u>. $n \geq 2$. It suffices to show that $[S_n(G)]^2$ contains a spanning eulerian subgraph. It follows immediately that if u is a vertex of G, then the degree of u in $[S_n(G)]^2$ is twice the degree of u in G; thus u has even degree in $[S_n(G)]^2$. Delete from $[S_n(G)]^2$ all edges incident only with vertices of $S_n(G)$ which are not vertices of G. In the resulting subgraph H of $[S_n(G)]^2$, each vertex which belongs to G has the same (even) degree as in $[S_n(G)]^2$, while every vertex of H which is not a vertex of G has degree two. Since H is connected, it follows that H is a spanning eulerian subgraph of $[S_n(G)]^2$.

<u>Case 2</u>. $n = 0$. If $G = K_2$, then $T([S_n(K_2)]^2) = K_3$, and the theorem is true; thus we assume G has order at least three. We prove here that G^2 contains a spanning eulerian subgraph. Certainly, if H is a spanning subgraph of G and H^2 contains a spanning eulerian subgraph then so does G^2. Hence in this case it is sufficient to prove that if T is any tree of order $p \geq 3$, then T^2 contains a spanning eulerian subgraph. We now proceed by induction on p.

For $p = 3$ and $p = 4$, the three trees T_i, $i = 1, 2, 3$, having these orders are shown in Figure 1, along with T_i^2 and a spanning eulerian subgraph H_i of T_i^2.

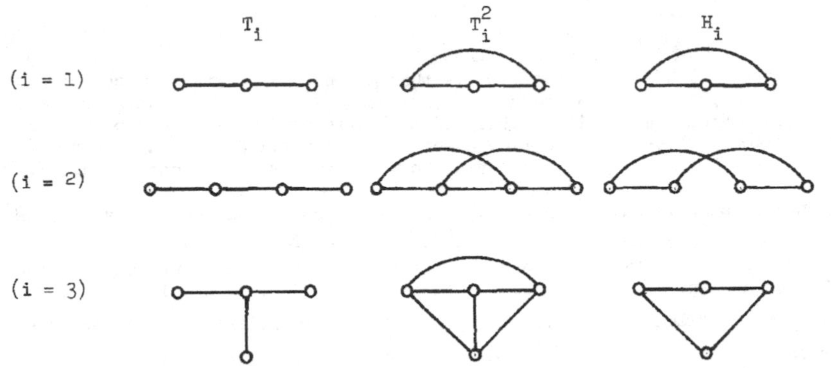

Figure 1

Let T be a tree with $p(\geq 5)$ vertices, and assume that the square of any tree having order at least three but less than p contains a spanning eulerian subgraph. Now every tree of order at least three contains a cutvertex w such that every vertex adjacent with w, with at most one exception, is an endvertex. Let v be such a cutvertex in T. We consider two subcases.

Subcase 1. $\deg v \geq 3$. Hence T contains two endvertices v_1 and v_2 adjacent with v. Since $F = T - v_1 - v_2$ is a tree (of order at least 3), F^2 contains a spanning eulerian subgraph F_1 by the induction hypothesis. Let $F_2 = \langle\{v, v_1, v_2\}\rangle$; then in T^2 the subgraph F_2^2 is a triangle (and is therefore eulerian). Since F_1 and F_2^2 are edge-disjoint eulerian subgraphs of T^2 having the vertex v in common, and each vertex of T is in F_1 or F_2^2, the graph T^2 contains a spanning eulerian subgraph.

Subcase 2. $\deg v = 2$. In this case v is adjacent to an endvertex u_1 and a cutvertex u_2 in T. Define $H = T - u_1 - v$ so that, by hypothesis, H^2 contains a spanning eulerian subgraph H_1. If we let $H_2 = \langle\{v, u_1, u_2\}\rangle$, then H_2^2 is a triangle in T^2. Because H_1 and H_2^2 are edge-disjoint eulerian subgraphs each with the vertex u_2, and since H_1 and H_2^2 span the vertex set of T, the graph T^2 contains a spanning eulerian subgraph.

This completes the proof.

The second case of the preceding result gives a proof of the following, which is interesting in its own right.

Corollary. For any nontrivial connected graph G, the graph $T(G^2)$ is hamiltonian.

The aforementioned corollary constitutes an improvement of the result that

$T(T(G))$ is hamiltonian for any non-trivial connected graph G. This result might then be considered a step towards verification of the Plummer Conjecture which states that the square of every 2-connected graph is hamiltonian. The Plummer Conjecture, if true, implies the corollary.

REFERENCE

1. M. Behzad and G. Chartrand, Total graphs and traversability, Proc. Edinburgh Math. Soc., 15 (1966), 117-120.

GENERALIZATIONS OF PRODUCT ISOMORPHISMS

S. M. Ulam, University of Colorado, Boulder, CO 80304

RESUME

Given two subsets A, B of the direct product $E^n = E \times E \times \ldots E$ of an abstract set E, the product isomorphism of $A \approx B$ was defined iff there exists a one-to-one transformation f(p) of E onto itself such that the transformation: $(p_1, p_2 \ldots p_n)$ goes into $(f(p_1), f(p_2), \ldots f(p_n))$ carries A onto B.

For (n = 2 this notion coincides with the isomorphism of graphs represented by sets of pairs: A, B.) Combinatorial and set theoretical properties of product isomorphisms were defined and studied in my master thesis in 1932.

In the present paper this notion is generalized in two different ways: We call 2 subsets A, B quasi-isomorphic if there exist one-to-one transformations $f_1, f_2 \ldots f_n$ of E onto itself, perhaps different from each other, such that the transformation $(p_1 \ldots p_n) \to (f_1(p_1) \ldots f_n(p_n))$, carries A onto B.

The second generalization is that of "equivalence by decomposition" (finite or countable). One calls two sets A, B equivalent by decomposition if $A = \bigcup A_i$, $A_i \cap A_j = 0$ for $i \neq j$, $B = \bigcup B_i$, $B_i \cap B_j$, $i \neq j$ and $A_i \approx B_i$ for all i (perhaps by different f's)

Some elementary set theoretical properties and enumeration results and problems are discussed. A fuller account of this work will appear elsewhere.

ON THE GENUS OF PRODUCTS OF GRAPHS

Arthur T. White, Western Michigan University, Kalamazoo, MI 49001

Given two graphs G and H with vertex sets $V(G)$ and $V(H)$ respectively, there are several graphical products which may be formed having the cartesian product $V(G) \times V(H)$ as the new vertex set. The purpose of this paper is to discuss the genus parameter, for four of these products, in the case where H is the graph K_2 and convenient conditions are placed upon G. A generalization, for one of the products discussed, will also be given.

The genus, $\gamma(G)$, of a graph G is the minimum genus among the genera of all closed orientable 2-manifolds M in which G can be imbedded. An imbedding of G in M is said to be minimal if M has genus $\gamma(G)$. The components of $M - G$ are called regions. If G has a minimal imbedding for which some region contains every vertex of G in its boundary, G is said to be outer-imbeddable. For example, K_5, K_6, $K_{3,3}$, and $K_{5,5}$, as well as all outer planar graphs, are outer-imbeddable. The first Betti number, $\beta(G)$, of a graph G is given by $\beta(G) = q - p + k$, where G has q edges, p vertices, and k components; $\beta(G)$ counts the number of independent cycles in G.

Let the two graphs G and H have vertex sets $V(G)$, $V(H)$ and edge sets $E(G)$, $E(H)$ respectively. Each of the four graphical products we are about to define has vertex set $V(G) \times V(H)$, so that the product in question will be determined by specifying its edge set. The following notation will simplify the definitions. For (u_i, v_j) and (u_k, v_m) in $V(G) \times V(H)$, the edge $(u_i,v_j)(u_k,v_m)$ in the product will be said to be of: type (1), if $u_i = u_k$ and $v_j v_m \in E(H)$; type (2), if $v_j = v_m$ and $u_i u_k \in E(G)$; type (3), if $u_i u_k \in E(G)$ with $v_j \neq v_m$ and $v_j v_m \notin E(H)$; type (4), if $u_i u_k \in E(G)$ and $v_j v_m \in E(H)$. The edge set of the cartesian product (see [7]), $G \times H$, consists of all edges of types (1) and (2). The edge set of the tensor product (or Kronecker product; see [9]), $G \otimes H$, consists of all edges of type (4), while the edge set of the strong cartesian product (see [7]), $G * H$, consists of all edges of types (1), (2), and (4). Finally, the edge set of the lexicographic product (or composition; see [8], [4]), $G(H)$, consists of all edges of types (1), (2), (3), and (4). Given graphs G_1 and G_2 with $V(G_1) = V(G_2)$ and $E(G_1) \cap E(G_2) = \emptyset$, the sum $G_1 + G_2$ has vertex set $V(G_1)$ and edge set $E(G_1) \cup E(G_2)$. It is apparent that, for arbitrary graphs G and H, the four products defined above are related as follows: $G * H = (G \times H) + (G \otimes H)$ and is a subgraph of $G(H)$. Furthermore, $G * H = G(H)$ if and only if H is complete.

For the special case $H = K_2$, we have the following results.

Theorem 1. (i) If G is bipartite, then $\gamma(G \otimes K_2) = 2\gamma(G)$.

(ii) If G is outer-imbeddable, then $\gamma(G \times K_2) = 2\gamma(G)$.

(iii) If G has no 3-cycles, then $\gamma(G * K_2) = \gamma(G(K_2)) = \beta(G)$.

Proofs: (i) For G bipartite, $G \otimes K_2$ is isomorphic to $2G$, two disjoint copies of G. By a result of Battle, Harary, Kodama, and Youngs [1], $\gamma(2G) = 2\gamma(G)$.

(ii) Since $2G$ is a subgraph of $G \times K_2$, $\gamma(G \times K_2) \geq \gamma(2G) = 2\gamma(G)$. Let G be minimally imbedded in M_1, with region R_1 containing every vertex of G in its boundary. Let M_2 be a second closed orientable 2-manifold of genus $\gamma(G)$, exterior to M_1, with G imbedded as a "mirror image" of the imbedding of G in M_1 . At this stage, all edges of type (2) in $G \times K_2$ have been accounted for. Now let D_i be an open disk interior to R_i, with simple closed boundary curve C_i, for $i = 1, 2$. Let T be a topological cylinder, with bases C_1 and C_2, such that $(M_1 \cup M_2) \cap T = C_1 \cup C_2$. Then $M = M_1 \cup M_2 \cup T$ is a closed orientable 2-manifold of genus $2\gamma(G)$; and all edges of type (1) in the cartesian product $G \times K_2$ can be added across the cylinder T, so that $\gamma(G \times K_2) \leq 2\gamma(G)$.

(iii) An argument involving the Euler formula shows that $\gamma(G * K_2) \geq \beta(G)$. A construction commencing (as in part (ii)) with two "mirror image" minimal imbeddings of G, adding tubes (cylinders) to accommodate edges of types (1) and (4), and resulting in a closed orientable 2-manifold of genus $\beta(G)$, can be developed to show that $\gamma(G * K_2) \leq \beta(G)$. We have already remarked that $G * K_2 = G(K_2)$.

Parts (i) and (iii) of Theorem 1 may be combined with the observation that $G \otimes K_2$ is a subgraph of $G(K_2)$ to show that, for G bipartite, $2\gamma(G) \leq \beta(G)$. This is hardly suprising, since it is equivalent to showing that the number of regions, for a minimal imbedding of G, is positive. In fact, Duke [3] has shown, for arbitrary graphs G of positive genus, that $2\gamma(G) \leq \beta(G) - 2$.

We note also that part (ii) of Theorem 1 generalizes the following result of Behzad and Mahmoodian [2]: if G is outerplanar , then $G \times K_2$ is planar. These authors also establish the converse of this result, in an argument employing forbidden subgraph characterizations of planar and outerplanar graphs. It is tempting to conjecture that the converse to Theorem 1, part (ii), also holds.

The constructive technique employed in the proof of Theorem 1, part (ii), has been extended in [10], wherein the genus is determined for several repeated cartesian products, each of which generalizes the graph associated with the 1-skeleton of the n-cube.

We next present a generalization of part (iii) of Theorem 1, wherein we regard the graph $G(K_2)$ as a composition, rather than as a strong cartesian product.

Theorem 2. Let G have p vertices of positive degree, q edges, k non-trivial components, and no 3-cycles. Let H have $2n$ $(n \geq 1)$ vertices and maximum degree at most one. Then $\gamma(G(H)) = k + n\,(nq - p)$.

Outline of proof: For each edge of G, there is a subgraph of $G(H)$ isomorphic to the complete bipartite graph $K_{2n,2n}$; the edges of these subgraphs are all those of types (2), (3), and (4) . We first construct an imbedding of $G(\overline{K}_{2n})$ (where $\overline{K}_{2n}$ is the complement of the complete graph K_{2n}), using the minimal imbedding given by Ringel [5] for $K_{2n,2n}$ q times. We then show that the type (1) edges can be added to the resulting closed orientable 2-manifold without increasing the genus and that the imbedding is minimal. It is then a simple matter to compute the genus. The details of this proof will be the subject of a subsequent paper.

We conclude with three corollaries to Theorem 2. Let C_s denote the cycle with s vertices. Ringel and Youngs have shown [6] that $\gamma(C_3(\overline{K}_m)) = \frac{(m-1)(m-2)}{2}$; this result combines with Theorem 2 to give the second corollary.

Corollary 2.1. Let G be a graph containing no 3-cycles. Then $\gamma(G(K_2)) = \gamma(G(\overline{K}_2)) = \beta(G)$.

Corollary 2.2.

$$\gamma(C_s(\overline{K}_{2n})) = \begin{cases} 1 + n(2n-3), & \text{if } s = 3 \\ 1 + ns(n-1), & \text{if } s \geq 4 \end{cases}$$

Corollary 2.3. Let H have $2n$ vertices $(n \geq 1)$ and maximum degree at most one. Then $\gamma(K_{r,s}(H)) = (nr - 1)(ns - 1)$.

REFERENCES

1. J. Battle, F. Harary, Y. Kodama, and J.W.T. Youngs, Additivity of the Genus of a Graph, Bull. Amer. Math. Soc. 68 (1962), 565-568.

2. M. Behzad and S.E. Mahmoodian, On Topological Invariants of the Product of Graphs, Canad. Math. Bull. 12 (1969), 157-166.

3. R. Duke, The Genus, Regional Number, and Betti Number of a Graph, Can. J. Math. 18 (1966), 817-822.

4. F. Harary, On the Group of the Composition of Two Graphs, Duke Math. J. 26 (1959), 29-34.

5. G. Ringel, Das Geschlecht des Vollständigen Paaren Graphen, Abh. Math. Sem. Univ. Hamburg 28 (1965), 139-150.

6. G. Ringel and J.W.T. Youngs, Das Geschlecht des Symmetrische Vollständige Drei-Farbaren Graphen, Comment. Math. Helv. (to appear).

7. G. Sabidussi, Graph Multiplication, Math. Z. 72 (1960), 446-457.

8. G. Sabidussi, The Lexicographic Product of Graphs, Duke Math. J. 28 (1961), 575-578.

9. P. Wiechsel, The Kronecker Product of Graphs, Proc. Amer. Math. Soc. 13 (1962), 47-52.

10. A.T. White, The Genus of Repeated Cartesian Products of Bipartite Graphs, Trans. Amer. Math. Soc. (to appear).

cture Notes in Physics

r erschienen/Already published

clusters crossing two cells within a window. There are two ways to form windows over the grid; we choose which according to an initial random bit. In the previous algorithm, clusters crossing two adjacent windows are not strictly forbidden but are discouraged in some sense.

In the new algorithm, the idea, roughly speaking, is to permit more clusters crossing windows. More specifically, call the grid point lying between two adjacent windows a *border*; generate a random bit for every border, where a 1 bit indicates an *open* border and a 0 bit indicates a *closed* border. Clusters crossing closed borders are still discouraged, but not clusters crossing open borders. (As it turns out, setting the probability of border opening/closing to 1/2 is indeed the best choice.)

The actual details of the algorithm are important and are carefully crafted. In the pseudocode below, $b(w, w')$ refers to the border indicator between windows w and w'. We say that a point *lies* in a cluster if inserting it to the cluster would not increase the length of the cluster, where the *length* of a cluster refers to the length of its smallest enclosing interval. We say that a point *fits* in a cluster if inserting it to the cluster would not cause the length to exceed 1.

RandBorder Algorithm: Partition the line into *windows* each of the form $[2i, 2i+2)$. With probability 1/2, shift all windows one unit to the right. For each two neighboring windows w and w' set $b(w, w')$ to a randomly drawn number from $\{0, 1\}$. For each new point p, find the window w containing p, and do the following:

1: **if** p fits in a cluster intersecting w **then**
2:     put p in the "closest" such cluster
3: **else if** p fits in a cluster u inside a neighboring window w' **then**
4:     **if** $b(w, w') = 1$ **then** put p in u
5:     **else if** w (completely) contains at least 1 cluster and
        w' (completely) contains at least 2 clusters
6:     **then** put p in u
7: **if** p is not put in any cluster **then** open a new cluster for p

---

Thus, a cluster is allowed to cross the boundary of two grid cells within a window freely, but it can cross the boundary of two adjacent windows only in two exceptional cases: when the corresponding border indicator is set to 1, or when the carefully specified condition in Line 5 arises (this condition is slightly different from the one in the previous algorithm). We will see the rationale for this condition during the analysis.

To see what the "closeness" exactly means in Line 2, we define the following two preference rules:

- Rule I. If p lies in a cluster u, then u is the closest cluster to p.
- Rule II. If p lies in a cell c, then any cluster intersecting c is closer to p than any cluster contained in a neighboring cell.

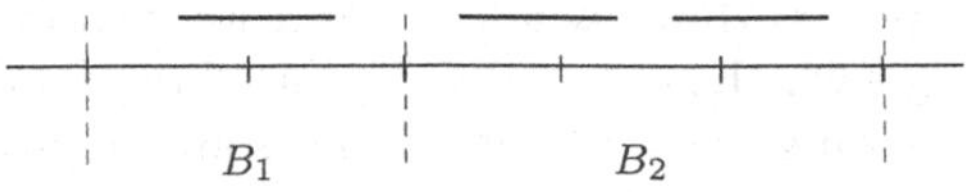

Fig. 1. Two blocks of sizes 2 and 3

The first preference rule prevents clusters from overlapping each other, and the second rule prevents clusters from unnecessarily crossing the boundary of two neighboring cells. The above preference rules and exceptional cases will be vital to the analysis.

Note that the random bits used for the border indicators can be easily generated on the fly as new borders are created.

3 Preliminaries for the Analysis

To prepare for the analysis, we first state a few definitions (borrowed from [1]).

Let σ be the input sequence. We denote by $\mathtt{opt}(\sigma)$ the optimal offline solution obtained by the following greedy algorithm: sort all points in σ from left to right; cover the leftmost point p and all points within unit distance to it by a unit interval started at p; and repeat the procedure for the remaining uncovered points. Obviously, the unit intervals obtained by this algorithm are disjoint.

We refer to a cluster as a *crossing cluster* if it intersects two adjacent grid cells, or as a *whole cluster* if it is contained completely in a grid cell.

For any real interval x (e.g., a grid cell or a group of consecutive cells), the *cost* of x denoted by $\mu(x)$ is defined to be the number of whole clusters contained in x plus half the number of clusters crossing the boundaries of x, in the solution produced by the RandBorder algorithm. We note that μ is additive, i.e., for two adjacent intervals x and y, $\mu(x \cup y) = \mu(x) + \mu(y)$.

A set of k consecutive grid cells containing $k-1$ intervals from $\mathtt{opt}(\sigma)$ is called a *block* of size k (see Fig. 1). We define $\rho(k)$ to be the expected competitive ratio of the RandBorder algorithm within a block of size k. In other words, $\rho(k)$ upper-bounds the expected value of $\mu(B)/(k-1)$ over all blocks B of size k.

In the following, a list of objects (e.g., grid cells or clusters) denoted by $\langle x_i, \ldots, x_j \rangle$ is always implicitly assumed to be ordered from left to right on the line. Moreover, $p_1 \ll p_2$ denotes the fact that point p_1 arrives before point p_2 in the input sequence.

We now establish some observations concerning the behavior of the RandBorder algorithm. Observations 1(ii) and (iii) are basically from [1] and have similar proofs (which are reproduced here for completeness' sake since the algorithm has changed); the other observations and subsequent lemmas are new and will be used multiple times in the analysis in the next section.

Observation 1.

(i) *Any interval in* $\mathtt{opt}(\sigma)$ *that does not cross a closed border can (completely) contain at most one whole cluster.*

(ii) *Any grid cell c can contain at most one whole cluster. Thus, we always have $\mu(c) \leq 1 + \frac{1}{2} + \frac{1}{2} = 2$.*
(iii) *If a grid cell c intersects a crossing cluster u_1 and a whole cluster u_2, then u_2 must be opened after u_1 has been opened, and after u_1 has become a crossing cluster.*

Proof. (i) Let u_1 and u_2 be two whole clusters contained in the said interval and suppose that u_1 is opened before u_2. Then all points of u_2 would be assigned to u_1, because Lines 2 and 4 precede Line 7. (ii) holds by the same argument, because Line 2 precedes Line 7.

For (iii), let p_1 be the first point of u_1 in c and p_1' be the first point of u_1 in a cell adjacent to c. Let p_2 be the first point of u_2. Among these three points, p_1 cannot be the last to arrive: otherwise, p_1 would be assigned to the whole cluster u_2 instead of u_1, because of Rule II. Furthermore, p_1' cannot be the last to arrive: otherwise, p_1 would be assigned to u_2 instead. So, p_2 must be the last to arrive. □

Observation 2. *Let u_1 be a whole cluster contained in a grid cell c, and let u_2 and u_3 be two clusters crossing the boundaries of c. Then*

(i) *u_1 and u_2 cannot be entirely contained in the same interval from $\mathtt{opt}(\sigma)$.*
(ii) *there are no two intervals I_1 and I_2 in $\mathtt{opt}(\sigma)$ such that $u_1 \cup u_2 \cup u_3 \subseteq I_1 \cup I_2$.*

Proof. (i) Suppose by way of contradiction that u_1 and u_2 are entirely contained in an interval I from $\mathtt{opt}(\sigma)$. Then by Observation 1(iii), u_1 is opened after u_2 has become a crossing cluster, but then the points of u_1 would be assigned to u_2 instead: a contradiction.

(ii) Suppose that $u_1 \cup u_2 \cup u_3 \subseteq I_1 \cup I_2$, where I_1 and I_2 are the two intervals from $\mathtt{opt}(\sigma)$ intersecting c. We now proceed as in part (i). By Observation 1(iii), u_1 is opened after u_2 and u_3 have become crossing clusters, but then the points of u_1 would be assigned to u_2 or u_3 instead: a contradiction. □

Lemma 1. *Let $B = \langle c_1, \ldots, c_k \rangle$ be a block of size $k \geq 2$, and S be the set of all odd-indexed (or even-indexed) cells in B. Then there exists a cell $c \in S$ such that $\mu(c) < 2$.*

Proof. Let $\langle I_1, \ldots, I_{k-1} \rangle$ be the $k-1$ intervals from $\mathtt{opt}(\sigma)$ in B, where each interval I_i intersects two cells c_i and c_{i+1} $(1 \leq i \leq k-1)$. Let O represent the set of all odd integers between 1 and k. We first prove the lemma for the odd-indexed cells.

Suppose by way of contradiction that for each $i \in O$, $\mu(c_i) = 2$. It means that for each $i \in O$, c_i intersects three clusters $\langle u_i^\ell, u_i, u_i^r \rangle$, where u_i is a whole cluster, and u_i^ℓ and u_i^r are two crossing clusters. We prove inductively that for each $i \in O$, $u_i \cap I_i \neq \emptyset$ and $u_i^r \cap I_{i+1} \neq \emptyset$.

Base Case: $u_1 \cap I_1 \neq \emptyset$ and $u_1^r \cap I_2 \neq \emptyset$.

The first part is trivial, because c_1 intersects just I_1, and hence, $u_1 \subseteq I_1$. The second part is implied by Observation 2(i), because u_1 and u_1^r cannot be entirely contained in I_1.

Inductive Step: $u_i \cap I_i \neq \emptyset \wedge u_i^r \cap I_{i+1} \neq \emptyset \Rightarrow u_{i+2} \cap I_{i+2} \neq \emptyset \wedge u_{i+2}^r \cap I_{i+3} \neq \emptyset$. Suppose by contradiction that $u_{i+2} \cap I_{i+2} = \emptyset$. Therefore, u_{i+2} must be entirely contained in I_{i+1}. On the other hand, $u_i^r \cap I_{i+1} \neq \emptyset$ implies that u_{i+2}^ℓ is entirely contained in I_{i+1}. But this is a contradiction, because u_{i+2} and u_{i+2}^ℓ are contained in the same interval, which is impossible by Observation 2(i). Now, suppose that $u_{i+2}^r \cap I_{i+3} = \emptyset$. Since $u_i^r \cap I_{i+1} \neq \emptyset$, and clusters do not overlap, u_{i+2}^ℓ, u_{i+2}, and u_{i+2}^r should be contained in $I_{i+1} \cup I_{i+2}$, which is impossible by Observation 2(ii).

Repeating the inductive step zero or more times, we end up at either $i = k$ or $i = k - 1$. If $i = k$, then $u_k \cap I_k \neq \emptyset$ which is a contradiction, because there is no I_k. If $i = k - 1$, then $u_{k-1}^r \cap I_k \neq \emptyset$ which is again a contradiction, because we have no I_k.

Both cases lead to contradiction. It means that there exists some $i \in O$ such that $\mu(c_i) < 2$. The proof for the even-indexed cells is similar. The only difference is that we need to prove the base case for $i = 2$, which is easy to get by Observations 2(i) and 2(ii). □

Lemma 2. *Let B be a block of size $k \geq 2$.*

(i) $\mu(B) \leq 2k - 1$.
(ii) *If all borders strictly inside B are open, then $\mu(B) \leq 2(k - 1)$.*

Proof. (i) is a direct corollary of Lemma 1, because there are at least two cells in B (one odd-indexed and one even-indexed) that have cost at most 3/2, and the other cells have cost at most 2.

(ii) is immediate from the fact that each block of size $k \geq 2$ contains exactly $k - 1$ intervals from $\texttt{opt}(\sigma)$, and that each of these $k - 1$ intervals has cost at most 2 by Observation 1(i). □

4 The Analysis

We are now ready to analyze the expected competitive ratio of our algorithm within a block of size $k \geq 2$.

Theorem 1. $\rho(2) = 27/16$.

Proof. Consider a block B of size 2, consisting of two cells $\langle c_1, c_2 \rangle$ (see Fig. 2). Let I be the single unit interval in B in $\texttt{opt}(\sigma)$. There are two possibilities.

Case 1: B falls completely in one window w. Let $\langle b_1, b_2 \rangle$ be the two border indicators at the boundaries of w. Let p_0 be the first point to arrive in I. W.l.o.g., assume p_0 is in c_2 (the other case is symmetric). We consider four subcases.

- Subcase 1.1: $\langle b_1, b_2 \rangle = \langle 0, 0 \rangle$. Here, both boundaries of B are closed. Thus, after a cluster u has been opened for p_0 (by Line 7), all subsequent points in I are put in the same cluster u. Note that the condition in Line 5 prevents points from the neighboring windows to join u and make crossing clusters. So, u is the only cluster in B, and hence, $\mu(B) = 1$.

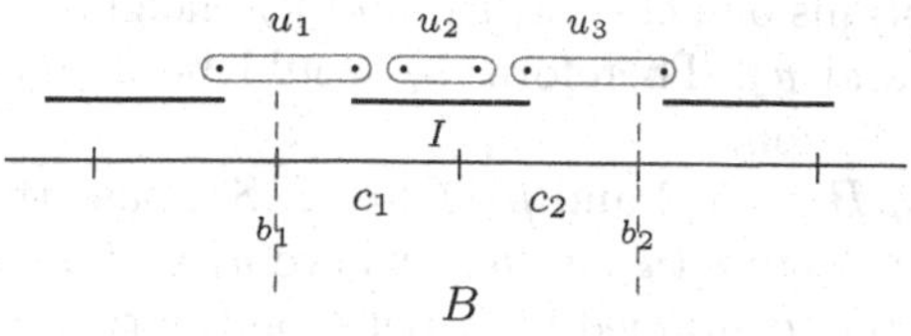

Fig. 2. Illustration of Subcase 1.3

- SUBCASE 1.2: $\langle b_1, b_2 \rangle = \langle 1, 0 \rangle$. When p_0 arrives, a new cluster u is opened, since p_0 is in c_2, the right border is closed, and w contains < 1 cluster at the time so that the condition in Line 5 fails. Again, all subsequent points in I are put in the same cluster, and points from the neighboring windows cannot join u and make crossing clusters. Hence, $\mu(B) = 1$.
- SUBCASE 1.3: $\langle b_1, b_2 \rangle = \langle 0, 1 \rangle$. We show that $\mu(B) < 2$. Suppose by contradiction that $\mu(B) = 2$. By Observation 1(i), I cannot contain two clusters entirely. Therefore, the only way to get $\mu(B) = 2$ is that I intersects three clusters $\langle u_1, u_2, u_3 \rangle$ (from left to right, as always), where u_1 and u_3 are crossing clusters, and u_2 is entirely contained in I (see Fig. 2). By a similar argument as in the proof of Observation 1(iii), u_2 is opened after u_1 and u_3 have become crossing clusters. Let p_1 be the first point of u_1 in w, and p_2 be the first point of u_1 in the neighboring window. We have two scenarios:
  - SUBSUBCASE 1.3.1: $p_1 \ll p_2$. In this case, cluster u_1 is opened for p_1. But p_2 cannot be put in u_1, because upon arrival of p_2, w contains < 2 clusters, and thus, the condition in line 5 does not hold.
  - SUBSUBCASE 1.3.2: $p_2 \ll p_1$. Here, cluster u_1 is opened for p_2. But p_1 cannot be put in u_1, because upon arrival of p_1, w contains < 1 cluster, and hence, the condition in line 5 does not hold.

  Both scenarios leads to contradiction. Therefore, $\mu(B) \leq 3/2$.
- SUBCASE 1.4: $\langle b_1, b_2 \rangle = \langle 1, 1 \rangle$. Here, Lemma 2(ii) implies that $\mu(B) \leq 2$.

Since each of the four subcases occurs with probability 1/4, we conclude that the expected value of $\mu(B)$ in Case 1 is at most $\frac{1}{4}(1 + 1 + \frac{3}{2} + 2) = \frac{11}{8}$.

CASE 2: B is split between two neighboring windows. Let b be the single border indicator inside B. Let $\mu_0(B)$ and $\mu_1(B)$ represent the value of $\mu(B)$ for the case that b is set to 0 and 1, respectively. It is clear by Lemma 2(ii) that $\mu_1(B) \leq 2$. We rule out two possibilities:

- SUBCASE 2.1: $\mu_0(B) = 3$. Since I cannot contain both a whole cluster and a crossing cluster by Observation 2(i), the only possible scenario is that c_1 intersects two clusters $\langle u_1, u_2 \rangle$, and c_2 intersects two clusters $\langle u_3, u_4 \rangle$, where u_1 and u_4 are crossing clusters, and u_2 and u_3 are whole clusters. Let p_1 be the first point in u_2 and p_2 be the first point in u_3. Suppose w.l.o.g. that $p_1 \ll p_2$. By Observation 1(iii), p_1 arrives after u_1 has been opened, and p_2 arrives after u_4 has been opened. But when p_2 arrives, the window containing it contains one cluster, u_4, and the neighboring window contains two clusters u_1 and u_2. Therefore, p_2 would be assigned to u_2 by Line 5 instead: a contradiction.

- SUBCASE 2.2: $\mu_0(B) = 5/2$ and $\mu_1(B) = 2$. Suppose that $\mu_1(B) = 2$. Then I intersects three clusters $\langle u_1, u_2, u_3 \rangle$, where u_1 and u_3 are crossing clusters, and u_2 is completely contained in I. Let t be the time at which u_1 becomes a crossing cluster, and let $\sigma(t)$ be the subset of input points coming up to time t. By a similar argument as in the proof of Observation 1(iii), any point in $I \cap c_1$ not contained in u_1 arrives after time t. Therefore, upon receiving the input sequence $\sigma(t)$, u_1 becomes a crossing cluster no matter whether the border between c_1 and c_2 is open or closed. Using the same argument we conclude that u_3 becomes a crossing cluster regardless of the value of b. Now consider the case where $b = 0$. Since both u_1 and u_3 remain crossing clusters, $\mu_0(B)$ must be an integer (1, 2, or 3) and cannot equal 5/2.

Ruling out these two subcases, we have $\mu_0(B) + \mu_1(B) \leq 4$ in all remaining subcases, and therefore, the expected value of $\mu(B)$ in this case is at most 2.

Since each of Cases 1 and 2 occurs with probability 1/2, we conclude that $\rho(2) \leq \frac{1}{2}(\frac{11}{8}) + \frac{1}{2}(2) = \frac{27}{16}$. (This bound is tight: to see this just consider the block $B = [2, 4)$, and the sequence of 8 points $\langle 1.5, 2.5, 0.5, 3.5, 4.5, 2.7, 3.2, 5.5 \rangle$ for which $E[\mu(B)] = \frac{27}{16}$.) □

Theorem 2. $\rho(3) \leq 17/8$.

Proof. Consider a block B of size 3, consisting of cells $\langle c_1, c_2, c_3 \rangle$, and let b be the single border indicator strictly inside B. We assume w.l.o.g. that c_1 and c_2 fall in the same window (the other scenario is symmetric). We consider two cases.

- CASE 1: $b = 0$. We rule out the following possibilities.
  - SUBCASE 1.1: $\mu(c_2) = 2$. Impossible by Lemma 1.
  - SUBCASE 1.2: $\mu(c_1) = \mu(c_3) = 2$. Impossible by Lemma 1.
  - SUBCASE 1.3: $\mu(c_1) = 2$ and $\mu(c_2) = \mu(c_3) = 3/2$. Here, B intersects six clusters $\langle u_1, \ldots, u_6 \rangle$, where u_1, u_3, u_6 are crossing clusters and u_2, u_4, u_5 are whole clusters. Let $\langle I_1, I_2 \rangle$ be the two unit intervals in B in $\mathtt{opt}(\sigma)$. By Observation 2(i), u_3 cannot be entirely contained in I_1. This implies that $u_4 \cup u_5 \subset I_2$. Now suppose w.l.o.g. that u_4 is opened after u_5. By Observation 1(iii), u_4 is the last to be opened after u_3, u_5, u_6. Consider any point p in u_4. Upon arrival of p, the window containing p contains at least one cluster, u_3, and the neighboring window contains two clusters u_5 and u_6. Therefore, by the condition in Line 5, the algorithm would assign p to u_5 instead of u_4, which is a contradiction.
  - SUBCASE 1.4: $\mu(c_1) = \mu(c_2) = 3/2$ and $\mu(c_3) = 2$. Similarly impossible.

  In all remaining subcases, $\mu(B)$ is at most $2 + \frac{3}{2} + 1 = \frac{9}{2}$ or $\frac{3}{2} + \frac{3}{2} + \frac{3}{2} = \frac{9}{2}$.
- CASE 2: $b = 1$. Here, Lemma 2(ii) implies that $\mu(B) \leq 4$.

Each of Cases 1 and 2 occurs with probability 1/2, therefore $\rho(3) \leq \frac{1}{2}(4 + \frac{9}{2})/2 = 17/8$. □

Theorem 3. $\rho(4) \leq 53/24$.

Proof. Consider a block B of size 4. We consider two easy cases.

- CASE 1: B falls completely in two windows. Let b be the single border indicator strictly inside B. Now, if $b = 1$, $\mu(B) \leq 6$ by Lemma 2(ii), otherwise, $\mu(B) \leq 7$ by Lemma 2(i). Therefore, the expected cost in this case is at most $\frac{1}{2}(6+7) = \frac{13}{2}$.
- CASE 2: B is split between three consecutive windows. Let $\langle b_1, b_2 \rangle$ be the two border indicators inside B. For the subcase where $\langle b_1, b_2 \rangle = \langle 1, 1 \rangle$ the cost is at most 6 by Lemma 2(ii), and for the remaining 3 subcases, the cost of B is at most 7 by Lemma 2(i). Thus, the expected cost in this case is at most $\frac{1}{4}(6) + \frac{3}{4}(7) = \frac{27}{4}$.

Since each of Cases 1 and 2 occurs with probability exactly 1/2, we conclude that $\rho(4) \leq \frac{1}{2}(\frac{13}{2} + \frac{27}{4})/3 = \frac{53}{24}$. □

Theorem 4. $\rho(k) \leq (2k-1)/(k-1)$ *for all* $k \geq 5$.

Proof. This is a direct implication of Lemma 2(i). □

5 The Combined Algorithm

The RandBorder algorithm as shown in the previous section has competitive ratio greater than 2 on blocks of size three and more. To overcome this deficiency, we need to combine RandBorder with another algorithm that works well for larger block sizes. A good candidate for this is the naïve grid algorithm:

Grid Algorithm: For each new point p, if the grid cell containing p contains a cluster, then put p in that cluster, else open a new cluster for p.

It is easy to verify that the Grid algorithm uses exactly k clusters on a block of size k. Therefore, the competitive ratio of this algorithm within a block of size k is $k/(k-1)$. We can now randomly combine the RandBorder algorithm with the Grid algorithm to obtain an expected competitive ratio strictly less than 2.

Combined Algorithm: With probability 8/15 run RandBorder, and with probability 7/15 run Grid.

Theorem 5. *The competitive ratio of the Combined algorithm is at most* 11/6 *against oblivious adversaries.*

Proof. The competitive ratios of RandBorder and Grid within blocks of size 2 are 27/16 and 2, respectively. Therefore, the expected competitive ratio of the Combined algorithm is $\frac{8}{15}(\frac{27}{16}) + \frac{7}{15}(2) = \frac{11}{6}$ within a block of size 2. For larger block sizes, the expected competitive ratio of Combined is always at most 11/6, as shown in Table 1. By summing over all blocks and exploiting the additivity of our cost function $\mu(\cdot)$, we see that the expected total cost of the solution produced by Combined is at most 11/6 times the size of $\texttt{opt}(\sigma)$ for every input sequence σ. □

Table 1. The competitive ratio of the algorithms within a block

Block Size	Grid	RandBorder	Combined
2	2	27/16	11/6
3	3/2	$\leq \frac{17}{8}$	$\leq 11/6$
4	4/3	$\leq \frac{53}{24}$	$\leq 9/5$
$k \geq 5$	$\frac{k}{k-1}$	$\leq \frac{2k-1}{k-1}$	$\leq \frac{23k-8}{15(k-1)}$

Remarks. Currently only a 4/3 randomized lower bound and a 3/2 deterministic lower bound are known for the one-dimensional problem [1]. Also, as a corollary to Theorem 5, we immediately get an upper bound of $(\frac{11}{12}) \cdot 2^d$ for the d-dimensional unit clustering problem under the L_∞ metric [1].

References

1. Chan, T.M., Zarrabi-Zadeh, H.: A randomized algorithm for online unit clustering. In: Erlebach, T., Kaklamanis, C. (eds.) WAOA 2006. LNCS, vol. 4368, pp. 121–131. Springer, Heidelberg (2007)
2. Charikar, M., Chekuri, C., Feder, T., Motwani, R.: Incremental clustering and dynamic information retrieval. SIAM J. Comput. 33(6), 1417–1440 (2004)
3. Cormen, T.H., Leiserson, C.E., Rivest, R.L., Stein, C.: Introduction to Algorithms, 2nd edn. MIT Press, Cambridge (2001)
4. Fotakis, D.: Incremental algorithms for facility location and k-median. In: Albers, S., Radzik, T. (eds.) ESA 2004. LNCS, vol. 3221, pp. 347–358. Springer, Heidelberg (2004)
5. Fowler, R.J., Paterson, M.S., Tanimoto, S.L.: Optimal packing and covering in the plane are NP-complete. Inform. Process. Lett. 12(3), 133–137 (1981)
6. Gonzalez, T.: Covering a set of points in multidimensional space. Inform. Process. Lett. 40, 181–188 (1991)
7. Gyárfás, A., Lehel, J.: On-line and First-Fit colorings of graphs. J. Graph Theory 12, 217–227 (1988)
8. Hochbaum, D.S., Maass, W.: Approximation schemes for covering and packing problems in image processing and VLSI. J. ACM 32, 130–136 (1985)
9. Kierstead, H.A., Qin, J.: Coloring interval graphs with First-Fit. SIAM J. Discrete Math. 8, 47–57 (1995)

10. Meyerson, A.: Online facility location. In: Proc. 42nd IEEE Sympos. Found. Comput. Sci., pp. 426–433 (2001)
11. Nielsen, F.: Fast stabbing of boxes in high dimensions. Theoret. Comput. Sci. 246, 53–72 (2000)
12. Tanimoto, S.L., Fowler, R.J.: Covering image subsets with patches. In: Proc. 5th International Conf. on Pattern Recognition, pp. 835–839 (1980)

Linear Time Algorithms for Finding a Dominating Set of Fixed Size in Degenerated Graphs

Noga Alon[1,*] and Shai Gutner[2,**]

[1] Schools of Mathematics and Computer Science, Tel-Aviv University, Tel-Aviv, 69978, Israel
noga@math.tau.ac.il.

[2] School of Computer Science, Tel-Aviv University, Tel-Aviv, 69978, Israel
gutner@tau.ac.il.

Abstract. There is substantial literature dealing with fixed parameter algorithms for the dominating set problem on various families of graphs. In this paper, we give a $k^{O(dk)}n$ time algorithm for finding a dominating set of size at most k in a d-degenerated graph with n vertices. This proves that the dominating set problem is fixed-parameter tractable for degenerated graphs. For graphs that do not contain K_h as a topological minor, we give an improved algorithm for the problem with running time $(O(h))^{hk}n$. For graphs which are K_h-minor-free, the running time is further reduced to $(O(\log h))^{hk/2}n$. Fixed-parameter tractable algorithms that are linear in the number of vertices of the graph were previously known only for planar graphs.

For the families of graphs discussed above, the problem of finding an induced cycle of a given length is also addressed. For every fixed H and k, we show that if an H-minor-free graph G with n vertices contains an induced cycle of size k, then such a cycle can be found in $O(n)$ expected time as well as in $O(n \log n)$ worst-case time. Some results are stated concerning the (im)possibility of establishing linear time algorithms for the more general family of degenerated graphs.

Keywords: H-minor-free graphs, degenerated graphs, dominating set problem, finding an induced cycle, fixed-parameter tractable algorithms.

1 Introduction

This paper deals with fixed-parameter algorithms for degenerated graphs. The degeneracy $d(G)$ of an undirected graph $G = (V, E)$ is the smallest number d for which there exists an acyclic orientation of G in which all the outdegrees are at most d. Many interesting families of graphs are degenerated (have bounded

* Research supported in part by a grant from the Israel Science Foundation, and by the Hermann Minkowski Minerva Center for Geometry at Tel Aviv University.

** This paper forms part of a Ph.D. thesis written by the author under the supervision of Prof. N. Alon and Prof. Y. Azar in Tel Aviv University.

G. Lin (Ed.): COCOON 2007, LNCS 4598, pp. 394–405, 2007.

degeneracy). For example, graphs embeddable on some fixed surface, degree-bounded graphs, graphs of bounded tree-width, and non-trivial minor-closed families of graphs.

There is an extensive literature dealing with fixed-parameter algorithms for the dominating set problem on various families of graphs. Our main result is a linear time algorithm for finding a dominating set of fixed size in degenerated graphs. This is the most general class of graphs for which fixed-parameter tractability for this problem has been established. To the best of our knowledge, linear time algorithms for the dominating set problem were previously known only for planar graphs. Our algorithms both generalize and simplify the classical bounded search tree algorithms for this problem (see, e.g., [2,13]).

The problem of finding induced cycles in degenerated graphs has been studied by Cai, Chan and Chan [8]. Our second result in this paper is a randomized algorithm for finding an induced cycle of fixed size in graphs with an excluded minor. The algorithm's expected running time is linear, and its derandomization is done in an efficient way, answering an open question from [8]. The problem of finding induced cycles in degenerated graphs is also addressed.

The Dominating Set Problem. The dominating set problem on general graphs is known to be $W[2]$-complete [12]. This means that most likely there is no $f(k) \cdot n^c$-algorithm for finding a dominating set of size at most k in a graph of size n for any computable function $f : \mathbb{N} \to \mathbb{N}$ and constant c. This suggests the exploration of specific families of graphs for which this problem is fixed-parameter tractable. For a general introduction to the field of parameterized complexity, the reader is referred to [12] and [14].

The method of bounded search trees has been used to give an $O(8^k n)$ time algorithm for the dominating set problem in planar graphs [2] and an $O((4g + 40)^k n^2)$ time algorithm for the problem in graphs of bounded genus $g \geq 1$ [13]. The algorithms for planar graph were improved to $O(4^{6\sqrt{34k}} n)$ [1], then to $O(2^{27\sqrt{k}} n)$ [17], and finally to $O(2^{15.13\sqrt{k}} k + n^3 + k^4)$ [15]. Fixed-parameter algorithms are now known also for map graphs [9] and for constant powers of H-minor-free graphs [10]. The running time given in [10] for finding a dominating set of size k in an H-minor-free graph G with n vertices is $2^{O(\sqrt{k})} n^c$, where c is a constant depending only on H. To summarize these results, fixed-parameter tractable algorithms for the dominating set problem were known for fixed powers of H-minor-free graphs and for map graphs. Linear time algorithms were established only for planar graphs.

Finding Paths and Cycles. The foundations for the algorithms for finding cycles, presented in this paper, have been laid in [4], where the authors introduce the color-coding technique. Two main randomized algorithms are presented there, as follows. A simple directed or undirected path of length $k - 1$ in a graph $G = (V, E)$ that contains such a path can be found in $2^{O(k)}|E|$ expected time in the directed case and in $2^{O(k)}|V|$ expected time in the undirected case. A simple directed or undirected cycle of size k in a graph $G = (V, E)$ that contains such a cycle can be found in either $2^{O(k)}|V||E|$ or $2^{O(k)}|V|^{\omega}$ expected time, where

$\omega < 2.376$ is the exponent of matrix multiplication. These algorithms can be derandomized at a cost of an extra $\log |V|$ factor. As for the case of even cycles, it is shown in [23] that for every fixed $k \geq 2$, there is an $O(|V|^2)$ algorithm for finding a simple cycle of size $2k$ in an undirected graph (that contains such a cycle). Improved algorithms for detecting given length cycles have been presented in [5] and [24]. The authors of [5] describe fast algorithms for finding short cycles in d-degenerated graphs. In particular, C_3's and C_4's can be found in $O(|E| \cdot d(G))$ time and C_5's in $O(|E| \cdot d(G)^2)$ time.

Finding Induced Paths and Cycles. Cai, Chan and Chan have recently introduced a new interesting technique they call *random separation* for solving fixed-cardinality optimization problems on graphs [8]. They combine this technique together with color-coding to give the following algorithms for finding an induced graph within a large graph. For fixed constants k and d, if a d-degenerated graph G with n vertices contains some fixed induced tree T on k vertices, then it can be found in $O(n)$ expected time and $O(n \log^2 n)$ worst-case time. If such a graph G contains an induced k-cycle, then it can be found in $O(n^2)$ expected time and $O(n^2 \log^2 n)$ worst-case time. Two open problems are raised by the authors of the paper. First, they ask whether the $\log^2 n$ factor incurred in the derandomization can be reduced to $\log n$. A second question is whether there is an $O(n)$ expected time algorithm for finding an induced k-cycle in a d-degenerated graph with n vertices. In this paper, we show that when combining the techniques of random separation and color-coding, an improved derandomization with a loss of only $\log n$ is indeed possible. An $O(n)$ expected time algorithm finding an induced k-cycle in graphs with an excluded minor is presented. We give evidence that establishing such an algorithm even for 2-degenerated graphs has far-reaching consequences.

Our Results. The main result of the paper is that the dominating set problem is fixed-parameter tractable for degenerated graphs. The running time is $k^{O(dk)}n$ for finding a dominating set of size k in a d-degenerated graph with n vertices. The algorithm is linear in the number of vertices of the graph, and we further improve the dependence on k for the following specific families of degenerated graphs. For graphs that do not contain K_h as a topological minor, an improved algorithm for the problem with running time $(O(h))^{hk}n$ is established. For graphs which are K_h-minor-free, the running time obtained is $(O(\log h))^{hk/2}n$. We show that all the algorithms can be generalized to the weighted case in the following sense. A dominating set of size at most k having minimum weight can be found within the same time bounds.

We address two open questions raised by Cai, Chan and Chan in [8] concerning linear time algorithms for finding an induced cycle in degenerated graphs. An $O(n)$ expected time algorithm for finding an induced k-cycle in graphs with an excluded minor is presented. The derandomization performed in [8] is improved and we get a deterministic $O(n \log n)$ time algorithm for the problem. As for finding induced cycles in degenerated graphs, we show a deterministic $O(n)$ time algorithm for finding cycles of size at most 5, and also explain why this is unlikely to be possible to achieve for longer cycles.

Techniques. We generalize the known search tree algorithms for the dominating set problem. This is enabled by proving some combinatorial lemmas, which are interesting in their own right. For degenerated graphs, we bound the number of vertices that dominate many elements of a given set, whereas for graphs with an excluded minor, our interest is in vertices that still need to be dominated and have a small degree.

The algorithm for finding an induced cycle in non-trivial minor-closed families is based on random separation and color-coding. Its derandomization is performed using known explicit constructions of families of (generalized) perfect hash functions.

2 Preliminaries

The paper deals with undirected and simple graphs, unless stated otherwise. Generally speaking, we will follow the notation used in [7] and [11]. For an undirected graph $G = (V, E)$ and a vertex $v \in V$, $N(v)$ denotes the set of all vertices adjacent to v (not including v itself). We say that v *dominates* the vertices of $N(v) \cup \{v\}$. The graph obtained from G by deleting v is denoted $G - v$. The subgraph of G induced by some set $V' \subseteq V$ is denoted by $G[V']$.

A graph G is *d-degenerated* if every induced subgraph of G has a vertex of degree at most d. It is easy and known that every d-degenerated graph $G = (V, E)$ admits an acyclic orientation such that the outdegree of each vertex is at most d. Such an orientation can be found in $O(|E|)$ time. A d-degenerated graph with n vertices has less than dn edges and therefore its average degree is less than $2d$.

For a directed graph $D = (V, A)$ and a vertex $v \in V$, the set of out-neighbors of v is denoted by $N^+(v)$. For a set $V' \subseteq V$, the notation $N^+(V')$ stands for the set of all vertices that are out-neighbors of at least one vertex of V'. For a directed graph $D = (V, A)$ and a vertex $v \in V$, we define $N_1^+(v) = N^+(v)$ and $N_i^+(v) = N^+(N_{i-1}^+(v))$ for $i \geq 2$.

An edge is said to be *subdivided* when it is deleted and replaced by a path of length two connecting its ends, the internal vertex of this path being a new vertex. A *subdivision* of a graph G is a graph that can be obtained from G by a sequence of edge subdivisions. If a subdivision of a graph H is the subgraph of another graph G, then H is a *topological minor* of G. A graph H is called a *minor* of a graph G if is can be obtained from a subgraph of G by a series of edge contractions.

In the parameterized dominating set problem, we are given an undirected graph $G = (V, E)$, a parameter k, and need to find a set of at most k vertices that dominate all the other vertices. Following the terminology of [2], the following generalization of the problem is considered. The input is a *black and white graph*, which simply means that the vertex set V of the graph G has been partitioned into two disjoint sets B and W of black and white vertices, respectively, i.e., $V = B \uplus W$, where $\uplus$ denotes disjoint set union. Given a black and white graph $G = (B \uplus W, E)$ and an integer k, the problem is to find a set of at most k vertices that dominate the black vertices. More formally, we ask whether there is a subset $U \subseteq B \uplus W$, such that $|U| \leq k$ and every vertex $v \in B - U$ satisfies $N(v) \cap U \neq \emptyset$. Finally we give a new definition, specific to this paper, for what it means to be a *reduced* black and white graph.

Definition 1. *A black and white graph $G = (B \uplus W, E)$ is called reduced if it satisfies the following conditions:*

- *W is an independent set.*
- *All the vertices of W have degree at least 2.*
- *$N(w_1) \neq N(w_2)$ for every two distinct vertices $w_1, w_2 \in W$.*

3 Algorithms for the Dominating Set Problem

3.1 Degenerated Graphs

The algorithm for degenerated graphs is based on the following combinatorial lemma.

Lemma 1. *Let $G = (B \uplus W, E)$ be a d-degenerated black and white graph. If $|B| > (4d+2)k$, then there are at most $(4d+2)k$ vertices in G that dominate at least $|B|/k$ vertices of B.*

Proof. Denote $R = \{v \in B \cup W | |(N_G(v) \cup \{v\}) \cap B| \geq |B|/k\}$. By contradiction, assume that $|R| > (4d+2)k$. The induced subgraph $G[R \cup B]$ has at most $|R|+|B|$ vertices and at least $\frac{|R|}{2} \cdot (\frac{|B|}{k} - 1)$ edges. The average degree of $G[R \cup B]$ is thus at least

$$\frac{|R|(|B|-k)}{k(|R|+|B|)} \geq \frac{\min\{|R|, |B|\}}{2k} - 1 > 2d.$$

This contradicts the fact that $G[R \cup B]$ is d-degenerated. □

Theorem 1. *There is a $k^{O(dk)}n$ time algorithm for finding a dominating set of size at most k in a d-degenerated black and white graph with n vertices that contains such a set.*

Proof. The pseudocode of algorithm *DominatingSetDegenerated*(G, k) that solves this problem appears below. If there is indeed a dominating set of size at most k, then this means that we can split B into k disjoint pieces (some of them can be empty), so that each piece has a vertex that dominates it. If $|B| \leq (4d+2)k$, then there are at most $k^{(4d+2)k}$ ways to divide the set B into k disjoint pieces. For each such split, we can check in $O(kdn)$ time whether every piece is dominated by a vertex. If $|B| > (4d+2)k$, then it follows from Lemma 1 that $|R| \leq (4d+2)k$. This means that the search tree can grow to be of size at most $(4d+2)^k k!$ before possibly reaching the previous case. This gives the needed time bound. □

Algorithm 1. *DominatingSetDegenerated(G, k)*

```
Input: Black and white d-degenerated graph G = (B ⊎ W, E), integers k, d
Output: A set dominating all vertices of B of size at most k or NONE if no
        such set exists
if B = ∅ then
  └ return ∅
else if k = 0 then
  └ return NONE
else if |B| ≤ (4d + 2)k then
  | forall possible ways of splitting B into k (possibly empty) disjoint pieces
  | B_1, ..., B_k do
  |   | if each piece B_i has a vertex v_i that dominates it then
  |   └   └ return {v_1, ..., v_k}
  └ return NONE
else
  | R ← {v ∈ B ∪ W | |(N_G(v) ∪ {v}) ∩ B| ≥ |B|/k}
  | forall v ∈ R do
  |   | Create a new graph G' from G by marking all the elements of N_G(v) as
  |   | white and removing v from the graph
  |   | D ← DominatingSetDegenerated(G', k − 1)
  |   | if D ≠ NONE then
  |   └   └ return D ∪ {v}
  └ return NONE
```

3.2 Graphs with an Excluded Minor

Graphs with either an excluded minor or with no topological minor are known to be degenerated. We will apply the following useful propositions.

Proposition 1. *[6,18] There exists a constant c such that, for every h, every graph that does not contain K_h as a topological minor is ch^2-degenerated.*

Proposition 2. *[19,21,22] There exists a constant c such that, for every h, every graph with no K_h minor is $ch\sqrt{\log h}$-degenerated.*

The following lemma gives an upper bound on the number of cliques of a prescribed fixed size in a degenerated graph.

Lemma 2. *If a graph G with n vertices is d-degenerated, then for every $k \geq 1$, G contains at most $\binom{d}{k-1}n$ copies of K_k.*

Proof. By induction on n. For $n = 1$ this is obviously true. In the general case, let v be a vertex of degree at most d. The number of copies of K_k that contain v is at most $\binom{d}{k-1}$. By the induction hypothesis, the number of copies of K_k in $G - v$ is at most $\binom{d}{k-1}(n-1)$. □

We can now prove our main combinatorial results.

Theorem 2. *There exists a constant $c > 0$, such that for every reduced black and white graph $G = (B \uplus W, E)$, if G does not contain K_h as a topological minor, then there exists a vertex $b \in B$ of degree at most $(ch)^h$.*

Proof. Denote $|B| = n > 0$ and $d = ch^2$ where c is the constant from Proposition 1. Consider the vertices of W in some arbitrary order. For each such vertex $w \in W$, if there exist two vertices $b_1, b_2 \in N(w)$, such that b_1 and b_2 are not connected, add the edge $\{b_1, b_2\}$ and remove the vertex w from the graph. Denote the resulting graph $G' = (B \uplus W', E')$. Obviously, $G'[B]$ does not contain K_h as a topological minor and therefore has at most dn edges. The number of edges in the induced subgraph $G'[B]$ is at least the number of white vertices that were deleted from the graph, which means that at most dn were deleted so far.

We now bound $|W'|$, the number of white vertices in G'. It follows from the definition of a reduced black and white graph that there are no white vertices in G' of degree smaller than 2. The graph G' cannot contain a white vertex of degree $h-1$ or more, since this would mean that the original graph G contained a subdivision of K_h. Now let w be a white vertex of G' of degree k, where $2 \le k \le h-2$. The reason why w was not deleted during the process of generating G' is because $N(w)$ is a clique of size k in $G'[B]$. The graph G' is a reduced black and white graph, and therefore $N(w_1) \neq N(w_2)$ for every two different white vertices w_1 and w_2. This means that the neighbors of each white vertex induce a different clique in $G'[B]$. By applying Lemma 2 to $G'[B]$, we get that the number of white vertices of degree k in G' is at most $\binom{d}{k-1}n$. This means that $|W'| \le \left[\binom{d}{1} + \binom{d}{2} + \cdots + \binom{d}{h-3}\right] n$. We know that $|W| \le |W'| + dn$ and therefore $|E| \le d(|B| + |W|) \le d\left[3d + \binom{d}{2} + \cdots + \binom{d}{h-3}\right] n$. Obviously, there exists a black vertex of degree at most $2|E|/n$. The result now follows by plugging the value of d and using the fact that $\binom{n}{k} \le (\frac{en}{k})^k$. □

Theorem 3. *There exists a constant $c > 0$, such that for every reduced black and white graph $G = (B \uplus W, E)$, if G is K_h-minor-free, then there exists a vertex $b \in B$ of degree at most $(c \log h)^{h/2}$.*

Proof. We proceed as in the proof of Theorem 2 using Proposition 2 instead of Proposition 1. □

Theorem 4. *There is an $(O(h))^{hk}n$ time algorithm for finding a dominating set of size at most k in a black and white graph with n vertices and no K_h as a topological minor.*

Proof. The pseudocode of algorithm $DominatingSetNoMinor(G, k)$ that solves this problem appears below. Let the input be a black and white graph $G = (B \uplus W, E)$. It is important to notice that the algorithm removes vertices and edges in order to get a (nearly) reduced black and white graph. This can be done in time $O(|E|)$ by a careful procedure based on the proof of Theorem 2 combined with radix sorting. We omit the details which will appear in the full version of the paper. The time bound for the algorithm now follows from Theorem 2. □

Algorithm 2. *DominatingSetNoMinor*(G, k)

Input: Black and white (K_h-minor-free) graph $G = (B \uplus W, E)$, integer k
Output: A set dominating all vertices of B of size at most k or $NONE$ if no such set exists

```
if B = ∅ then
  └ return ∅
else if k = 0 then
  └ return NONE
else
  │ Remove all edges of G whose two endpoints are in W
  │ Remove all white vertices of G of degree 0 or 1
  │ As long as there are two different vertices w1, w2 ∈ W with
  │ N(w1) = N(w2), |N(w1)| < h − 1, remove one of them from the graph
  │ Let b ∈ B be a vertex of minimum degree among all vertices in B
  │ forall v ∈ N_G(b) ∪ {b} do
  │   │ Create a new graph G′ from G by marking all the elements of N_G(v) as
  │   │ white and removing v from the graph
  │   │ D ← DominatingSetNoMinor(G′, k − 1)
  │   │ if D ≠ NONE then
  │   │   └ return D ∪ {v}
  │   └
  └ return NONE
```

Theorem 5. *There is an* $(O(\log h))^{hk/2}n$ *time algorithm for finding a dominating set of size at most* k *in a black and white graph with* n *vertices which is* K_h*-minor-free.*

Proof. The proof is analogues to that of Theorem 4 using Theorem 3 instead of Theorem 2. □

3.3 The Weighted Case

In the weighted dominating set problem, each vertex of the graph has some positive real weight. The goal is to find a dominating set of size at most k, such that the sum of the weights of all the vertices of the dominating set is as small as possible. The algorithms we presented can be generalized to deal with the weighted case without changing the time bounds. In this case, the whole search tree needs to be scanned and one cannot settle for the first valid solution found.

Let $G = (B \uplus W, E)$ be the input graph to the algorithm. In algorithm 1 for degenerated graphs, we need to address the case where $|B| \leq (4d + 2)k$. In this case, the algorithm scans all possible ways of splitting B into k disjoint pieces $B_1, \ldots, B_k$, and it has to be modified, so that it will always choose a vertex with minimum weight that dominates each piece. In algorithm 2 for graphs with an excluded minor, the criterion for removing white vertices from the graph is modified so that whenever two vertices $w_1, w_2 \in W$ satisfy $N(w_1) = N(w_2)$, the vertex with the bigger weight is removed.

4 Finding Induced Cycles

4.1 Degenerated Graphs

Recall that $N_i^+(v)$ is the set of all vertices that can be reached from v by a directed path of length exactly i. If the outdegree of every vertex in a directed graph $D = (V, A)$ is at most d, then obviously $|N_i^+(v)| \leq d^i$ for every $v \in V$ and $i \geq 1$.

Theorem 6. *For every fixed $d \geq 1$ and $k \leq 5$, there is a deterministic $O(n)$ time algorithm for finding an* **induced** *cycle of length k in a d-degenerated graph on n vertices.*

Proof. Given a d-degenerated graph $G = (V, E)$ with n vertices, we orient the edges so that the outdegree of all vertices is at most d. This can be done in time $O(|E|)$. Denote the resulting directed graph $D = (V, A)$. We can further assume that $V = \{1, 2, \ldots, n\}$ and that every directed edge $\{u, v\} \in A$ satisfies $u < v$. This means that an out-neighbor of a vertex u will always have an index which is bigger than that of u. We now describe how to find cycles of size at most 5.

To find cycles of size 3 we simply check for each vertex v whether $N^+(v) \cap N_2^+(v) \neq \emptyset$. Suppose now that we want to find a cycle $v_1 - v_2 - v_3 - v_4 - v_1$ of size 4. Without loss of generality, assume that $v_1 < v_2 < v_4$. We distinguish between two possible cases.

- $v_1 < v_3 < v_2 < v_4$: Keep two counters C_1 and C_2 for each pair of vertices. For every vertex $v \in V$ and every unordered pair of distinct vertices $u, w \in N^+(v)$, such that u and w are not connected, we raise the counter $C_1(\{u, w\})$ by one. In addition to that, for every vertex $x \in N^+(v)$ such that $u, w \in N^+(x)$, the counter $C_2(\{u, w\})$ is incremented. After completing this process, we check whether there are two vertices for which $\binom{C_1(\{u,w\})}{2} - C_2(\{u, w\}) > 0$. This would imply that an induced 4-cycle was found.
- $v_1 < v_2 < v_3 < v_4$ or $v_1 < v_2 < v_4 < v_3$: Check for each vertex v whether the set $\{v\} \cup N^+(v) \cup N_2^+(v) \cup N_3^+(v)$ contains an induced cycle.

To find an induced cycle of size 5, a more detailed case analysis is needed. It is easy to verify that such a cycle has one of the following two types.

- There is a vertex v such that $\{v\} \cup N^+(v) \cup N_2^+(v) \cup N_3^+(v) \cup N_4^+(v)$ contains the induced cycle.
- The cycle is of the form $v - x - u - y - w - v$, where $x \in N^+(v)$, $u \in N^+(x) \cap N^+(y)$, and $w \in N^+(v) \cap N^+(y)$. The induced cycle can be found by defining counters in a similar way to what was done before. We omit the details. □

The following simple lemma shows that a linear time algorithm for finding an induced C_6 in a 2-degenerated graph would imply that a triangle (a C_3) can be found in a general graph in $O(|V| + |E|) \leq O(|V|^2)$ time. It is a long standing open question to improve the natural $O(|V|^\omega)$ time algorithm for this problem [16].

Lemma 3. *Given a linear time algorithm for finding an induced C_6 in a 2-degenerated graph, it is possible to find triangles in general graphs in $O(|V|+|E|)$ time.*

Proof. Given a graph $G = (V, E)$, subdivide all the edges. The new graph obtained G' is 2-degenerated and has $|V| + |E|$ vertices. A linear time algorithm for finding an induced C_6 in G' actually finds a triangle in G. By assumption, the running time is $O(|V| + |E|) \leq O(|V|^2)$. □

4.2 Minor-Closed Families of Graphs

Theorem 7. *Suppose that G is a graph with n vertices taken from some non-trivial minor-closed family of graphs. For every fixed k, if G contains an* ***induced*** *cycle of size k, then it can be found in $O(n)$ expected time.*

Proof. There is some absolute constant d, so that G is d-degenerated. Orient the edges so that the maximum outdegree is at most d and denote the resulting graph $D = (V, E)$. We now use the technique of random separation. Each vertex $v \in V$ of the graph is independently removed with probability $1/2$, to get some new directed graph D'. Now examine some (undirected) induced cycle of size k in the original directed graph D, and denote its vertices by U. The probability that all the vertices in U remained in the graph and all vertices in $N^+(U) - U$ were removed from the graph is at least $2^{-k(d+1)}$.

We employ the color-coding method to the graph D'. Choose a random coloring of the vertices of D' with the k colors $\{1, 2, \ldots, k\}$. For each vertex v colored i, if $N^+(v)$ contains a vertex with a color which is neither $i-1$ nor $i+1 \pmod{k}$, then it is removed from the graph. For each induced cycle of size k, its vertices will receive distinct colors and it will remain in the graph with probability at least $2k^{1-k}$.

We now use the $O(n)$ time algorithm from [4] to find a multicolored cycle of length k in the resulting graph. If such a cycle exists, then it must be an induced cycle. Since k and d are constants, the algorithm succeeds with some small constant probability and the expected running time is as needed. □

The next theorem shows how to derandomize this algorithm while incurring a loss of only $O(\log n)$.

Theorem 8. *Suppose that G is a graph with n vertices taken from some non-trivial minor-closed family of graphs. For every fixed k, there is an $O(n \log n)$ time deterministic algorithm for finding an* ***induced*** *cycle of size k in G.*

Proof. Denote $G = (V, E)$ and assume that G is d-degenerated. We derandomize the algorithm in Theorem 7 using an $(n, dk+k)$-family of perfect hash functions. This is a family of functions from $[n]$ to $[dk + k]$ such that for every $S \subseteq [n]$, $|S| = dk + k$, there exists a function in the family that is 1-1 on S. Such a family of size $e^{dk+k}(dk + k)^{O(\log(dk+k))} \log n$ can be efficiently constructed [20]. We think of each function as a coloring of the vertices with the $dk + k$ colors

$C = \{1, 2, \ldots, dk + k\}$. For every combination of a coloring, a subset $L \subseteq C$ of k colors and a bijection $f : L \to \{1, 2, \ldots, k\}$ the following is performed. All the vertices that got a color from $c \in L$ now get the color $f(c)$. The other vertices are removed from the graph.

The vertices of the resulting graph are colored with the k colors $\{1, 2, \ldots, k\}$. Examine some induced cycle of size k in the original graph, and denote its vertices by U. There exists some coloring c in the family of perfect hash functions for which all the vertices in $U \cup N^+(U)$ received different colors. Now let L be the k colors of the vertices in the cycle U and let $f : L \to [k]$ be the bijection that gives consecutive colors to vertices along the cycle. This means that for this choice of c, L, and f, the induced cycle U will remain in the graph as a multicolored cycle, whereas all the vertices in $N^+(U) - U$ will be removed from the graph.

We proceed as in the previous algorithm. Better dependence on the parameters d and k can be obtained using the results in [3]. □

5 Concluding Remarks

- The algorithm for finding a dominating set in graphs with an excluded minor, presented in this paper, generalizes and improves known algorithms for planar graphs and graphs with bounded genus. We believe that similar techniques may be useful in improving and simplifying other known fixed-parameter algorithms for graphs with an excluded minor.
- An interesting open problem is to decide whether there is a $2^{O(\sqrt{k})}n^c$ time algorithm for finding a dominating set of size k in graphs with n vertices and an excluded minor, where c is some absolute constant that does not depend on the excluded graph. Maybe even a $2^{O(\sqrt{k})}n$ time algorithm can be achieved.

References

1. Alber, J., Bodlaender, H.L., Fernau, H., Kloks, T., Niedermeier, R.: Fixed parameter algorithms for DOMINATING SET and related problems on planar graphs. Algorithmica 33(4), 461–493 (2002)
2. Alber, J., Fan, H., Fellows, M.R., Fernau, H., Niedermeier, R., Rosamond, F.A., Stege, U.: A refined search tree technique for dominating set on planar graphs. J. Comput. Syst. Sci. 71(4), 385–405 (2005)
3. Alon, N., Cohen, G.D., Krivelevich, M., Litsyn, S.: Generalized hashing and parent-identifying codes. J. Comb. Theory, Ser. A 104(1), 207–215 (2003)
4. Alon, N., Yuster, R., Zwick, U.: Color-coding. Journal of the ACM 42(4), 844–856 (1995)
5. Alon, N., Yuster, R., Zwick, U.: Finding and counting given length cycles. Algorithmica 17(3), 209–223 (1997)
6. Bollobás, B., Thomason, A.: Proof of a conjecture of Mader, Erdös and Hajnal on topological complete subgraphs. Eur. J. Comb. 19(8), 883–887 (1998)
7. Bondy, J.A., Murty, U.S.R.: Graph theory with applications. American Elsevier Publishing, New York (1976)

8. Cai, L., Chan, S.M., Chan, S.O.: Random separation: A new method for solving fixed-cardinality optimization problems. In: Bodlaender, H.L., Langston, M.A. (eds.) IWPEC 2006. LNCS, vol. 4169, pp. 239–250. Springer, Heidelberg (2006)
9. Demaine, E.D., Fomin, F.V., Hajiaghayi, M., Thilikos, D.M.: Fixed-parameter algorithms for (k, r)-center in planar graphs and map graphs. ACM Transactions on Algorithms 1(1), 33–47 (2005)
10. Demaine, E.D., Fomin, F.V., Hajiaghayi, M., Thilikos, D.M.: Subexponential parameterized algorithms on bounded-genus graphs and H-minor-free graphs. Journal of the ACM 52(6), 866–893 (2005)
11. Diestel, R.: Graph theory, 3rd edn. Graduate Texts in Mathematics, vol. 173. Springer, Heidelberg (2005)
12. Downey, R.G., Fellows, M.R.: Parameterized complexity. Monographs in Computer Science. Springer, Heidelberg (1999)
13. Ellis, J.A., Fan, H., Fellows, M.R.: The dominating set problem is fixed parameter tractable for graphs of bounded genus. J. Algorithms 52(2), 152–168 (2004)
14. Flum, J., Grohe, M.: Parameterized complexity theory. Texts in Theoretical Computer Science. An EATCS Series. Springer, Heidelberg (2006)
15. Fomin, F.V., Thilikos, D.M.: Dominating sets in planar graphs: branch-width and exponential speed-up. In: Proceedings of the Fourteenth Annual ACM-SIAM Symposium on Discrete Algorithms, pp. 168–177. ACM Press, New York (2003)
16. Itai, A., Rodeh, M.: Finding a minimum circuit in a graph. SIAM Journal on Computing 7(4), 413–423 (1978)
17. Kanj, I.A., Perkovic, L.: Improved parameterized algorithms for planar dominating set. In: Diks, K., Rytter, W. (eds.) MFCS 2002. LNCS, vol. 2420, pp. 399–410. Springer, Heidelberg (2002)
18. Komlós, J., Szemerédi, E.: Topological cliques in graphs II. Combinatorics, Probability & Computing 5, 79–90 (1996)
19. Kostochka, A.V.: Lower bound of the Hadwiger number of graphs by their average degree. Combinatorica 4(4), 307–316 (1984)
20. Naor, M., Schulman, L.J., Srinivasan, A.: Splitters and near-optimal derandomization. In: 36th Annual Symposium on Foundations of Computer Science, pp. 182–191 (1995)
21. Thomason, A.: An extremal function for contractions of graphs. Math. Proc. Cambridge Philos. Soc. 95(2), 261–265 (1984)
22. Thomason, A.: The extremal function for complete minors. J. Comb. Theory, Ser. B 81(2), 318–338 (2001)
23. Yuster, R., Zwick, U.: Finding even cycles even faster. SIAM Journal on Discrete Mathematics 10(2), 209–222 (1997)
24. Yuster, R., Zwick, U.: Detecting short directed cycles using rectangular matrix multiplication and dynamic programming. In: Proceedings of the Fifteenth Annual ACM-SIAM Symposium on Discrete Algorithms, pp. 254–260 (2004)

Single-Edge Monotonic Sequences of Graphs and Linear-Time Algorithms for Minimal Completions and Deletions*

Pinar Heggernes and Charis Papadopoulos

Department of Informatics, University of Bergen, N-5020 Bergen, Norway
{pinar,charis}@ii.uib.no

Abstract. We study graph properties that admit an increasing, or equivalently decreasing, sequence of graphs on the same vertex set such that for any two consecutive graphs in the sequence their difference is a single edge. This is useful for characterizing and computing minimal completions and deletions of arbitrary graphs into having these properties. We prove that threshold graphs and chain graphs admit such sequences. Based on this characterization and other structural properties, we present linear-time algorithms both for computing minimal completions and deletions into threshold, chain, and bipartite graphs, and for extracting a minimal completion or deletion from a given completion or deletion. Minimum completions and deletions into these classes are NP-hard to compute.

1 Introduction

A graph property $\mathcal{P}$ is called *monotone* if it is closed under any edge or vertex removal. Equivalently, a property is monotone if it is closed under taking subgraphs that are not necessarily induced. A property is *hereditary* if it is closed under taking induced subgraphs. Every monotone property is hereditary but the converse is not true. For example bipartiteness and planarity are monotone properties, as they are characterized through forbidden subgraphs that are not necessarily induced, whereas perfectness is a hereditary but not monotone property. Some of the most well-studied graph properties are monotone [1,3] or hereditary [14]. If a given monotone (hereditary) property is equivalent to belonging to a graph class then this graph class is called monotone (hereditary).

In this work, we introduce *sandwich monotonicity* of graph properties and graph classes. We say that a graph property $\mathcal{P}$ is *sandwich monotone* if $\mathcal{P}$ satisfies the following: Given two graphs $G_1 = (V, E)$ and $G_2 = (V, E \cup F)$ that both satisfy $\mathcal{P}$, the edges in F can be ordered $f_1, f_2, \ldots, f_{|F|}$ such that when single edges of F are added to G_1 one by one in this order (or removed from G_2 in the reverse order), the graph obtained at each step satisfies $\mathcal{P}$. Every monotone property is clearly sandwich monotone as well. However every hereditary

* This work is supported by the Research Council of Norway through grant 166429/V30.

G. Lin (Ed.): COCOON 2007, LNCS 4598, pp. 406–416, 2007.

property is not necessarily sandwich monotone. Here, we are interested in identifying non-monotone hereditary graph classes that are sandwich monotone. Until now, sandwich monotonicity of only two such classes has been shown: chordal graphs [28] and split graphs [15], and it has been an open question which other graph classes are sandwich monotone [2]. Simple examples exist to show that the hereditary classes of perfect graphs, comparability graphs, permutation graphs, cographs, trivially perfect graphs, and interval graphs are not sandwich monotone. In this extended abstract we show that threshold graphs and chain graphs are sandwich monotone.

Our main motivation for studying sandwich monotonicity comes from the problem of adding edges to or deleting edges from a given arbitrary graph so that it satisfies a desired property. For example, a *chordal completion* is a chordal supergraph on the same vertex set, and a *bipartite deletion* is a spanning bipartite subgraph of an arbitrary graph. Completions and deletions into other graph classes are defined analogously. A completion (deletion) is *minimum* if it has the smallest possible number of added (deleted) edges. Unfortunately minimum completions and deletions into most interesting graph classes, including threshold, chain, and bipartite graphs, are NP-hard to compute [13,7,22,25,29]. However, minimum completions (deletions) are a subset of minimal completions (deletions), and hence we can search for minimum among the set of minimal. A completion (deletion) is *minimal* if no subset of the added (deleted) edges is enough to give the desired property when added to (deleted from) the input graph.

Given as input an arbitrary graph, there are two problems related to minimal completions (deletions) : 1. Computing a minimal completion (deletion) of the given input graph into the desired graph class, 2. Extracting a minimal completion (deletion) which is a subgraph (supergraph) of a given arbitrary completion (deletion) of the input graph into the desired class. A solution of problem 2 requires a characterization of minimal completions (deletions) into a given class, and readily gives a solution of problem 1. A solution of problem 1 might generate only a subset of all possible minimal completions (deletions), and does not necessarily solve problem 2. Solution of problem 2 in polynomial time is known only for completions into chordal [4], split [15], and interval [18] graphs, and for deletions into chordal [9], split [16], and planar [10,20] graphs. As an example of usefulness of a solution of problem 2, various characterizations of minimal chordal completions [21,6] have made it possible to design approximation algorithms [25] and fast exact exponential time algorithms [11] for computing minimum chordal completions. A solution of problem 2 also allows the computation of minimal completions that are not far from minimum in practice [5].

For a graph property $\mathcal{P}$ that is sandwich monotone, problem 2 of extracting a minimal completion from a given completion of an input graph into $\mathcal{P}$ has always a polynomial time solution if $\mathcal{P}$ can be recognized in polynomial time. Given G and a supergraph G_2 of G satisfying $\mathcal{P}$, if G_2 is not a minimal completion, then a minimal completion G_1 exists sandwiched between G and G_2. Hence by trying one by one all edges of G_2 that are not in G for removal, one obtains a minimal extraction algorithm with a number of iterations that is quadratic in the number of edges appearing in G_2 but not in G. Similarly, problem 2 of extracting a minimal deletion from a given deletion can also be solved with a number of iterations that is quadratic in the number of deleted edges.

In this extended abstract, by showing that threshold graphs and chain graphs are sandwich monotone, we thus establish that minimal completions and deletions of arbitrary graphs into these graph classes can be computed in polynomial time. Even more interesting, we give linear-time algorithms for minimal completions into threshold graphs and minimal deletions into bipartite and chain graphs. This does not follow from sandwich monotonicity in general. Furthermore, we solve the extraction version of these problems (problem 2 above) in linear time for these graph classes. Linear-time minimal completion algorithms have been known only into two classes previously; split [15] and proper interval [27] graphs. The only linear-time minimal extraction algorithm known is the one into split graphs [15]. Linear-time minimal deletion algorithms are known only into split [16] and planar graphs (which are monotone) [10,20].

Notation and Terminology: We consider simple undirected graphs. For a graph $G = (V, E)$, $V(G) = V$ and $E(G) = E$, with $n = |V|$ and $m = |E|$. For $S \subseteq V$, the *subgraph of G induced by S* is denoted by $G[S]$. Moreover, we denote by $G - S$ the graph $G[V \setminus S]$ and by $G - v$ the graph $G[V \setminus \{v\}]$. We distinguish between *subgraphs* and *induced subgraphs*. By a *subgraph* of G we mean a graph G' on the same vertex set containing a subset of the edges of G, and we denote it by $G' \subseteq G$. If G' contains a proper subset of the edges of G, we write $G' \subset G$. We write $G - uv$ to denote the graph $(V, E \setminus \{uv\})$.

The *neighborhood* of a vertex x of G is $N_G(x) = \{v \mid xv \in E\}$. The *degree* of x in G is $d_G(x)$. For $S \subseteq V$ $N_G(S) = \bigcup_{x \in S} N_G(x) \setminus S$. The size of a largest clique in G is $\omega(G)$. A chordless cycle on k vertices is denoted by C_k and a chordless path on k vertices is denoted by P_k. The graph consisting of only two disjoint edges is denoted by $2K_2$. The *complement* $\overline{G}$ of a graph G consists of all vertices and all non-edges of G. For a graph class $\mathcal{C}$, the class *co-*$\mathcal{C}$ is the set of graphs $\overline{G}$ for which $G \in \mathcal{C}$. A class $\mathcal{C}$ is *self-complementary* if $\mathcal{C} =$ co-$\mathcal{C}$.

Given an arbitrary graph $G = (V, E)$ and a graph class $\mathcal{C}$, a $\mathcal{C}$ *completion* of G is a graph $H = (V, E \cup F)$ such that $H \in \mathcal{C}$, and H is a *minimal* $\mathcal{C}$ completion of G if $(V, E \cup F')$ fails to be in $\mathcal{C}$ for every $F' \subset F$. Similarly, $H = (V, E \setminus D)$ is a $\mathcal{C}$ *deletion* of G if $H \in \mathcal{C}$, and H is minimal if $(V, E \setminus D')$ fails to be in $\mathcal{C}$ for every $D' \subset D$. The edges added to the original graph in order to obtain a completion are called *fill edges*, and the edges removed from the original graph in order to obtain a deletion are called *deleted edges*.

2 Minimal Completions and Deletions into Sandwich Monotone Graph Classes

Definition 1. *A graph class* $\mathcal{C}$ *is* sandwich monotone *if the following is true for any pair of graphs* $G = (V, E)$ *and* $H = (V, E \cup F)$ *in* $\mathcal{C}$ *with* $E \cap F = \emptyset$*: There is* *an ordering* $f_1, f_2, \ldots, f_{|F|}$ *of the edges in* F *such that in the sequence of graphs* $G = G_0, G_1, \ldots, G_{|F|} = H$, *where* G_{i-1} *is obtained by removing edge* f_i *from* G_i, *(or equivalently,* G_i *is obtained by adding edge* f_i *to* G_{i-1}*), every graph belongs to* $\mathcal{C}$*.*

Minimal completions and deletions into sandwich monotone graph classes have the following algorithmically useful characterization, as observed on chordal graphs [28] and split graphs [15] previously.

Observation 1. *Let* $\mathcal{C}$ *be a graph class and let* $\mathcal{P}$ *be the property of belonging to* $\mathcal{C}$*. The following are equivalent:*

(i) $\mathcal{C}$ *is sandwich monotone.*
(ii) co-$\mathcal{C}$ *is sandwich monotone.*
(iii) A $\mathcal{C}$ *completion is minimal if and only if no single fill edge can be removed without destroying the property* $\mathcal{P}$*.*
(iv) A $\mathcal{C}$ *deletion is minimal if and only if no single deleted edge can be added without destroying the property* $\mathcal{P}$*.*

Observation 2. *Let* $\mathcal{C}$ *be a sandwich monotone graph class. Given a polynomial time algorithm for the recognition of* $\mathcal{C}$*, there is a polynomial time algorithm for extracting a minimal* $\mathcal{C}$ *completion (deletion) of an arbitrary graph* G *from any given* $\mathcal{C}$ *completion (deletion) of* G*.*

Even though we know that minimal $\mathcal{C}$ completions and deletions can be computed in polynomial time for a sandwich monotone graph class $\mathcal{C}$, the actual running time is not necessarily practical. In the following sections, we give linear-time algorithms for computing and extracting minimal completions into threshold, and deletions into bipartite and chain graphs.

For the sandwich monotone graph classes previously studied for completions and deletions, linear-time algorithms exist for computing and extracting minimal split completions [15] and deletions [16], and minimal planar deletions [10,20]. As a comparison, although chordal graphs are sandwich monotone and they can be recognized in linear time, the best known running time is $O(n^{2.376})$ for a minimal chordal completion algorithm [19], and $O(\Delta m)$ for a minimal chordal deletion algorithm [9], where Δ is the largest degree in the input graph.

Minimal Bipartite Deletions: A graph is *bipartite* if its vertex set can be partitioned into two independent sets, called a *bipartition.* The bipartition of a bipartite graph is unique if and only if the graph is connected. Bipartite graphs are exactly the class of graphs that do not contain cycles of odd length. It is well known that simple modifications of breadth-first search (BFS) can be used to find an odd cycle in a graph or provide a bipartition of it in linear time. Bipartite graphs are monotone, and hence also sandwich monotone. Computing minimum bipartite deletions is NP-hard problem [13].

If an arbitrary input graph G is connected then there exists a connected bipartite deletion of G, since G has a spanning tree, and trees are bipartite. If G is not connected, then the following result can be applied to each connected component of G separately.

Theorem 1. *Let $H = (V, E \setminus D)$ be a bipartite deletion of an arbitrary graph $G = (V, E)$ with $D \subseteq E$. A minimal bipartite deletion $H' = (V, E \setminus D')$ of G, such that $D' \subseteq D$ can be computed in $O(n + m)$ time.*

Corollary 1. *Any minimal bipartite deletion of an arbitrary graph can be computed in $O(n + m)$ time.*

3 Minimal Threshold Completions

A graph $G = (V, E)$ is called a *threshold graph* if there exist nonnegative real numbers w_v for $v \in V$, and t such that for every $I \subseteq V$, $\sum_{v \in I} w_v \leq t$ if and only if I is an independent set [8,14,23].

A graph is a *split graph* if its vertex set can be partitioned into a clique K and an independent set S, in which case (S, K) is a (not necessarily unique) *split partition* of the graph. Split graphs can be recognized and a split partition can be computed in linear time [14]. It is known that a graph is split if and only if it does not contain any vertex subset inducing $2K_2$, C_4, or C_5 [12]. Hence the next theorem states that a graph is threshold if and only if it is split and does not contain any vertex set inducing a P_4.

Theorem 2 ([8]). *A graph is a threshold graph if and only if it does not contain any vertex set inducing $2K_2$, C_4, or P_4.*

Consequently, in a disconnected threshold graph there is at most one connected component that contains an edge. An ordering $v_1, v_2, \ldots, v_{|S|}$ of a subset $S \subseteq V(G)$ of vertices is called *nested neighborhood ordering* if it has the property that $(N_G(v_1) \setminus S) \subseteq (N_G(v_2) \setminus S) \subseteq \ldots \subseteq (N_G(v_{|S|}) \setminus S)$.

Theorem 3 ([23]). *A graph is a threshold graph if and only if it is split and the vertices of the independent set have a nested neighborhood ordering in any split partition.*

Threshold graphs can be recognized and a nested neighborhood ordering can be computed in linear time, since sorting the vertices of the independent set by their degrees readily gives such an ordering for threshold graphs [14]. It is NP-hard to compute minimum threshold completions of arbitrary graphs; even split graphs [26].

A split partition of a threshold graph is never unique [14]. Here we define a *threshold partition* (S, K) of a threshold graph in a unique way. Note first that all vertices of degree more than $\omega(G) - 1$ must belong to K, and all vertices of degree less than $\omega(G) - 1$ must belong to S. The set of vertices of degree exactly $\omega(G) - 1$ is either an independent set or a clique [15]. If this set is a clique, we place all of these vertices in K, and if it is an independent set we place them in S. We refine the sets S and K further as follows: $(S_0, S_1, S_2, \ldots, S_\ell)$ is a partition of S such that S_0 is the set of isolated vertices, and $N(S_1) \subset N(S_2) \subset \ldots \subset N(S_\ell)$, where ℓ is as large as possible. Hence all vertices in S_i have the same degree for $0 \leq i \leq \ell$. This also defines a partition $(K_1, K_2, \ldots, K_\ell)$ of K, where $K_1 = N(S_1)$

and $K_i = N(S_i) \setminus N(S_{i-1})$ for $2 \leq i \leq \ell$. The remaining vertices of K form another set $K_{\ell+1} = K \setminus N(S_\ell)$. Again, all vertices in K_i have the same degree for $1 \leq i \leq \ell + 1$. By definition, a graph is threshold if and only if its vertex set admits such a threshold partition. Moreover the threshold partition of a threshold graph is unique and all sets S_i and K_i, $1 \leq i \leq \ell$, are non-empty except possibly the sets S_0 and $K_{\ell+1}$. If $K_{\ell+1} = \emptyset$ then S_ℓ contains at least two vertices, and if $K_{\ell+1} \neq \emptyset$ then $K_{\ell+1}$ contains at least two vertices [14].

Lemma 1. *Let G be a threshold graph with threshold partition $((S_0, \ldots, S_\ell), (K_1, \ldots, K_{\ell+1}))$ and let uv be an edge satisfying the following: either $u \in S_i$ and $v \in K_i$ for some $i \in \{1, \ldots, \ell\}$, or $u, v \in K_{\ell+1}$. Then $G - uv$ is a threshold graph.*

Proof. Assume first that $u \in S_i$ and $v \in K_i$ for some i. Removing uv from G results in a split graph since we remove and edge between the clique and the independent set of the split partition. In the graph $G' = G - uv$ we have that $N_{G'}(S_{i-1}) \subseteq N_{G'}(u) \subset N_{G'}(S_i \setminus \{u\})$. Thus the new independent set has a nested neighborhood ordering also in G', and hence G' is threshold by Theorem 3.

Assume now that $u, v \in K_{\ell+1}$. We describe the threshold partition of G': If $K_{\ell+1}$ contains more than two vertices then the new threshold partition has the sets (K_i, S_i), $1 \leq i \leq \ell$, unchanged, and it also contains the new sets $K'_{\ell+1} = K_{\ell+1} \setminus \{u, v\}$ and $S'_{\ell+1} = \{u, v\}$. If $K_{\ell+1}$ has exactly two vertices, then every set remains as before, except $K_{\ell+1}$ which is removed and S_ℓ which now becomes $S'_\ell = S_\ell \cup \{u, v\}$. It is easy to check that removal of uv results in exactly the described threshold partition for G' and thus G' is a threshold graph.

For simplicity, we will call an edge uv of G that satisfies the condition in Lemma 1 a *candidate edge* of G.

Lemma 2. *Let $G = (V, E)$ and $G' = (V, E \cup F)$ be two threshold graphs such that $F \cap E = \emptyset$ and $F \neq \emptyset$. At least one edge in F is a candidate edge of G'.*

Proof. Let $((S'_0, \ldots, S'_\ell), (K'_1, \ldots, K'_{\ell+1}))$ be the threshold partition of G'. Assume for a contradiction that F does not contain any candidate edge, and let $uv \in F$. Since uv cannot be a candidate edge, it is of one of the following three types: (i) $u, v \in K'_i$, for some i satisfying $1 \leq i \leq \ell$, (ii) $u \in K'_i$ and $v \in K'_j$, for some i and j satisfying $1 \leq i < j \leq \ell + 1$, or (iii) $u \in K'_i$ and $v \in S'_j$, for some i and j satisfying $1 \leq i < j \leq \ell$. (Recall that there are no edges uv in G' with $u \in K'_i$ and $v \in S'_j$ where $j < i$ by the definition of threshold partition.)

If uv is of type (i), then since K'_i is non-empty, there is a vertex $x \in S'_i$ such that ux and uv are edges of G'. Since there are no candidate edges in F, $ux, uv \in E$. Assume first that $i \neq \ell$. Then there is a vertex $y \in K'_\ell$ such that uy and vy are edges of G' and xy is not an edge of G'. If both uy and vy belong to E, then $\{u, x, v, y\}$ induces a C_4 in G. If exactly one of uy and vy belongs to E, say uy, and the other belongs to F, then $\{y, u, x, v\}$ induces a P_4 in G. If both uy and vy belong to F, then there is a vertex $z \in S'_l$ such that $\{x, u, z, y\}$ induces a $2K_2$ in G, since zy is a candidate edge of G' and hence cannot be in F. Hence all possibilities lead to a contradiction since G is a threshold graph and cannot contain any of the mentioned induced subgraphs. If $i = \ell$ and $K'_{\ell+1} \neq \emptyset$ then there are at least two vertices y and z in $K'_{\ell+1}$ that can substitute the role of y and z as in the case $i \neq \ell$, since yz is a candidate edge of G' and hence must belong to E. If $i = \ell$ and $K'_{\ell+1} = \emptyset$ then we know that there are at least two vertices $y, z \in S'_\ell$, and that $uy, uz, vy, vz \in E$ (since they are candidate edges). Hence $\{u, y, v, z\}$ induces a C_4 in G, contradicting that G is threshold.

If uv is of type (ii), assume first that $j \neq \ell + 1$. Then we know that there is a vertex $x \in S'_i$ and a vertex $y \in S'_j$ such that $ux, vy \in E$ since they are candidate edges. We know that xv is not an edge of G'. If $yu \in E$ then $\{x, u, y, v\}$ induces a P_4 in G, and if $yu \in F$ then the same set induces a $2K_2$ in G, contradicting in both cases that G is threshold. If $j = \ell+1$ then we know that there is at least one more vertex $z \neq v$ in $K'_{\ell+1}$ where $vz \in E$ (since it is a candidate edge). If uz belongs to F then $\{v, z, u, x\}$ induces a $2K_2$ in G. Otherwise this set induces a P_4 in G, because zx and vx are not edges in G or G', contradicting that G is threshold.

If uv is of type (iii), then we know that there are vertices $x \in S'_i$ and $y \in K'_j$ such that $ux, vy \in E$ by the same arguments as before. If $uy \in E$ then $\{x, u, y, v\}$ induces a P_4 in G, and if $uy \in F$, then this set induces a $2K_2$ in G. Hence by by Theorem 2, we reach a contradiction in all possible cases. Consequently, either G is not threshold, or F must contain a candidate edge, and the proof is complete.

Theorem 4. *Threshold graphs are sandwich monotone.*

Proof. Given G and G' as in the premise of Lemma 2, we know by Lemmas 1 and 2 that there is an edge $f \in F$ such that $G' - f$ is threshold. Now the same argument can be applied to G and $G' = G' - f$ with $F = F \setminus \{f\}$ repeatedly, until G' becomes equal to G.

Theorem 5. *Let G be an arbitrary graph and H be a threshold completion of G. H is a minimal threshold completion of G if and only if no fill edge is a candidate edge of H.*

Proof. If H is a minimal threshold completion of G, then there cannot be any fill edge that is a candidate edge of H, because otherwise the removal of this edge would result in a threshold graph by Lemma 1, contradicting that H is minimal. If H is not a minimal threshold completion of G, then there exists a minimal threshold completion M of G such that $E(G) \subseteq E(M) \subset E(H)$. By Lemma 2 there is an edge $e \in E(H) \setminus E(M)$ that is a candidate edge of H.

Now given any threshold completion H of an arbitrary graph G, let us consider the problem of extracting a minimal threshold completion H' of G such that $G \subseteq H' \subseteq H$.

Theorem 6. *Let $H = (V, E \cup F)$ be a threshold completion of an arbitrary graph $G = (V, E)$ with $F \cap E = \emptyset$. A minimal threshold completion $H' = (V, E \cup F')$ of G, such that $F' \subseteq F$, can be computed in $O(n + m + |F|)$ time.*

For the proof of Theorem 6 we give Algorithm Extr_Min_Threshold which describes a way of computing H'.

Algorithm. Extr_Min_Threshold

Input: A graph $G = (V, E)$ and a threshold graph $H = (V, E \cup F)$ with $F \cap E = \emptyset$.
Output: A minimal threshold completion $H' = (V, E \cup F')$ of G such that $F' \subseteq F$.

$H' = H$; Unmark all vertices of H';
Let $(S = (S_0, S_1, \ldots, S_\ell)$, $K = (K_1, K_2, \ldots, K_{\ell+1}))$ be the threshold partition of H';
While there is an unmarked vertex v such that $d_{H'}(v) \leq \omega(H') - 1$ **do**
  Pick an unmarked vertex v of minimum $d_{H'}(v)$;
  If $v \in S$ **then** compute the index j for which $v \in S_j$;
  Else $j = \ell + 1$;
  Find a vertex $u \in N_G(v)$ of minimum $d_{H'}(u)$;
  Compute $U = \{u' \in N_G(v) \mid d_{H'}(u') = d_{H'}(u)\}$;
  Compute the index i for which $U \subseteq K_i$;
  Remove all edges between v and $(K_i \setminus U) \cup K_{i+1} \cup \cdots \cup K_j$ from H';
  Update the threshold-partition of H' and mark v;
Return H';

Next we show how to obtain a minimal threshold completion H of G directly. The motivation for this is that we now compute H in time linear in the size of G. The idea behind our approach is the following: Compute a minimal split completion of G using the algorithm of [15], and then compute a minimal threshold completion of the computed split completion by giving the vertices of the independent set a nested neighborhood ordering. Computing a minimal split completion G' with split partition (K, S) of G can be done in $O(n + m)$ time by the algorithm given in [15]. After that we compute an order of the vertices of S such that $d(v_1) \geq d(v_2) \geq \ldots \geq d(v_{|S|})$. The important point is that $d_{G'}(v) = d_G(v)$ for each $v \in S$, since all fill edges of G' have both endpoints in K. Then for each vertex v_i we make it adjacent to the vertices of $N(\{v_{i+1}, v_{i+2}, \ldots, v_{|S'|}\})$, starting from v_1. We show in the proof of the following theorem that this indeed results in a minimal threshold completion of G.

Theorem 7. *A minimal threshold completion of an arbitrary graph G can be computed in $O(n + m)$ time.*

4 Minimal Chain Deletions

Yannakakis introduced *chain graphs* and defined a bipartite graph to be a chain graph if one of the sides of the bipartition has nested neighborhood ordering [30]. He also showed that one side has this property if and only if both sides have the property. Chain graphs can be recognized in linear time [23]. It is also known that a graph is a chain graph if and only if it does not contain a vertex set inducing $2K_2$, C_3, or C_5 as an induced subgraph [23]. Computing a minimum chain deletion of a bipartite graph is an NP-hard problem [29].

Theorem 8 ([23]). *A graph G is a chain graph if and only if G is bipartite and turning one side of the bipartition into a clique gives a threshold graph for any bipartition of G.*

By Theorem 8, a chain graph G can have at most one connected component that contains an edge. Isolated vertices can belong to any side of the bipartition. We will here define a unique way of partitioning the vertices of a chain graph that we call *chain partition*, similar to threshold partition. Define X_0 to be the set of all isolated vertices of G. The remaining vertices induce a connected bipartite graph which thus has a unique bipartition (X, Y). Partition X into $(X_1, X_2, \ldots, X_\ell)$ where $N_G(X_1) \subset N_G(X_2) \subset \ldots \subset N_G(X_\ell)$, and ℓ is as large as possible. Hence vertices of X_i have the same neighborhood, for each i. This defines a partition of Y into $(Y_1, Y_2, \ldots, Y_\ell)$, where $Y_i = N_G(X_i) \setminus N_G(X_{i-1})$, $2 \leq i \leq \ell$. Observe that each set X_i contains at least one vertex which implies that the set Y_i is also a non-empty set. If there are only isolated vertices in G, we let $\ell = 0$. The chain partition is unique (upon exchanging X with Y).

Theorem 9. *Chain graphs are sandwich monotone.*

Proof. Let $G = (V, E)$ and $G' = (V, E \cup F)$ be two chain graphs such that $E \cap F = \emptyset$ and $F \neq \emptyset$. We will show that there is an edge $f \in F$ that can be removed from G' so that the resulting graph remains a chain graph, from which the result follows by induction on the edges in F.

Let $((X'_0, X'_1, \ldots, X'_\ell), (Y'_1, Y'_2, \ldots, Y'_\ell))$ be the chain-partition of G'. First we prove that if F contains an edge xy such that $x \in X'_i$ and $y \in Y'_i$ for some $i \in \{1, \ldots, \ell\}$ then removing xy from G' results in a chain graph. Let G'' be the graph that we obtain after removing xy; that is, $G'' = G' - xy$. In G'' we have that $N_{G''}(X'_{i-1}) \subset N_{G''}(x) \subset N_{G''}(X'_i \setminus \{x\}) \subset N_{G''}(X'_{i+1})$. Thus the nested neighborhood ordering is maintained which implies that G'' is a chain graph.

Now we prove that F contains at least one edge with one endpoint in X'_i and one endpoint in Y'_i, for some i satisfying $1 \leq i \leq \ell$. Assume for a contradiction that there are no edges in F with one endpoint in X'_i and the other in Y'_i. Hence for every edge $xy \in F$, $x \in X'_j$ and $y \in Y'_i$, where $1 \leq i < j \leq \ell$. Since X'_j and Y'_i are non-empty, X'_i and Y'_j are non-empty, too by definition. Let $x' \in X'_i$ and $y' \in Y'_j$. Edges xy' and $x'y$ cannot belong to F and hence they are edges in E by our assumption. Edge $x'y'$ is not an edge of G' by the definition of chain partition, and hence it is not an edge of G either. Since xy is not an edge of G, $\{x, y', x', y\}$ induces a $2K_2$ in G contradicting that G is a chain graph.

Lemma 3. *Let $G = (V, E)$ be an arbitrary graph, and let H be a chain deletion of G with chain partition $(X = (X_0, X_1, \ldots, X_\ell),\ Y = (Y_1, \ldots, Y_\ell))$. H is a minimal chain deletion of G if and only if no deleted edge uv has the following property: $u \in X_{i-1}$ and $v \in Y_i$ for some $1 \leq i \leq \ell$, or $u \in X_\ell$ and $v \in X_0$.*

Next we consider the problem of extracting a minimal chain deletion from any chain deletion H of an arbitrary graph G. We briefly describe such an algorithm. Let (X, Y) be the chain partition of H. At the beginning we unmark all vertices of X and at each step we pick an unmarked vertex $v \in X$ of largest degree in H. We find deleted edges incident to v not having the properties given in Lemma 3. If the deleted edges can be added to H without destroying the chain property then we add them and proceed with the next unmarked vertex of X of largest degree in H.

Theorem 10. *Let $H = (V, E \setminus D)$ be a chain deletion of an arbitrary graph $G = (V, E)$ with $D \subseteq E$. A minimal chain deletion $H' = (V, E \setminus D')$ of G, such that $D' \subseteq D$ can be computed in time $O(n + m)$.*

Corollary 2. *Any minimal chain deletion of an arbitrary graph graph can be computed in $O(n + m)$ time.*

5 Concluding Remarks and Open Questions

In this extended abstract we proved sandwich monotonicity of threshold and chain graphs. Moreover, we gave linear-time algorithms for computing minimal threshold completions, minimal bipartite deletions, and minimal chain deletions; and for extracting these from any given threshold completion, bipartite deletion, or chain deletion, respectively. In fact, the results presented in this extended abstract, by careful but not difficult argumentation, lead to linear-time algorithms also for computing **minimal threshold deletions** and **minimal co-bipartite completions** of arbitrary graphs, and **minimal chain completions** of bipartite graphs (see also [24]); as well as extracting these from any given such completion or deletion. Details of these results are left to a full version [17].

It has been shown recently that minimum weakly chordal completions of arbitrary graphs are NP-hard to compute [7]. We would like to know whether weakly chordal graphs are sandwich monotone. Also, we repeat the open question of [2]: are chordal bipartite graphs sandwich monotone? In addition, we would like to know whether minimum chordal bipartite completions of bipartite graphs are NP-hard to compute.

References

1. Alon, N., Shapira, A.: Every monotone graph property is testable. In: Proceedings of STOC 2005 - 37th Annual Symposium on Theory of Computing, pp. 128–137 (2005)
2. Bakonyi, M., Bono, A.: Several results on chordal bipartite graphs. Czechoslovak Math. J. 46, 577–583 (1997)
3. Balogh, J., Bolobás, B., Weinreich, D.: Measures on monotone properties of graphs. Disc. Appl. Math. 116, 17–36 (2002)
4. Blair, J., Heggernes, P., Telle, J.A.: A practical algorithm for making filled graphs minimal. Theoretical Computer Science 250, 125–141 (2001)
5. Bodlaender, H.L., Koster, A.M.C.A.: Safe separators for treewidth. Discrete Math. 306, 337–350 (2006)
6. Bouchitté, V., Todinca, I.: Treewidth and minimum fill-in: Grouping the minimal separators. SIAM J. Comput. 31, 212–232 (2001)
7. Burzyn, P., Bonomo, F., Durán, G.: NP-completeness results for edge modification problems. Disc. Appl. Math. 154, 1824–1844 (2006)
8. Chvátal, V., Hammer, P.L.: Set-packing and threshold graphs. Univ. Waterloo Res. Report, CORR 73–21 (1973)
9. Dearing, P.M., Shier, D.R., Warner, D.D.: Maximal chordal subgraphs. Disc. Appl. Math. 20, 181–190 (1988)

10. Djidjev, H.: A linear algorithm for finding a maximal planar subgraph. SIAM J. Disc. Math. 20, 444–462 (2006)
11. Fomin, F.V., Kratsch, D., Todinca, I.: Exact (exponential) algorithms for treewidth and minimum fill-in. In: Díaz, J., Karhumäki, J., Lepistö, A., Sannella, D. (eds.) ICALP 2004. LNCS, vol. 3142, pp. 568–580. Springer, Heidelberg (2004)
12. Földes, S., Hammer, P.L.: Split graphs. Congressus Numer. 19, 311–315 (1977)
13. Garey, M., Johnson, D., Stockmeyer, L.: Some simplified NP-complete graph problems. Theoretical Computer Science 1, 237–267 (1976)
14. Golumbic, M.C.: Algorithmic Graph Theory and Perfect Graphs, 2nd edn. Annals of Discrete Mathematics, vol. 57. Elsevier, Amsterdam (2004)
15. Heggernes, P., Mancini, F.: Minimal split completions of graphs. In: Correa, J.R., Hevia, A., Kiwi, M. (eds.) LATIN 2006. LNCS, vol. 3887, pp. 592–604. Springer, Heidelberg (2006)
16. Heggernes, P., Mancini, F.: A completely dynamic algorithm for split graphs. Reports in Informatics 334, University of Bergen, Norway (2006)
17. Heggernes, P., Papadopoulos, C.: Single-edge monotonic sequences of graphs and linear-time algorithms for minimal completions and deletions. Reports in Informatics 345, University of Bergen, Norway (2007)
18. Heggernes, P., Suchan, K., Todinca, I., Villanger, Y.: Characterizing minimal interval completions: Towards better understanding of profile and pathwidth. In: Thomas, W., Weil, P. (eds.) STACS 2007. LNCS, vol. 4393, pp. 2007–2024. Springer, Heidelberg (2007)
19. Heggernes, P., Telle, J.A., Villanger, Y.: Computing minimal triangulations in time $O(n^\alpha \log n) = o(n^{2.376})$. SIAM J. Disc. Math. 19, 900–913 (2005)
20. Hsu, W.-L.: A linear time algorithm for finding a maximal planar subgraph based on PC-trees. In: Wang, L. (ed.) COCOON 2005. LNCS, vol. 3595, pp. 787–797. Springer, Heidelberg (2005)
21. Kaplan, H., Shamir, R., Tarjan, R.E.: Tractability of parameterized completion problems on chordal, strongly chordal, and proper interval graphs. SIAM J. Comput. 28(5), 1906–1922 (1999)
22. Kashiwabara, T., Fujisawa, T.: An NP-complete problem on interval graphs. IEEE Symp. of Circuits and Systems, pp. 82–83. IEEE Computer Society Press, Los Alamitos (1979)
23. Mahadev, N., Peled, U.: Threshold graphs and related topics. Annals of Discrete Mathematics 56. North Holland, Amsterdam (1995)
24. Meister, D.: Recognition and computation of minimal triangulations for AT-free claw-free and co-comparability graphs. Disc. Appl. Math. 146, 193–218 (2005)
25. Natanzon, A., Shamir, R., Sharan, R.: Complexity classification of some edge modification problems. Disc. Appl. Math. 113, 109–128 (2001)
26. Peng, S.-L., Chen, C.-K.: On the interval completion of chordal graphs. Disc. Appl. Math. 154, 1003–1010 (2006)
27. Rapaport, I., Suchan, K., Todinca, I.: Minimal proper interval completions. In: Fomin, F.V. (ed.) WG 2006. LNCS, vol. 4271, pp. 217–228. Springer, Heidelberg (2006)
28. Rose, D., Tarjan, R.E., Lueker, G.: Algorithmic aspects of vertex elimination on graphs. SIAM J. Comput. 5, 266–283 (1976)
29. Yannakakis, M.: Computing the minimum fill-in is NP-complete. SIAM J. Alg. Disc. Meth. 2, 77–79 (1981)
30. Yannakakis, M.: Node deletion problems on bipartite graphs. SIAM J. Comput. 10, 310–327 (1981)

On the Hardness of Optimization in Power Law Graphs*

Alessandro Ferrante[1], Gopal Pandurangan[2], and Kihong Park[2]

[1] Dipartimento di Informatica ed Applicazioni "R.M. Capocelli", University of Salerno, Via Ponte don Melillo - 84084 Fisciano (SA), Italy
ferrante@dia.unisa.it
[2] Department of Computer Science, Purdue University, West Lafayette, IN 47907, USA
{gopal,park}@cs.purdue.edu

Abstract. Our motivation for this work is the remarkable discovery that many large-scale real-world graphs ranging from Internet and World Wide Web to social and biological networks exhibit a power-law distribution: the number of nodes y_i of a given degree i is proportional to $i^{-\beta}$ where $\beta > 0$ is a constant that depends on the application domain. There is practical evidence that combinatorial optimization in power-law graphs is easier than in general graphs, prompting the basic theoretical question: Is combinatorial optimization in power-law graphs easy? Does the answer depend on the power-law exponent β? Our main result is the proof that many classical NP-hard graph-theoretic optimization problems remain NP-hard on power law graphs for certain values of β. In particular, we show that some classical problems, such as CLIQUE and COLORING, remains NP-hard for all $\beta \geq 1$. Moreover, we show that all the problems that satisfy the so-called "optimal substructure property" remains NP-hard for all $\beta > 0$. This includes classical problems such as MIN VERTEX-COVER, MAX INDEPENDENT-SET, and MIN DOMINATING-SET. Our proofs involve designing efficient algorithms for constructing graphs with prescribed degree sequences that are tractable with respect to various optimization problems.

1 Overview and Results

The elegant theory of NP-hardness serves as an important cornerstone in understanding the difficulty of solving various combinatorial optimization problems in graphs. A natural and relevant question is whether such hardness results on combinatorial problems are applicable to "real-world" graphs since such graphs possess certain well-defined special properties which may very well render them tractable. Our motivation for this work is the remarkable discovery that many large-scale real-world graphs ranging from Internet and World Wide Web to social and biological networks exhibit a power-law distribution. In such networks, the number of nodes y_i of a given degree i is proportional to $i^{-\beta}$ where $\beta > 0$

* Work partially supported by funds for research from MIUR ex 60% 2005.

G. Lin (Ed.): COCOON 2007, LNCS 4598, pp. 417–427, 2007.

is a constant that depends on the application domain. Power-law degree distribution has been observed in the Internet ($\beta = 2.1$), World Wide Web ($\beta = 2.1$), social networks (movie actors graph with $\beta = 2.3$, citation graph with $\beta = 3$), and biological networks (protein domains with $\beta = 1.6$, protein-protein interaction graphs with $\beta = 2.5$). In most real-world graphs, β ranges between 1 and 4 (see [3] for a comprehensive list). Thus, power-law graphs have emerged as a partial answer to the perennial search for representative real-world graphs in combinatorial optimization.

There is practical evidence that combinatorial optimization in real-world power law graphs is easier than in general graphs. For example, experiments in Internet measurement graphs (power law with $\beta = 2.1$) show that a simple greedy algorithm that exploits the power law property yields a very good approximation to the MINIMUM VERTEX COVER problem (much better than random graphs with no power law) [11,12]. Gkantsidis, Mihail, and Saberi [8] argue that the performance of the Internet suggests that multi-commodity flow can be routed more efficiently (i.e., with near-optimal congestion) in Internet graphs than in general graphs. Eubank et al. [6] show that in power-law social networks, a simple and natural greedy algorithm that again exploits the power-law property (choose enough high-degree vertices) gives a $1+o(1)$ approximation to the DOMINATING SET problem. There is also similar practical evidence that optimization in power-law biological networks is easier [9]. All these results on disparate problems on various real-world graphs motivate a coherent and systematic algorithmic theory of optimization in power law graphs (and in general, graphs with prescribed degree sequences).

In this work, we study the following theoretical questions: What are the implications of power-law degree distribution to the algorithmic complexity of NP-hard optimization problems? Can the power-law degree distribution property alone be sufficient to design polynomial-time algorithms for NP-hard problems on power-law graphs? And does the answer depend on the exponent β?

A number of power law graph models have been proposed in the last few years to capture and/or explain the empirically observed power-law degree distribution in real-world graphs. They can be classified into two types. The first takes a power-law degree sequence and generates graph instances with this distribution. The second type arises from attempts to explain the power-law starting from basic assumptions about a growth evolution. Both approaches are well motivated and there is a large literature on both (e.g., [4,1,2]). Following Aiello, Chung, and Lu [1,2], we adopt the first approach, and use the following model for (undirected) power-law graphs (henceforth called ACL model): the number of vertices y_i with degree i is roughly given[1] by $y_i = e^\alpha / i^\beta$, where e^α is a normalization constant (so that the total number of vertices sum to the size of the graph, thus α determines the size). While the above model is potentially less accurate than the detailed modeling approach of the second type, it has the advantage of being robust and general [1]: the structural properties that are true in this model will be true for all graphs with the given degree sequence.

[1] Our model is defined precisely in Section 2.

Investigating the complexity of problems in power law graphs (in particular, the ACL model) involves an important subtlety. The ACL model allows graphs with self-loops and multi-edges. However, many real-world networks, such as Internet domain graphs, are simple undirected power-law graphs. Thus, we restrict ourselves to simple undirected power-law graphs (no multi-edges or self-loops). In this paper we study the complexity of many classical graph problems in the ACL model. In particular, we first show that problems such as COLORING and CLIQUE remains NP-hard in simple power-law graphs of the the ACL model for all $\beta \geq 1$, and then we show that all the graph problems that satisfy an "optimal substructure" property (such as MINIMUM VERTEX COVER, MAXIMUM INDEPENDENT SET and MINIMUM DOMINATING SET) remain NP-hard on simple power law graphs of the ACL model for all $\beta > 0$. This property essentially states that the optimal solution for a problem on given graph is the union of the optimal (sub-)solutions on its maximal connected components. A main ingredient in our proof is a technical lemma that guarantees that any arbitrary graph can be "embedded" in a suitably large (but polynomial in the size of the given graph) graph that conforms to the prescribed power-law degree sequence. This lemma may be of independent interest and can have other applications as well e.g., in showing hardness of *approximation* of problems in power-law graphs. Another contribution is constructions of graphs with prescribed degree sequences that admit polynomial time algorithms. These constructions are useful in showing the NP-hardness of certain optimization problems that do not satisfy the optimal substructure property. In particular, we will use them to show the NP-hardness of CLIQUE and COLORING for all $\beta \geq 1$.

Our results show that the worst-case complexity of power law graphs is hard with respect to many important graph optimization problems. However, experimental evidence shows that optimization is considerably easier in real-world power-law graphs. This suggests that real-world graphs are not "worst-case" instances of power-law graphs, but rather typical instances which may be well modeled by power law *random graph* models (e.g., [1,6,3,4,8,10]). Combinatorial optimization is generally easier in random graphs and hence from an optimization perspective this somewhat justifies using power law random graphs to model real-world power law graphs. We believe that further investigation, both in the modeling of real-world graphs and in the optimization complexity of real-world graphs and their models, is needed to gain a better understanding of this important issue.

2 Notations and Definitions

In this section we introduce some notations and definitions that we will use throughout the paper. For all $x, y \in \mathbb{N}$ with $x \leq y$, we will use $[x, y]$ to denote $\{x, x+1, \ldots, y\}$ and $[x]$ to denote $[1, x]$.

Given a graph, we will refer to two types of sequence of integers: *y*-degree sequence and *d*-degree sequence. The first type lists the number of vertices with a certain degree (i.e., the degree distribution) and the latter lists the degrees of the vertices in non-increasing order (i.e., the degree sequence of the graph in non-increasing order). More formally, we can define the y-degree sequence as follows. Given a graph $G = (V, E)$ with maximum degree m, the y-degree sequence is the sequence $Y^G = \langle y_1^G, \ldots, y_m^G \rangle$ where $y_i = |\{v \in V : degree(v) = i\}|$, $i \in [m]$. Given a graph of n vertices, the d-degree sequence will be denoted by $D^G = \langle d_1^G, \ldots, d_n^G \rangle$, where d_i^G's are the vertex degrees in non-increasing order. When the referred graph is clear from the context, we will use only Y and D to denote the y- and d-degree sequence respectively. (We note that we don't allow vertices with zero degree (i.e., singletons) in G. This is not really a issue, because we will deal with problems in which singletons can be treated separately from the rest of the graph to obtain a (optimum) solution to the problem with "minor effects" on the running time.)

Given a sequence of integers $S = \langle s_1, \ldots, s_m \rangle$, we define the following operator that expands S in a new non increasing sequence of integers.

Definition 1 (Expansion). *Let $S = \langle s_1, \ldots, s_n \rangle$ be a sequence of integers and $j \in [n]$. Then we define*

$$\mathbb{EXP}(S) = \langle \overbrace{n, \ldots, n}^{s_n}, \ldots, \overbrace{1, \ldots, 1}^{s_1} \rangle.$$

Note that the expansion operation converts a y-degree sequence into a d-sequence. In the rest of the paper, given two degree sequences $S = \langle s_1, \ldots, s_n \rangle$ and $T = \langle t_1, \ldots, t_m \rangle$ with $n \geq m$, we will denote $S - T = \langle x_1, \ldots, x_n \rangle$ with $x_i = s_i - t_i$ if $i \in [m]$ and $x_i = s_i$ otherwise.

The ACL model of power-law graphs introduced in [1] have a particular kind of y-degree sequence which we henceforth call (β, α)-degree sequence and is defined as follows.

Definition 2 ((β, α)-degree sequence). *Given $\alpha, \beta \in \mathbb{R}^+$, the y-degree sequence of a graph $G = (V, E)$ is a (β, α)-degree sequence (denoted by $Y^{(\beta,\alpha)} = \langle y_1^{(\beta,\alpha)}, \ldots, y_m^{(\beta,\alpha)} \rangle$) if $m = \lfloor e^{\alpha/\beta} \rfloor$ and, for $i \in [m]$*

$$y_i = \begin{cases} \left\lfloor \frac{e^\alpha}{i^\beta} \right\rfloor & \text{if } i > 1 \text{ or } \sum_{k=1}^m \left\lfloor \frac{e^\alpha}{k^\beta} \right\rfloor \text{ is even} \\ \lfloor e^\alpha \rfloor + 1 & \text{otherwise.} \end{cases}$$

In the rest of the paper, given a sequence of integers $S = \langle s_1, \ldots, s_k \rangle$, we will define $tot(S) = \sum_{i=1}^k s_i$ and $w(S) = \sum_{i=1}^k i s_i$. Note that if S is the y-degree sequence of a graph, then $w(S)$ is the total degree of the graph, whereas if S is the d-degree sequence of a graph, then $tot(S)$ is the total degree of the graph.

Our aim is to study the NP-hardness of graph-theoretic optimization problems when they are restricted to ACL power-law graphs with a fixed β, in particular, simple graphs belonging to this class. (Of course, showing hardness results for this class implies hardness for arbitrary power law graphs as well.) Formally, we define such graphs as:

Definition 3 (β-graph). *Given $\beta \in \mathbb{R}^+$, a graph $G = (V, E)$ is a β-graph if it is simple and there exists $\alpha \in \mathbb{R}^+$ such that the y-degree sequence of G is a (β, α)-degree sequence.*

3 NP-Hardness of CLIQUE AND COLORING

In this section we introduce a general technique to prove the NP-hardness of some optimization problems. The main idea of the proof is the following. Given an arbitrary graph G, it is possible to construct a simple graph G_1 which contains G as a set of maximal connected components. Let $G_2 = G_1 \backslash G$ be the remaining graph. Obviously, G_2 is simple and if we can show that we can efficiently (i.e., in polynomial time) compute the optimal solution in G_2 then this essentially gives us the result. However, it is a priori not obvious how to design an efficient algorithm given a particular problem. The key idea we will use here is that we have the choice of constructing G_1 (and hence G_2) and thus we can construct the graph in such a way that it admits an efficient algorithm. If we construct the graph in a careful way, it will be possible to design a polynomial time algorithm that finds the optimal.

Below we illustrate this idea by showing the NP-completeness of certain problems, including CLIQUE AND COLORING, in β-graphs for $\beta \geq 1$. Our idea here is to make G_2 to be a *simple bipartite* graph. Since bipartite graphs are 2-colorable and have a maximum clique of size 2, this immediately gives the reduction. Obviously, the main difficulty is in constructing the bipartite graph. We first need the following definitions.

Definition 4 (Contiguous Sequence). *A sequence* $D = \langle d_1, \ldots, d_n \rangle$ *with maximum value* m *is* contiguous *if* $y_i^D > 0$ *for all* $i \in [m]$, *where* $y_i^D = |\{j \in [n] \text{ s.t. } d_j = i\}|$.

Definition 5 (Bipartite-Eligible Sequence). *A sequence* $D = \langle d_1, \ldots, d_n \rangle$ *with maximum value* m *is* bipartite-eligible *if it is contiguous and* $m \leq \lfloor n/2 \rfloor$.

Given a simple graph $G = (V, E)$, for every vertex $u \in V$ we will denote $NEIG(u) = \{v \in V \backslash \{u\} \text{ s.t. } (u, v) \in E\}$.

Lemma 1. *Let* $D = \langle d_1, \ldots, d_n \rangle$ *be a sequence. If* D *is non increasing and bipartite-eligible and* $tot(D)$ *is even, then it is possible to construct in time* $O(n^2)$ *a simple bipartite graph* $G = (V, E)$ *such that* $D^G = D$.

Proof. First note that since D is non increasing and bipartite-eligible, $d_1 \leq \lfloor n/2 \rfloor$. We build the graph iteratively by adding some edges to certain vertices. Define the *residual* degree of a vertex as its final degree minus its "current" degree. Initially all the vertices have degree 0. To build the graph we use the following algorithm:

1. let $d(s_i)$ and $d(t_i)$ be the *residual degree* of the i-th vertex of S and T;
2. $E \leftarrow \emptyset$; $S \leftarrow \emptyset$; $T \leftarrow \emptyset$; $tot(S) \leftarrow 0$; $tot(S) \leftarrow 0$; $k \leftarrow |S|$; $l \leftarrow |T|$;
3. **while** $i \leq n$ **do**
   (a) **while** $i \leq n$ **and** $tot(S) \leq tot(T)$ **do**
      i. $S \leftarrow S \cup \{u \mid u \notin S\}$; $k \leftarrow k + 1$; $d(s_k) \leftarrow d_i$; $tot(S) \leftarrow tot(S) + d_i$;
   (b) **while** $i \leq n$ **and** $tot(T) \leq tot(S)$ **do**
      i. $T \leftarrow T \cup \{v \mid v \notin T\}$; $l \leftarrow l + 1$; $d(t_l) \leftarrow d_i$; $tot(T) \leftarrow tot(T) + d_i$;

4. **while** $tot(S) > 0$ **do**
   (a) **SORT** S and T separately in non increasing order of the residual degree;
   (b) **for** $i \leftarrow 1$ **to** $d(s_1)$ **do**
      i. $E \leftarrow E \cup \{(s_1, t_i)\}$; $d(s_1) \leftarrow d(s_1) - 1$; $d(t_i) \leftarrow d(t_i) - 1$; $tot(S) \leftarrow tot(S) - 1$; $tot(T) \leftarrow tot(T) - 1$;
   (c) **for** $i \leftarrow 2$ **to** $d(t_1) + 1$ **do**
      i. $E \leftarrow E \cup \{(t_1, s_i)\}$; $d(t_1) \leftarrow d(t_1) - 1$; $d(s_i) \leftarrow d(s_i) - 1$; $tot(T) \leftarrow tot(T) - 1$; $tot(S) \leftarrow tot(S) - 1$;
5. **return** $G = (S \cup T, E)$;

Note that the entire loop 3) requires $O(n^2)$ time to be completed. Moreover, in every iteration of the loop 4), at least one vertex is completed and will be no longer considered in the algorithm. Therefore, the loop 4) is completed in $O(n^2)$ time and the algorithm has complexity $O(n^2)$.

Now we prove that the algorithm correctly works. We first introduce some notations. The residual degree of the set S (T respectively) after the *SORT* instruction of the round i is denoted by $R_i(S)$ ($R_i(T)$ respectively). The number of vertices with positive residual degree (*non full* vertices) in S (T) is denoted by $N_i(S)$ ($N_i(T)$). The set S is $s_1^i, \ldots, s_h^i$ and the set T is $t_1^i, \ldots, t_k^i$.

The proof is by induction on the round i. More exactly, we prove the following invariant: *After the SORT instruction we have: (i)* $R_i(S) = R_i(T)$ *and (ii)* $N_i(T) \geq d(s_1^i)$ *and* $N_i(S) \geq d(t_1^i)$.

It is easy to see that if this invariant holds, then the algorithm correctly builds a bipartite graph. We start proving the base ($i = 1$) by showing that the above two conditions hold.

1. Let $tot_j(S)$ and $tot_j(T)$ be the total degree of the sets S and T after the insertion of the j-th vertex. We first show that $|tot_j(S) - tot_j(T)| \leq d_{j+1}$ for all $j \in [2, n-1]$. This is obvious for $j = 2$ since the sequence is non increasing and contiguous. Let us suppose that this is true until $j - 1$ and let us show it for j.

   Without loss of generality, let us suppose that the j-th vertex is assigned to T. Then this implies that $tot_{j-1}(S) \geq tot_{j-1}(T)$ and by induction $tot_{j-1}(S) - tot_{j-1}(T) \leq d_j$ and, therefore, $tot_j(S) - tot_j(T) \leq 0$.

   Now we can complete the proof of the bases. w.l.o.g. let us suppose that the last one vertex is assigned to T. Then we have $R_1(S) \geq R_1(T) - 1$. But from the preceding proof we also know that $R_1(S) \leq R_1(T)$ and, from the fact that the last one vertex has degree 1 and that the total degree of D is even, we have the claim.
2. Since the degree sequence is contiguous and after the insertion we have $tot(S) = tot(T)$, it is easy to see that after the insertion we have $-1 \leq |S| - |T| \leq 1$. From this and from the hypothesis $d_1 \leq \lfloor n/2 \rfloor$ the claim follows.

Let us suppose that the invariant is true until $i - 1$ and let us prove it for i.

1. We have $R_i(S) = R_{i-1}(S) - d(s_1^{i-1}) - (d(t_1^{i-1}) - 1) = R_{i-1}(T) - d(t_1^{i-1}) - (d(s_1^{i-1}) - 1) = R_i(T)$ as claimed.

2. The case $d(s_1^i) = 0$ is trivial, therefore let us suppose that $d(s_1^i) \geq 1$. If $d(s_2^{i-1}) = 1$, then $d(s_1^i) = 1$ since the degrees in S are non increasing. Moreover, from item (1) we have $R_i(T) = R_i(S) \geq 1$ and this completes this case.

   If $d(s_2^{i-1}) > 1$, then we have two cases. If $d(t_2^{i-1}) = 1$, from item (1) and the fact that $d(t_j^i) \leq 1$ for all j we simply have the claim. On the other hand, if $d(s_2^{i-1}) > 1$, we have $N_i(T) = N_{i-1}(T) \geq d(s_1^{i-1}) \geq d(s_1^i)$. □

The following lemma shows that for $\beta \geq 1$ it is possible to embed a simple graph G in a polynomial-size β-graph G_1 such that G is a set of maximal connected components of G_1 and $G_2 = G_1 \backslash G$ is bipartite-eligible.

Lemma 2. *Let $G = (V, E)$ be a simple graph with n_1 vertices and $\beta \geq 1$. Let $\alpha_0 = \max\{4\beta, \beta \ln n_1 + \ln(n_1 + 1)\}$. Then, for all $\alpha \geq \alpha_0$ the sequence $D = \mathbb{EXP}(Y^{(\beta,\alpha)} - Y^G)$ is contiguous and bipartite-eligible.*

Proof. Let n_2 be the number of elements in D and $\alpha \geq \alpha_0$. We have

$$n_2 \geq \sum_{i=1}^{\lfloor e^{\alpha/\beta} \rfloor} \left\lfloor \frac{e^\alpha}{i^\beta} \right\rfloor - n_1 > e^\alpha \sum_{i=1}^{\lfloor e^{\alpha/\beta} \rfloor} \frac{1}{i^\beta} - \lfloor e^{\alpha/\beta} \rfloor - n_1 \geq e^\alpha \int_{i=1}^{\lfloor e^{\alpha/\beta} \rfloor + 1} \frac{1}{i^\beta} - \lfloor e^{\alpha/\beta} \rfloor - n_1.$$

If $\beta = 1$, then we have

$$n_2 \geq \alpha e^\alpha - e^\alpha - n_1 \geq 4e^\alpha - 2e^\alpha + 1 \geq 2m + 1.$$

and if $\beta > 1$ we have

$$n_2 \geq \frac{e^\alpha}{\beta - 1} - e^{\alpha/\beta} - n_1 \geq 4e^{\alpha/\beta} - 2e^{\alpha/\beta} + 1 \geq 2m + 1.$$

Moreover $y_{n_1}^{(\beta,\alpha)} \geq \left\lfloor \frac{e^\alpha}{n_1^\beta} \right\rfloor > \frac{e^\alpha}{n_1^\beta} - 1 \geq \frac{n_1^{\beta+1} + n_1^\beta}{n_1^\beta} - 1 = n_1$, that is $\mathbb{EXP}(Y)$ is contiguous. Therefore, $\mathbb{EXP}(Y)$ is bipartite-eligible and this completes the proof of this lemma. □

We finally show the NP-completeness of certain problems in β-graphs with $\beta \geq 1$. The following definition is useful to introduce the class of problems we analyze in what follows.

Definition 6 (c-Oracle). *Let P be an optimization problem and $c > 0$ a constant. A c-oracle for the problem P is a polynomial-time algorithm $A_c^P(I)$ which takes in input an instance I of P and correctly returns an optimum solution for P given that on the instance I the problem has an optimum solution with size at most c.*

The following theorem shows the NP-completeness of a particular class of decision problems defined using the c-oracle in β-graphs with $\beta \geq 1$.

Theorem 1. *Let $\beta \geq 1$. Let P be a graph decision problem such that its optimization version obeys the following properties:*

1. *$OPT(G) = \max_{1 \leq i \leq k} OPT(C_i)$ (where C_i are the maximal connected components of G),*
2. *exists a constant $c > 0$ such that for all bipartite simple graphs H it holds $|OPT(H)| \leq c$ and*
3. *it admits a c-oracle.*

If P is NP-complete in general graphs, then it is NP-complete in β-graphs too.

Proof. From Lemmas 2 and 1, it is possible to construct, in time $poly(|G|)$, a β-graph G_1 embedding G such that $|G_1| = poly(|G|)$, G is a set of maximal connected components of G_1 and $G_2 = G_1 \backslash G$ is a simple bipartite graph. Since $OPT(G_1) = \max_k \{OPT(C_k)\}$, $|OPT(G_2)| \leq c$ and the optimization version of P admits a c-oracle, it is easy to see that P can be reduced in polynomial time to β-P (where β-P is P restricted to β-graphs). □

Since CLIQUE and COLORING satisfy all conditions of Theorem 1 with $c = 2$, we easily obtain the following corollary.

Corollary 1. *CLIQUE, and COLORING are NP-Complete in β-graphs for all $\beta \geq 1$.*

4 Hardness of Optimization Problems with Optimal Substructure

We show that if an optimization problem is NP-hard on (simple) general graphs (i.e., computing a solution in polynomial time is hard) and it satisfies the following "optimal substructure" property, then it is NP-hard on β-graphs also. We state this property as follows. Let P be an optimization problem which takes a graph as input. For *every* input G, the following should be true: every optimum solution of P on G should contain an optimum solution of P on each of G's maximal connected components. To illustrate with an example, it is easy to see that MINIMUM VERTEX COVER problem satisfies this property: an optimal vertex cover on any graph G should contain within it an optimal vertex cover of its maximal connected components. On the other hand, MINIMUM COLORING does not satisfy the above property, since the optimal coloring of a graph need not contain an optimal coloring of all its maximal connected components. We first need some definitions. We say that a sequence D is *graphic* if there exists a simple graph G such that $D^G = D$.

Definition 7 (Eligible Sequence). *A sequence of integers $S = \langle s_1, \ldots, s_n \rangle$ is eligible if $s_1 \geq \cdots \geq s_n$ and, for all $k \in [n]$, $f_S(k) \geq 0$, where*

$$f_S(k) = k(k-1) + \sum_{i=k+1}^{n} \min\{k, s_i\} - \sum_{i=1}^{k} s_i.$$

The following result due to Havel and Hakimi ([5]) gives a straightforward algorithm to construct a simple graph from a graphic degree sequence.

Lemma 3 ([5]). *A sequence of integers $D = \langle d_1, \ldots, d_n \rangle$ is graphic if and only if it is non-increasing, and the sequence of values $D' = \langle d_2 - 1, d_3 - 1, \ldots, d_{d_1+1} - 1, d_{d_1+2}, \ldots, d_n \rangle$ when sorted in non-increasing order is graphic.*

In the next technical lemma (whose complete proof appears in the full version [7]), we introduce a new sufficient condition for a sequence of integers to be eligible.

Lemma 4. *Let $Y^{(1)}$ and $Y^{(2)}$ be two degree sequences with m_1 and m_2 elements respectively such that (i) $y_j^{(1)} \leq y_j^{(2)}$ for all $j \in [m_1]$, and (ii) $D^{(1)} = \mathbb{EXP}(Y^{(1)})$ and $D^{(2)} = \mathbb{EXP}(Y^{(2)})$ are contiguous. If $D^{(1)}$ is eligible then $D^{(2)}$ is eligible.*

Proof sketch. Let $Y^{(1)}$ and $Y^{(2)}$ be two degree sequences with m_1 and m_2 elements respectively such that *(i)* $y_j^{(1)} \leq y_j^{(2)}$ for all $j \in [m_1]$, and *(ii)* $D^{(1)} = \mathbb{EXP}(Y^{(1)})$ and $D^{(2)} = \mathbb{EXP}(Y^{(2)})$ are contiguous.

We have to show that "if $D^{(1)}$ is eligible then $D^{(2)}$ is eligible". Let us note that the transformation from the degree sequence $Y^{(1)}$ to the degree sequence $Y^{(2)}$ (and hence from $D^{(1)}$ to $D^{(2)}$) can be seen as a sequence of rounds of the following type: in every step a vertex with degree d is transformed into a vertex with degree $(d+1)$ and the global sequence is rearranged with respect to the relation $y_j^{(1)} \leq y_j^{(2)}$ for all $j \in [m_1]$. In other words, to transform $Y^{(1)}$ to $Y^{(2)}$ (and hence $D^{(1)}$ to $D^{(2)}$) we can execute the following simple algorithm[2]:

1. $S^{(0)} \leftarrow D^{(1)}$; $i \leftarrow 0$
2. **while** $S^{(i)} \neq D^{(2)}$ **do**
   (a) **for** $j \leftarrow m_2$ **downto** 2 **do**
      i. **if** $|\{x \in S^{(i+1)} \text{ s.t. } x = j\}| < y_j^{(2)}$ **and** $|\{x \in S^{(i+1)} \text{ s.t. } x = j-1\}| > 0$ **then**
         A. $k \leftarrow \min\{x \in |S^{(i+1)} \text{ s.t. } s_x^{(i+1)} = j-1\}$
         B. $s_k^{(i+1)} \leftarrow s_k^{(i+1)} + 1$
   (b) $S^{(i+1)} \leftarrow S_{(i+1)} + x$
   (c) $i \leftarrow i + 1$

Let $n_2 = |D^{(2)}|$. From definition of eligibility, $D^{(2)}$ is eligible if $f_{D^{(2)}}(k) > 0$ for all $k \in [n_2]$. It can be showed that at the end of each iteration of the while loop of the previous algorithms, if $f_{S^{(i-1)}}(k) \geq 0$ for all $i \in [n^{(i-1)}]$ then $f_{S^{(i)}}(k) \geq 0$ for all $i \in [n^{(i)}]$, where $n^{(i)} = |S^{(i)}|$. Since $f_{S^{(0)}}(k) \geq 0$ for $k \in n^{(0)}$, this completes the proof of this lemma. □

The previous lemma is useful to show the following key lemma (*Embedding Lemma*) that shows that it is possible to quickly construct a β-graph with a certain property.

[2] In the rest of the paper, given a sequence $S = \langle s_1, \ldots, s_n \rangle$ and an integer x we will use the notation $S + x = \langle s_1, \ldots, s_n, x \rangle$ to denote the concatenation of S with the integer x.

Lemma 5 (Embedding Lemma). *Let $G = (V, E)$ be a simple undirected graph and $\beta \in \mathbb{R}^+$. Then there exists a simple undirected graph $G_1 = (V_1, E_1)$ such that G is a set of maximal connected components of G_1, $|V_1| = poly(|V|)$ and G_1 is a β-graph. Furthermore, given G, we can construct G_1 in time polynomial in the size of G.*

Proof. Let $n_1 = |V|$. From Lemma 3, we have only to show that there exist $\alpha_0 = O(\ln n_1)$ such that for all $\alpha \geq \alpha_0$, the degree sequence $D = \mathbb{EXP}(Y) = Y^{(\beta,\alpha)} - Y^G$ is graphic, that is, from Lemma 3 such that D is eligible. For $\beta \geq 1$ the proof directly comes from Lemmas 2 and 1. Let us complete the proof for $0 < \beta < 1$.

Note that, $y_i^{(1,\alpha)} \leq y_i^{(\beta,\alpha)}$ and $\lfloor e^{\alpha/\beta} \rfloor \geq \lfloor e^{\alpha} \rfloor$ for $0 < \beta < 1$ and $i \in \lfloor e^{\alpha} \rfloor$ and, from Lemma 2, $\mathbb{EXP}(Y^{(1,\alpha)} - Y^G)$ is contiguous for $\alpha \geq \max\{4, \ln n_1 + \ln(n_1+1)\}$.

Therefore, from Lemma 4, the sequence $\mathbb{EXP}(Y^{(\beta,\alpha)} - Y^G)$ is eligible for $0 < \beta < 1$ and $\alpha \geq \max\{4, \ln n_1 + \ln(n_1 + 1)\}$ and this completes the proof of this lemma. □

Now we are ready to show the main theorem of this section.

Theorem 2. *Let P be an optimization problem on graphs with the optimal substructure property. If P is NP-hard on (simple) general graphs, then it is also NP-hard on β-graphs for all $\beta > 0$.*

Proof. We show that we can reduce the problem of computing an optimal solution on general graphs to computing an optimal solution on β-graphs and this reduction takes polynomial time. Let $G = (V, E)$ be a simple undirected graph. Lemma 5 says that we can construct (in time polynomial in the size of G) a simple undirected graph $G_1 = (V_1, E_1)$ such that G is a set of maximal connected components of G_1, and G_1 is a β-graph with $|V_1| = poly(|V|)$. Since P has the optimal substructure property and G is a set of maximal connected components of G_1, this implies that an optimum solution for the graph G can be computed easily from an optimal solution for G_1. □

5 Concluding Remarks and Open Problems

We showed a general technique for establishing NP-hardness and NP-completeness of a large class of problems in power-law graphs. Our technique of "embedding" any arbitrary (given) graph into a polynomial-sized power-law graph is quite general and can have other applications, e.g., in showing hardness of *approximation* in power-law graphs (which is the next important question, now that we have established hardness). On the positive side, one may investigate approximation algorithms that exploit the power law property to get better approximation ratios compared to general graphs. Another interesting and relevant direction is to investigate the hardness or easiness of non-trivial restrictions of the ACL model.

We conclude by mentioning some open problems that follow directly from our work. We showed NP-hardness of CLIQUE and COLORING only for power law graphs with $\beta \geq 1$. We believe that a different construction might show that these problems are NP-Complete for all $\beta > 0$. It will also be interesting to investigate the complexity of node- and edge-deletion problems, that is a general and important class of problems defined in [13]. We finally note that our technique does not directly imply hardness in *connected* power-law graphs. We conjecture that our techniques can be extended to show these results.

References

1. Aiello, W., Chung, F.R.K., Lu., L.: A Random Graph Model for Massive Graphs. In: Proceedings of STOC 2000, pp. 171–180. ACM Press, New York (2000)
2. Aiello, W., Chung, F.R.K., Lu, L.: A random graph model for power law graphs. In Experimental Mathematics 10, 53–66 (2000)
3. Barabasi, A.: Emergence of Scaling in Complex Networks. In: Bornholdt, S., Schuster, H. (eds.) Handbook of Graphs and Networks, Wiley, Chichester (2003)
4. Bollobas, B., Riordan, O.: Mathematical Results on Scale-free Random Graphs. In: Bornholdt, S., Schuster, H. (eds.) Handbook of Graphs and Networks (2003)
5. Bondy, J.A., Murty, U.S.R.: Graph Theory with Applications. North-Holland, Amsterdam (1976)
6. Eubank, S., Kumar, V.S.A., Marathe, M.V., Srinivasan, A., Wang, N.: Structural and Algorithmic Aspects of Massive Social Networks. In: Proceedings of 15th ACM-SIAM Symposium on Discrete Algorithms (SODA 2004), pp. 711–720. ACM Press, New York (2004)
7. Ferrante, A., Pandurangan, G., Park, K.: On the Hardness of Optimization in Power-Law Graphs, `http://www.cs.purdue.edu/homes/gopal/papers-by-date.html`
8. Gkantsidis, C., Mihail, M., Saberi, A.: Throughput and Congestion in Power-Law Graphs. In: Proceedings of SIGMETRICS 2003, pp. 148–159. ACM Press, New York (2003)
9. Koyuturk, M., Grama, A., Szpankowski, W.: Assessing significance of connectivity and conservation in protein interaction networks. In: Apostolico, A., Guerra, C., Istrail, S., Pevzner, P., Waterman, M. (eds.) RECOMB 2006. LNCS (LNBI), vol. 3909, pp. 45–49. Springer, Heidelberg (2006)
10. Mihail, M., Papadimitriou, C., Saberi, A.: On Certain Connectivity Properties of the Internet Topology. In: Proc. of FOCS 2003, pp. 28–35. IEEE Computer Society Press, Los Alamitos (2003)
11. Park, K., Lee, H.: On the effectiveness of route-based packet filtering for distributed DoS attack prevention in power-law internets. In: Proceedings of SIGCOMM 2001, pp. 15–26. ACM Press, New York (2001)
12. Park, K.: The Internet as a complex system. In: Park, K., Willinger, W. (eds.) The Internet as a Large-Scale Complex System. Santa Fe Institute Studies on the Sciences of Complexity, Oxford University Press, Oxford (2005)
13. Yannakakis, M.: Node- and Edge-Deletion NP-Complete Problems. In: Proceedings of STOC 1978. SIAM 1978, San Diego, California, pp. 253–264 (1978)

Can a Graph Have Distinct Regular Partitions?

Noga Alon[1,⋆], Asaf Shapira[2], and Uri Stav[3]

[1] Schools of Mathematics and Computer Science, Raymond and Beverly Sackler Faculty of Exact Sciences, Tel Aviv University, Tel Aviv, Israel
nogaa@tau.ac.il
[2] Microsoft Research, Redmond, WA
asafico@microsoft.com
[3] School of Computer Science, Raymond and Beverly Sackler Faculty of Exact Sciences, Tel Aviv University, Tel Aviv, Israel
uristav@tau.ac.il.

Abstract. The regularity lemma of Szemerédi gives a concise approximate description of a graph via a so called regular-partition of its vertex set. In this paper we address the following problem: can a graph have two "distinct" regular partitions? It turns out that (as observed by several researchers) for the standard notion of a regular partition, one can construct a graph that has very distinct regular partitions. On the other hand we show that for the stronger notion of a regular partition that has been recently studied, all such regular partitions of the same graph must be very "similar".

En route, we also give a short argument for deriving a recent variant of the regularity lemma obtained independently by Rödl and Schacht ([11]) and Lovász and Szegedy ([9],[10]), from a previously known variant of the regularity lemma due to Alon et al. [2]. The proof also provides a deterministic polynomial time algorithm for finding such partitions.

1 Introduction

We start with some of the basic definitions of regularity and state the regularity lemmas that we refer to in this paper. For a comprehensive survey on the regularity lemma the reader is referred to [7]. For a set of vertices $A \subseteq V$, we denote by $E(A)$ the set of edges of the graph induced by A in G, and by $e(A)$ the size of $E(A)$. Similarly, if $A \subseteq V$ and $B \subseteq V$ are two vertex sets, then $E(A, B)$ stands for the set of edges of G connecting vertices in A and B, and $e(A, B)$ denotes the number of ordered pairs (a, b) such that $a \in A$, $b \in B$ and ab is an edge of G. Note that if A and B are disjoint this is simply the number of edges of G that connect a vertex of A with a vertex of B, that is $e(A, B) = |E(A, B)|$. The *edge density* of the pair (A, B) is defined as $d(A, B) = e(A, B)/|A||B|$. When several graphs on the same set of vertices are involved, we write $d_G(A, B)$ to specify the graph to which we refer.

⋆ Research supported in part by a grant from the Israel Science Foundation, by the Hermann Minkowski Minerva Center for Geometry at Tel Aviv University, and by a USA-Israeli BSF grant.

G. Lin (Ed.): COCOON 2007, LNCS 4598, pp. 428–438, 2007.

Definition 1 (ϵ-Regular Pair). *A pair (A, B) is* ϵ-regular, *if for any two subsets $A' \subseteq A$ and $B' \subseteq B$, satisfying $|A'| \geq \epsilon|A|$ and $|B'| \geq \epsilon|B|$, the inequality $|d(A', B') - d(A, B)| \leq \epsilon$ holds.*

A partition $\mathcal{A} = \{V_i \ : \ 1 \leq i \leq k\}$ of the vertex set of a graph is called an *equipartition* if $|V_i|$ and $|V_j|$ differ by no more than 1 for all $1 \leq i < j \leq k$ (so in particular each V_i has one of two possible sizes). For the sake of brevity, we will henceforth use the term partition to denote an equipartition. We call the number of sets in a partition (k above) the *order* of the partition.

Definition 2 (ϵ-Regular partition). *A partition $\mathcal{V} = \{V_i \ : \ 1 \leq i \leq k\}$ of $V(G)$ for which all but at most $\epsilon\binom{k}{2}$ of the pairs (V_i, V_j) are ϵ-regular is called an* ϵ-regular partition *of $V(G)$.*

The Regularity Lemma of Szemerédi can be formulated as follows.

Lemma 1 ([13]). *For every m and $\epsilon > 0$ there exists an integer $T = T_1(m, \epsilon)$ with the following property: Any graph G on $n \geq T$ vertices, has an ϵ-regular partition $\mathcal{V} = \{V_i \ : \ 1 \leq i \leq k\}$ with $m \leq k \leq T$.*

The main drawback of the regularity-lemma is that the bounds on the integer T, and hence on the order of $\mathcal{V}$, have an enormous dependency on $1/\epsilon$. The current bounds are towers of exponents of height $O(1/\epsilon^5)$. This means that the regularity measure (ϵ in Lemma 1) is very large compared to the inverse of the order of the partition (k in Lemma 1). In some cases, however, we would like the regularity measure between the pairs to have some (strong) relation to the order of the partition. This leads to the following definition.

Definition 3 (f-Regular partition). *For a function $f : N \mapsto (0, 1)$, a partition $\mathcal{V} = \{V_i \ : \ 1 \leq i \leq k\}$ of $V(G)$ is said to be* f-regular *if all pairs (V_i, V_j), $1 \leq i < j \leq k$, are $f(k)$-regular.*

Note that as opposed to Definition 2, in the above definition, the order of the partition and the regularity measure between the sets of the partition go "hand in hand" via the function f. One can (more or less) rephrase Lemma 1 as saying that every graph has a $(\log^*(k))^{-1/5}$-regular partition [1]. Furthermore, Gowers [6] showed that this is close to being tight. Therefore, one cannot guarantee that a general graph has an f-regular partition for a function f approaching zero faster than roughly $1/\log^*(k)$. One should thus look for certain variants of this notion and still be able to show that any graph has a similar partition.

A step in this direction was first taken by Alon, Fischer, Krivelevich and Szegedy [2] who proved a stronger variant of the regularity lemma. See Lemma 2 below for the precise statement. The following is yet another variant of the regularity lemma that was recently proved independently by Rödl and Schacht [11] and by Lovász [9] (implicitly following a result of Lovász and Szegedy in [10]).

[1] This is not accurate because Definition 3 requires *all* pairs to be $f(k)$-regular, while Lemma 1 guarantees that only *most* pairs are regular.

This lemma does not guarantee that for any f we can find an f-regular partition of any given graph. Rather, it shows that any graph is "close" to a graph that has an f-regular partition.

Theorem 1 ([11], [9]). *For every m, $\epsilon > 0$ and non-increasing function $f : N \mapsto (0,1)$, there is an integer $T = T_1(f, \epsilon, m)$ so that given a graph G with at least T vertices, one can add-to/remove-from G at most ϵn^2 edges and thus get a graph G' that has an f-regular partition of order k, where $m \leq k \leq T$.*

Our first result in this paper is a new short proof of the above theorem. The proof is a simple application of the variant of the regularity lemma of [2] mentioned above. Basing the proof on this method provides both explicit bounds and a polynomial time algorithm for finding the partition and the necessary modifications. Section 2 consists of the proof of Theorem 1 and in Section 3 we describe a deterministic polynomial time algorithm for finding a regular partition and a set of modifications that are guaranteed by this theorem.

We now turn to the second result of this paper. In many cases, one applies the regularity lemma on a graph G, to get an ϵ-regular partition $\mathcal{V} = \{V_i : 1 \leq i \leq k\}$ and then defines a weighted complete graph on k vertices $\{1, \ldots, k\}$, in which the weight of the edge connecting vertices (i, j) is $d(V_i, V_j)$. This relatively small weighted graph, sometimes called the *regularity-graph* of G, carries a lot of information on G. For example, it can be used to approximately count the number of copies of any fixed small graph in G, and to approximate the size of the maximum-cut of G. A natural question, which was suggested to us by Madhu Sudan [12], is how different can two regularity-graphs of the same graph be. We turn to define what it means for two regularity graphs, or equivalently for two regular partitions, to be ϵ-isomorphic.

Definition 4 (ϵ-Isomorphic). *We say that two partitions $\mathcal{U} = \{U_i : 1 \leq i \leq k\}$ and $\mathcal{V} = \{V_i : 1 \leq i \leq k\}$ of a graph G are ϵ-isomorphic if there is a permutation $\sigma : [k] \mapsto [k]$, such that for all but at most $\epsilon\binom{k}{2}$ pairs $1 \leq i < j \leq k$, we have $|d(U_i, U_j) - d(V_{\sigma(i)}, V_{\sigma(j)})| \leq \epsilon$.*

We first show that if one considers the standard notion of an ϵ-regular partitions (as in Definition 2), then ϵ-regular partitions of the same graph are not necessarily similar. In fact, as the following theorem shows, even $f(k)$-regular partitions of the same graph, where $f(k) = 1/k^{\delta}$, are not necessarily similar. A variant of this theorem has been proved by Lovász [9].

Theorem 2. *Let $f(k) = 1/k^{1/4}$. For infinitely many k, and for every $n > n_2(k)$ there is a graph $G = (V, E)$ on n vertices with two f-regular partitions of order k that are not $\frac{1}{4}$-isomorphic.*

The proof of Theorem 2 provides explicit examples. We note that an inexplicit probabilistic proof shows that the assertion of the theorem holds even for $f(k) = \Theta(\frac{\log^{1/3} k}{k^{1/3}})$. See Section 4 for more details.

Using the terminology of Definition 2, the above theorem and its proof can be restated as saying that for any (small) $\epsilon > 0$ and all large enough $n > n_0(\epsilon)$, there exists an n vertex graph that has two ϵ-regular partitions of order ϵ^{-4}, that are not $\frac{1}{4}$-similar. Therefore, ϵ-regular partitions of the same graph may be very far from isomorphic.

Recall now that Theorem 1 guarantees that for any function f, any graph can be slightly modified in a way that the new graph admits an f-regular partition. As the following theorem shows, whenever $f(k) < 1/k^2$ all the regular partitions of the new graph must be close to isomorphic.

Theorem 3. *Let $f(k)$ be any function satisfying $f(k) \leq \min\{1/k^2, \frac{1}{2}\epsilon\}$, and suppose $\mathcal{U}$ and $\mathcal{V}$ are two f-regular partitions of some graph G. Then $\mathcal{U}$ and $\mathcal{V}$ are ϵ-isomorphic.*

This theorem illustrates the power of f-regular partitions, showing that (for $f(k) < 1/k^2$) they enjoy properties that do not hold for usual regular partitions. Observe that the above results imply that when, e.g., $f(k) > 1/k^{\frac{1}{4}}$, then two f-regular partitions of the same graph are not necessarily similar, whereas whenever $f(k) < 1/k^2$ they are. It may be interesting to find a tight threshold for f that guarantees ϵ-isomorphism between f-regular partitions of the same graph. It should also be interesting to find a similar threshold assuring that partitions of two *close* graphs are similar.

2 Proof of Theorem 1

In this section we show how to derive Theorem 1 from a variant of the regularity lemma due to Alon et al. [2]. Before we get to the proof we observe the following three simple facts. First, a standard probabilistic argument shows that for every δ and η, and for every large enough $n > n_0(\delta)$ there exists a δ-regular pair (A, B) with $|A| = |B| = n$ and $d(A, B) = \eta$. [2] The additional two facts we need are given in the following two claims, where we use the notation $x = y \pm \epsilon$ to denote the fact that $y - \epsilon \leq x \leq y + \epsilon$.

Claim 2.1. *Let δ and γ be fixed positive reals and let $n > n_0(\delta, \gamma)$ be a large enough integer. Suppose (A, B) is a δ-regular pair satisfying $d(A, B) = \eta \pm \gamma$ and $|A| = |B| = n$. Then, one can add or remove at most $2\gamma n^2$ edges from (A, B) and thus turn it into a 3δ-regular pair satisfying $d(A, B) = \eta \pm \delta$.*

Proof. Let us assume that $d(A, B) = \eta + \gamma$. The general case where $\eta - \gamma \leq d(A, B) \leq \eta + \gamma$ is similar. Suppose we delete each of the edges connecting A and B with probability $\frac{\gamma}{\eta+\gamma}$. Clearly the expected value of $d(A, B)$ after these modifications is η and assuming n is large enough, we get from a standard application of Chernoff's bound that the probability that the new density deviates from η by more than δ is at most $\frac{1}{4}$. Also, the expected number of edges removed is γn^2 and again, if n is large enough, the probability that we removed more than

[2] Here and throughout the rest of the paper, we say that $d(A, B) = \eta$ if $|e(A, B) - \eta|A||B|\,| \leq 1$. This avoids rounding problems arising from the fact that $\eta|A||B|$ may be non-integral.

$2\gamma n^2$ edges is at most $\frac{1}{4}$. Consider now two subsets $A' \subseteq A$ and $B' \subset B$ each of size δn. As (A, B) was initially δ-regular we initially had $d(A', B') = (\eta+\gamma) \pm \delta$. As each edge is removed with probability $\frac{\gamma}{\eta+\gamma}$ the expected value of $d(A', B')$ after these modifications is $\eta \pm \frac{\delta\eta}{\eta+\gamma} = \eta \pm \delta$. By Chernoff's bound we get that for large enough n, for every such pair (A', B') the probability that $d(A', B')$ deviates from $\eta \pm \delta$ by more than δ is bounded by 2^{-4n}. As there are less than 2^{2n} choices for (A', B') we get that with probability at least $\frac{3}{4}$ all pairs (A', B') have density $\eta \pm 2\delta$. To recap, we get that with probability at least $\frac{1}{4}$ we made at most $2\gamma n^2$ modifications, $d(A, B) = \eta \pm \delta$ and $d(A', B') = \eta \pm 2\delta$, implying that (A, B) is 3δ-regular. □

Claim 2.2. *Let (A, B) be a pair of vertex sets with $|A| = |B| = n$. Suppose A and B are partitioned into subsets $A_1, \ldots, A_l$ and $B_1, \ldots, B_l$ such that all pairs (A_i, B_j) are $\frac{1}{4}\delta^2$-regular and satisfy $d(A_i, B_j) = d(A, B) \pm \frac{1}{4}\delta$. Then (A, B) is δ-regular.*

Proof. Consider two subsets $A' \subseteq A$ and $B' \subseteq B$ of size δn each, and set $A'_i = A' \cap A_i$ and $B'_i = B' \cap B_i$. The number of pairs $(a \in A', b \in B')$, where $a \in A'_i$, $b \in B'_j$, and either $|B'_j| < \frac{1}{4}\delta^2 |B_j|$ or $|A'_i| < \frac{1}{4}\delta^2 |A_i|$ is bounded by $\frac{1}{2}\delta^3 n^2$. Therefore, the possible contribution of such pairs to $d(A', B')$ is bounded by $\frac{1}{2}\delta$.

Consider now the pairs (A'_i, B'_j) satisfying $|B'_j| \geq \frac{1}{4}\delta^2 |B_j|$ and $|A'_i| \geq \frac{1}{4}\delta^2 |A_i|$. As (A_i, B_j) is $\frac{1}{4}\delta^2$-regular we have $d(A'_i, B'_j) = d(A_i, B_j) \pm \frac{1}{4}\delta$. As $d(A_i, B_j) = d(A, B) \pm \frac{1}{4}\delta$ we conclude that $d(A'_i, B'_j) = d(A, B) \pm \frac{1}{2}\delta$. As the pairs discussed in the preceding paragraph can change $d(A', B')$ by at most $\frac{1}{2}\delta$, we conclude that $d(A', B') = d(A, B) \pm \delta$, as needed. □

The following is the strengthened version of the regularity lemma, due to Alon et al. [2], from which we will deduce Theorem 1.

Lemma 2 ([2]). *For every integer m and function $f : N \mapsto (0, 1)$ there exists an integer $T = T_2(m, f)$ with the following property: If G is a graph with $n \geq T$ vertices, then there exists a partition $\mathcal{A} = \{V_i \ : \ 1 \leq i \leq k\}$ and a refinement $\mathcal{B} = \{V_{i,j} \ : \ 1 \leq i \leq k, 1 \leq j \leq l\}$ of $\mathcal{A}$ that satisfy:*

1. *$|\mathcal{A}| = k \geq m$ but $|\mathcal{B}| = kl \leq T$.*
2. *For all $1 \leq i < i' \leq k$, for all $1 \leq j, j' \leq l$ but at most $f(k)l^2$ of them, the pair $(V_{i,j}, V_{i',j'})$ is $f(k)$-regular.*
3. *All $1 \leq i < i' \leq k$ but at most $f(0)\binom{k}{2}$ of them are such that for all $1 \leq j, j' \leq l$ but at most $f(0)l^2$ of them, $|d(V_i, V_{i'}) - d(V_{i,j}, V_{i',j'})| < f(0)$ holds.*

Proof of Theorem 1: Given a graph G, an integer m, a real ϵ and some function $f : N \mapsto (0, 1)$ as an input to Theorem 1, let us apply Lemma 2 with the function $f'(k) = \min\{f^2(k)/12, \epsilon/8\}$ and with $m' = m$. By Lemma 2, if G has more than $T = T_2(m', f')$ vertices, then G has two partitions $\mathcal{A} = \{V_i \ : \ 1 \leq i \leq k\}$ and $\mathcal{B} = \{V_{i,j} \ : \ 1 \leq i \leq k, 1 \leq j \leq l\}$ satisfying the three assertions of the lemma.

We claim that we can make less than ϵn^2 modifications in a way that all pairs (V_i, V_j) will become $f(k)$-regular.

We start by considering the pairs $(V_{i,j}, V_{i',j'})$, with $i < i'$, which are not $f'(k)$-regular. Every such pair is simply replaced by an $f'(k)$-regular bipartite graph of density $d(V_{i,j}, V_{i',j'})$. Such a pair exists by the discussion at the beginning of this section. The number of edge modifications needed for each such pair is at most $(n/kl)^2$ and by the second assertion of Lemma 2 we get that the total number of modifications we make at this stage over all pairs (V_i, V_j) is bounded by $\binom{k}{2} \cdot f'(k)l^2 \cdot (n/kl)^2 \leq \frac{\epsilon}{8}n^2$.

We now consider the pairs $(V_i, V_{i'})$ that do not satisfy the third assertion of Lemma 2, that is, those for which there are more than $f'(0)l^2$ pairs $1 \leq j, j' \leq l$ satisfying $|d(V_i, V_{i'}) - d(V_{i,j}, V_{i',j'})| \geq f'(0)$. For every such pair $(V_i, V_{i'})$ we simply remove all edges connecting V_i and $V_{i'}$. As by the third assertion there are at most $f'(0)\binom{k}{2} < \frac{\epsilon}{8}k^2$ such pairs, the total number of edge modifications we make is bounded by $\frac{\epsilon}{8}n^2$.

We finally consider the pairs $(V_i, V_{i'})$ that satisfy the third assertion of Lemma 2. Let us denote $d = d(V_i, V_{i'})$. We start with pairs $(V_{i,j}, V_{i',j'})$ satisfying $|d - d(V_{i,j}, V_{i',j'})| \geq f'(0)$. Each such pair is replaced with an $f'(k)$-regular pair of density d. As there are at most $f'(0)l^2 \leq \frac{\epsilon}{8}l^2$ such pairs in each pair (V_i, V_j), the total number of modifications made in the whole graph due to such pairs is bounded by $\frac{\epsilon}{8}n^2$. Let us now consider the pairs $(V_{i,j}, V_{i',j'})$ satisfying $|d - d(V_{i,j}, V_{i',j'})| \leq f'(0)$. If $d(V_{i,j}, V_{i',j'}) = d \pm f'(k)$ we do nothing. Otherwise, we apply Claim 2.1 on $(V_{i,j}, V_{i',j'})$ with $\eta = d$, $\gamma = |d - d(V_{i,j}, V_{i',j'})|$ and $\delta = f'(k)$. Note that here we are guaranteed to have $\gamma \leq f'(0) \leq \frac{1}{8}\epsilon$. Claim 2.1 guarantees that we can make at most $2\gamma(n/kl)^2 \leq \frac{1}{4}\epsilon(n/kl)^2$ modifications and thus turn $(V_{i,j}, V_{i',j'})$ into a $3f'(k)$-regular pair with density $d \pm f'(k)$. The total number of modifications over the entire graph is bounded by $\frac{\epsilon}{4}n^2$.

To conclude, the overall number of modifications we have made in the above stages is less than ϵn^2, as needed. Moreover, at this stage all the pairs $(V_{i,j}, V_{i',j'})$ satisfy $|d(V_{i,j}, V_{i',j'}) - d(V_i, V_{i'})| \leq f'(k) \leq \frac{1}{4}f(k)^2$ and they are all $\frac{1}{4}f^2(k)$-regular. Therefore, by Claim 2.2 all pairs (V_i, V_j) are $f(k)$-regular, as needed. □

3 Deterministic Algorithmic Version of Theorem 1

As mentioned before, we show that it is also possible to obtain an algorithmic version of Theorem 1. Here is a rough sketch, following the proof of Theorem 1 step by step. As described in [2], one can obtain the partition of Lemma 2 in polynomial time. In order to find the modifications that make it f-regular, the random graphs can be replaced by appropriate pseudo-random bipartite graphs. The last ingredient we need is an algorithm for finding the modifications to a bipartite graph (A, B) that are guaranteed by Claim 2.1. The algorithm we describe here combines the use of conditional probabilities (see, e.g., [3]) with a certain local condition that ensures regularity. We first describe such a condition.

Given a bipartite graph on a pair of vertex sets (A,B) we denote by $d_{C_4}(A,B)$ the density of four-cycles in (A,B), namely, the number of copies of C_4 divided by $\binom{|A|}{2}\binom{|B|}{2}$. A pair (A,B) is said to be *ϵ-quad-regular* if $d_{C_4}(A,B) = d^4(A,B) \pm \epsilon$. This local condition indeed ensures ϵ-regularity, as detailed in the following Lemma. The proof of the lemma appears in [5] and is based on the results of [1].

Lemma 3 ([5]). *Let (A,B) be a bipartite graph on A and B where $|A| = |B| = n$ and $\delta > 0$. Then:*

1. *If (A,B) is $\frac{1}{4}\delta^{10}$-quad-regular then it is δ-regular.*
2. *If (A,B) is δ-regular then it is 8δ-quad-regular.*

We shall design a deterministic algorithm for the following slightly weaker version of Claim 2.1.

Claim 3.1. *There is a deterministic polynomial time algorithm that given a $\frac{1}{200}\delta^{20}$-regular pair (A,B) with n vertices in each part (with n large enough) and $d(A,B) = \eta \pm \gamma$, modifies up to $2\gamma n^2$ edges and thus turns the bipartite graph into a 2δ-regular pair with edge density $d'(A,B) = \eta \pm \delta$.*

Note that the polynomial loss in the regularity measure with respect to Claim 2.1 can be evened by modifying the definition of f' in the proof of Theorem 1 so that $f'(k) = \min\{f^2(k)/8, \epsilon^{20}/2000\}$. Hence Claim 3.1 indeed implies an algorithm for finding the modifications and partition guaranteed by Theorem 1.

Proof of Claim 3.1: Assume $d(A,B) = \eta + \gamma$ and $\gamma > \delta$. The case $d(A,B) = \eta - \gamma$ can be treated similarly.

Consider an arbitrary ordering of the edges of (A,B) and a random process in which each edge is deleted independently with probability $\frac{\gamma}{\eta+\gamma}$. We first consider this setting and later show that a sequence of deterministic choices of the deletions can be applied so that the resulting graph satisfies the desired properties.

Define the indicator random variable X_i, $1 \leq i \leq t = \eta n^2$, for the event of *not* deleting the i'th edge. Denote the number of four cycles in (A,B) by $s = d_{C_4}(A,B)\binom{n}{2}^2$, and arbitrarily index them by $1,\ldots,s$. For every C_4 in (A,B) define the indicator Y_i, $1 \leq i \leq s$, for the event of its survival (i.e., none of its edges being deleted). Also let $X = \sum_{i=1}^{t} X_i$ and $Y = \sum_{i=1}^{s} Y_i$ which account for the numbers of edges and four-cycles (respectively) at the end of this process. Now define the following conditional expectations for $i = 0,1,\ldots,t$.

$$f_i(x_1,\ldots,x_i) = \mathbb{E}\left[n^4(X-\eta n^2)^2 + (Y - \eta^4\tbinom{n}{2}^2)^2 \mid X_1 = x_1,\ldots,X_i = x_i\right] \quad (1)$$

where the expectation in (1) is taken over a uniform independent choice of $X_{i+1},\ldots,X_t$.

We first obtain an upper bound on f_0. Since $X \sim B((\eta+\gamma)n^2, \frac{\eta}{\eta+\gamma})$, hence $\mathbb{E}[(X-\eta n^2)^2] = V(X) = O(n^2)$ and thus the first term in the expression for f_0 is $O(n^6)$. The expectation of the second term is

$$\mathbb{E}[(Y - \eta^4\tbinom{n}{2}^2)^2] = \mathbb{E}[Y^2] - 2\mathbb{E}[Y]\eta^4\tbinom{n}{2}^2 + \eta^8\tbinom{n}{2}^4$$

For the linear term we have $\mathbb{E}[Y] = \sum_{i=1}^{s} \mathbb{E}[Y_i] = s(\frac{\eta}{\eta+\gamma})^4$. As for the quadratic term, for any pair $1 \le i < j \le s$ of four-cycles which share no common edge, the corresponding Y_i and Y_j are independent and hence $\mathbb{E}[Y_i Y_j] = (\frac{\eta}{\eta+\gamma})^8$. There are only $O(n^6)$ non-disjoint pairs of C_4s, thus $\mathbb{E}[Y^2] = \mathbb{E}[\sum_{1\le i,j\le s} Y_i Y_j] = s^2(\frac{\eta}{\eta+\gamma})^8 \pm O(n^6)$. By Lemma 3, $d_{C_4}(A,B) = (\eta+\gamma)^4 \pm \frac{1}{25}\delta^{20}$ and so $s = ((\eta+\gamma)^4 \pm \frac{1}{25}\delta^{20})\binom{n}{2}^2$. Therefore, we conclude that

$$\begin{aligned}\mathbb{E}[(Y - \eta^4\tbinom{n}{2}^2)^2] &= s^2(\tfrac{\eta}{\eta+\gamma})^8 \pm O(n^6) - 2s(\tfrac{\eta}{\eta+\gamma})^4\eta^4\tbinom{n}{2}^2 + \eta^8\tbinom{n}{2}^4 \\ &\le \tfrac{1}{5}\delta^{20}\tbinom{n}{2}^4 + O(n^6)\end{aligned}$$

This implies that altogether, for a large enough n, $f_0 \le \frac{1}{4}\delta^{20}\binom{n}{2}^4$.

However, each $f_i(x_1,\dots,x_i)$ is a convex combination of $f_{i+1}(x_1,\dots,x_i,0)$ and $f_{i+1}(x_1,\dots,x_i,1)$. Thus, there is some choice of a value x_{i+1} for X_{i+1} such that $f_{i+1}(x_1,\dots,x_{i+1}) \le f_i(x_1,\dots,x_i)$. Therefore, choosing an x_{i+1} that minimizes f_{i+1} sequentially for $i = 0,\dots,t-1$ results in an assignment of $(x_1,\dots,x_t)$ such that $f_t(x_1,\dots,x_t) \le f_0 \le \frac{1}{4}\delta^{20}\binom{n}{2}^4$. In order to apply this process, one needs to be able to efficiently compute f_i. But this is straightforward, since for any partial assignment of values to the X_is, the mutual distribution of any pair Y_i, Y_j can be calculated in time $O(1)$. Therefore, since there are at most $O(n^8)$ pairs of four-cycles, computing the expected value of the sum in (1) requires $O(n^8)$ operations. Repeating this for each edge accumulates to $O(n^{10})$.[3]

To complete the proof of the claim, we only need to show that the modifications we obtained above, namely such that $(x_1,\dots,x_t)$ satisfy $f_t(x_1,\dots,x_t) \le \frac{1}{4}\delta^{20}\binom{n}{2}^4$, are guaranteed to satisfy the conditions of the claim. Indeed, in this case, each of the two addends which sum up to f_t is bounded by $\frac{1}{4}\delta^{20}\binom{n}{2}^4$. By the first addend, the new edge density d' is $d' = \eta \pm \frac{1}{2}\delta^{10}$. Thus, with much room to spare, the conditions on the edge density and the number of modifications are fulfilled. Note that it also follows that $d'^4 = \eta^4 \pm 3\delta^{10}$ (for, e.g., $\delta < \frac{1}{4}$), and the second addend implies that the new four-cycles density is $\eta^4 \pm \frac{1}{2}\delta^{10} = d'^4 \pm 4\delta^{10}$. By Lemma 3 the pair is now $4^{1/5}\delta$-regular, and hence the modified graph attains all the desired properties. □

Remark 1. Another possible proof of Claim 3.1 can be obtained by using an appropriate 8-wise independent space for finding $(x_1,\dots,x_t)$ such that f_t attains at most its expected value.

4 Isomorphism of Regular Partitions

In this section we prove Theorems 2 and 3. In order to simplify the presentation, we omit all floor and ceiling signs whenever these are not crucial. We

[3] Note that each edge effects only at most $O(n^6)$ pairs of four cycles, thus the complexity can easily be reduced to $O(n^8)$, and in fact can be further reduced by a more careful implementation.

start with the proof of Theorem 2. The basic ingredient of the construction is a pseudo-random graph which satisfies the following conditions.

Lemma 4. *Let k be a square of a prime power, then there exists a graph $F = (V, E)$ on $|V| = k$ vertices such that*

1. *F is $\lfloor k/2 \rfloor$-regular, and hence* $d(V, V) = \frac{\lfloor k/2 \rfloor}{k}$
2. *For any pair of vertex sets A and B, if $|A| \geq k^{\frac{3}{4}}$ and $|B| \geq k^{\frac{3}{4}}$, then* $d(A, B) = d(V, V) \pm k^{-\frac{1}{4}}$

Proof: We use some known pseudo-random graphs as follows, see the survey [8] for further definitions and details. An *(n, d, λ)-graph* is a d-regular graph on n vertices all of whose eigenvalues, except the first one, are at most λ in their absolute values. It is well known that if λ is much smaller than d, then such graphs have strong pseudo-random properties. In particular, (see, e.g., [3], Chapter 9), in this case for any two sets of vertices A and B of G: $d(A, B) = \frac{d}{n} \pm \lambda(|A||B|)^{-\frac{1}{2}}$. Thus, it is easy to verify that a $(k, \lfloor \frac{k}{2} \rfloor, \sqrt{k})$-graph would satisfy the assertions of the lemma.

There are many known explicit constructions of (n, d, λ)-graphs. Specifically, we use the graph constructed by Delsarte and Goethals and by Turyn (see [8]). In this graph the vertex set $V(G)$ consists of all elements of the two dimensional vector space over $GF(q)$, where q is a prime power, so G has $k = q^2$ vertices. To define the edges of G we fix a set L of $\frac{q+1}{2}$ lines through the origin. Two vertices x and y of the graph G are adjacent if $x - y$ is parallel to a line in L. It is easy to check that this graph is $\frac{(q+1)(q-1)}{2} = \frac{q^2-1}{2}$ -regular. Moreover, because it is a strongly regular graph, one can compute its eigenvalues precisely and show that besides the first one they all are either $-\frac{q+1}{2}$ or $\frac{q-1}{2}$. Therefore, indeed, we obtain an $(k, \lfloor \frac{k}{2} \rfloor, \lambda)$-graph with $\lambda < \sqrt{k}$ as necessary. □

Proof of Theorem 2: We construct our example as follows. Pick a graph F on k vertices $V(F) = \{1, \ldots, k\}$ which satisfies the conditions of Lemma 4. Suppose $n \geq k^2$. The graph on n vertices G will be an $\frac{n}{k}$ blow-up of F: every vertex of F is replaced by an independent set of size $\frac{n}{k}$, and each edge is replaced by a complete bipartite graph connecting the corresponding independent sets. Every non-edge corresponds to an empty bipartite graph between the parts. Let $\mathcal{U} = \{U_i \ : \ 1 \leq i \leq k\}$ be the partition of $V(G)$ where U_i is an independent set which corresponds to the vertex i in F. It follows from the construction that for any $1 \leq i < j \leq k$ the edge density of (U_i, U_j) is either 0 or 1, and (U_i, U_j) is ϵ-regular for any $\epsilon > 0$. The second partition $\mathcal{V}$ is generated by arbitrarily splitting every U_i into k equal-sized sets $W_{i,t}$, $1 \leq t \leq k$, and setting $V_t = \bigcup_{i=1}^{k} W_{i,t}$. Note that for any $1 \leq i < j \leq k$ the edge density $d_G(V_i, V_j)$ is exactly $d_F(V(F), V(F))$. Yet by Lemma 4, $d_F(V(F), V(F)) = \frac{2e(F)}{k^2} = \frac{\lfloor k/2 \rfloor}{k}$, which for $k \geq 2$ is strictly between $\frac{1}{4}$ and $\frac{3}{4}$. Hence $\mathcal{U}$ and $\mathcal{V}$ are not $\frac{1}{4}$-similar, as $|d(U_i, U_j) - d(V_{i'}, V_{j'})| > \frac{1}{4}$ for all pairs $i < j$ and $i' < j'$.

Thus, we complete the proof of the theorem by showing that all pairs (V_i, V_j) are $k^{-\frac{1}{4}}$-regular. Suppose, towards a contradiction and without loss of generality, that there are subsets $A \subseteq V_1$ and $B \subseteq V_2$ such that $|A| \geq k^{-\frac{1}{4}}|V_1|$, $|B| \geq k^{-\frac{1}{4}}|V_2|$ and $|d(A,B) - d(V_1,V_2)| > k^{-\frac{1}{4}}$.

For any $1 \leq i \leq k$ we denote $A_i = A \cap W_{i,1}$ and $B_i = B \cap W_{i,2}$. For any vertex $x \in A$, let the *fractional degree* of x with respect to B be defined by $d_B(x) = e(\{x\}, B)/|B|$. Note that $d(A,B) = \frac{1}{|A|}\Sigma_{x\in A} d_B(x)$ and that if x_1 and x_2 come from the same $W_{i,1}$, then $d_B(x_1) = d_B(x_2)$. Therefore, $d(A,B)$ is a convex combination

$$d(A,B) = \sum_{i=1}^{k} \frac{|A_i|}{|A|} d_B(x \in W_{i,1})$$

of (at most) k possible fractional degrees of vertices in A, which come from different sets $W_{i,1}$.

First assume that $d(A,B) > d(V_1,V_2) + k^{-\frac{1}{4}}$. We sort the vertices of A by their fractional degrees with respect to B, and consider a subset $\hat{A}$ of V_1 which consists of the union of the $k^{\frac{3}{4}}$ sets $W_{i,1}$ which have the highest fractional degrees with respect to B. Since $|A| > k^{-\frac{1}{4}}|V_1| = |\hat{A}|$, it follows that $d(\hat{A}, B) \geq d(A,B)$. Similarly, by considering the fractional degrees of the vertices of B with respect to the new subset $\hat{A}$, we may obtain a subset $\hat{B}$ of V_2 such that $d(\hat{A},\hat{B}) \geq d(\hat{A},B) \geq d(A,B) > d(V_1,V_2) + k^{-\frac{1}{4}}$. It also follows that both $\hat{A}$ and $\hat{B}$ are unions of sets $W_{i,1}$ and $W_{i,2}$ respectively. Thus, the edge density $d(\hat{A},\hat{B})$ is exactly the edge density of the corresponding vertex sets in F (both of size $k^{\frac{3}{4}}$). By Lemma 4, we get that $d(\hat{A},\hat{B}) \leq d_F(V(F),V(F)) + k^{-\frac{1}{4}} = d_G(V_1,V_2) + k^{-\frac{1}{4}}$, which leads to a contradiction and completes the proof of Theorem. 2. The case where $d(A,B) < d(V_1,V_2) - k^{-\frac{1}{4}}$ can be treated similarly. □

Remark 2. By using the random graph $G(k, \frac{1}{2})$ one could establish an inexplicit probabilistic proof for an analog of Lemma 4. The proof applies standard Chernoff bounds on the number of edges between *any* pair of *small* vertex sets. This extends the result for any $k > 2$ and with a stronger regularity constraint. Repeating the proof of Theorem 2 with such a graph F implies that Theorem 2 holds even for $f(k) = \Theta(\frac{\log^{1/3} k}{k^{1/3}})$.

We conclude this section with the proof of Theorem 3.

Proof of Theorem 3: Consider two f-regular partitions $\mathcal{U} = \{U_i : 1 \leq i \leq k\}$ and $\mathcal{V} = \{V_i : 1 \leq i \leq k\}$ of order k. Let $W_{i,j}$ denote $V_i \cap U_j$. Consider a matrix A where $A_{i,j} = \frac{|W_{i,j}|}{|V_i|}$ is the fraction of vertices of V_i in U_j, and note that A is doubly stochastic, that is, the sum of entries in each column and row is precisely 1. A well known (and easy) theorem of Birkhoff [4] guarantees that A is a convex combination of (less than) k^2 permutation matrices. In other words, there are k^2 permutations $\sigma_1, \ldots, \sigma_{k^2}$ of the elements $\{1, \ldots, k\}$, and k^2 reals $0 \leq \lambda_1, \ldots, \lambda_{k^2} \leq 1$ such that $\sum_t \lambda_t = 1$ and $A = \sum_t \lambda_t A_{\sigma_t}$, where A_σ is the permutation matrix corresponding to σ. Let λ_p be the largest of these k^2 coefficients. Clearly $\lambda_p \geq 1/k^2$, and observe that as A is a convex combinations of the matrices A_{σ_t}, this means that for every $1 \leq i \leq k$ we have $|W_{i,\sigma_p(i)}| \geq \frac{1}{k^2}|V_i|$ and similarly $|W_{i,\sigma_p(i)}| \geq \frac{1}{k^2}|U_{\sigma_p(i)}|$. As both $\mathcal{V}$ and $\mathcal{U}$ are assumed to be $f(k)$-regular and $f(k) \leq \min\{1/k^2, \epsilon/2\}$, this guarantees that for all $1 \leq i < j \leq k$ we have

$$|d(V_i, V_j) - d(U_{\sigma_p(i)}, U_{\sigma_p(j)})| \leq$$

$$|d(V_i, V_j) - d(W_{i,\sigma_p(i)}, W_{j,\sigma_p(j)})| + |d(W_{i,\sigma_p(i)}, W_{j,\sigma_p(j)}) - d(U_{\sigma_p(i)}, U_{\sigma_p(j)})| \leq \epsilon \,,$$

completing the proof. □

Acknowledgments. We would like to thank Madhu Sudan for a conversation that initiated this study, and Laci Lovász for fruitful discussions.

References

1. Alon, N., Duke, R.A., Lefmann, H., Rödl, V., Yuster, R.: The algorithmic aspects of the Regularity Lemma. In: Proc. 33rd IEEE FOCS, Pittsburgh, pp. 473–481. IEEE, Los Alamitos (1992) Also: J. of Algorithms 16, 80–109 (1994)
2. Alon, N., Fischer, E., Krivelevich, M., Szegedy, M.: Efficient testing of large graphs. In: Proc. of the 40 IEEE FOCS, pp. 656–666. IEEE, Los Alamitos (1999) Also: Combinatorica 20, 451–476 (2000)
3. Alon, N., Spencer, J.H.: The Probabilistic Method, 2nd edn. Wiley, New York (2000)
4. Birkhoff, G.: Three observations on linear algebra. Univ. Nac. Tucumán. Rev. Ser. A 5, 147–151 (1946)
5. Fischer, E., Matsliach, A., Shapira, A.: A hypergraph approach for finding small regular partitions. (preprint 2007)
6. Gowers, T.: Lower bounds of tower type for Szemerédi's uniformity lemma. GAFA 7, 322–337 (1997)
7. Komlós, J., Simonovits, M.: Szemerédi's Regularity Lemma and its applications in graph theory. In: Miklós, D., Sós, V.T., Szönyi, T. (eds.) Combinatorics, Paul Erdös is Eighty, Budapest. János Bolyai Math. Soc., vol. II, pp. 295–352 (1996)
8. Krivelevich, M., Sudakov, B.: Pseudo-random graphs. In: Györi, E., Katona, G.O.H., Lovász, L. (eds.) More sets, graphs and numbers. Bolyai Society Mathematical Studies, vol. 15, pp. 199–262.
9. Lovász, L.: Private communication (2006)
10. Lovász, L., Szegedy, B.: Szemerédi's lemma for the analyst, GAFA (to appear)
11. Rödl, V., Schacht, M.: Regular partitions of hypergraphs, Combinatorics, Probability and Computing (to appear)
12. Sudan, M.: Private communication (2005)
13. Szemerédi, E.: Regular partitions of graphs. In: Bermond, J.C., Fournier, J.C., Las Vergnas, M., Sotteau, D. (eds.) Proc. Colloque Inter. CNRS, pp. 399–401 (1978)

Algorithms for Core Stability, Core Largeness, Exactness, and Extendability of Flow Games*

Qizhi Fang[1], Rudolf Fleischer[2], Jian Li[2,**], and Xiaoxun Sun[3]

[1] Department of Mathematics, Ocean University of China, Qingdao, China
qfang@ouc.edu.cn
[2] Department of Computer Science and Engineering
Shanghai Key Laboratory of Intelligent Information Processing
Fudan University, Shanghai, China
rudolf,lijian83@fudan.edu.cn
[3] Department of Mathematics and Computing, University of Southern Queensland
w0072830@mail.connect.usq.edu.au

Abstract. In this paper, we give linear time algorithms to decide core stability, core largeness, exactness, and extendability of flow games on uniform networks (all edge capacities are 1). We show that a uniform flow game has a stable core if and only if the network is a balanced DAG (for all non-terminal vertices, indegree equals outdegree), which can be decided in linear time. Then we show that uniform flow games are exact, extendable, and have a large core if and only if the network is a balanced directed series-parallel graph, which again can be decided in linear time.

1 Introduction

In 1944, von Neumann and Morgenstern [19] introduced the concept of *stable sets* to analyse bargaining situations in cooperative games. In 1968, Lucas [13] described a ten-person game without a stable set. Deng and Papadimitriou [5] pointed out that deciding the existence of a stable set for a given cooperative game is not known to be computable. Jain and Vohra [8] recently showed that it is decidable whether a balanced game has a stable core. When the game is convex, the core is always a stable set. In general, however, the core and the stable set are not related. Hence the question arises: when do the core and the stable set coincide, and how can we decide core stability?

Several sufficient conditions for core stability have been discussed in the literature. Sharkey [14] introduced the concepts of subconvexity of a game and core largeness. He showed that convexity implies subconvexity which implies core largeness which implies core stability. In an unpublished paper, Kikuta and Shapley [12] studied another concept, later called extendability of the game by

* The work described in this paper was supported by NCET, NSFC (No. 10371114s and No. 70571040/G0105) and partially supported by NSFC (No. 60573025).
** The work was partially done when this author was visiting The Hong Kong University of Science and Technology.

G. Lin (Ed.): COCOON 2007, LNCS 4598, pp. 439–447, 2007.

Gellekom *et al.* [18], and proved that it is necessary for core largeness and sufficient for core stability.

However, only few results are known about core stability and related concepts for concrete cooperative games. Solymosi and Raghavan [15] studied these concepts for assignment games, and Bietenhader and Okamoto [1] studied them for minimum coloring games on perfect graphs.

In this paper we study flow games, introduced by Kalai and Zemel [10,11], which arise from the profit distribution problem related to the maximum flow in networks. We give the first linear time algorithms to decide core stability, core largeness, exactness, and extendability of uniform flow games (all edge capacities are 1). We obtain these efficient algorithms by characterizing structural properties of those networks that have flow games with the desired properties. These characterizations might be useful in other contexts.

We show that a uniform flow game has a stable core if and only if the network is a balanced DAG (for all non-terminal vertices, indegree equals outdegree). This can easily be tested in linear time, so we get a linear time algorithm to decide core stability, which improves a previous algorithm with runtime $O(|V|^2 \cdot |E|^2)$ [16]. We also show that uniform flow games are exact, extendable, and have a large core if and only if the network is a balanced directed series-parallel graph. Again, this can be tested in linear time [17], so we also get a linear time algorithm to decide exactness, extendability, and core largeness of uniform flow games. In [16], Sun *et al.* established the equivalence of these three properties and proved them to be equivalent to a certain structural property of the network (see Section 2.2.4) but left it as an open problem to design an efficient algorithm to decide this property. Note that core largeness, exactness, and extendability of a flow game all imply core stability (because flow games are totally balanced).

This paper is organized as follows. In Section 2 we define cooperative games and review some results on the core of flow games and its stability. In Section 3, we characterize those networks that have uniform flow games with these properties, leading to linear time algorithms to decide them. We conclude with some open problems in Section 4.

2 Definitions

2.1 Graphs

A *flow network* is a directed graph $G = (V, E; \omega)$, where V is the vertex set, E is the edge set, and $\omega : E \rightarrow \mathbb{R}^+$ is the edge capacity function. Let s (the *source*) and t (the *sink*) be two distinct vertices of G. W.l.o.g., we assume that every edge in E lies on some simple (s,t)-path. G is a *uniform flow network* if all edge capacities are equal. By scaling the capacities we can w.l.o.g. assume that the edge capacities are equal to one in this case.

If S and T partition the vertex set into two parts such that $s \in S$ and $t \in T$, then the set C of edges going from a node in S to a node in T is an *(s,t)-cut*. The *capacity* of C is the sum of its edge capacities. C is a *minimal* (s,t)-cut if no proper subset of C is an (s,t)-cut. C is a *minimum* (s,t)-cut if it has smallest capacity among all (s,t)-cuts. We say a uniform network G is *cut-normal* if every minimal (s,t)-cut is already a minimum (s,t)-cut.

A directed graph is a *DAG (directed acyclic graph)* if it does not contain a directed cycle. A *2-terminal DAG* is a DAG with two vertices s and t such that the indegree of s and the outdegree of t are zero and every other vertex appears in at least one simple (s,t)-path. A 2-terminal DAG is *balanced* if the indegree of every vertex other than s and t equals its outdegree.

A *2-terminal directed series-parallel graph (2-DSPG)* is a directed graph with two distinguished vertices (*terminals*) s and t that is obtained inductively as follows. A basic 2-DSPG consists of two terminals s and t, connected by a directed edge (s,t). If G_1 and G_2 are 2-DSPGs with terminals s_i and t_i, $i=1,2$, then we can combine them in series by identifying t_1 with s_2 to obtain a 2-DSPG with terminals s_1 and t_2, or in parallel by identifying s_1 with s_2 and t_1 with t_2 to obtain a 2-DSPG with terminals the combined vertex s_1/s_2 and the combined vertex t_1/t_2.

2.2 Cooperative Games

A *cooperative (profit) game* $\Gamma = (N, v)$ consists of a player set $N = \{1, 2, \cdots, n\}$ and a profit function $v : 2^N \to \mathbb{R}$ with $v(\emptyset) = 0$. A *coalition* S is a non-empty subset of N, and $v(S)$ represents the profit that can be achieved by the players in S without help of other players. The central problem in a cooperative game is how to allocate the total profit $v(N)$ among the individual players in a 'fair' way. An *allocation* is a vector $x \in \mathbb{R}^n$ with $x(N) = v(N)$, where $x(S) = \sum_{i\in S} x_i$ for any $S \subseteq N$.

Different requirements for fairness, stability and rationality lead to different optimal allocations which are generally called *solution concepts*. The core is an important solution concept.

An allocation $x \in \mathbb{R}^n$ is called an *imputation* if $x_i \geq v(\{i\})$ for all players $i \in N$. Every player is happy in this case because he gets at least as much as he could expect from the profit function of the game. We denote by $X(\Gamma)$ the set of imputations of Γ.

The *core* $C(\Gamma)$ of Γ is the set of imputations where no coalition S has an incentive to split off from the grand coalition N and go their own way. Formally, $C(\Gamma) = \{x \in \mathbb{R}^n \mid x(N) = v(N) \text{ and } x(S) \geq v(S) \text{ for all } S \subseteq N\}$. A game is *balanced* if its core is not empty.

For a subset $S \subseteq N$, the *induced subgame* (S, v_S) on S has profit function $v_S(T) = v(T)$ for each $T \subseteq S$. A cooperative game Γ is called *totally balanced* if all its subgames are balanced, i.e., all its subgames have non-empty cores.

2.3 Core Stability, Core Largeness, Extendability, and Exactness

In their classical work on game theory, von Neumann and Morgenstern [19] introduced the stable set which is very useful for the analysis of bargaining situations. Suppose that x and y are imputations. We say that x *dominates* y if there is a coalition S such that $x(S) \leq v(S)$ and $x_i > y_i$ for each $i \in S$. A set $\mathcal{F}$ of imputations is *stable* if no two imputations in $\mathcal{F}$ dominate each other, and any imputation not in $\mathcal{F}$ is dominated by at least one imputation in $\mathcal{F}$. In particular, the core of a game is stable if for any non-core imputation y there is a core imputation x dominating y, i.e., $x(S) = v(S)$ and $x_i > y_i$ for each $i \in S$.

There are three other concepts closely related to the core stability. Let $\Gamma = (N, v)$ be an n-player cooperative game. The core of Γ is *large* if for every $y \in \mathbb{R}^n$ satisfying $y(S) \geq v(S)$, for all $S \subseteq N$, there exists a core imputation x such that $x \leq y$. Γ is *extendable* if for every $S \subseteq N$ and every core imputation y of the subgame (S, v_S) there exists a core imputation $x \in C(\Gamma)$ such that $x_i = y_i$ for all $i \in S$. Γ is called *exact* if for every $S \subset N$ there exists $x \in C(\Gamma)$ such that $x(S) = v(S)$.

Kikuta and Shapley [12] showed that a balanced game with a large core is extendable, and an extendable balanced game has a stable core. Sharkey [14] showed that a totally balanced game with a large core is exact. Biswas *et al.* [2] pointed out that extendability also implies exactness. Note that flow games are totally balanced.

2.4 Flow Games

Flow games were introduced by Kalai and Zemel [10,11]. Consider a flow network $G = (V, E; \omega)$. In the corresponding *flow game* $\Gamma = (E, v)$ on G each player controls one edge. The profit $v(S)$ of a coalition $S \subseteq E$ is the value of a maximum (s,t)-flow that only uses edges controlled by S. The flow game is *uniform* if the underlying network is uniform.

In this paper, we focus on uniform flow games. These games belong to the class of packing and covering games introduced by Deng *et al.* [4]. Kalai and Zemel [11] showed that flow games are totally balanced. Fang *et al.* [7] proved that the problem of testing membership in the core of a flow game is *co-$\mathcal{NP}$*-complete. Deng *et al.* [4] showed that the core of a uniform flow game is exactly the convex hull of the indicator vectors of the minimum (s,t)-cuts of the network, which can be computed in polynomial time.

An edge $e \in E$ is called a *dummy edge* if $v(E \setminus \{e\}) = v(E)$, i.e., removal of e does not change the value of the maximum (s,t)-flow. Sun *et al.* [16] showed that a uniform flow game has a stable core if and only if the network does not contain dummy edges. Based on this structural property they designed an $O(|V|^2 \cdot |E|^2)$ time algorithm to decide core stability of uniform flow games. In Section 3 we will see that we can recognize graphs without dummy edges much faster. Note that dummy edges also play a role in the efficient computation of the nucleolus of flow games [3].

Sun *et al.* [16] also showed that for uniform flow games the concepts of exactness, extendability, and core largeness are equivalent, and that they are equivalent to the property of the network that every (s,t)-cut contains a minimum (s,t)-cut. It is easy to see that this is equivalent to being cut-normal.

Lemma 1. *A 2-terminal DAG with terminals s and t is cut-normal if and only if every (s,t)-cut contains a minimum (s,t)-cut.* □

In next section we show that the cut-normal uniform flow networks are exactly the balanced directed serial-parallel graphs with two terminals. This immediately implies a linear time algorithm to decide whether a uniform flow game is exact, extendable, and has a large core.

3 Efficient Algorithms

3.1 Core Stability

Let $G = (V, E)$ be a uniform flow network with source s and sink t. In this section we will give a linear time algorithm to decide core stability of the flow game on G.

Theorem 2. *G contains no dummy edge if and only if G is a balanced DAG.*

Proof. **If:** Suppose G is a balanced DAG. Let f be a maximum integer (s,t)-flow (which is also a maximum (s,t)-flow). Then, f pushes either zero or one unit of flow along each edge. Consider the subgraph G' of G of all edges $e \in E$ with $f(e) = 0$. Since G is balanced and f satisfies the flow conservation property in every vertex except s and t, G' is also balanced. As a subgraph of G it is also acyclical. Thus, if G' is not empty, there must be a simple (s,t)-path in G'. But then we can push one more unit of flow along this path, contradicting the maximality of f. Thus, G' is empty, i.e., G contains no dummy edge.

Only if: The flow conservation property and the fact that all edges have capacity one imply that for each maximum flow f at least one edge incident to an unbalanced vertex is not used by f. Thus, every unbalanced graph contains dummy edges.

Suppose G contains a directed cycle C. Since we may w.l.o.g. assume that a maximum (s,t)-flow f does not have flow flowing around in cycles, at least one edge of C will not be used by f, i.e., it is a dummy edge. □

Corollary 3. *We can decide core stability of uniform flow games in linear time.*

Proof. Sun *et al.* [16] have shown that a uniform flow game has a stable core if and only if the network does not contain dummy edges. □

3.2 Extendability, Exactness, and Core Largeness

In this section we will give a linear time algorithm to decide exactness, extendability, and core largeness of uniform flow games. Note that these properties imply core stability. In view of Theorem 2 we can therefore w.l.o.g. assume in this subsection that flow networks are balanced DAGs.

Two graphs are *homeomorphic* if they can be made isomorphic by inserting new vertices of degree two into edges, i.e., substituting directed edges by directed paths (which does not change the topology of the graph). Let H denote the graph shown on the left side of Fig. 1.

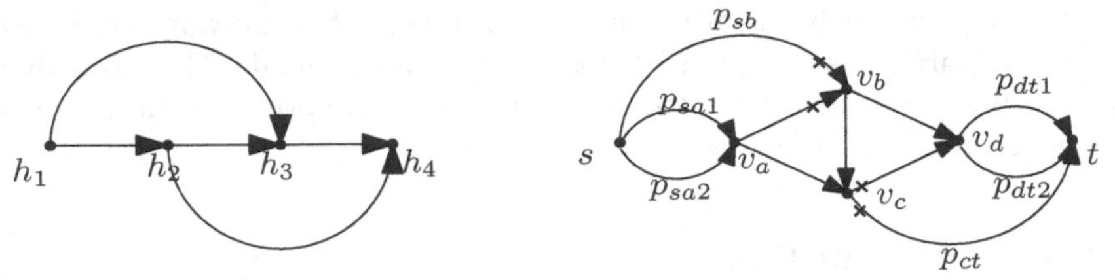

Fig. 1. Left: the graph H; Right: the graph G_3

Theorem 4. *[6,9,17] A 2-terminal DAG is a 2-DSPG if and only if it does not contain a subgraph homeomorphic to H.* □

We now give a characterization of balanced 2-DSPGs.

Theorem 5. *A balanced 2-terminal DAG is cut-normal if and only if it is a balanced 2-DSPG.*

Proof. **If:** It is easy to see that balanced 2-DSPGs can be generated inductively as 2-DSPGs with the additional constraint that a combination in series step (where we identify the two terminals t_1 and s_2) requires the indegree of t_1 being equal to the outdegree of s_2.

We now prove the claim by induction on $|E|$. If $|E| = 1$, the graph is (s,t)-normal. Suppose the statement holds for all balanced 2-DSPGs with fewer than $|E|$ edges. If G was generated from $G_1 = (V_1, E_1)$ and $G_2 = (V_2, E_2)$ by combination in parallel, then any minimal (s,t)-cut C in G consists of a minimal (s_1,t_1)-cut $C \cap E_1$ in G_1 and a minimal (s_2,t_2)-cut $C \cap E_2$ in G_2. By inductive hypothesis, these are minimum cuts. Thus, C is a minimum (s,t)-cut in G.

If G was generated from G_1 and G_2 by combination in series, then a minimal (s,t)-cut C in G is either a minimal (and thus minimum) (s_1,t_1)-cut in G_1 or a minimal (and thus minimum) (s_2,t_2)-cut in G_2. Since the indegree of t_1 equals the outdegree of s_2 and cutting all edges incident to t_1 (s_2) defines a minimal (s_1,t_1)-cut in G_1 (minimal (s_2,t_2)-cut in G_2), a minimum (s_1,t_1)-cut in G_1 has the same capacity as a minimum (s_2,t_2)-cut in G_2. Thus, C is a minimum (s,t)-cut in G.

Only if: Let $G = (V, E)$ be a balanced 2-terminal DAG with terminals s and t. Let $V = \{s = v_1, v_2, \ldots, v_n = t\}$ be a topological ordering of G.

Let k denote the outdegree of s. If G is not a 2-DSPG, then we can construct a minimal cut of size larger than k, contradicting the assumption that G is cut-normal. By Theorem 4, G contains a subgraph homeomorphic to H which we denote by $H(a,b,c,d,P_{ab},P_{bc},P_{cd},P_{ac},P_{bd})$, where v_a is the node corresponding to h_1, v_b the node corresponding to h_2, v_c the node corresponding to h_3, v_d the node corresponding to h_4, and the vertex-disjoint (except at their endpoints) paths P_{ab}, P_{bc}, P_{cd}, P_{ac} and P_{bd} correspond to the edges (h_1,h_2), (h_2,h_3), (h_3,h_4), (h_1,h_3), and (h_2,h_4) in H, respectively. Among all subgraphs homeomorphic to H we choose one, G_H, with largest b.

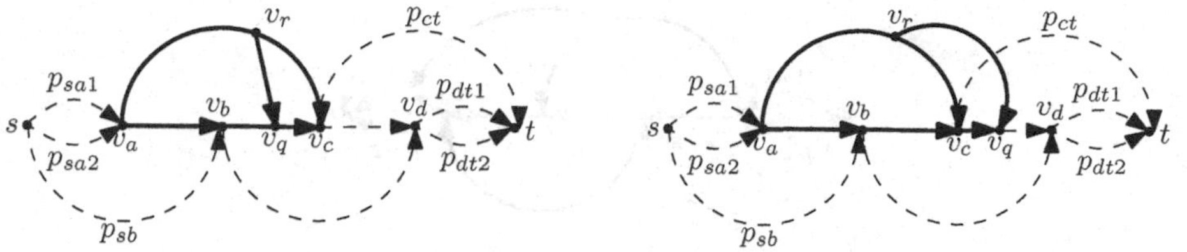

Fig. 2. Cases (1) and (2) in the proof of Theorem 5

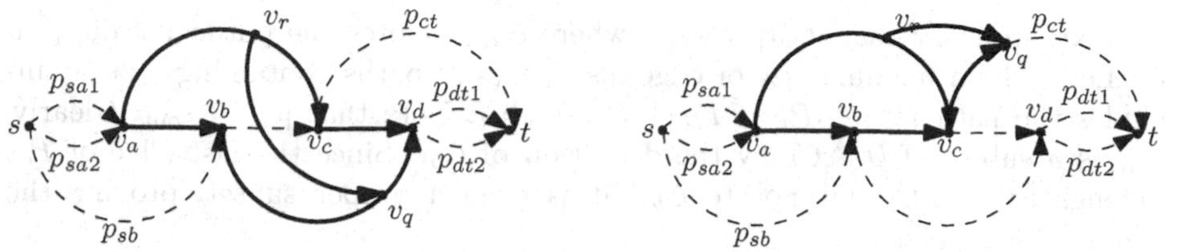

Fig. 3. Cases (3) and (4) in the proof of Theorem 5

G_H is not balanced, but since G is balanced we can find in G six pairwise edge-disjoint paths (they are also edge-disjoint with G_H) P_{sa1}, P_{sa2}, P_{sb}, P_{ct}, P_{dt1}, and P_{dt2} that we can add to G_H to obtain a balanced graph G_3 (see Fig. 1, right side). Note that G_3 is the union of three edge-disjoint (s,t)-paths $P_{sa1} + P_{ab} + P_{bd} + P_{dt1}$, $P_{sa2} + P_{ac} + P_{cd} + P_{dt2}$, and $P_{sb} + P_{bc} + P_{ct}$ (we use $P_1 + P_2$ to denote the concatenation of paths P_1 and P_2).

Consider the (s,t)-cut C_b in G induced by the partition of V into $\{v_1, \ldots, v_{b-1}\}$ and $\{v_b, \ldots, v_n\}$. Since G is a DAG and all vertices lie on some (s,t)-path, this is a minimal cut. If $|C_b| > k$, we have a minimal cut that is not a minimum cut. So assume $|C_b| = k$. We partition the edges of C_b into two classes, C_{b1} and C_{b2}. An edge e belongs to C_{b1} if every (s,t)-path containing e passes through v_c, otherwise it belongs to C_{b2}.

We will now show that C_{b1} is not empty. Specifically, we show it contains the edge $e = P_{ac} \cap C_b$. Assume there is an (s,t)-path P_e containing e but not v_c. Let v_r be the largest node in $P_{ac} \cap P_e$. Let v_q be the first node corresponding to a node in G_3 that we encounter on P_e when starting to walk at v_r (such a node exists because P_e ends at $t \in G_3$). Let P_{rq} denote the path from v_r to v_q.

Since $e \in C_b$, $r \geq b$. Actually, $r > b$ because v_b is not on P_{ac}. And $r < c$ because $v_r \in P_{ac}$ and $v_c \notin P_e$. But then there exists another $H(a, r, v_{c'}, \ldots)$ in G, a contradiction because $r > b$. To see this, we distinguish five cases, depending on the location of v_q (see Fig. 2–4). Let $P_{xy}(i,j)$ denote the subpath of P_{xy} from v_i to v_j, where $x, y \in \{a, b, c, d\}$.

1. $v_q \in P_{bc}$: $H(a, r, q, c, P_{ac}(a,r), P_{rq}, P_{bc}(q,c), P_{ab} + P_{bc}(b,q), P_{ac}(r,c))$.
2. $v_q \in P_{cd}$: $H(a, r, c, q, P_{ac}(a,r), P_{ac}(r,c), P_{cd}(c,q), P_{ab} + P_{bc}, P_{rq})$.
3. $v_q \in P_{bd}$: $H(a, r, q, d, P_{ac}(a,r), P_{rq}, P_{bd}(q,d), P_{ab} + P_{bd}(b,q), P_{ac}(r,c) + P_{cd})$.
4. $v_q \in P_{ct}$: $H(a, r, c, q, P_{ac}(a,r), P_{ac}(r,c), P_{ct}(c,q), P_{ab} + P_{bc}, P_{rq})$.
5. $v_q \in P_{dt1}$ or $v_q \in P_{dt2}$: $H(a, r, d, q, P_{ac}(a,r), P_{ac}(r,c) + P_{cd}, P_{dq}, P_{ab} + P_{bd}, P_{rq})$.

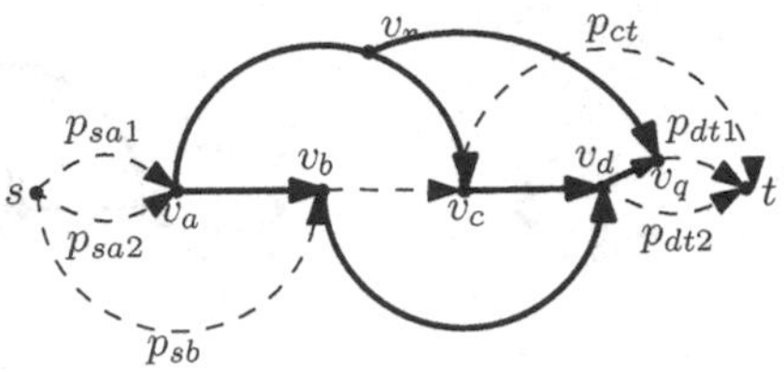

Fig. 4. Case (5) in the proof of Theorem 5

Next, we show that $|C_{b1}| < c_{\text{out}}$, where c_{out} denotes the outdegree of v_c in G. Let U be a maximal set of edge-disjoint (s,t)-paths containing v_c that includes the path $P_{sa1} + P_{ab} + P_{bc} + P_{cd} + P_{dt1}$. Note that $|U| \leq c_{\text{out}}$. Clearly, C_{b1} is a subset of $U \cap C_b$ by the definition of C_{b1}. Since the last edge of P_{ab} belongs to $U \cap C_b$ but not to C_{b1}, it is even a proper subset, proving the claim.

Let C be the set C_{b2} plus all edges outgoing from v_c. C is an (s,t)-cut because every (s,t)-path must either contain an edge in C_{b2} or the node v_c. C is a minimal (s,t)-cut because every outgoing edge of v_c is necessary because C_{b1} is not empty, and every edge in C_{b2} is necessary by definition. The size of C is $|C| = |C_{b2}| + c_{\text{out}} > |C_{b2}| + |C_{b1}| = |C_b| = k$. Thus, C is not a minimum (s,t)-cut, a contradiction. Therefore, the assumption that G is not a 2-DSPG must be wrong. □

Corollary 6. *We can test in linear time whether a balanced uniform flow network is cut-normal. Consequently, we can decide exactness, extendability, and core largeness of uniform flow games in linear time.*

Proof. We can test in linear time whether a DAG is balanced and whether it is a 2-DSPG [17]. By Theorem 5 this is equivalent to being cut-normal. By Lemma 1, a uniform flow network is cut-normal if and only if every (s,t)-cut contains a minimum (s,t)-cut. Sun *et al.* [16] have shown that this is equivalent to the flow game being exact, extendable, and having a large core. □

4 Open Problems

In this paper, we gave structural characterizations of exact, extendable, large-core and stable-core uniform flow games that can be tested in linear time. Currently, little is known about core stability of flow games on networks with arbitrary capacities. Although it is *co-$\mathcal{NP}$*-complete to decide whether an imputation belongs to the core this does not rule out the possibility that core stability can be decided efficiently. We leave it as an open problem.

Acknowledgements

We would like to thank Mordecai Golin for his helpful discussions.

References

1. Bietenhader, T., Okamoto, Y.: Core stability of minimum coloring games. In: Hromkovič, J., Nagl, M., Westfechtel, B. (eds.) WG 2004. LNCS, vol. 3353, pp. 389–401. Springer, Heidelberg (2004)
2. Biswas, A.K., Parthasarathy, T., Potters, J.A.M., Voorneveld, M.: Large cores and exactness. Game and Economic Beheavior 28, 1–12 (1999)
3. Deng, X., Fang, Q., Sun, X.: Finding nucleolus of flow game. In: Proceedings of the 17th Annual ACM-SIAM Symposium on Discrete Algorithms (SODA'06), pp. 124–131. ACM Press, New York (2006)
4. Deng, X., Ibaraki, T., Nagamochi, H.: Algorithmic aspects of the core of combinatorial optimization games. Mathematics of Operations Research 24, 751–766 (1999)
5. Deng, X., Papadimitriou, C.H.: On the complexity of cooperative solution concepts. Mathematics of Operations Research 19, 257–266 (1994)
6. Duffin, R.: Topology of series-parallel networks. Journal of Mathematical Analysis and Applications 10, 303–318 (1965)
7. Fang, Q., Zhu, S., Cai, M., Deng, X.: Membership for core of LP games and other games. In: Wang, J. (ed.) COCOON 2001. LNCS, vol. 2108, pp. 247–256. Springer, Heidelberg (2001)
8. Jain, K., Vohra, R.V.: On stability of the core. Manuscript (2006), `http://www.kellogg.northwestern.edu/faculty/vohra/ftp/newcore.pdf`
9. Jakoby, A., Liśkiewicz, M., Reischuk, R.: Space efficient algorithms for directed series-parallel graphs. Journal of Algorithms 60(2), 85–114 (2006)
10. Kalai, E., Zemel, E.: Totally balanced games and games of flow. Mathematics of Operations Research 7, 476–478 (1982)
11. Kalai, E., Zemel, E.: Generalized network problems yielding totally balanced games. Operations Research 30, 998–1008 (1982)
12. Kikuta, K., Shapley, L.S.: Core stability in n-person games. Manuscript (1986)
13. Lucas, W.F.: A game with no solution. Bulletin of the American Mathematical Society 74, 237–239 (1968)
14. Sharkey, W.W.: Cooperative games with large cores. International Journal of Game Theory 11, 175–182 (1982)
15. Solymosi, T., Raghavan, T.E.S.: Assignment games with stable cores. International Journal of Game Theory 30, 177–185 (2001)
16. Sun, X., Fang, Q.: Core Stability of Flow Games. In: Akiyama, J., Chen, W.Y.C., Kano, M., Li, X., Yu, Q. (eds.) CJCDGCGT 2005. LNCS, vol. 4381, pp. 189–199. Springer, Heidelberg (2007)
17. Valdes, J., Tarjan, R.E., Lawler, E.L.: The recognition of series-parallel digraphs. In: Proceedings of the 11th Annual ACM Symposium on Theory of Computing (STOC'79), pp. 1–12. ACM Press, New York (1979)
18. van Gellekom, J.R.G., Potters, J.A.M., Reijnierse, J.H.: Prosperity properties of TU-games. International Journal of Game Theory 28, 211–227 (1999)
19. von Neumann, J., Morgenstern, O.: Theory of Games and Economic Behaviour. Princeton University Press, Princeton (1944)

Computing Symmetric Boolean Functions by Circuits with Few Exact Threshold Gates

Kristoffer Arnsfelt Hansen

Department of Computer Science
University of Aarhus*
arnsfelt@daimi.au.dk

Abstract. We consider constant depth circuits augmented with few exact threshold gates with arbitrary weights. We prove strong (up to exponential) size lower bounds for such circuits computing symmetric Boolean functions. Our lower bound is expressed in terms of a natural parameter, the *balance*, of symmetric functions. Furthermore, in the quasi-polynomial size setting our results provides an *exact* characterization of the class of symmetric functions in terms of their balance.

1 Introduction

A central program of circuit complexity is that aimed towards strong size lower bounds for larger and larger classes of constant depth circuits. Such lower bounds have been obtained for constant depth circuits consisting of AND and OR gates ($\mathbf{AC^0}$ circuits) [1,2,3,4]. If we however allow gates performing modular counting or a majority vote, we obtain classes of circuits ($\mathbf{ACC^0}$ and $\mathbf{TC^0}$) for which no strong size lower bounds are known. The present work belongs to two distinct lines of research both aimed at increasing the understanding of these classes thereby approaching such lower bounds.

The first of these lines of research considers $\mathbf{AC^0}$ circuits augmented with just few additional more powerful gates. A series of papers [5,6,7,8] considers circuits with few MAJ gates, resulting in the result that such circuits with $n^{o(1)}$ MAJ gates require size $2^{n^{\Omega(1)}}$ to compute the MOD_m function for any m. More recently have lower bounds been obtained for circuits with few modular counting gates and symmetric gates [9,10,11].

The second line of research concerns classifying which symmetric Boolean functions a class of circuits can compute. This naturally relies heavily on strong lower bounds for the class of circuits in question, but provides much more information than just a lower bound for a particular Boolean function. These results are usually stated in terms of properties of the *value vector* of a symmetric Boolean function, being composed of the outputs of the function for inputs of the $n+1$ different weights. Fagin et al. [12] prove that the symmetric functions computed by $\mathbf{AC^0}$ circuits of polynomial size are precisely those whose value vector

* Currently at The University of Chicago, supported by a Villum Kann Rasmussen Post.Doc. fellowship.

G. Lin (Ed.): COCOON 2007, LNCS 4598, pp. 448–458, 2007.

is constant except within both ends of poly-logarithmic length. Interestingly, the class remains the same when allowing quasi-polynomial size circuits. Zhang et al. [13] prove that the symmetric functions computed by quasi-polynomial size $\mathbf{THR} \circ \mathbf{AC^0}$ circuits ($\mathbf{AC^0}$ circuits with an arbitrary threshold gate at the output) are precisely those whose value vector has only a poly-logarithmic number of *sign changes* (positions in the value vector where there is a change of values). Lu [14] prove that the symmetric functions computed by quasi-polynomial size $\mathbf{AC^0}$ circuits augmented with MOD_q gates, where $q = p^k$ is a prime power, are precisely those whose value vector is *periodic* with poly-logarithmic period $p^{t(n)} = \log^{O(1)} n$, except within both ends of poly-logarithmic length.

Beigel [7] showed that a quasi-polynomial size $\mathbf{AC^0}$ circuit augmented with a poly-logarithmic number of MAJ gates can be converted into a quasi-polynomial size $\mathbf{MAJ} \circ \mathbf{AC^0}$ circuit computing the same function, which implies that the characterization of Zhang et al. extends to this class of circuits.

A natural next step would be to consider circuits with few threshold gates, however here no strong lower bounds are known for computing simple functions such as symmetric Boolean functions. (For more complicated functions, some lower bounds can be obtained by the results of [9,10] combined with a multiparty protocol for evaluating threshold gates due to Nisan [15]). In this paper we instead consider circuits with few exact threshold gates computing symmetric Boolean functions. This class of circuits is simpler than the class of circuits with few threshold gates but seems to be incomparable to the class of circuits with few majority gates.

We state our lower bounds in terms of a notion of *balance* $b(f)$ of a symmetric Boolean function f, defined as the minimum number of 0's or 1's in the value vector of f. Our main result is the following general lower bound statement.

Theorem 1. *There is a constant $c > 0$ such that any depth h circuit containing s exact threshold gates and computing a symmetric function f must have size at least*

$$\frac{1}{n2^{s+1}} 2^{\frac{1}{14}\left(c\frac{b(f)}{s\log\left(\frac{n}{b(f)}\right)}\right)^{\frac{1}{h}}} .$$

A notable special case to single out is quasi-polynomial size circuits. In this case we obtain a complete characterization of the symmetric functions computable.

Theorem 2. *A symmetric Boolean function can be computed by a constant depth $\mathbf{AC^0}$ circuit of size $2^{\log^{O(1)} n}$ containing $\log^{O(1)} n$ exact threshold gates if and only if it has balance $\log^{O(1)} n$.*

2 Preliminaries

2.1 Constant Depth Circuits

We consider circuits built from families of unbounded fanin gates. Inputs are allowed to be Boolean variables and their negations as well as the constants 0 and 1. In addition to AND, OR and NOT gates we consider the following variants of *threshold* functions. Let $x_1, \ldots, x_n$ be n Boolean inputs. The majority function, MAJ, is 1 if and only if $\sum_{i=1}^n x_i \geq \frac{n}{2}$. Similarly, the *exact* majority function, EMAJ, is 1 if and only if $\sum_{i=1}^n x_i = \frac{n}{2}$. Introducing weights we get a much richer class of functions. Let $w \in \mathbf{R}^n$ and let t be any real number. The threshold function with weights w and threshold t, $\mathrm{THR}_{w,t}$ is 1 if and only if $\sum_{i=1}^n w_i x_i \geq t$. Similarly, the *exact* threshold function with weights w and threshold t, $\mathrm{ETHR}_{w,t}$ is 1 if and only if $\sum_{i=1}^n w_i x_i = t$. For both these functions it is easy to see that we can in fact assume that the weights and threshold are integer valued, although this is not needed for our results.

Additionally, for a positive integer m, let MOD_m be the function that outputs 1 if and only if $\sum_{i=1}^n x_i \not\equiv 0 \pmod{m}$.

Let **AND** and **OR** denote the families of unbounded fanin AND and OR gates. Let **MAJ**, **ETHR**, **THR** denote the families of MAJ, $\mathrm{ETHR}_{w,t}$ and $\mathrm{THR}_{w,t}$ gates, for arbitrary w and t. If G is a family of boolean gates and $\mathcal{C}$ is a family of circuits we let $G \circ \mathcal{C}$ denote the class of circuits consisting of a G gate taking circuits from $\mathcal{C}$ as inputs.

$\mathbf{AC^0}$ is the class of functions computed by constant depth circuits built from AND, OR and NOT gates. We define the larger class $\mathbf{TC^0}$ as the class of functions computed by constant depth circuits built from AND, OR and MAJ gates. When there is no special restrictions on the number of threshold gates, we can exchange MAJ gates in this definition with either of the other three variants of threshold gates defined above without changing the class of functions computed.

2.2 The Switching Lemma

A *restriction* on a set V of boolean variables is a map $\rho : V \to \{0, 1, \star\}$. It acts on a boolean function $f : V \to \{0, 1\}$, creating a new boolean function f_ρ on the set of variables for which $\rho(x) = \star$, obtained by substituting $\rho(x)$ for $x \in V$ whenever $\rho(x) \neq \star$. Let R_n^l denote the set of all restriction ρ leaving l of n variables free.

A decision tree is a binary tree, where the internal nodes are labeled by variables and leafs are labeled by either 0 or 1. On a given input x, its value is the value of the leaf reached by starting at the root, and at any internal node labeled by x_i proceeding to the left child if $x_i = 0$ and to the right child otherwise. We will use the following version of Håstads Switching Lemma due to Beame [16].

Lemma 1. *Let f be a DNF formula in n variables with terms of length at most r. Let $l = pn$ and pick ρ uniformly at random from R_n^l. Then the probability that f_ρ does not have a decision tree of depth d is less than $(7pr)^d$.*

The advantage of using this version of the switching lemma is that it directly gives us a decision tree. If we convert a decision tree into a DNF, we in fact obtain a *disjoint* DNF, i.e. a DNF where all terms are *mutually contradictory*. We can view it as a sum of terms, instead as an OR of AND's. This view allows us to absorb the sum into a threshold gate. We elaborate on this after the next proposition.

Using the switching lemma several times, allows one to obtain decision trees for every gate in a given $\mathbf{AC^0}$ circuit.

Proposition 1. *Let C be a* $\mathbf{AC^0}$ *circuit of depth h and size S. Let d be any positive integer, and choose a restriction $\rho \in R_n^{n_h}$, where $n_h = \frac{n}{14(14d)^{h-1}}$. Then with probability greater than $1 - S2^{-d}$, is every function computed by any gate of C computed by a decision tree of depth less than d, after applying the restriction ρ.*

Proof. First, whenever an input to a NOT gate is computed by a decision tree of depth less than d, the function computed by the NOT gate is also computed by a decision tree of depth less than d, by simply negating the constants at the leaves of the decision tree. Henceforth we will only consider OR and AND gates.

Define, for $0 \leq i \leq h-1$, $n_{i+1} = \frac{n}{14(14d)^i}$. We choose the restriction $\rho \in R_n^{n_h}$ at random as a composition of randomly chosen restrictions $\rho_1, \dots, \rho_h$, where $\rho_1 \in R_n^{n_1}$ and $\rho_{i+1} \in R_{n_i}^{n_{i+1}}$ for $1 \leq i \leq h-1$. Consider an AND or OR gate at level 1 of C. We can consider the inputs of the gate to be terms or clauses of size 1. This means that we can consider the gate to be a CNF with clauses of size 1 or a DNF with terms of size 1. Then by Lemma 1, after applying ρ_1, the probability that the function computed by the gate is not computed by a decision tree of depth less than d is at most $(7\frac{n_1}{n})^d = 2^{-d}$.

Now assume that after having applied the restrictions $\rho_1, \dots, \rho_i$, every function computed by a gate at level i is also computed by a decision tree of depth at most d. Consider now a gate at level $i+1$. If the gate is an OR gate we rewrite every input to the gate as DNF's with terms of size at most d. The OR gates of these can then be merged with the OR gate at level $i+1$, thus obtaining a DNF with terms of size at most d computing the same function. Similarly, if the gate is an AND gate we rewrite every input to the gate as CNF's with clauses of size at most t. The AND gates of these can then be merged with the AND gate at level $i+1$, thus obtaining a CNF with clauses of size at most d for the same function. Then by using Lemma 1 again, the probability that the function computed by the gate at level $i+1$ is not computed by a decision tree of depth less than d is less than $(7\frac{n_{i+1}}{n_i}d)^d = 2^{-d}$.

To conclude, since the probability of error at a single gate is less than 2^{-d}, the total probability of error is less than $S2^{-d}$ and the result follows.

Consider now an $\mathbf{ETHR} \circ \mathbf{AC^0}$ circuit C. After successful use of Proposition 1 we have a restriction ρ such that after applying ρ, every function computed at an input to the output gate of C is computed be a decision tree of depth at most d, and hence by disjoint DNF's with terms of size at most d. Assume that the output gate evaluates whether $\sum_{i=1}^m w_i x_i = t$, and let t_j^i be the terms of the i'th DNF. Then we have that the output of C_ρ evaluates whether $\sum_{i=1}^m w_i(\sum_j(\prod_{x_k \in t_j^i} x_k \cdot \prod_{\neg x_k \in t_j^i}(1-x_k))) = t$. This means we have a real polynomial $P(x)$ of degree at most d such $P(x) \neq 0$ if and only if $C_\rho(x) = 1$ for all x.

2.3 Representation by Polynomials

Let P be a real polynomial on n variables and f a Boolean function on n variables. We say that P is a *weak equality* representation f, if P is not identically 0 and whenever $P(x) \neq 0$ for $x \in \{0,1\}^n$ we must have $f(x) = 1$. (Thus our definition is a real valued analogue of the notion of weak representation over $\mathbf{Z}_m$ as defined by Green [17]).

Let f be a symmetric Boolean function on n variables. Since f then only depends on the *weight* of its input, i.e. $\sum_{i=1}^{n} x_i$, we will identify f with a function $f : \{0,\ldots,n\} \to \{0,1\}$. By $v(f)$ we denote the string $f(0)f(1)\cdots f(n) \in \{0,1\}^{n+1}$, called the *value vector* of f.

For a string $s \in \{0,1\}^*$, let $n_0(s)$ denote the number of 0's in a, and likewise let $n_1(s)$ denote the number of 1's in s. Let further $n_0(f) = n_0(v(f))$ and $n_1(f) = n_1(v(f))$. Define the *balance* $b(f)$ of f by $b(f) = \min(n_0(f), n_1(f))$.

Using the symmetrization technique of Minsky and Papert [18] we can prove lower bounds on the degree of a polynomial representing a symmetric Boolean function, in terms of the number of 0's of its value vector.

Proposition 2. *Let P be a polynomial, not identically* 0*, that is a weak equality representation of a symmetric Boolean function f on n variables. Then the degree of P must be at least $\frac{n_0(f)}{2}$.*

Proof. Let d be the degree of P. Define

$$Q(x) = \sum_{\sigma \in S_n} (P(x_{\sigma(1)},\ldots,x_{\sigma(n)}))^2 \ .$$

Observe that Q is a symmetric polynomial of degree $2d$ that is also a weak equality representation of f. We thus have a polynomial H of degree $2d$ such that $H(\sum_{i=1}^{n} x_i) = Q(x)$ for all $x \in \{0,1\}^n$. This polynomial H is not identically 0, and if $H(w) \neq 0$ we must have $f(w) = 1$. Since H must have at least $n_0(f)$ roots it then follows that $2d \geq n_0(f)$.

For our purposes we will for a given function f need lower bounds for polynomials that represents *either* f or its negation $\neg f$. We get precisely such a lower bound from Proposition 2 in terms of the balance of f.

Corollary 1. *Let f be any symmetric Boolean function. If P is a polynomial that is either a weak equality representation of f or $\neg f$, the degree of P must be at least $\frac{b(f)}{2}$.*

Proof. For representing f, Proposition 2 implies that the degree of P must be at least $\frac{n_0(f)}{2}$, and similarly for representing $\neg f$ the degree of P must be at least $\frac{n_0(\neg f)}{2}$. Since $n_1(f) = n_0(\neg f)$ the result follows.

3 Circuit Lower Bounds

By use of the switching lemma we will be able to obtain lower bounds of the size of circuits with few exact threshold gates computing a Boolean function f depending on the degree required for a polynomial to represent either f_ρ or $\neg f_\rho$, where ρ is a random restriction. If f_ρ is a symmetric function we can use Corollary 1 to express the lower bound in terms of the balance of f_ρ. To express the lower bound solely in terms of a symmetric f, we will carefully choose a restriction f' of f, for which we can relate the balance of f'_ρ to that of f, for a suitable random restriction ρ.

By an *interval* of the value vector of a symmetric Boolean function f we mean a contiguous substring. We will denote such an interval by $[a, b]$, where a and b are the endpoints of the interval. When applying a restriction ρ to a symmetric Boolean function f, the function f_ρ obtained is also a symmetric Boolean function, and its value vector is an interval of the value vector of f. Conversely, every interval of the value vector of f is the value vector of a restriction of f. Let $f_I = f_{[a,b]}$ be the Boolean function whose value vector is the interval $I = [a, b]$ of the value vector of f.

With these definitions in place, we can outline the properties we desire of f'. We would like to have $f' = f_I$ for an interval $I = [a, b]$, such that a smaller interval I' in the middle of I has as large a balance as possible. This will mean that we can ensure that $f_{I'}$ is a further restriction of f'_ρ and thus will the balance of f'_ρ be at least as large as the balance of $f_{I'}$.

This is precisely the approach that Zhang et al.[13] took using *sign changes* instead of balance to prove their lower bound. However, while our approach here is similar, we have to be more careful. The reason is as follows. If we divide an interval containing a certain number of sign changes into two parts then one of the parts must contain at least half of the sign changes. For balance, the same thing clearly does not hold. As a simple example consider an interval where the first half is all 0 and the other half is all 1. Although this interval has maximal balance, each of the two halves have balance 0. What is true instead, is that there is *some* subinterval of half the length with half the balance. We will establish that we can find the function f' we desire in the following series of lemmas.

Lemma 2. *Let $s \in \{0,1\}^*$. Suppose that $n_0(s[a, a + l]) \geq n_1(s[a, a + l])$ and $n_1(s[c, c + l]) \geq n_0(s[c, c + l])$ for $a \leq c$. Then there exists b such that $a \leq b \leq c$ and $n_0(s[b, b + l]) \leq n_1(s[b, b + l]) \leq n_0(s[b, b + l]) + 1$.*

Proof. Define $f : \{a, \ldots, c\} \rightarrow \mathbf{Z}$ by $f(d) = n_1(s[d, d+l]) - n_0(s[d, d+l])$. Observe that $f(b + 1) - f(b) \in \{-2, 0, 2\}$. Since by assumption $f(a) \leq 0$ and $f(c) \geq 0$ there exists $a \leq b \leq c$ such that $0 \leq f(b) \leq 1$, which implies $n_0(s[b, b + l]) \leq n_1(s[b, b + l]) \leq n_0(s[b, b + l]) + 1$.

Lemma 3. *Let f be a symmetric Boolean function on n variables. Let $n' = 2^k - 1$, where $2^k < n + 1 \leq 2^{k+1}$. Then there is a restriction f' of f on n' variables such that $b(f') \geq \frac{b(f)-1}{2}$.*

Proof. Assume without loss of generality that $n_0(f) \leq n_1(f)$. We will then find an interval I of length at most 2^k of the value vector of f such that $n_1(f_I) \geq n_0(f_I) \geq \frac{n_0(f)-1}{2}$. This interval can then be extended to an interval I' of length 2^k such that $\min(n_0(f_{I'}), n_1(f_{I'})) \geq n_0(f_I)$. Thus for $f' = f_{I'}$ on n' variables we have $b(f') \geq \frac{b(f)-1}{2}$.

Consider the two intervals $I_1 = [0, \lfloor \frac{n+1}{2} \rfloor - 1]$ and $I_2 = [\lfloor \frac{n+1}{2} \rfloor, 2\lfloor \frac{n+1}{2} \rfloor - 1]$. Note that each of these are of length $\lfloor \frac{n+1}{2} \rfloor \leq 2^k$. Since at most one index is left out, we must have either $n_0(f_{I_1}) \geq \frac{n_0(f)-1}{2}$ or $n_0(f_{I_2}) \geq \frac{n_0(f)-1}{2}$. Assume without loss of generality that $n_0(f_{I_1}) \geq \frac{n_0(f)-1}{2}$. If also $n_1(f_{I_1}) \geq n_0(f_{I_1})$ we may choose $I = I_1$. Thus assume $n_0(f_{I_1}) > n_1(f_{I_1})$. Now we must have $n_1(f_{I_2}) \geq n_0(f_{I_2})$, as otherwise we would have $n_0(f) \geq n_0(f_{I_1}) + n_0(f_{I_2}) \geq n_1(f_{I_1}) + n_1(f_{I_2}) + 2 \geq n_1(f) + 1$. Then using Lemma 2 we can find an interval I_3 of length $\lfloor \frac{n+1}{2} \rfloor \leq 2^k$ such that $n_0(f_{I_3}) \leq n_1(f_{I_3}) \leq n_0(f_{I_3}) + 1$. We then have $n_0(f) \leq \lfloor \frac{n+1}{2} \rfloor = n_0(f_{I_3}) + n_1(f_{I_3}) \leq 2n_0(f_{I_3}) + 1$, and we may choose $I = I_3$.

Lemma 4. *Let $s \in \{0,1\}^{2^k}$. If $n_0(s) \geq n_1(s)$ then there is a substring s' of s of length 2^{k-1} such that $n_0(s') \geq n_1(s')$. Analogously if $n_1(s) \geq n_0(s)$ then there is a substring s' of s of length 2^{k-1} such that $n_1(s') \geq n_0(s')$.*

Proof. We prove the case $n_0(s) \geq n_1(s)$. We then have $n_0(s[0, 2^{k-1} - 1]) + n_0(s[2^{k-1}, 2^k - 1]) \geq n_1(s[0, 2^{k-1} - 1]) + n_1(s[2^{k-1}, 2^k - 1])$ and thus we must have $n_0(s[0, 2^{k-1} - 1]) \geq n_1(s[0, 2^{k-1} - 1])$ or $n_0(s[2^{k-1}, 2^k - 1]) \geq n_1(s[2^{k-1}, 2^k - 1])$. We can then let $s' = s[0, 2^{k-1} - 1]$ or $s' = s[2^{k-1}, s^k - 1]$.

Lemma 5. *Let f be a symmetric Boolean function on n variables, where $n = 2^k - 1$. Let $2^h \leq b(f) < 2^{h+1}$, and assume that $h > 2$. Then there exist an interval I of length $2^{k'}$, where $k' \geq h - 2$, such that I is a subinterval of the interval $[2^{k'}, n - 2^{k'}]$ and $b(f_I) \geq \frac{b(f)}{4 \log_2\left(\frac{8(n+1)}{b(f)}\right)}$.*

Proof. Let $b = 2^{h-2}$, and assume without loss of generality that $n_0(f_{[b,n-b]}) \leq n_1(f_{[b,n-b]})$. Consider the intervals $I_i^L = [b2^i, b2^{i+1} - 1]$ and $I_i^R = [n - b2^{i+1} + 1, n - b2^i]$ for $i = 0, \ldots, k - h$. Note that these $2(k - h + 1)$ intervals precisely cover the interval $[b, n - b]$.

Suppose first that $n_1(I_i^L) \geq n_0(I_i^L)$ and $n_1(I_i^R) \geq n_0(I_i^R)$ for all i. Then we may select I among the I_i^L and I_i^R intervals obtaining $n_1(f_I) \geq n_0(f_I) \geq \frac{n_0(f_{[b,n-b]})}{2(k-h+1)} \geq \frac{b(f)-2b}{2(k-h+1)} \geq \frac{b(f)}{4 \log_2\left(\frac{4(n+1)}{b(f)}\right)}$.

Otherwise we must have two intervals I_1 and I_2 chosen from the I_i^L and I_i^R intervals, such that $n_0(f_{I_1}) \geq n_1(f_{I_1})$ and $n_1(f_{I_2}) \leq n_0(f_{I_2})$, (By assumption $n_0(f_{[b,n-b]}) \leq n_1(f_{[b,n-b]})$, so we can not have that $n_1(I_i^L) < n_0(I_i^L)$ and $n_1(I_i^R) < n_0(I_i^R)$ for all i). Suppose I_1 is of length 2^{k_1} and I_2 is of length 2^{k_2}, and let $k = \min(k_1, k_2)$. Note then that I_1 and I_2 are subintervals of the interval $[2^k, n - 2^k]$.

Using Lemma 4 we may find subintervals I_1' of I_1 and I_2' of I_2 of length k, that are also subintervals of the interval $[2^k, n - 2^k]$, such that $n_0(f_{I_1'}) \geq n_1(f_{I_1'})$ and $n_1(f_{I_2'}) \leq n_0(f_{I_2'})$.

Then using Lemma 2 we can find an interval I of length $2^k \geq b$, which is a subinterval of the interval $[2^k, n - 2^k]$, such that $n_0(f_I) \leq n_1(f_I) \leq n_0(f_I) + 1$, which means in fact, $n_0(f_I) = n_1(f_I)$. Thus we have $b(f_I) \geq \frac{b}{2} > \frac{b(f)}{16} = \frac{b(f)}{4 \log_2\left(\frac{8(n+1)}{\frac{n+1}{2}}\right)} \geq \frac{b(f)}{4 \log_2\left(\frac{8(n+1)}{b(f)}\right)}$.

Corollary 2. *Let f be a symmetric Boolean function on n variables, where $n = 2^k - 1$. Assume that $b(f) \geq 8$ and let m be such that $2^m \leq \frac{b(f)}{8}$. Then there exist exist an interval I' of length $2^{k'}$, for some $k' > m$, such that the interval I'' of length $2^{k'-m}$ in the middle of I' satisfies $b(f_{I''}) \geq \frac{b(f)}{2^{m+2}\log_2\left(\frac{8(n+1)}{b(f)}\right)}$.*

Proof. Let I be the interval given by Lemma 5 and divide it into interval $I_1, \ldots, I_{2^m}$, each of length $2^{k'-m}$. We will find an subinterval I'' of I such that $b(f_{I''}) \geq \frac{b(f_I)}{2^m}$. Then since I is a subinterval of $[2^{k'}, n - 2^{k'}]$, we can let I' be the interval of length $2^{k'}$ with I'' in the middle. Assume without loss of generality that $n_0(f_I) \leq n_1(f_I)$. If $n_0(f_{I_i}) \leq n_1(f_{I_i})$ for all i, then for some j we have $n_1(f_{I_j}) \geq n_0(f_{I_j}) \geq \frac{n_0(f_I)}{2^m}$, and we can let $I'' = I_j$. Otherwise, we have $n_0(f_{I_i}) \leq n_1(f_{I_i})$ and $n_1(f_{I_j}) > n_0(f_{I_j})$, for some i and j. By Lemma 2 we then have a subinterval I'' of I such that $n_0(f_{I''}) = n_1(f_{I''})$, and thus $b(f_{I''}) \geq \frac{b(f_I)}{2^m}$.

By combining Lemma 3 and Corollary 2 we finally obtain the following proposition we will use in the proof of our main lower bound.

Proposition 3. *Let f be a symmetric Boolean function on n variables. Assume that m is an integer such that $2^m \leq \frac{b(f)}{16}$. Then there exist an interval I of length 2^k for some $k > m$ such that the interval I' of length 2^m in the middle of I satisfies $b(f_{I'}) \geq \frac{b(f)-1}{2^{m+3}\log_2\left(\frac{16n}{b(f)-1}\right)}$*

With this proposition in place we can give the proof of Theorem 1.

Proof. (Theorem 1) Assume that C is a depth h circuit containing s exact threshold gates $g_1, \ldots, g_s$ computing f. Assume there is no path from the output of g_j to an input of g_i if $i < j$. Define $b = \frac{b(f)-1}{8\log_2\left(\frac{16n}{b(f)-1}\right)}$ and $d = \frac{1}{14}\left(\frac{b}{4(s+1)}\right)^{\frac{1}{h}}$. Assume the size of C is $S = 2^{d-(s+1)-\log_2 n}$. For $\alpha \in \{0,1\}^s$ let C_i^α be the $\mathbf{ETHR} \circ \mathbf{AC^0}$ subcircuit of C with g_i as output, where every g_j for $j < i$ is replaced by the constant α_j. Similarly let C^α be the $\mathbf{AC^0}$ circuit obtained from C when every g_i is replaced by α_i.

Now let m be an integer such that $2^m \geq 14(14t)^{h-1} > 2^{m-1}$. Using Proposition 3 we have a restriction of f to a Boolean function f' on $n' = 2^k - 1$ variables such the subinterval I' of length 2^{k-m} in the middle of the value vector of f' satisfies $b(f_{I'}) \geq \frac{b}{2^m}$.

Pick a random restriction $\rho \in R_{n'}^{n_h}$, where $n_h = \frac{n'}{14(14d)^{h-1}}$ and apply it simultaneously to the circuits C^α and the circuits obtained from C_i^α by removing the output gate. Using Proposition 1 we then have that after applying ρ, except with probability at most $2^{s+1}S2^{-d} = \frac{1}{n}$ is every function computed by any gate of these circuits computed by a decision tree of depth at most d. By the discussion following Proposition 1 we obtain real polynomials P_i^α and Q^α of degree less then d, such that for all x, $C_{i,\rho}(x) = 1$ if and only if $P_i^\alpha(x) = 0$ and $C_\rho^\alpha(x) = Q^\alpha(x)$. Pick a *maximal* set G of the ETHR gates that are 0 at the same time for some input x to C_ρ, and define α such that $\alpha_i = 1$ if and only if $g_i \in G$.

The probability that the number of variables assigned 0 and 1 by ρ differ by at most 1 is at least $\frac{1}{n'-n_d}\binom{n'-n_d}{\frac{n'-n_d}{2}}$. Using Stirling's approximation this is $\Omega\left((n'-n_d)^{-\frac{1}{2}}\right) = \Omega(n^{-\frac{1}{2}})$. Thus for sufficiently large n we can assume that ρ gives the polynomials as above and the number of variables assigned 0 and 1 by ρ differ by at most 1. We then have that I' is a subinterval of the interval defined by ρ.

Now, if there exists x such that all gates in G evaluate to 0 and at the same time $C_\rho(x) = 1$, then the polynomial $Q^\alpha(x)\prod_{g_i \in G} P_i^\alpha(x)$ is a weak equality representation of f'_ρ. Otherwise the polynomial $\prod_{g_i \in G} P_i^\alpha(x)$ is a weak equality representation of $\neg f'_\rho$. The correctness of this claim follows from the maximality of the set G.

These polynomials are of degree less than $(s+1)t$ and since $b(f'_\rho) \geq \frac{b}{2^m}$, we must have $(s+1)t \geq \frac{b}{2^{m+1}}$ using Corollary 1.

However we have

$$t(s+1) = \frac{1}{14}\left(\frac{b}{4(s+1)}\right)^{\frac{1}{h}}(s+1)$$
$$= \frac{b}{4 \cdot 14\left(\frac{b}{4(s+1)}\right)^{\frac{h-1}{h}}} = \frac{b}{4 \cdot 14(14d)^{h-1}} < \frac{b}{2^{m+1}}$$

thus contracting the existence of C.

For *concrete* symmetric Boolean function slightly better lower bounds can be obtained by avoiding the use of Proposition 3. For example for the MAJ function the subinterval in the middle satisfies the requirements of the proof. For the MOD_2 function things are even better, since *any* subinterval satisfies the requirements. We just state the bounds since the proof is similar to the proof of Theorem 1.

Theorem 3. *Let C be a depth h $\mathbf{AC^0}$ circuit containing s exact threshold gates. If C computes the* MAJ *function or the* MOD_2 *function the size of C must be at least*

$$\frac{1}{n2^{s+1}}2^{\frac{1}{14}\left(\frac{n}{4(s+1)}\right)^{\frac{1}{h}}} \quad \text{and} \quad \frac{1}{2^{s+1}}2^{\frac{1}{14}\left(\frac{n}{4(s+1)}\right)^{\frac{1}{h}}}$$

respectively.

Turning to the proof of Theorem 2, one half of theorem follows readily from Theorem 1. The other half follows from the following proposition.

Proposition 4. *Any symmetric function f can be computed by a depth 3 $\mathbf{AC^0}$ circuit of size $b(f)+2$ augmented with $b(f)$ exact threshold gates.*

Proof. If $b(f) = n_1(f)$ we can compute f by a $\mathbf{OR} \circ \mathbf{ETHR}$ circuit where there is an exact threshold gate corresponding to every 1 in the value vector of f. If $b(f) = n_0(f)$ we construct the circuit as before for $\neg f$ and then add a negation gate at the output.

4 Conclusion

We have obtained strong lower bounds for circuits with few exact threshold gates with arbitrary weights computing symmetric Boolean functions. With our results we have essentially reached the best we could hope for with our current techniques: the approach taken in this paper does not provide the possibility to prove lower bounds with a superlinear amount exact threshold gates. Furthermore we would obviously also need to consider non-symmetric functions since every symmetric function can be computed with a linear number of exact threshold gates.

However, we find that further exploration of the power of exact threshold in constant depth circuits could prove to be very fruitful. In general, while circuits with threshold have been extensively studied, especially depth 2 and 3 circuits, very little research have considered circuits with (weighted) exact threshold gates, even though such a study is likely to provide insight into circuits with threshold gates. Many lower bounds are known for various classes of circuits with a (weighted or unweighted) threshold gate at the output. Interestingly, most of these lower bounds holds equally well with an exact threshold gate at the output, since the lower bounds actually holds with a Boolean function at the output that is determined by sign of a very small degree (e.g constant) polynomial. This allows one to simulate an exact threshold gate, since $p(x) = 0$ if and only if $(p(x))^2 \leq 0$.

For a concrete question, proving lower bounds for depth 2 threshold circuits is well known as being a notoriously difficult problem. Is the same true for depth 2 exact threshold circuits?

References

1. Furst, M., Saxe, J.B., Sipser, M.: Parity, circuits, and the polynomial-time hierarchy. Mathematical Systems Theory 17(1), 13–27 (1984)
2. Ajtai, M.: Σ_1^1-formulae on finite structures. Annals of Pure and Applied Logic 24, 1–48 (1983)
3. Håstad, J.: Computational limitations of small-depth circuits. MIT Press, Cambridge (1987)
4. Yao, A.C.C.: Separating the polynomial–time hierarchy by oracles. In: Proceedings 26st Annual Symposium on Foundations of Computer Science, pp. 1–10. IEEE Computer Society Press, Los Alamitos (1985)
5. Aspnes, J., Beigel, R., Furst, M.L., Rudich, S.: The expressive power of voting polynomials. Combinatorica 14(2), 135–148 (1994)
6. Beigel, R., Reingold, N., Spielman, D.A.: PP is closed under intersection. Journal of Computer and System Sciences 50(2), 191–202 (1995)
7. Beigel, R.: When do extra majority gates help? Polylog(n) majority gates are equivalent to one. Computational Complexity 4(4), 314–324 (1994)
8. Barrington, D.A.M., Straubing, H.: Complex polynomials and circuit lower bounds for modular counting. Computational Complexity 4(4), 325–338 (1994)

9. Chattopadhyay, A., Hansen, K.A.: Lower bounds for circuits with few modular and symmetric gates. In: Caires, L., Italiano, G.F., Monteiro, L., Palamidessi, C., Yung, M. (eds.) ICALP 2005. LNCS, vol. 3580, pp. 994–1005. Springer, Heidelberg (2005)
10. Viola, E.: Pseudorandom bits for constant depth circuits with few arbitrary symmetric gates. In: Proceedings of the 20th Annual IEEE Conference on Computational Complexity, pp. 198–209. IEEE Computer Society Press, Los Alamitos (2005)
11. Hansen, K.A.: Lower bounds for circuits with few modular gates using exponential sums. Technical Report 79, Electronic Colloquium on Computational Complexity (2006)
12. Fagin, R., Klawe, M.M., Pippenger, N.J., Stockmeyer, L.: Bounded-depth, polynomial-size circuits for symmetric functions. Theoretical Computer Science 36(2–3), 239–250 (1985)
13. Zhang, Z.L., Barrington, D.A.M., Tarui, J.: Computing symmetric functions with AND/OR circuits and a single MAJORITY gate. In: Enjalbert, P., Wagner, K.W., Finkel, A. (eds.) STACS 93. LNCS, vol. 665, pp. 535–544. Springer, Heidelberg (1993)
14. Lu, C.J.: An exact characterization of symmetric functions in $qAC^0[2]$. Theoretical Computer Science 261(2), 297–303 (2001)
15. Nisan, N.: The communication complexity of threshold gates. In: Miklós, D., Szönyi, T., S., V.T. (eds.) Combinatorics, Paul Erdös is Eighty. Bolyai Society. Mathematical Studies 1, vol. 1, pp. 301–315 (1993)
16. Beame, P.: A switching lemma primer. Technical Report UW-CSE-95-07-01, Department of Computer Science and Engineering, University of Washington (1994), Available online at `www.cs.washington.edu/homes/beame`
17. Green, F.: A complex-number fourier technique for lower bounds on the mod-m degre. Computational Complexity 9(1), 16–38 (2000)
18. Minsky, M., Papert, S.: Perceptrons - An Introduction to Computational Geometry. MIT Press, Cambridge (1969)

On the Complexity of Finding an Unknown Cut Via Vertex Queries

Peyman Afshani, Ehsan Chiniforooshan, Reza Dorrigiv, Arash Farzan, Mehdi Mirzazadeh, Narges Simjour, and Hamid Zarrabi-Zadeh

School of Computer Science, University of Waterloo
Waterloo, Ontario, N2L 3G1, Canada
{pafshani,echinifo,rdorrigiv,afarzan, mmirzaza,nsimjour,hzarrabi}@cs.uwaterloo.ca

Abstract. We investigate the problem of finding an unknown cut through querying vertices of a graph G. Our complexity measure is the number of submitted queries. To avoid some worst cases, we make a few assumptions which allow us to obtain an algorithm with the worst case query complexity of $O(k) + 2k \log \frac{n}{k}$ in which k is the number of vertices adjacent to cut-edges. We also provide a matching lowerbound and then prove if G is a tree our algorithm can asymptotically achieve the information theoretic lowerbound on the query complexity. Finally, we show it is possible to remove our extra assumptions but achieve an approximate solution.

1 Introduction

Consider a graph G together with a partition of its set of vertices, $V(G)$, into two sets, A and B. Here, we study the problem of finding the sets A and B by only asking queries about the vertices of G. In other words, the algorithm has only access to the graph G and an oracle which given a vertex v will tell the algorithm whether $v \in A$ or $v \in B$. Although we study this problem from a theoretical point of view, we can establish connections to the existing concepts and problems studied in machine learning.

In the standard learning problems, the learner is given a collection of labeled data items, which is called the *training data*. The learner is required to find a "hypothesis", using the training data and thus predict the labels of all (or most of) the data items, even those not seen by the learner algorithm. In this context, labeling the data points is considered to be an expensive operation. Thus, reducing the size of the training data is one of the important objectives. Semi-supervised learning attempts to accomplish this by using additional information about the whole data set. Recently, new models for semi-supervised learning have emerged which use spectral or graph techniques. We can name the work of Blum et al. [1], in which they built a graph and proposed several strategies to weigh the edges of the graph and proved finding a minimum cut corresponds to several of the previously employed learning algorithms based on Random Markov Fields. This is supported by the fact that it is common to restrict the set of possible labels

G. Lin (Ed.): COCOON 2007, LNCS 4598, pp. 459–469, 2007.

to $\{+, -\}$ [7] which implies a cut in graph G. We need to mention that spectral clustering techniques (for instance [9,8]) which are closely related to cuts have also been used in context of learning (for instance see [3,5]). We refer the reader to a line of papers in this area [1,10,4,2].

Other concepts related to the problem studied here are *Query learning* and *active learning*. Basically, under these assumptions the learner algorithm is allowed to interactively ask for the label of any data item. Thus, we can claim the problem studied in this paper has strong connection to the existing topics in machine learning; in short, our problem can be described as actively learning an unknown cut in a graph G.

In the next section, we define the problem precisely and obtain a simple lower bound on the number of queries. Then, in Section 3 we develop an algorithm that can solve a stronger version of our problem. We prove that the number of queries needed by our algorithm matches the lower bound. We discuss the problem for trees as a special family of graphs in Section 4. Finally, we relax the balancedness assumption and develop an ϵ-approximation algorithm instead of an exact algorithm in Section 5.

2 Preliminaries

Given a graph G, here we choose to represent the cut using a *labeling* $l : V(G) \rightarrow \{+, -\}$ which is the assignment of $+$ or $-$ to the vertices of G. Our goal is to design an algorithm which through querying labels of vertices can detect *all* the cut-edges. Clearly, the challenge is to minimize the number of queries or otherwise n queries can trivially solve the problem. Thus, we measure the complexity of the problem by the number of submitted queries and the parameters involved are the number of vertices, n, number of edges in the cut, k, and number of vertices adjacent to cut-edges, k'.

Notice that if graph G is a k-regular graph and all the vertices except one vertex v are labeled $+$, then it is easy to see that the algorithm must perform n queries in the worst case to find the single vertex v. This (the unbalanced cut having undesirable properties) is a common phenomenon which also appears in the spectral and clustering techniques. For instance, the definitions of normalized cut and ratio cut both factor in a form of balancedness condition. Here, we require a different form of balancedness: suppose we remove all the cut-edges in the graph. We call a labeling of a graph *α-balanced*, if each connected component in the new graph has at least αn vertices. Now we formally define our first problem:

Definition 1 (Problem A). *Suppose a graph G with an unknown α-balanced labeling of cut-size k is given. Use the structure of G together with the value of α to find this labeling using a small number of queries.*

Now, we show how to reduce the above problem to a seemingly easier problem by a probabilistic reduction. For a cut $\mathcal{C}$, define $G \setminus \mathcal{C}$ as the graph constructed from G by deleting all cut-edges of $\mathcal{C}$. Given a graph G, a *hint-set* is a set S of vertices of G such that S has at least one vertex from each connected component of $G \setminus \mathcal{C}$.

Definition 2 (Problem B). *Given an input graph G and a hint-set for an unknown labeling of G, find the labeling using a small number of queries.*

To show that Problem A can be reduced to Problem B, we select $\frac{c \log n}{\alpha}$ vertices uniformly at random and query the labels of the selected vertices. Since a connected component C contains at least αn vertices, the probability that a randomly selected vertex is not in C is at most $1-\alpha$. Hence, the probability that all selected vertices fall outside C is at most

$$(1-\alpha)^{\frac{c \log n}{\alpha}} \leq ((1-\alpha)^{\frac{1}{\alpha}})^{c \log n} \leq (\frac{1}{n})^c$$

The number of connected components is α^{-1} thus, with high probability we have obtained a hint-set and reduced the Problem A to Problem B.

In addition to this probabilistic reduction, a non-probabilistic one, though with an exponential running time, can also be proposed for the situation in which an upper bound k on the number of cut-edges of the labeling is given. As follows from a theorem proved by Kleinberg [6], for any graph G, parameter α, and integer $k < \frac{\alpha^2}{20}|V(G)|$, any subset of size $O(\frac{1}{\alpha}\log\frac{2}{\alpha})$, with probability at least $\frac{1}{2}$, is a hint-set for all α-balanced cuts of G with at most k cut-edges. So, we may examine each subset of vertices of G against every α-balanced cut of G with at most k cut-edges to find a subset that is a hint-set for all possible input labellings; then we will have an equivalent instance of problem B.

2.1 The Lower Bound

The balancedness condition prevents the problem from having a huge and trivial lowerbound. The idea behind our lowerbound is to construct a large number of balanced cuts on a fixed hint-set S.

Lemma 1. *Treating α as a constant, any algorithm that solves Problem A or B needs $k \log \frac{n}{k} - O(k)$ queries in the worst case, where n is the size of the graph and k is the cut-size of the labeling.*

Proof. Let G consist of k paths, each of length n/k, connected to each other through their endpoints as depicted in Figure 1. Suppose that the vertices in the

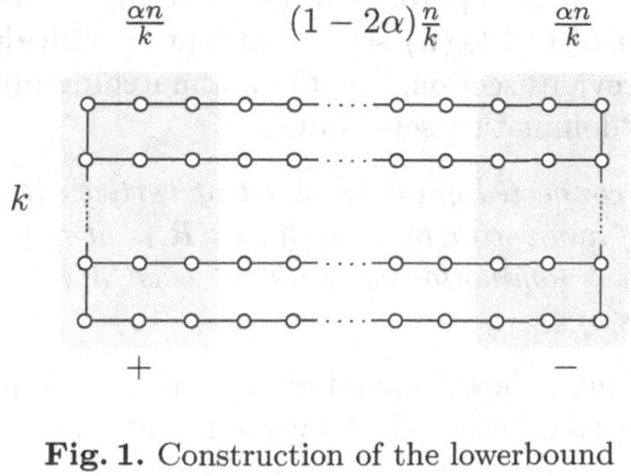

Fig. 1. Construction of the lowerbound

first (respectively, the last) α portion of each path are labeled with + (respectively, with −). The cut we are looking for is formed by k edges from the middle $(1-2\alpha)$ portion of the paths, one from each path. There are $\left((1-2\alpha)(\frac{n}{k})\right)^k$ choices for selecting these edges. Thus, any algorithm for finding the cut in this graph needs $\log((1-2\alpha)\frac{n}{k})^k = k\log\frac{n}{k} - O(k)$ queries in the worst case. □

3 An Optimal Algorithm

First we examine a very special case in which the graph G is a path and the cut is a single edge. This special case will come handy in the solution for the general case.

3.1 Algorithm for Paths with Cut-Size One

In this case the cut essentially is just one edge and all vertices to one side of it are all labeled + and the vertices on the other side are all labeled −. The solution is direct and is similar to the binary search algorithm. We start from both ends and query the labels of the endpoints. These will have opposite labels. Then we query the label of the midpoint and depending on the answer our search will be confided to one side of the path. We continue this binary search which eventually will find the cut using $O(\log n)$ queries.

Notice this binary search approach can still be used to find a cut-edge provided we start from two vertices with opposite labels.

3.2 Algorithm for Balanced Cuts

In this section, we develop an algorithm which matches the lowerbound proved in Section 2.1. We also focus on Problem B and assume a hint set $S = \{v_1, \ldots, v_c\}$ is given. First, the algorithm uses c queries to find out labels of vertices in S. It then computes a sequence $G_0 = G, G_1, \ldots, G_k = G \setminus \mathcal{C}$ of subgraphs of G where, for $1 \le i \le k$, $G_i = G_{i-1} - e_i$ for a cut-edge e_i of $\mathcal{C}$. To find a cut-edge e_i in G_{i-1}, it selects vertices u and v of S that are in the same connected component of G_i but have different labels. Then a "binary search" on a path between u and v in G_i is used to find a cut-edge. If computed naively, this path can have $\Omega(n)$ nodes, causing the algorithm to perform too many queries ($\Omega(\log n)$ in the worst case) at each step. Thus this naive approach will only result in the bound $O(k\log n)$. Although this bound of $O(k\log n)$ seems efficient, it still does not match the lowerbound in the previous section. To obtain a matching upperbound, we need one last ingredient: "domination sets" of G_i.

Definition 3. *In a connected graph G, a set of vertices R is an* r-domination set, *if the distance of any vertex in G to the set R is at most r. We use $f(r)$ to denote the size of an r-domination set with the least number of vertices among all r-domination sets of G.*

The next two lemma shows how domination sets are used to find a path of length at most $2r+1$ connecting two vertices with different labels in G_{i-1}.

Lemma 2. *Given a graph G and an r-domination set R of G, for every two vertices u and v of the same connected component of G, one can construct a walk W from u to v such that every $2r+1$ consecutive vertices in W contain at least one vertex from R.*

Proof. Let $P = (u = p_1, p_2, \ldots, p_l = v)$ be an arbitrary path between u and v in G. Since R is an r-domination, for each vertex $p_i \in P$ there is a vertex $r_i \in R$ such that there is a path P_i of length at most r connecting p_i to r_i. We define P'_i to be the reverse of P_i, for all $1 \leq i \leq l$, and W to be the walk $(P_1, P'_1, P_2, P'_2, \ldots, P_l, P'_l)$. Intuitively, the walk W starts walking along the vertices in P and at each vertex p_i, it first goes to r_i and then returns to p_i. So, each segment of size $2r+1$ of W contains at least one vertex from R. □

Lemma 3. *Suppose l is a labeling of a graph G, R is an r-domination set in G, and u_1 and u_2 are vertices with $l(u_1) \neq l(u_2)$. There are vertices v_1 and v_2 of distance at most $2r+1$ in $R \cup \{u_1, u_2\}$ such that $l(v_1) \neq l(v_2)$.*

Proof. Consider the walk W form u_1 to u_2 described in Lemma 2 and let $r_1, r_2, \ldots, r_p$ be the vertices of R appearing in W in order from u_1 to u_2. Also, define $r_0 = u_1$ and $r_{p+1} = u_2$. By Lemma 2, there are at most $2r$ vertices in W between r_i and r_{i+1}, exclusive, for $0 \leq i \leq p$. Since $l(r_0) \neq l(r_{p+1})$, there is an $0 \leq i \leq p$ such that $l(r_i) \neq l(r_{i+1})$. The correctness of the lemma follows by setting $v_1 = r_i$ and $v_2 = r_{i+1}$. □

Our algorithm will use the above two lemmas to iteratively find and extract the cut edges. Since at each step of the algorithm we will remove a cut edge, we must be able to update the r-domination set under edge deletions. Furtunately, this can be done trivially as the next lemma shows. We omit the prove since the proof is mostly intuitive.

Lemma 4. *If R is an r-domination set in G and $uv \in E$, then $R' = R \cup \{u, v\}$ is an r-domination set in $G' = G - uv$.*

The following property of r-domination sets allow the algorithm to asymptotically achieve the query complexity of the existing lower bound.

Lemma 5. *In any n-vertex connected graph, $f(r) \leq 2n/r$.*

Proof. This upper bound can be achieved by a naive greedy algorithm. Start with one vertex as the initial domination set R and progressively add vertices to R in steps. In each step, find a vertex with distance more than r to R and add it to R (if there is no such vertex, we are finished). We claim $|R| \leq 2n/r$.

We define the *d-neighborhood* of a vertex u, denoted by $N_u(d)$, as the set of all vertices within distance d of u. If v and w are in the greedily-selected r-domination set R, then $N_v(r/2)$ and $N_w(r/2)$ have an empty intersection. Hence, there are $|R|$ $\frac{r}{2}$-neighborhoods formed around vertices of R that are pairwise disjoint. Since, there are at least $r/2$ vertices (including v) in $N_v(r/2)$, for any vertex v, $r|R|/2$ is a lower bound on the number of vertices n. This immediately implies that $|R| \leq 2n/r$. □

Algorithm FindCutEdges(graph G, hint-set S, integers r and κ)

Find an r-domination set R for G using the greedy approach
Query labels of vertices in S and R
Set $C = \emptyset$
while *there are* u_1, u_2 *in* S *in the same component of* G *with* $l(u_1) \neq l(u_2)$ **do**
    Find a path P between two vertices v_1 and v_2 in $R \cup \{u_1, u_2\}$ with length at most $2r + 1$ such that $l(v_1) \neq l(v_2)$
    Use binary search to find an edge $e = wx$ of P such that $l(w) \neq l(x)$
    Set $C = C \cup e$, $G = G - e$, and $R = R \cup \{w, x\}$
    if $|C| > \kappa$ **then** Fail;
Return C

Fig. 2. Algorithm for finding the cut-edges of an unknown labeling of a graph

The greedy algorithm above seems naive as it can be far from achieving the optimal value of $f(r)$. However we prove, in the next lemma, that the size of its output can be upper bounded by the function f in some way.

Lemma 6. *Suppose, on an input graph G, the greedy algorithm, as described in Lemma 5, comes up with an r-domination set of size $Greedy(r)$. Then,*

$$f(r) \leq Greedy(r) \leq f(r/2).$$

Proof. It is obvious that $f(r) \leq Greedy(r)$, as $f(r)$ is the minimum size of a domination set and $Greedy(r)$ is the size one such set. To prove $Greedy(r) \leq f(r/2)$, it is sufficient to consider the $\frac{r}{2}$-neighborhoods formed around the selected vertices in Lemma 5. For a set to be an eligible $\frac{r}{2}$-domination set, it has to include at least one vertex from each of these $\frac{r}{2}$-neighborhoods; otherwise the center would be at a distance more than $r/2$ from all the vertices in the domination set. Since these neighborhoods are pairwise disjoint (see the proof of Lemma 5), the size of any eligible $\frac{r}{2}$-domination set must be at least the number of $\frac{r}{2}$-neighborhoods which is exactly $Greedy(r)$. Hence, $f(r/2) \geq Greedy(r)$. □

Consider the algorithm shown in Figure 2. It accepts additional parameters r and κ where κ is enforced to be an upper-bound on the cut-size. The algorithm starts by constructing an r-domination set R of G. Then, it performs $|R| + |S|$ queries to find out the labels of the vertices in R and in S. The algorithm runs in at most κ steps and in the i-th step it constructs the graph G_i. We use G_i, R_i, and C_i, respectively, to denote values of variables G, R, and C at the beginning of the i-th iteration of the while loop ($i \geq 0$).

At the beginning of the iteration i, we have a partially computed cut C_i, a graph $G_i = G \setminus C_i$, and an r-domination set R_i for G_i. Also, the algorithm knows the labels of vertices in S and R_i. Next, it uses Lemma 3 to find a path P of length at most $2r+1$ between two vertices with different labels. Hence, the algorithm uses at most $\lceil \log{(2r+1)} \rceil$ queries to find a new cut-edge e in G_i. The edge e is removed from the graph and the r-domination set is updated based on

Lemma 4. If $\kappa \geq k$, removing all edges of $\mathcal{C}$ one by one in this way, the algorithm finds the solution using $|S| + Greedy(r) + \kappa \log r$ queries. The algorithm fails, when $\kappa < k$, before submitting more than $|S| + Greedy(r) + \kappa \log r$ queries.

Notice that we can compute $Greedy(r)$ for different values of r in advance and choose the value which minimizes the query complexity. Then, the number of queries will be at most $|S| + O(\kappa) + \min_r\{Greedy(r) + \kappa \log r\}$ which is at most

$$|S| + O(\kappa) + \min_r\{f(r/2) + \kappa \log r\} = |S| + O(\kappa) + \min_r\{f(r) + \kappa \log r\}$$

according to Lemma 6.

Let U be the set of endpoints of all cut-edges. We can further reduce the number of queries by noticing that every edge between a vertex u with positive label and a vertex v with negative label must be a cut-edge. Once we remove all such trivial cut-edges, the next step of the algorithm will find a new vertex $v \in U$. This implies we can bound the number of steps by $k' = |U|$. The final theorem is as follows:

Theorem 1. *For a graph G, a labeling l, and an integer κ, one can use $|S| + O(\kappa) + \min_r\{f(r) + \kappa \log r)\}$ queries to discover if $\kappa < k'$ and to solve the problem B when $\kappa \geq k'$, where S is a hint-set for l.*

According to Lemma 5, $f(2n/\kappa) \leq \kappa$. Thus, if we run the algorithm of Theorem 1 with parameters $\kappa = 1, \kappa = 2, \kappa = 4, \ldots$ and $r = 2n/\kappa$, until κ becomes as large as k', we get the following theorem.

Corollary 1. *For a graph G and a labeling l, the problem B can be solved using $|S| + O(k') + 2k' \log \frac{n}{k'}$ queries, where k' is the number of vertices adjacent to cut-edges and S is a hint-set for l.*

Finally, we must mention that our probabilistic reduction of problem A to problem B used $\frac{c \log n}{\alpha}$ queries which is always asymptotically smaller than the above query complexity and thus we can safely omit this term.

4 The Tightness of the Bounds

The result of the Corollary 1 is tight with respect to parameters n and k'. However, for specific graphs better bounds might be possible. Note that given a graph G and a hit-set S one available lowerbound is the logarithm of the number of cuts for which S is the hint-set. Clearly, this number depends entirely on the structure of graph G. Same can be said about the result of Theorem 1. For instance, if G is a full binary tree then we have $f(r) = \theta(\frac{n}{2^r})$ which means the query complexity of Theorem 1 is in fact $\kappa \log \log \frac{n}{\kappa} + O(\kappa)$ with a matching asymptotic lowerbound for this particular tree. In this section we generalize this observation for all trees. In other words, we prove if G is a tree then the result of Theorem 1 is asymptotically tight by providing a matching lowerbound which entirely depends on the structure of G and thus throughout this section we always assume G is a tree.

We use the output S of the greedy algorithm of Lemma 5 to construct many of labellings for G, all with the same hint-set S. Consider an arbitrary r and the r-domination set S returned by greedy algorithm. This special r-domination set has the property that for every $u, v \in S$, $dist(u,v) \geq r+1$. We call any r-domination set with this property a *distributed r-domination set.* For a vertex $v \in S$ let N_v be the $\lceil \frac{r}{2} \rceil$-neighborhood of v. If $u \neq v$, then the edge-sets of the graphs induced by N_u and N_v do not intersect, that is, $E(G[N_u]) \cap E(G[N_v]) = \emptyset$. Suppose $v_0, \ldots v_{m-1}$ is an ordering of the vertices of S. We have assumed G is a tree and thus there is a unique path connecting v_i to v_j. Define P_{ij} to be the portion of this path which falls inside N_{v_i}. We have $|P_{ij}| \geq 1 + \lceil r/2 \rceil$, for every $0 \leq i < j \leq m-1$. The fact that $E(G[N_{v_{i_1}}]) \cap E(G[N_{v_{i_2}}]) = \emptyset$ implies that the edge sets of the paths $P_{i_1 j_1}$ and $P_{i_2 j_2}$ do not intersect, for $i_1 \neq i_2$. This results in the following lemma.

Lemma 7. *If S is a distributed r-domination set of size m in a tree G and $k < m$ be an arbitrary integer then, there are at least $\lceil \frac{r}{2} \rceil^k$ cuts each having k cut-edges such that S is a hint-set for every one of them.*

Proof. Consider the notation above and the following algorithm:

Set $C = \emptyset$
while $|C| < k$ **do**
- Choose the lexicographically smallest pair (i,j), $0 \leq i < j \leq m-1$ such that v_i and v_j are in the same component of $G \setminus C$.
- Nondeterministically select an edge of P_{ij} and add it to C

Since $k \leq m-1$, the selection of the pair (i,j) in each execution of the body of the while loop is feasible. As each path P_{ij} has $\lceil \frac{r}{2} \rceil$ edges and the body of the while loop is executed k times, there are $\lceil \frac{r}{2} \rceil^k$ possibilities, in overall, for nondeterministic choices of the algorithm. Moreover, each time that an edge is deleted from $G \setminus C$ and a connected component of $G \setminus C$ is split into two, each new connected component still includes a vertex of S (either v_i or v_j). Therefore, S is a hint-set for the cut specified by any set C generated by this algorithm.

It remains to show that all the $\lceil \frac{r}{2} \rceil^k$ sets C generated by the algorithm are distinct. Consider two different executions $\mathcal{E}_1$ and $\mathcal{E}_2$ of the algorithm and consider the first point that the algorithm makes different decisions in $\mathcal{E}_1$ and in $\mathcal{E}_2$. Suppose $\mathcal{E}_1$ chooses an edge e_1 of P_{ij} while $\mathcal{E}_2$ selects a different edge e_2 of P_{ij}. Without loss of generality, assume e_1 appears before e_2 in P_{ij}. The edge e_2 is in $\lceil \frac{r}{2} \rceil$-neighborhood of v_i and after adding e_1 to C in $\mathcal{E}_1$, e_2 is not in the same component as v_i anymore; so $\mathcal{E}_1$ will never add e_2 to C. Thus, the value of C in $\mathcal{E}_2$ will be different form the value of C in $\mathcal{E}_1$. So, each execution of the algorithm generates a distinct set of edges. □

Next theorem uses this set of cuts to give a lowerbound.

Theorem 2. *For every tree G with n vertices and integer $k > 0$, there is an integer r such that any algorithm solving problem B must perform $f(r) + k \log r - O(k)$ queries.*

Proof. Due to the term $-O(k)$, we can assume $k \leq \frac{n}{2}$. Define m_t as the maximum size of any distributed t-domination set. The size of the maximum independent set of a tree is at least $\frac{n}{2}$ which implies $m_1 \geq \frac{n}{2}$. As r increases, m_r must decrease, and in particular $m_n = 1$. We know $k \leq \lceil \frac{r}{2} \rceil$ and thus there is an integer $t > 1$ such that $f(t+1) \leq m_{t+1} \leq k < m_t$. According to Lemma 7 any algorithm solving problem B must perform $k \log \frac{t}{2}$ queries and a simple calculation shows that $f(t+1) + k\log(t+1) - O(k) \leq k \log t$ which means $f(r) + k \log r - O(k)$ is a lowerbound for the number of queries for $r = t+1$. □

5 Relaxing the Balancedness Assumption

In this section we show how to remove the balancedness assumption at the expense of obtaining an approximation algorithm. Thus, we present an ε-approximation algorithm that performs $k \ln(3/\varepsilon) + k \log(n/k) + O(k)$ queries on a given graph with n vertices and cut-size k and reports a labeling which with high probability has at most εn vertices mislabeled.

The algorithm is fundamentally same as before. We select a random set of vertices uniformly as the hint-set and perform Algorithm 2 just once for $\kappa = k$ and $r = n/k$. Nevertheless, the probabilistic analysis presented in Section 2 holds no more; the connected components here can be arbitrarily small so there could be components which do not contain any vertex from the hint set.

We call the components that have a vertex representative in the sample as the *represented* components and the rest as the *unrepresented*. Firstly, we claim that if we select $(k+1)\ln(\varepsilon/3)$ sample vertices, with high probability, the number of vertices in an unrepresented component is at most εn. The argument is probabilistic; we compute the expected number of vertices in unrepresented components. We denote by $n_1, \ldots, n_t$ the sizes of connected components $C_1, \ldots, C_t$ in the initial graph respectively. Using basic probability arguments, the probability that the component C_i is unrepresented is at most $\left(1 - \frac{n_i}{n}\right)^{(k+1)\ln(\varepsilon/3)}$. Therefore, the expected total number of vertices in unrepresented components is

$$\mathbf{E} = \sum_{i=1}^{t} n_i \left(1 - \frac{n_i}{n}\right)^{(k+1)\ln(\varepsilon/3)} \leq n \sum_{i=1}^{t} \frac{n_i}{n} e^{-\frac{n_i}{n}(k+1)\ln(\varepsilon/3)}.$$

As the function xe^{-cx} is convex for any fixed c, the maximum happens when $\frac{n_i}{n}$ are equal and thus $\mathbf{E} \leq n \left(t \frac{1}{t} e^{-\frac{1}{t}(k+1)\ln(\varepsilon/3)}\right) \leq n\varepsilon/3$. The last inequality is due to the fact that k cannot be less than $t-1$. By using Markov inequality, one can assert that the probability that there are more than εn vertices in unreported components is less than $1/3$.

Secondly, we observe that by a run of Algorithm 2, labels of vertices of represented components are reported correctly. This fact follows from the correctness of the algorithm, and the observation that it never reports an edge as a cut-edge if it is not indeed a cut-edge. The algorithm might miss some cut-edges in contrast to the previous setting, as some components are unrepresented, and so the labeling of such components can be reported arbitrarily.

Combining the two latter facts, we conclude that the total number of vertices whose labels are misreported is less than εn for an arbitrary small $\varepsilon < 1$.

6 Conclusion

In this paper we studied the query learning problem, while we assumed both labeled and unlabeled data were available to us and we had a similarity graph constructed on data items. The problem was discussed in the following two settings: We studied the case where we are given a hint set of vertices in which there exists at least one vertex from each connected component of the same label. We gave the lower bound $k \log n/k - O(k)$ on the number of queries required for this case. We provided an algorithm that finds the optimal hypothesis using at most $O(k \log n/k)$ queries, which matched our lower bound for general graphs. Hence, our proposed algorithm is optimal in terms of the number of queries.

In the second setting, we showed that if the labeling is α-balanced, i.e. each connected component of the vertices with the same label has at least α fraction of all vertices, we could find a hint set with high probability using $\frac{c \log n}{\alpha}$ random queries. Note that our previous lower bound works for the α-balanced graphs, too. This gives an evidence of the optimality of our algorithm for general graphs in the latter setting.

Although our algorithm for the first setting was proved to be optimal for general graphs, we may have different lower bounds for special families of graphs. In Section 4, we investigated the problem for trees and proved for tree our algorithm is asymptotically optimal. One natural question is whether it is possible to extend this result to other classes of graphs or not.

Finally, we considered the problem for non-balance cuts. We developed an ε-approximation algorithm for this case. However, we assumed that the number of cut-edges is given to the algorithm. Thus, the following problem remained open:

Open Problem: *Suppose that G is a graph, l is a labeling of G with k cut-edges, and $0 < \varepsilon \leq 1$ is a real number. Does there exist an ε-approximation algorithm that runs in polynomial time and submits at most $O(\text{poly}(1/\varepsilon) k \log n/k)$ queries without knowing k in advance?*

References

1. Blum, A., Chawla, S.: Learning from labeled and unlabeled data using graph mincuts. In: Proceedings of the Eighteenth International Conference on Machine Learning, pp. 19–26. Morgan Kaufmann Publishers, San Francisco (2001)
2. Blum, A., Lafferty, J., Rwebangira, M.R., Reddy, R.: Semi-supervised learning using randomized mincuts. In: Proceedings of the twenty-first international conference on Machine learning, p. 13. ACM Press, New York (2004)
3. Joachims, T.: Transductive learning via spectral graph partitioning. In: Twentieth International Conference on Machine Learning (2003)

4. Joachims, T.: Transductive learning via spectral graph partitioning. In: Proceedings of the International Conference on Machine Learning, pp. 290–297 (2003)
5. Kamvar, S., Klein, D., Manning, C.: Spectral learning. In: International Joint Conference On Artificial Intelligence (2003)
6. Kleinberg, J.: Detecting a network failure. In: Proceedings of the Forty-First Annual Symposium on Foundations of Computer Science, p. 231. IEEE Computer Society Press, Los Alamitos (2000)
7. Mitchell, T.: Machine Learning. McGraw-Hill, New York (1997)
8. Ng, A., Jordan, M., Weiss, Y.: On spectral clustering: Analysis and an algorithm. In: Advances in Neural Information Processing Systems (2001)
9. Shi, J., Malik, J.: Normalized cuts and image segmentation. IEEE Trans. Pattern Anal. Mach. Intell. 22(8), 888–905 (2000)
10. Zhu, X., Ghahramani, Z., Lafferty, J.: Semi-supervised learning using Gaussian fields and harmonic functions. In: Proceedings of the Twentieth International Conference on Machine Learning, pp. 912–919 (2003)

“Resistant” Polynomials and Stronger Lower Bounds for Depth-Three Arithmetical Formulas*

Maurice J. Jansen and Kenneth W. Regan**

Dept. of CSE, University at Buffalo (SUNY), 201 Bell Hall, Buffalo, NY 14260-2000
Tel.: (716) 645-3180 x114; Fax: 645-3464
regan@cse.buffalo.edu.

Abstract. We derive quadratic lower bounds on the $*$-complexity of sum-of-products-of-sums ($\Sigma\Pi\Sigma$) formulas for classes of polynomials f that have too few partial derivatives for the techniques of Shpilka and Wigderson [10,9]. This involves a notion of “resistance” which connotes full-degree behavior of f under any projection to an affine space of sufficiently high dimension. They also show stronger lower bounds over the reals than the complex numbers or over arbitrary fields. Separately, by applying a special form of the Baur-Strassen Derivative Lemma tailored to $\Sigma\Pi\Sigma$ formulas, we obtain sharper bounds on $+,*$-complexity than those shown for $*$-complexity by Shpilka and Wigderson [10], most notably for the lowest-degree cases of the polynomials they consider.

Keywords: Computational complexity, arithmetical circuits, lower bounds, constant depth formulas, partial derivatives.

1 Introduction

In contrast to the presence of exponential size lower bounds on constant-depth Boolean circuits for majority and related functions [3,13,7], and depth-3 arithmetical circuits over finite fields [5,6], Shpilka and Wigderson [10] observed that *over fields of characteristic zero* (which are infinite), super-quadratic lower bounds are not known even for constant-depth *formulas*. Indeed they are unknown for unbounded fan-in, depth 3 formulas that are sums of products of affine linear functions, which they call $\Sigma\Pi\Sigma$ formulas. These formulas have notable *upper-bound power* because they can carry out forms of Lagrange interpolation. As they ascribed to M. Ben-Or, $\Sigma\Pi\Sigma$ formulas can compute the elementary symmetric polynomials S_n^k (defined as the sum of all degree-k monomials in n variables, and analogous to majority and threshold-k Boolean functions) in size $O(n^2)$ independent of k. Thus $\Sigma\Pi\Sigma$ formulas present a substantial challenge for lower bounds, as well as being a nice small-scale model to study.

Shpilka and Wigderson defined the *multiplicative size* of an arithmetical (circuit or) formula ϕ to be the total fan-in to multiplication gates. We denote this

* Part of this work by both authors was supported by NSF Grant CCR-9821040.
** Corresponding author.

G. Lin (Ed.): COCOON 2007, LNCS 4598, pp. 470–481, 2007.

by $\ell^*(\phi)$, and write $\ell(\phi)$ for the total fan-in to all gates, i.e. + gates as well. The best known lower bound for general arithmetical circuits has remained for thirty years the $\Omega(n \log n)$ lower bound on ℓ^* by the "Degree Method" of Strassen [11] (see also [1,2]). However, this comes nowhere near the exponential lower bounds conjectured by Valiant [12] for the permanent and expected by many for other NP-hard arithmetical functions. For polynomials f of total degree $n^{O(1)}$, the method is not even capable of $\Omega(n^{1+\epsilon})$ circuit lower bounds, not for any $\epsilon > 0$. Hence it is notable that [10] achieved better lower bounds on $\ell_3^*(f)$, where the subscript-3 refers to $\Sigma\Pi\Sigma$ formulas. These were $\Omega(n^2)$ for $f = S_n^k$ when $k = \Theta(n)$, $n^{2-\epsilon_k}$ for S_n^k with small values of k, and $\Omega(N^2/\operatorname{polylog}(N))$ for the determinant, with $N = n^2$. However, $\Omega(n^2)$ is the best this can do for $\Sigma\Pi\Sigma$ formulas. Shpilka [9] got past this only in some further-restricted cases, and also considered a depth-2 model consisting of an arbitrary symmetric function of sums. This barrier provides another reason to study the $\Sigma\Pi\Sigma$ model, in order to understand the obstacles and what might be needed to surpass them.

The techniques in [8,10,9] all depend on the set of dth-order partial derivatives of f being large. This condition fails for functions such as $f(x_1, \ldots, x_n) = x_1^n + \ldots + x_n^n$, which has only n dth-order partials for any d. We refine the analysis to show the sufficiency of f behaving like a degree-r polynomial on any affine subspace A of sufficiently high dimension (for this f, $r = n$ and any affine line suffices). Our technical condition is that for every polynomial g of total degree at most $r - 1$ and every such A, *there exists* a d-th order partial of $f - g$ that is non-constant on A. This enables us to prove an absolutely sharp n^2 bound on $\ell_3^*(f)$ for this f computed over the real or rational numbers, and a lower bound of $n^2/2$ over any field of characteristic zero. Note the absence of "O, Ω" notation. We prove similar tight bounds for sums of powered monomial blocks, powers of inner-products, and functions depending on ℓ_p-norm distance from the origin, and also replicate the bounds of [10,9] for symmetric polynomials. Even in the last case, we give an example where our simple existential condition may work deeper than the main question highlighted in [9] on the maximum dimension of subspaces A on which S_n^k vanishes.

In Section 5 we prove lower bounds on $+, *$ complexity $\ell_3(f)$ that are significantly higher (but still sub-quadratic) than those given for $\ell_3^*(f)$ in [10] when the degree r of f is small. This is done intuitively by exploiting a closed-form application of the Baur-Strassen "Derivative Lemma" [1] to $\Sigma\Pi\Sigma$ formulas, showing that f and all of its n first partial derivatives can be computed with only a constant-factor increase in ℓ and ℓ^* over $\Sigma\Pi\Sigma$ formulas for f.

2 Preliminaries

A $\Sigma\Pi\Sigma$-formula is an arithmetic formula consisting of four consecutive layers: a layer of inputs, next a layer of addition gates, then a layer of multiplication gates, and finally the output sum gate. The gates have unbounded fan-in from the previous layer (only), and individual wires may carry arbitrary constants from the underlying field. Given a $\Sigma\Pi\Sigma$-formula we can write $p = \sum_{i=1}^{s} M_i$, where $M_i = \Pi_{j=1}^{d_i} l_{i,j}$, and $l_{i,j} = c_{i,j,1}x_1 + c_{i,j,2}x_2 + \ldots + c_{i,j,n}x_n + c_{i,j,0}$. Here d_i is the in-degree of the ith multiplication gate, and $c_{i,j,k}$ is nonzero iff there is a wire from x_k to the addition gate computing $l_{i,j}$.

Let $X = (x_1, \ldots, x_n)$ be an n-tuple of variables. For any affine linear subspace $A \subset F^n$, we can always find a set of variables $B \subset X$, and affine linear forms l_b in the variables $X \setminus B$, for each $b \in B$, such that A is the set of solutions of $\{x_b = l_b : b \in B\}$. This representation is not unique. The set B is called a *base* of A. The size $|B|$ always equals the co-dimension of A. In the following, we always assume some base B of A to be fixed. Any of our numerical "progress measures" used to prove lower bounds will not depend on the choice of a base.

Following Shpilka and Wigderson [10], for polynomial $f \in F[x_1, \ldots, x_n]$, the *restriction of f to A* is defined to be the polynomial obtained by substitution of l_b for variable x_b for each $b \in B$, and is denoted by $f_{|A}$. For a set of polynomials W, define $W_{|A} = \{f_{|A} \mid f \in W\}$. For a linear form $l = c_1x_1 + \ldots + c_nx_n + c_0$, we denote $l^h = c_1x_1 + \ldots + c_nx_n$. For a set S of linear forms, $S^h = \{l^h : l \in S\}$.

3 Resistance of Polynomials

We state our new definition in the weakest and simplest form that suffices for the lower bounds, although the functions in our applications all meet the stronger condition of Lemma 1 below.

Definition 1. *A polynomial f in variables $x_1, x_2, \ldots, x_n$ is (d, r, k)-*resistant *if for any polynomial $g(x_1, x_2, \ldots, x_n)$ of degree at most $r-1$, for any affine linear subspace A of co-dimension k, there exists a dth order partial derivative of $f - g$ that is non-constant on A.*

For a multiset X of size d with elements taken from $\{x_1, x_2, \ldots, x_n\}$, we will use the notation $\frac{\partial^d f}{\partial X}$ to indicate the dth-order derivative with respect to the variables in X. As our applications all have $r = \deg(f)$, we call f simply (d, k)-*resistant* in this case. Then the case $d = 0$ says that f itself has full degree on any affine A of co-dimension k, and in most cases corresponds to the non-vanishing condition in [10]. We separate our notion from [10] in applications and notably in the important case of the elementary symmetric polynomials in Section 4.4 below.

The conclusion of Definition 1 is not equivalent to saying that some $(d+1)$st-order partial of $f - g$ is non-vanishing on A, because the restriction of this partial on A need not be the same as a first-partial of the restriction of the dth-order partial to A. Moreover, (d, k)-resistance need not imply $(d-1, k)$-resistance, even for $d, k = 1$: consider $f(x, y) = xy$ and A defined by $x = 0$.

Theorem 1. *Suppose $f(x_1, x_2, \ldots, x_n)$ is (d, r, k)-resistant, then*

$$\ell_3^*(f) \geq r\frac{k+1}{d+1}.$$

Proof. Consider a $\Sigma\Pi\Sigma$-formula that computes f. Remove all multiplication gates that have degree at most $r-1$. Doing so we obtain a $\Sigma\Pi\Sigma$ formula $\mathcal{F}$ computing $f-g$, where g is some polynomial of degree at most $r-1$. Say $\mathcal{F}$ has s multiplication gates. Write: $f-g=\sum_{i=1}^{s} M_i$, where $M_i=\Pi_{j=1}^{d_i} l_{i,j}$ and $l_{i,j}=c_{i,j,1}x_1+c_{i,j,2}x_2+\ldots+c_{i,j,n}x_n+c_{i,j,0}$. The degree of each multiplication gate in $\mathcal{F}$ is at least r, i.e. $d_i \geq r$, for each $1 \leq i \leq s$. Now select a set S of input linear forms using the following algorithm:

$S=\emptyset$
for $i=1$ to s **do**
  repeat $d+1$ times:
    if $(\exists j \in \{1,2,\ldots,d_i\})$: $S^h \cup \{l^h_{i,j}\}$ is a set of independent vectors
      then $S=S\cup\{l_{i,j}\}$

Let A be the set of common zeroes of the linear forms in S. Since S^h is an independent set, A is affine linear of co-dimension $|S| \leq (d+1)s$.

We claim that if at a multiplication gate M_i we picked strictly fewer than $d+1$ linear forms, then any linear form that was not picked is constant on A. Namely, each linear form l that was not picked had l^h already in the span of S^h, for the set S built up so far. Hence we can write $l=c+l^h=c+\sum_{g\in S} c_g g^h$, for certain scalars c_g. Since each g^h is constant on A, we conclude l is constant on A. This settles the claim, and yields that for each multiplication gate either

1. $(d+1)$ input linear forms vanish on A, or
2. fewer than $(d+1)$ linear forms vanish on A, with all others constant on A.

For each multiset X of size d with elements from $\{x_1,x_2,\ldots,x_n\}$, the dth order partial derivative $\partial^d(f-g)/\partial X$ is in the linear span of the set

$$\left\{\prod_{\substack{j=1\\ j\notin J}}^{d_i} l_{ij} \,:\, 1\leq i\leq s,\ J\subseteq\{1,2,\ldots,d_i\},\ |J|=d\right\}$$

This follows from the sum and product rules for derivatives and the fact that a first order derivative of an individual linear form l_{ij} is a constant. Consider $1 \leq i \leq s$ and $J \subseteq \{1,2,\ldots,d_i\}$ with $|J|=d$. If item 1. holds for the multiplication gate M_i, then $\prod_{\substack{j=1\\ j\notin J}}^{d_i} l_{ij}$ vanishes on A, since there must be one l_{ij} that vanishes on A that was not selected, given that $|J|=d$. If item 2 holds for M_i, then this product is constant on A.

Hence, we conclude that $\partial^d(f-g)/\partial X$ is constant on A. Since f is (d,r,k)-resistant, we must have that the co-dimension of A is at least $k+1$. Hence $(d+1)s \geq k+1$. Since each gate in $\mathcal{F}$ is of degree at least r, we obtain $\ell_3^*(\mathcal{F}) \geq r\frac{k+1}{d+1}$. Since $\mathcal{F}$ was obtained by removing zero or more multiplication gates from a $\Sigma\Pi\Sigma$-formula computing f, we have proven the statement of the theorem. □

To prove lower bounds on resistance, we supply the following lemma:

Lemma 1. *Over fields of characteristic zero, for any $d \leq r$, $k > 0$, and any polynomial $f(x_1, x_2, \ldots, x_n)$, if for every affine linear subspace A of co-dimension k, there exists some dth order partial derivative of f such that*

$$\deg\left(\left(\frac{\partial^d f}{\partial X}\right)_{|A}\right) \geq r - d + 1, \qquad \textit{then } f \textit{ is } (d, r+1, k)\textit{-resistant.}$$

Proof. Assume for every affine linear subspace A of co-dimension k, there exists some dth order partial derivative derivative of f such that

$$\deg\left(\left(\frac{\partial^d f}{\partial X}\right)_{|A}\right) \geq r - d + 1.$$

Let g be an arbitrary polynomial of degree r. Then

$$\left(\frac{\partial^d f - g}{\partial X}\right)_{|A} = \left(\frac{\partial^d f}{\partial X} - \frac{\partial^d g}{\partial X}\right)_{|A} = \left(\frac{\partial^d f}{\partial X}\right)_{|A} - \left(\frac{\partial^d g}{\partial X}\right)_{|A}.$$

The term $\left(\frac{\partial^d f}{\partial X}\right)_{|A}$ has degree at least $r - d + 1$, whereas the term $\left(\frac{\partial^d g}{\partial X}\right)_{|A}$ can have degree at most $r - d$. Hence $\deg(\left(\frac{\partial^d f - g}{\partial X}\right)_{|A}) \geq r - d + 1 \geq 1$. Since over fields of characteristic zero, syntactically different polynomials define different mappings, we conclude $\frac{\partial^d f - g}{\partial X}$ must be non-constant on A. □

The main difference between Lemma 1 and the original Definition 1 appears to be the order of quantifying the polynomial "g" of degree $r - 1$ out front in the former, whereas analogous considerations in the lemma universally quantify it later (making a stronger condition). We have not found a neat way to exploit this difference in any prominent application, however.

4 Applications

4.1 Sum of nth Powers Polynomial

Consider $f = \sum_{i=1}^n x_i^n$. By repeated squaring for each x_i^n, one obtains $\Sigma\Pi$ circuits (not formulas) of size $O(n \log n)$. All arithmetical circuits require size $\Omega(n \log n)$ for f [1]. The expression for f yields a $\Sigma\Pi\Sigma$ *formula* ϕ with n multiplication gates of degree n, with n^2 wires in the top linear layer fanning in to them. This works over any field, but makes $\ell(\phi) = \ell^*(\phi) = n^2$. We prove that this is close to optimal.

Theorem 2. *Over fields of characteristic zero, any $\Sigma\Pi\Sigma$-formula for $f = \sum_{i=1}^n x_i^n$ has multiplicative size at least $n^2/2$.*

Proof. By Theorem 1 it suffices to show f is $(1, n-1)$-resistant. Let g be an arbitrary polynomial of degree $n-1$. Letting $g_1, \ldots, g_n$ denote the first order partial derivatives of g, we get that the *ith* partial derivative of $f - g$ equals $nx_i^{n-1} - g_i(x_1, \ldots, x_n)$. Note that the g_i's are of total degree at most $n-2$.

We claim there is no affine linear subspace of dimension greater than zero on which all $\partial f / \partial x_i$ are constant. Consider an arbitrary affine line $x_i = c_i + d_i t$ parameterized by a variable t, where c_i and d_i are constants for all $i \in [n]$, and with at least one d_i nonzero. Then $\frac{\partial(f-g)}{\partial x_i}$ restricted to the line is given by $n(c_i + d_i t)^{n-1} - h_i(t)$, for some univariate polynomials $h_i(t)$ of degree $\leq n-2$. Since there must exist *some* i such that d_i is nonzero, we know some partial derivative restricted to the affine line is parameterized by a univariate polynomial of degree $n-1$, and thus, given that the field is of characteristic zero, is not constant for all t. □

In case the underlying field is the real numbers **R** and n is even, we can improve the above result to prove an absolutely tight n^2 lower bound.

Theorem 3. *Over the real numbers, for even n, any $\Sigma\Pi\Sigma$-formula for $f = \sum_{i=1}^{n} x_i^n$ has multiplicative size at least n^2.*

Proof. Since f is symmetric we can assume without loss of generality that A has the base representation $x_{k+1} = l_1(x_1, \ldots, x_k)$, $\ldots$, $x_n = l_{n-k}(x_1, \ldots, x_k)$. Then

$$f_{|A} = x_1^n + \ldots x_k^n + l_1^n + \ldots + l_{n-k}^n.$$

Hence $f_{|A}$ must include the term x_1^n, since each l_j^n has a non-negative coefficient for the term x_1^n and n is even. Thus via Lemma 1 we conclude that over the real numbers f is $(0, n-1)$-resistant. Hence, by Theorem 1 we get that $\ell_3^*(f) \geq \deg(f)\frac{n}{1} = n^2$. □

Let us note that $f = \sum_{i=1}^{n} x_i^n$ is an example of a polynomial that has few d-th partial derivatives, namely only n of them regardless of d. This renders the partial derivatives technique of Shpilka and Wigderson [10]—which we will describe and extend in the next section—not directly applicable.

4.2 Blocks of Powers Polynomials

Let the underlying field have characteristic zero, and suppose $n = m^2$ for some m. Consider the "m blocks of m powers" polynomial $f = \sum_{i=1}^{m} \prod_{j=(i-1)m+1}^{im} x_j^m$. The straightforward $\Sigma\Pi\Sigma$-formula for f, that computes each term/block using a multiplication gate of degree n, is of multiplicative size $n^{3/2}$. We will show this is tight.

Proposition 1. *The blocks of powers polynomial f is $(0, m-1)$-resistant.*

Proof. Consider an affine linear space of co-dimension $m-1$. For any base B of A, restriction to A consists of substitution of the $m-1$ variables in B by linear forms in the remaining variables X/B. This means there is at least one term/block

$B_i := \prod_{j=(i-1)m+1}^{im} x_j^m$ of f whose variables are disjoint from B. This block B_i remains the same under restriction to A. Also, for every other term/block there is at least one variable that is not assigned to. As a consequence, B_i cannot be canceled against terms resulting from restriction to A of other blocks. Hence $\deg(f_{|_A}) = \deg(f)$. Hence by Lemma 1 we have that f is $(0, m-1)$-resistant. □

Corollary 1. *For the blocks of powers polynomial f, $\ell_3^*(f) \geq nm = n^{3/2}$.*

Alternatively, one can observe that by substitution of a variable y_i for each variable appearing in the ith block one obtains from a $\Sigma\Pi\Sigma$-formula $\mathcal{F}$ for f a formula for $f' = \sum_{i=1}^{m} y_i^n$ of the same size as $\mathcal{F}$. Theorem 2 generalizes to show that $\ell_3^*(f') \geq \frac{1}{2}n^{3/2}$, which implies $\ell_3^*(f) \geq \frac{1}{2}n^{3/2}$.

4.3 Polynomials Depending on Distance to the Origin

Over the real numbers, $d_2(x) = x_1^2 + x_2^2 + \cdots + x_n^2$ is the square of the Euclidean distance of the point x to the origin. Polynomials f of the form $q(d_2(x))$ where q is a single-variable polynomial can be readily seen to have high resistance. Only the leading term of q matters. For example, consider $f = (x_1^2 + x_2^2 + \cdots + x_n^2)^m$. On any affine line L in $\mathbf{R}^n$, $\deg(f_{|_L}) = 2m$. Therefore, by Lemma 1, over the reals, f is $(0, n-1)$-resistant. Hence by Theorem 1 we get that

Proposition 2. *Over the real numbers, $\ell_3^*((x_1^2 + x_2^2 + \cdots + x_n^2)^m) \geq 2mn$.*

Observe that by reduction this means that the "mth-power of an inner product polynomial", defined by $g = (x_1y_1 + x_2y_2 + \cdots + x_ny_n)^m$, must also have $\Sigma\Pi\Sigma$-size at least $2mn$ over the reals numbers. Results for l_p norms, $p \neq 2$, are similar.

4.4 The Case of Symmetric Polynomials

The special case of $(0, k)$-resistance is implicitly given by Shpilka [9], at least insofar as the sufficient condition of Lemma 1 is used for the special case $d = 0$ in which no derivatives are taken. For the elementary symmetric polynomial S_n^r of degree $r \geq 2$ in n variables, Theorem 4.3 of [9] implies (via Lemma 1) that S_n^r is $(0, n - \frac{n+r}{2})$-resistant. Shpilka proves for $r \geq 2$, $\ell_3^*(S_n^r) = \Omega(r(n-r))$, which can be verified using Theorem 1: $\ell_3^*(S_n^r) \geq (r+1)(n - \frac{n+r}{2}) = \Omega(r(n-r))$.

The symmetric polynomials S_n^k collectively have the "telescoping" property that every dth-order partial is (zero or) the symmetric polynomial S_{n-d}^{k-d} on an $(n-d)$-subset of the variables. Shpilka [9] devolves the analysis into the question, "What is the maximum dimension of a linear subspace of $\mathbf{C}^n$ on which S_n^r vanishes?" In Shpilka's answer, divisibility properties of r come into play as is witnessed by Theorem 5.9 of [9]. To give an example case of this theorem, one can check that S_9^2 vanishes on the 3-dimensional linear space given by

$$\{(x_1, \omega x_1, \omega^2 x_1, x_2, \omega x_2, \omega^2 x_2, x_3, \omega x_3, \omega^2 x_3) : x_1, x_2, x_3 \in \mathbf{C}\},$$

where ω can be selected to be either primitive 3rd root of unity. Let

$$\rho_0(f) = \max\{k : \text{ for any linear } A \text{ of codim. } k,\ f_{|_A} \neq 0\}.$$

Shpilka proved for $r > n/2$, that $\rho_0(S_n^r) = n - r$, and for $r \geq 2$, that $\frac{n-r}{2} < \rho_0(S_n^r) \leq n-r$. For S_9^2 we see via divisibility properties of d that the value for ρ_0 can get less than the optimum value, although the $\frac{n-r}{2}$ lower bound suffices for obtaining the above mentioned $\ell_3^*(S_n^r) = \Omega(r(n-r))$ lower bound. We have some indication from computer runs using the polynomial algebra package *Singular* [4] that the "unruly" behavior seen for ρ_0 because of divisibility properties for $r \leq n/2$ can be made to go away by considering the following notion:

$$\rho_1(f) = \max\left\{k : \text{ for any linear } A \text{ of codim. } k, \text{ there exists } i, \left(\frac{\partial f}{\partial x_i}\right)_{|A} \neq 0\right\}$$

One can still see from the fact that S_n^r is homogeneous and using Lemma 1 and Theorem 1 that $\ell_3^*(S_n^r) \geq \frac{r\cdot(\rho_1(S_n^r)+1)}{2}$. Establishing the exact value of $\rho_1(S_n^r)$, which we conjecture to be $n+1-r$ at least over the rationals, seems at least to simplify obtaining the $\ell_3(S_n^r) = \Omega(d(n-d))$ lower bound. In the full version we prove that for $r \geq 2$, $\rho_1(S_{n+1}^{r+1}) \geq \rho_0(S_n^{r-1})$.

For another example, S_6^3 is made to vanish at dimension 3 not by any subspace that zeroes out 3 co-ordinates but rather by $A = \{\,(u, -u, w, -w, y, -y) : u, w, y \in \mathbf{C}\,\}$. Now add a new variable t in defining $f = S_7^4$. The notable fact is that f 1-resists the dimension-3 subspace A' obtained by adjoining $t = 0$ to the equations for A, upon existentially choosing to derive by a variable other than t, such as u. All terms of $\partial f/\partial u$ that include t vanish, leaving 10 terms in the variables v, w, x, y, z. Of these, 4 pairs cancel under the equations $x = -w$, $z = -y$, but the leftover $vwx + vyz$ part equates to $uw^2 + uy^2$, which not only doesn't cancel but also dominates any contribution from the lower-degree g. Gröbner basis runs using *Singular* imply that S_7^4 is $(1, 4)$-resistant over $\mathbf{C}$ as well as the rationals and reals, though we have not yet made this a consequence of a general resistance theorem for all S_n^r.

Hence our $(1, k)$-resistance analysis for S_7^4 is not impacted by the achieved upper bound of 3 represented by A. Admittedly the symmetric polynomials f have $O(n^2)$ upper bounds on $\ell_3(f)$, so our distinction in this case does not directly help surmount the quadratic barrier. But it does show promise of making progress in our algebraic understanding of polynomials in general.

5 Bounds for +,*-Complexity

The partial derivatives technique used by Shpilka and Wigderson [10] ignores the wires of the formula present in the first layer. In the following we show how to account for them. As a result we get a sharpening of several lower bounds, though not on ℓ_3^* but on total formula size. We employ the concepts and lemmas from [10]. For $f \in F[x_1, \ldots, x_n]$, let $\partial_d(f)$ be the set of all dth order *formal* partial derivatives of f w.r.t. variables from $\{x_1, \ldots, x_n\}$. For a set of polynomials $A = \{f_1, \ldots, f_t\}$ span(A) $= \{\sum_{i=1}^t c_i f_i \mid c_i \in F\}$. Write dim$[A]$ as shorthand for dim[span(A)]. Note span$(f_1, \ldots, f_t)_{|A} =$ span$(f_{1|A}, \ldots, f_{t|A})$, and that dim$[W_{|A}] \leq$ dim$[W]$. The basic inequality from [10] then becomes:

Proposition 3. $\dim[\partial_d(c_1 f_1 + c_2 f_2)_{|A}] \leq \dim[\partial_d(f_1)_{|A}] + \dim[\partial_d(f_2)_{|A}]$.

We refine two main results in [10] $*$-complexity into results with tighter bounds but for $+,*$-complexity. In each case we compare old and new versions.

Theorem 4 ([10]). *Let $f \in F[x_1, \ldots, x_n]$. Suppose for integers d, D, κ it holds that for every affine subspace A of co-dimension κ, $\dim(\partial_d(f)_{|A}) > D$. Then $\ell_3^*(f) \geq \min(\frac{\kappa^2}{d}, \frac{D}{\binom{\kappa+d}{d}})$;*

Theorem 5 (new). *Let $f \in F[x_1, \ldots, x_n]$. Suppose for integers d, D, κ it holds that for every affine subspace A of co-dimension κ, $\sum_{i=1}^n \dim[\partial_d(\frac{\partial f}{\partial x_i})_{|A}] > D$. Then $\ell_3(f) \geq \min(\frac{\kappa^2}{d+2}, \frac{D}{\binom{\kappa+d}{d}})$.*

Proof. Consider a minimum-size $\Sigma\Pi\Sigma$-formula for f with multiplication gates $M_1, \ldots, M_s$. We have that $f = \sum_{i=1}^s M_i$, where for $1 \leq i \leq s$, $M_i = \Pi_{j=1}^{d_i} l_{i,j}$ and $l_{i,j} = c_{i,j,1}x_1 + c_{i,j,2}x_2 + \ldots + c_{i,j,n}x_n + c_{i,j,0}$, for certain constants $c_{i,j,k} \in F$. Computing the partial derivative of f w.r.t. variable x_k we get

$$\frac{\partial f}{\partial x_k} = \sum_{i=1}^{s} \sum_{j=1}^{d_i} c_{i,j,k} \frac{M_i}{l_{i,j}}. \tag{1}$$

Let $S = \{i : \dim[M_i^h] \geq \kappa\}$. If $|S| \geq \frac{\kappa}{d+2}$, then $\ell_3(f) \geq \frac{\kappa^2}{d+2}$. Suppose $|S| < \frac{\kappa}{d+2}$. If $S = \emptyset$, then let A be an arbitrary affine subspace of co-dimension κ. Otherwise, construct an affine space A as follows. Since $|S|(d+2) < \kappa$, and since for each $j \in S$, $\dim[M_i^h] \geq \kappa$, it is possible to pick $d+2$ input linear forms $l_{j,1}, \ldots, l_{j,d+2}$ of each multiplication gate M_j with $j \in S$, such that $\{l_{j,1}^h, \ldots, l_{j,d+2}^h | j \in S\}$ is a set of $|S|(d+2) < \kappa$ independent homogeneous linear forms. Define

$$A = \{x : \ l_{i,j}(x) = 0, \text{ for any } i \in S, \ j \in [d+2]\}.$$

We have that the co-dimension of A is at most κ. W.l.o.g. assume the co-dimension of A equals κ. For each $i \in S$, d+2 linear forms of M_i vanish on A. This implies that $\dim[\partial_d(\frac{M_i}{l_{i,j}})_{|A}] = 0.$ for any $i \in S$. For any $i \notin S$, by Proposition 2.3 in [10], $\dim[\partial_d(\frac{M_i}{l_{i,j}})_{|A}] < \binom{\kappa+d}{d}$. Let $D_k = \dim[\partial_d(\frac{\partial f}{\partial x_k})_{|A}]$. By Proposition 3 and equation (1),

$$D_k \leq \sum_{i \notin S} \sum_{\substack{j \\ c_{i,j,k} \neq 0}} \dim\left[\partial_d\left(\frac{M_i}{l_{i,j}}\right)_{|A}\right].$$

Hence there must be at least $\frac{D_k}{\binom{\kappa+d}{d}}$ terms on the r.h.s., i.e. there are at least that many wires from x_k to gates in the first layer. Hence in total the number of wires to the first layer is at least $\sum_{i=1}^n \frac{D_i}{\binom{\kappa+d}{d}} > \frac{D}{\binom{\kappa+d}{d}}$. □

Theorem 6 ([10]). *Let $f \in F[x_1, \ldots, x_n]$. Suppose for integers d, D, κ it holds that for every affine subspace A of co-dimension κ, $\dim(\partial_d(f_{|A})) > D$. Then for every $m \geq 2$, $\ell_3^*(f) \geq \min(\kappa m, \frac{D}{\binom{m}{d}})$.*

Theorem 7 (new). *Let $f \in F[x_1, \ldots, x_n]$. Suppose for integers d, D, κ with $d \geq 1$, it holds that for every affine subspace A of co-dimension κ, $\sum_{i=1}^{n} \dim[\partial_d(\frac{\partial f}{\partial x_i}_{|A})] > D$. Then for every $m \geq 2$, $\ell_3(f) \geq \min(\frac{1}{2}\kappa m, \frac{D}{\binom{m-1}{d}})$.*

The proof of Theorem 7 is analogous to above and appears in the full version.

In [10] it was proved that for $d \leq \log n$, $\ell_3^*(S_n^{2d}) = \Omega(\frac{n^{\frac{2d}{d+2}}}{d})$. Note that for $d = 2$, this lower bound is only $\Omega(n)$. We can apply Theorem 5 to prove the following stronger lower bound on the total formula size of S_n^{2d}. In particular for $d = 2$, we get an $\Omega(n^{\frac{4}{3}})$ bound.

Theorem 8. *For $1 \leq d \leq \log n$, $\ell_3(S_n^{2d}) = \Omega(\frac{n^{\frac{2d}{d+1}}}{d})$.*

Proof. For any affine subspace A of co-dimension κ and $d \geq 2$ we have that

$$\sum_{i=1}^{n} \dim\left[\partial_{d-1}\left(\frac{\partial S_n^{2d}}{\partial x_i}\right)_{|A}\right] \geq \dim[\partial_d(S_n^{2d})_{|A}] \geq \binom{n-\kappa}{d}.$$

The latter inequality follows from Lemma 4.4 in [10]. Applying Theorem 5 we get that

$$\ell_3(S_n^{2d}) \geq \min\left(\frac{\kappa^2}{d+1}, \frac{\binom{n-\kappa}{d}}{\binom{\kappa+d-1}{d-1}}\right) = \min\left(\frac{\kappa^2}{d+1}, \frac{\binom{n-\kappa}{d}}{\binom{\kappa+d}{d}}\frac{\kappa+d}{d}\right). \quad (2)$$

Set $\kappa = \frac{1}{9}n^{\frac{d}{d+1}}$. Then we have that

$$\frac{\binom{n-\kappa}{d}}{\binom{\kappa+d}{d}}\frac{\kappa+d}{d} \geq \left(\frac{n-\kappa}{\kappa+d}\right)^d \frac{\kappa+d}{d} \geq \left(\frac{8/9n}{2/9n^{\frac{d}{d+1}}}\right)^d \frac{\kappa+d}{d} = 4^d n^{\frac{d}{d+1}} \frac{\kappa+d}{d} \geq \frac{4^d}{9d} n^{\frac{2d}{d+1}} \geq n^{\frac{2d}{d+1}}.$$

Hence (2) is at least $\min(\frac{n^{\frac{2d}{d+1}}}{81(d+1)}, n^{\frac{2d}{d+1}}) = \Omega(\frac{n^{\frac{2d}{d+1}}}{d})$. □

Corollary 2. $\ell_3(S_n^4) = \Omega(n^{4/3})$.

Shpilka and Wigderson defined the "product-of-inner-products" polynomial over $2d$ variable sets of size n (superscript indicate different variables, each variable has degree one) by $PIP_n^d = \prod_{i=1}^{d} \sum_{j=1}^{n} x_j^i y_j^i$.

Theorem 9. *For any constant $d > 0$, $\ell_3(PIP_n^d) = \Omega(n^{\frac{2d}{d+1}})$.*

Proof. Let $f = PIP_n^d$. Essentially we have that $\frac{\partial f}{\partial x_j^i} = y_j^i PIP_n^{d-1}$, where the PIP_n^{d-1} must be chosen on the appropriate variable set. Let A be an arbitrary affine linear subspace of co-dimension κ. Then

$$\begin{aligned}\sum_{i=1}^{d}\sum_{j=1}^{n} \dim\left[\partial_{d-1}\left(\frac{\partial f}{\partial x_j^i}_{|A}\right)\right] &= \sum_{i=1}^{d}\sum_{j=1}^{n} \dim[\partial_{d-1}(y_j^i PIP_n^{d-1}{}_{|A})] \\ &\geq (dn-\kappa)\dim[\partial_{d-1}(PIP_n^{d-1}{}_{|A})]\end{aligned}$$

The last inequality follows because at least $dn - \kappa$ of the y-variables are not assigned to with the restriction to A. From Lemma 4.9 in [10] one gets

$$\dim[\partial_{d-1}(PIP_n^{d-1}|_A) \geq n^{d-1} - 2^{2d-1}\kappa n^{d-2}.$$

Using Theorem 7 we get

$$\ell_3(f) \geq \min\left(\frac{\kappa^2}{2}, \frac{(dn-\kappa)(n^{d-1} - 2^{2d-1}\kappa n^{d-2})}{\binom{\kappa-1}{d-1}}\right).$$

Taking $\kappa = n^{\frac{d}{d+1}}$, one gets for constant d that $\ell_3(PIP_n^d) = \Omega(n^{\frac{2d}{d+1}})$. □

For comparison, in [10] one gets $\ell_3^*(PIP_n^d) = \Omega(n^{\frac{2d}{d+2}})$.

6 Conclusion

We have taken some further steps after Shpilka and Wigderson [10,9], obtaining absolutely tight (rather than asymptotically so) multiplicative size lower bounds for some natural functions, and obtaining somewhat improved bounds on $+, *$-size for low-degree symmetric and product-of-inner-product polynomials. However, these may if anything enhance the feeling from [10,9] that the concepts being employed may go no further than quadratic for lower bounds. One cannot after all say that a function $f(x_1, \ldots, x_n)$ is non-vanishing on an affine-linear space of co-dimension more than n. The quest then is for a mathematical invariant that scales beyond linear with the number of degree-d-or-higher multiplication gates in the formula.

Acknowledgments. We thank Avi Wigderson for comments on a very early version of this work, and referees of later versions for very helpful criticism.

References

1. Baur, W., Strassen, V.: The complexity of partial derivatives. Theor. Comp. Sci. 22, 317–330 (1982)
2. Bürgisser, P., Clausen, M., Shokrollahi, M.A.: Algebraic Complexity Theory. Springer, Heidelberg (1997)
3. Furst, M., Saxe, J., Sipser, M.: Parity, circuits, and the polynomial-time hierarchy. Math. Sys. Thy. 17, 13–27 (1984)
4. Greuel, G.-M., Pfister, G., Schönemann, H.: Singular 3.0. A Computer Algebra System for Polynomial Computations, Centre for Computer Algebra, University of Kaiserslautern (2005), `http://www.singular.uni-kl.de`
5. Grigoriev, D., Karpinski, M.: An exponential lower bound for depth 3 arithmetic circuits. In: Proc. 30th Annual ACM Symposium on the Theory of Computing, pp. 577–582. ACM Press, New York (1998)
6. Grigoriev, D., Razborov, A.A.: Exponential lower bounds for depth 3 algebraic circuits in algebras of functions over finite fields. Applicable Algebra in Engineering, Communication, and Computing 10, 465–487 (1998) (preliminary version FOCS 1998)

7. Håstad, J.: Almost optimal lower bounds for small-depth circuits. In: Micali, S. (ed.) Randomness and Computation. Advances in Computing Research, vol. 5, pp. 143–170. JAI Press, Greenwich, CT (1989)
8. Nisan, N., Wigderson, A.: Lower bounds on arithmetic circuits via partial derivatives. Computational Complexity 6, 217–234 (1996)
9. Shpilka, A.: Affine projections of symmetric polynomials. J. Comp. Sys. Sci. 65, 639–659 (2002)
10. Shpilka, A., Wigderson, A.: Depth-3 arithmetic formulae over fields of characteristic zero. Computational Complexity 10, 1–27 (2001)
11. Strassen, V.: Berechnung und Programm II. Acta Informatica 2, 64–79 (1973)
12. Valiant, L.: The complexity of computing the permanent. Theor. Comp. Sci. 8, 189–201 (1979)
13. Yao, A.: Separating the polynomial-time hierarchy by oracles. In: Proc. 26th Annual IEEE Symposium on Foundations of Computer Science, pp. 1–10. IEEE Computer Society Press, Los Alamitos (1985)

An Improved Algorithm for Tree Edit Distance Incorporating Structural Linearity*

Shihyen Chen and Kaizhong Zhang

Department of Computer Science,
The University of Western Ontario, London, Ontario, Canada, N6A 5B7
{schen,kzhang}@csd.uwo.ca

Abstract. An ordered labeled tree is a tree in which the nodes are labeled and the left-to-right order among siblings is significant. The edit distance between two ordered labeled trees is the minimum cost of transforming one tree into the other by a sequence of edit operations. Among the best known tree edit distance algorithms, the majority can be categorized in terms of a framework named cover strategy. In this paper, we investigate how certain locally linear features may be utilized to improve the time complexity for computing the tree edit distance. We define structural linearity and present a method incorporating linearity which can work with existing cover-strategy based tree algorithms. We show that by this method the time complexity for an input of size n becomes $\mathcal{O}(n^2 + \phi(\mathcal{A}, \widetilde{n}))$ where $\phi(\mathcal{A}, \widetilde{n})$ is the time complexity of any cover-strategy algorithm $\mathcal{A}$ applied to an input size $\widetilde{n}$, with $\widetilde{n} \leq n$, and the magnitude of $\widetilde{n}$ is reversely related to the degree of linearity. This result is an improvement of previous results when $\widetilde{n} < n$ and would be useful for situations in which $\widetilde{n}$ is in general substantially smaller than n, such as RNA secondary structure comparisons in computational biology.

Keywords: Tree edit distance, dynamic programming, RNA secondary structure comparison.

1 Introduction

An ordered labeled tree is a tree in which the nodes are labeled and the left-to-right order among siblings is significant. Trees can represent many phenomena, such as grammar parses, image descriptions and structured texts, to name a couple. In many applications where trees are useful representations of objects, the need for comparing trees frequently arises.

The tree edit distance metric was introduced by Tai [7] as a generalization of the string edit distance problem [9]. Given two trees T_1 and T_2, the tree edit distance between T_1 and T_2 is the minimum cost of transforming one tree into the other, with the sibling and ancestor orders preserved, by a sequence of edit operations on the nodes (relabeling, insertion and deletion) as shown in Figure 1.

* Research supported partially by the Natural Sciences and Engineering Research Council of Canada under Grant No. OGP0046373 and a grant from MITACS, a Network of Centres of Excellence for the Mathematical Sciences.

G. Lin (Ed.): COCOON 2007, LNCS 4598, pp. 482–492, 2007.

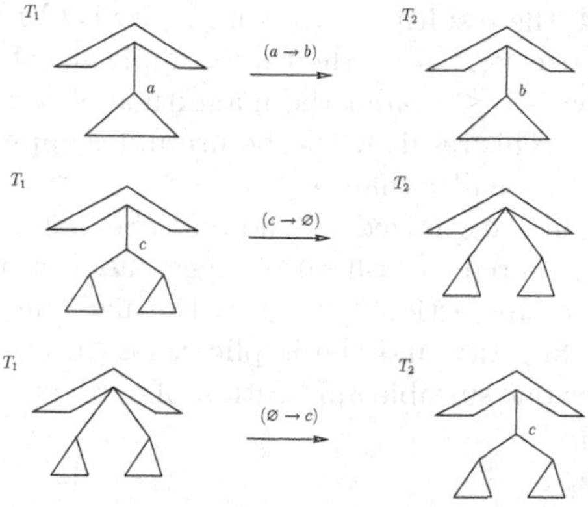

Fig. 1. Tree edit operations. From top to bottom: relabeling, deletion and insertion.

Among the known algorithms of comparable results such as in [1,2,4,10], the majority [2,4,10] can be categorized in terms of a generalized framework by the name of *cover strategy* [3] which prescribes the direction by which a dynamic program builds up the solution. Briefly, a tree is decomposed into a set of disjoint paths which, in this paper, we refer to as *special paths*. Each special path is associated with a subtree such that the special path coincides with a path of the subtree which runs from the root to a leaf. The dynamic program proceeds in a bottom-up order with respect to the special paths such that for any node i on a special path, the subtrees hanging off the portion of the special path no higher than i have been processed before the node i is reached. When there are subtrees hanging off on both sides of a special path, the decision as to which side takes precedence is referred to as *strategy*. In the Zhang-Shasha algorithm [10], the special paths are chosen to be the leftmost paths. In Klein's algorithm [4] as well as that of Demaine *et al.* [2], the special paths are chosen such that every node on a special path is the root of a largest subtree over its sibling subtrees. These special paths are referred to as *heavy paths* [6]. Examples of leftmost paths and heavy paths are shown in Figure 2.

In these algorithms, no consideration is given to any structural characteristics which may exist in the tree. In this paper, we investigate the possibility of utilizing certain linear features within the trees to speed up the computation of the tree edit distance. We show that by embedding a procedure in any cover-

Fig. 2. Left: Decomposition of a tree into leftmost paths (in bold). Right: Decomposition of a tree into heavy paths (in bold).

strategy algorithm $\mathcal{A}$, the resulting time complexity is $\mathcal{O}(n^2 + \phi(\mathcal{A}, \widetilde{n}))$ where n is the original input size, $\phi(\mathcal{A}, \widetilde{n})$ is the time complexity of algorithm $\mathcal{A}$ applied to an input size $\widetilde{n}$, with $\widetilde{n} \leq n$, and the magnitude of $\widetilde{n}$ is reversely related to the degree of linearity. This result would be useful for applications in which $\widetilde{n}$ is in general substantially smaller than n.

The rest of the paper is organized as follows. In Section 2, we define structural linearity and give a new representation of trees based on reduction of the tree size due to the linearity. In Section 3, we show the algorithmic aspects as a result of incorporating the linearity and the implications on the time complexity. In Section 4, we describe one suitable application of our result. We give concluding remarks in Section 5.

2 Preliminaries

2.1 Notations

Given a tree T, we denote by $t[i]$ the ith node in the left-to-right post-order numbering. The index of the leftmost leaf of the subtree rooted at $t[i]$ is denoted by $l(i)$. We denote by $F[i, j]$ the ordered sub-forest of T induced by the nodes indexed i to j inclusive. The subtree rooted at $t[i]$ in T is denoted by $T[i]$, i.e., $T[i] = F[l(i), i]$. The sub-forest induced by removing $t[i]$ from $T[i]$ is denoted by $F[i]$, i.e., $F[i] = F[l(i), i-1]$. When referring to the children of a specific node, we adopt a subscript notation in accordance with the left-to-right sibling order. For example, the children of $t[i]$, from left to right, may be denoted by $(t[i_1], t[i_2], \cdots, t[i_k])$.

2.2 Linearity

Definition 1 (V-Component). *Given a tree T with left-to-right post-order and pre-order numberings, a path π of T is a v-component (i.e., vertically linear component) if all of the following conditions hold.*

- *Both the post-order and the pre-order numberings along this path form a sequence of continuous indices.*
- *No other path containing π satisfies the above condition.*

Definition 2 (V-Reduction). *The v-reduction on a tree is to replace every v-component in the tree by a single node.*

Definition 3 (H-Component). *Given a tree T and another tree $\widetilde{T}$ obtained by a v-reduction on T, any set of connected components of T corresponding to a set of leaves $\widetilde{L}$ in $\widetilde{T}$ form an h-component (i.e., horizontally linear component) if all of the following conditions hold.*

- $|\widetilde{L}| \geq 2$.
- *All the leaves in $\widetilde{L}$ share the same parent.*
- *A left-to-right post-order or pre-order numbering on $\widetilde{T}$ produces a sequence of continuous indices for $\widetilde{L}$.*
- *No other set of leaves containing $\widetilde{L}$ satisfy the above conditions.*

Definition 4 (H-Reduction). *The h-reduction on a tree is to replace every h-component in the tree by a single node.*

A tree possesses vertical (horizontal) linearity if it contains any v-component (h-component). A tree is v-reduced if it is obtained by a v-reduction only. A tree is vh-reduced if it is obtained by a v-reduction followed by an h-reduction.

Note that a reduced tree is just a compact representation of the original tree. The edit distance of two reduced trees is the same as the edit distance of the original trees. In the case when a tree does not possess any linearity as defined above, the reduced tree is the same as the original tree.

In Figure 3, we give an example showing the v-components and h-components of a tree and the corresponding reduced trees. Note that an h-component can also contain v-components.

Each node in a v-reduced tree corresponds to either a v-component or a single node in the corresponding full tree. Given a v-reduced tree $\widetilde{T}$, we define

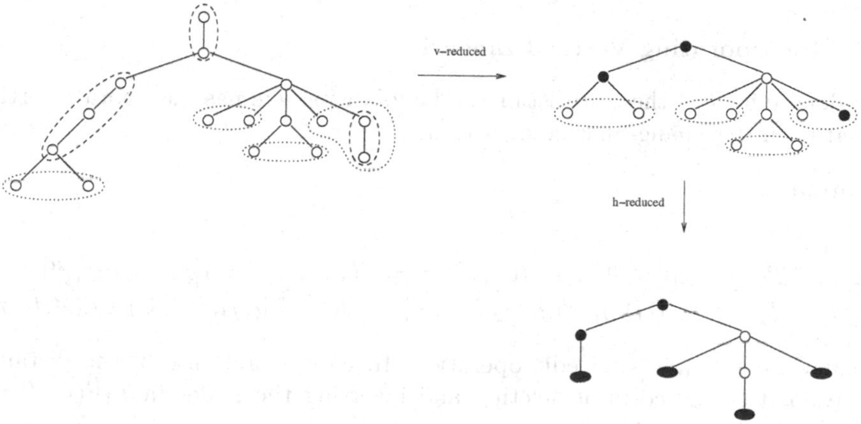

Fig. 3. The v-components (in dashed enclosures) and the h-components (in dotted enclosures). Also shown are the reduced trees as a result of reduction. The parts of the original tree affected by reduction are represented by black nodes in the reduced tree.

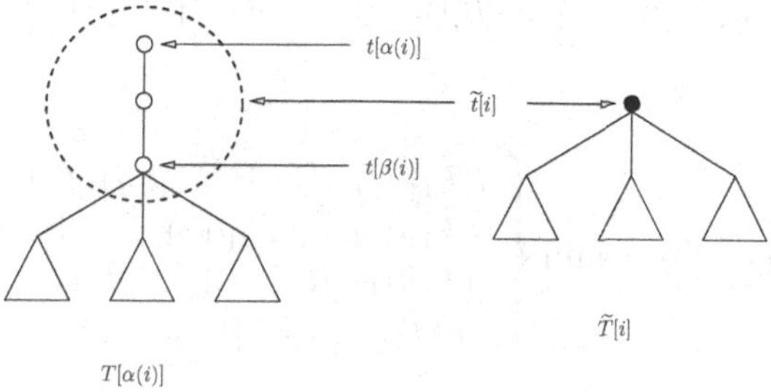

Fig. 4. A partial view of the mapping of nodes between a tree (left) and its v-reduced tree (right)

two functions $\alpha()$ and $\beta()$ which respectively map a node $\widetilde{t}[i]$ to the highest indexed node $t[\alpha(i)]$ and the lowest indexed node $t[\beta(i)]$ of the corresponding v-component in the full tree T. In the special case when $\widetilde{t}[i]$ corresponds to a single node in T, $t[\alpha(i)] = t[\beta(i)]$. An example of this mapping is given in Figure 4. When $\widetilde{T}$ is h-reduced to yield $\widehat{T}$, $\alpha()$ and $\beta()$ apply in the same way for the mapping from $\widehat{T}$ to $\widetilde{T}$.

3 Algorithm

In this section, we show how to incorporate vertical linearity in the Zhang-Shasha algorithm. We will also show that the method can be incorporated in all the cover-strategy algorithms.

Due to space limitation, we shall not discuss the incorporation of horizontal linearity in this paper. The method involves adapting techniques from matrix searching and will be given elsewhere in the future.

3.1 Incorporating Vertical Linearity

We denote by $d(,)$ the edit distance. The following lemmas incorporate vertical linearity in the Zhang-Shasha algorithm.

Lemma 1

1. $d(\varnothing, \varnothing) = 0$.
2. $\forall i \in \widetilde{T}_1$, $\forall i' \in [l(i), i]$, $d(\widetilde{F}_1[l(i), i'], \varnothing) = d(\widetilde{F}_1[l(i), i'-1], \varnothing) + d(\widetilde{t}_1[i'], \varnothing)$.
3. $\forall j \in \widetilde{T}_2$, $\forall j' \in [l(j), j]$, $d(\varnothing, \widetilde{F}_2[l(j), j']) = d(\varnothing, \widetilde{F}_2[l(j), j'-1]) + d(\varnothing, \widetilde{t}_2[j'])$.

Proof Case 1 requires no edit operation. In case 2 and case 3, the distances correspond to the costs of deleting and inserting the nodes in $\widetilde{F}_1[l(i), i']$ and $\widetilde{F}_2[l(j), j']$, respectively. □

Lemma 2. *$\forall (i, j) \in (\widetilde{T}_1, \widetilde{T}_2)$, $\forall i' \in [l(i), i]$ and $\forall j' \in [l(j), j]$, if $l(i') = l(i)$ and $l(j') = l(j)$,*

$$d(\widetilde{F}_1[l(i), i'], \widetilde{F}_2[l(j), j']) = d(\widetilde{T}_1[i'], \widetilde{T}_2[j']) \ ;$$

otherwise,

$$d(\widetilde{F}_1[l(i), i'], \widetilde{F}_2[l(j), j']) = \min \left\{ \begin{array}{l} d(\widetilde{F}_1[l(i), i'-1], \widetilde{F}_2[l(j), j']) + d(\widetilde{t}_1[i'], \varnothing), \\ d(\widetilde{F}_1[l(i), i'], \widetilde{F}_2[l(j), j'-1]) + d(\varnothing, \widetilde{t}_2[j']), \\ d(\widetilde{F}_1[l(i), l(i')-1], \widetilde{F}_2[l(j), l(j')-1]) \\ +d(\widetilde{T}_1[i'], \widetilde{T}_2[j']) \end{array} \right\} .$$

Proof. The condition "$l(i') = l(i)$ and $l(j') = l(j)$" implies that the two forests are simply two trees and the equality clearly holds. We now consider the other condition in which "$l(i') \neq l(i)$ or $l(j') \neq l(j)$". If $t_1[\alpha(i')] = t_1[\beta(i')]$ and

$t_2[\alpha(j')] = t_2[\beta(j')]$, the formula holds as a known result. Otherwise, at least one of $\widetilde{t}_1[i']$ and $\widetilde{t}_2[j']$ corresponds to a v-component in $(T_1[\alpha(i)], T_2[\alpha(j)])$. Consider the connected components in $(T_1[\alpha(i)], T_2[\alpha(j)])$ corresponding to $(\widetilde{t}_1[i'], \widetilde{t}_2[j'])$. There are two cases to consider: either (1) there is no occurrence of node-to-node match between the connected components; or (2) there is at least one occurrence of node-to-node match between the connected components. In case 1, one of the components must be entirely deleted which implies that either $\widetilde{t}_1[i']$ must be deleted or $\widetilde{t}_2[j']$ must be inserted. In case 2, in order to preserve the ancestor-descendent relationship $\widetilde{T}_1[i']$ and $\widetilde{T}_2[j']$ must be matched. □

Note. In Lemma 2 for the condition "$l(i') \neq l(i)$ or $l(j') \neq l(j)$" the value of $d(\widetilde{T}_1[i'], \widetilde{T}_2[j'])$ would already be available if implemented in a bottom-up order, since it involves a subproblem of $d(\widetilde{F}_1[l(i), i'], \widetilde{F}_2[l(j), j'])$ and would have been computed. For the condition "$l(i') = l(i)$ and $l(j') = l(j)$", however, we encounter the problem involving $(\widetilde{T}_1[i'], \widetilde{T}_2[j'])$ for the first time and must compute its value.

We show how to compute $d(\widetilde{T}_1[i'], \widetilde{T}_2[j'])$ in the following lemmas.

Lemma 3. $\forall u \in [\beta(i'), \alpha(i')]$,

$$d(T_1[u], F_2[\beta(j')]) = \min \left\{ \begin{array}{l} d(F_1[u], F_2[\beta(j')]) + d(t_1[u], \varnothing), \\ \min_{j'_1 \le q \le j'_l} \{d(T_1[u], T_2[\alpha(q)]) - d(\varnothing, T_2[\alpha(q)])\} \\ +d(\varnothing, F_2[\beta(j')]) \end{array} \right\}.$$

Proof. This is the edit distance between the tree $T_1[u]$ and the forest $F_2[\beta(j')]$. There are two cases. In the first case, $t_1[u]$ is constrained to be deleted and the remaining substructure $F_1[u]$ is matched to $F_2[\beta(j')]$. In the second case, $t_1[u]$ is constrained to be matched to a node somewhere in $F_2[\beta(j')]$. This is equivalent to stating that $T_1[u]$ is constrained to be matched to a subtree in $F_2[\beta(j')]$. The question thus becomes finding a subtree in $F_2[\beta(j')]$ to be matched to $T_1[u]$ so as to minimize the distance between $T_1[u]$ and $F_2[\beta(j')]$ under such constraint. This can be done by considering the set of all combinations in which exactly one tree in $F_2[\beta(j')]$ is matched to $T_1[u]$ while the remainder of $F_2[\beta(j')]$ is deleted. The minimum in this set is the edit distance for the second case. □

Lemma 4. $\forall v \in [\beta(j'), \alpha(j')]$,

$$d(F_1[\beta(i')], T_2[v]) = \min \left\{ \begin{array}{l} d(F_1[\beta(i')], F_2[v]) + d(\varnothing, t_2[v]), \\ \min_{i'_1 \le p \le i'_k} \{d(T_1[\alpha(p)], T_2[v]) - d(T_1[\alpha(p)], \varnothing)\} \\ +d(F_1[\beta(i')], \varnothing) \end{array} \right\}.$$

Proof. This is symmetric to that of Lemma 3. □

Lemma 5. $\forall u \in [\beta(i'), \alpha(i')]$ *and* $\forall v \in [\beta(j'), \alpha(j')]$,

$$d(T_1[u], T_2[v]) = \min \left\{ \begin{array}{l} d(F_1[u], T_2[v]) + d(t_1[u], \varnothing), \\ d(T_1[u], F_2[v]) + d(\varnothing, t_2[v]), \\ d(F_1[u], F_2[v]) + d(t_1[u], t_2[v]) \end{array} \right\}.$$

Proof. This is a known result for the tree-to-tree edit distance. □

Note. In the computation for every $d(\widetilde{T}_1[i'], \widetilde{T}_2[j'])$, we save the values for $d(T_1[u], T_2[\alpha(j')])$ $\forall u \in [\beta(i'), \alpha(i')]$ and $d(T_1[\alpha(i')], T_2[v])$ $\forall v \in [\beta(j'), \alpha(j')]$. This ensures that when $d(T_1[u], F_2[\beta(j')])$ in Lemma 3 and $d(F_1[\beta(i')], T_2[v])$ in Lemma 4 are evaluated in a bottom-up order the values of the terms involving $d(T_1[u], T_2[\alpha(q)])$ and $d(T_1[\alpha(p)], T_2[v])$ would be available.

Lemma 6. $d(\widetilde{T}_1[i'], \widetilde{T}_2[j']) = d(T_1[\alpha(i')], T_2[\alpha(j')])$.

Proof. The result follows from the tree definitions. □

3.2 The New Algorithm

For every node i of tree T, we designate a child of i, if any, to be its *special child*, denoted by $sc(i)$. Note that in the Zhang-Shasha algorithm $sc(i)$ is the leftmost child of i whereas in a different cover-strategy the choice of $sc(i)$ may be different. Denote by $p(i)$ the parent of i. We define a set of nodes, called *key roots*, for tree T as follows.

$$keyroots(T) = \{k \mid k = root(T) \text{ or } k \neq sc(p(k))\} \ .$$

This is a generalized version of the *LR_keyroots* used in [10] and is suitable for any known decomposition strategy as in [2,4,10]. Referring to Figure 2, in every special path the highest numbered node in a left-to-right post-order is a key root.

We now give the new algorithm in Algorithms 1 and 2. Algorithm 1 contains the main loop which repeatedly calls Algorithm 2 to compute $d(\widetilde{T}_1[i], \widetilde{T}_2[j])$ where (i, j) are key roots in $(\widetilde{T}_1, \widetilde{T}_2)$.

Theorem 1. *The new algorithm correctly computes* $d(T_1, T_2)$.

Proof. The correctness of all the computed values in Algorithm 2 follows from the lemmas. By Lemma 6, $d(\widetilde{T}_1, \widetilde{T}_2) = d(T_1, T_2)$ when (i', j') are set to be the roots of $(\widetilde{T}_1, \widetilde{T}_2)$. Since these roots are key roots, $d(T_1, T_2)$ is always computed by Algorithm 1. □

Algorithm 1. Computing $d(\widetilde{T}_1, \widetilde{T}_2)$

Input: $(\widetilde{T}_1, \widetilde{T}_2)$
Output: $d(\widetilde{T}_1[i], \widetilde{T}_2[j])$, where $1 \leq i \leq |\widetilde{T}_1|$ and $1 \leq j \leq |\widetilde{T}_2|$

```
Sort keyroots(T̃1) and keyroots(T̃2) in increasing order into arrays K1 and K2,
    respectively
for i' ← 1, ··· , |keyroots(T̃1)| do
    for j' ← 1, ··· , |keyroots(T̃2)| do
        i ← K1[i']
        j ← K2[j']
        Compute d(T̃1[i], T̃2[j]) by Algorithm 2
    end for
end for
```

Algorithm 2. Computing $d(\widetilde{T}_1[i], \widetilde{T}_2[j])$

$d(\varnothing, \varnothing) \leftarrow 0$
for $i' \leftarrow l(i), \cdots, i$ **do**
  $d(\widetilde{F}_1[l(i), i'], \varnothing) \leftarrow d(\widetilde{F}_1[l(i), i'-1], \varnothing) + d(\widetilde{t}_1[i'], \varnothing)$
end for
for $j' \leftarrow l(j), \cdots, j$ **do**
  $d(\varnothing, \widetilde{F}_2[l(j), j']) \leftarrow d(\varnothing, \widetilde{F}_2[l(j), j'-1]) + d(\varnothing, \widetilde{t}_2[j'])$
end for
for $i' \leftarrow l(i), \cdots, i$ **do**
  for $j' \leftarrow l(j), \cdots, j$ **do**
    if $l(i') = l(i)$ and $l(j') = l(j)$ **then**
      for $u \leftarrow \beta(i'), \cdots, \alpha(i')$ **do**
        $d(T_1[u], F_2[\beta(j')]) \leftarrow$
        $$\min \begin{cases} d(F_1[u], F_2[\beta(j')]) + d(t_1[u], \varnothing), \\ \min_{j'_1 \le q \le j'_l} \{d(T_1[u], T_2[\alpha(q)]) - d(\varnothing, T_2[\alpha(q)])\} \\ +d(\varnothing, F_2[\beta(j')]) \end{cases}$$
      end for
      for $v \leftarrow \beta(j'), \cdots, \alpha(j')$ **do**
        $d(F_1[\beta(i')], T_2[v]) \leftarrow$
        $$\min \begin{cases} d(F_1[\beta(i')], F_2[v]) + d(\varnothing, t_2[v]), \\ \min_{i'_1 \le p \le i'_k} \{d(T_1[\alpha(p)], T_2[v]) - d(T_1[\alpha(p)], \varnothing)\} \\ +d(F_1[\beta(i')], \varnothing) \end{cases}$$
      end for
      for $u \leftarrow \beta(i'), \cdots, \alpha(i')$ **do**
        for $v \leftarrow \beta(j'), \cdots, \alpha(j')$ **do**
          $$d(T_1[u], T_2[v]) \leftarrow \min \begin{cases} d(F_1[u], T_2[v]) + d(t_1[u], \varnothing), \\ d(T_1[u], F_2[v]) + d(\varnothing, t_2[v]), \\ d(F_1[u], F_2[v]) + d(t_1[u], t_2[v]) \end{cases}$$
        end for
      end for
      $d(\widetilde{T}_1[i'], \widetilde{T}_2[j']) \leftarrow d(T_1[\alpha(i')], T_2[\alpha(j')])$
    else
      $d(\widetilde{F}_1[l(i), i'], \widetilde{F}_2[l(j), j']) \leftarrow$
      $$\min \begin{cases} d(\widetilde{F}_1[l(i), i'-1], \widetilde{F}_2[l(j), j']) + d(\widetilde{t}_1[i'], \varnothing), \\ d(\widetilde{F}_1[l(i), i'], \widetilde{F}_2[l(j), j'-1]) + d(\varnothing, \widetilde{t}_2[j']), \\ d(\widetilde{F}_1[l(i), l(i')-1], \widetilde{F}_2[l(j), l(j')-1]) + d(\widetilde{T}_1[i'], \widetilde{T}_2[j']) \end{cases}$$
    end if
  end for
end for

Theorem 2. *The new algorithm runs in* $\mathcal{O}(n^2 + \widetilde{n}^4)$ *time and* $\mathcal{O}(n^2)$ *space, where* n *is the original input size and* $\widetilde{n} \le n$.

Proof. We first consider the time complexity. The v-reduced trees can be built in linear time in a preprocess. Identifying and sorting the key roots can be done in linear time. As well, all the values associated with insertion or deletion of a subtree or a sub-forest, as appearing in Lemma 3 and Lemma 4, can be obtained beforehand in linear time during a tree traversal. The Zhang-Shasha algorithm has a worst-case running time of $\mathcal{O}(\widetilde{n}^4)$ for an input size $\widetilde{n}$. Referring to Algorithm 2, we consider the block from line 11 to 21 which concerns computation involving the (i', j') pairs on leftmost paths, based on Lemmas 3, 4 and 5. This is the part of the algorithm that structurally differs from its counterpart in the Zhang-Shasha algorithm. For each such (i', j') pair, this part takes $\mathcal{O}((\alpha(i') - \beta(i') + 1) \times (j'_l - j'_1 + 1) + (\alpha(j') - \beta(j') + 1) \times (i'_k - i'_1 + 1) + (\alpha(i') - \beta(i') + 1) \times (\alpha(j') - \beta(j') + 1))$. All the subproblems of (T_1, T_2) associated with these (i', j') pairs are disjoint. Summing over all these pairs, we can bound the complexity by $\mathcal{O}(n^2)$. Hence, the overall time complexity is $\mathcal{O}(n^2 + \widetilde{n}^4)$.

We now consider the space complexity. We use three different arrays: the full-tree array, the reduced-tree array and the reduced-forest array. The reduced-forest array is a temporary array and its values can be rewritten during the computation of the reduced-forest distances. The other two arrays are permanent arrays for storing tree distances. The space for the reduced-tree array and the reduced-forest array is bounded by $\mathcal{O}(\widetilde{n}^2)$. The space for the full-tree array is bounded by $\mathcal{O}(n^2)$. Hence, the space complexity is $\mathcal{O}(n^2)$. □

Theorem 3. *Given (T_1, T_2) of maximum size n, the edit distance $d(T_1, T_2)$ can be computed in $\mathcal{O}(n^2 + \phi(\mathcal{A}, \widetilde{n}))$ time where $\phi(\mathcal{A}, \widetilde{n})$ is the time complexity of any cover-strategy algorithm $\mathcal{A}$ applied to an input size $\widetilde{n}$, with $\widetilde{n} \leq n$.*

Proof. Since a v-component is consisted of a simple path, there is only one way a dynamic program can recurse along this path regardless which strategy is used. Hence, Lemmas 3 to 6 are valid for all cover strategies. Lemmas 1 and 2, after a proper adjustment of the subtree orderings in each forest to adapt to the given strategy, are also valid. The theorem is implied from Theorems 1 and 2 when the lemmas are properly embedded in any cover-strategy algorithm. □

4 Application

We describe one application which would benefit from our result, namely RNA secondary structure comparison. RNA is a molecule consisted of a single strand of nucleotides (abbreviated as A, C, G and U) which folds back onto itself by means of hydrogen bonding between distant complementary nucleotides, giving rise to a so-called secondary structure. The secondary structure of an RNA molecule can be topologically represented by a tree. An example is depicted in Figure 5. In this representation, an internal node represents a pair of complementary nucleotides interacting via hydrogen bonding. When a number of such pairs are stacked up, they form a local structure called *stem*, which corresponds to a v-component in the tree representation. The secondary structure plays an important role in the functions of RNA [5]. Therefore, comparing the secondary structures of RNA molecules can help understand their comparative functions. One way to compare two trees is to compute the edit distance between them.

To gain an impression, we list the size reductions for a set of selected tRNA molecules in Table 1. $|T|$ is the size of the original tree. $|\widetilde{T}|$ is the size of the

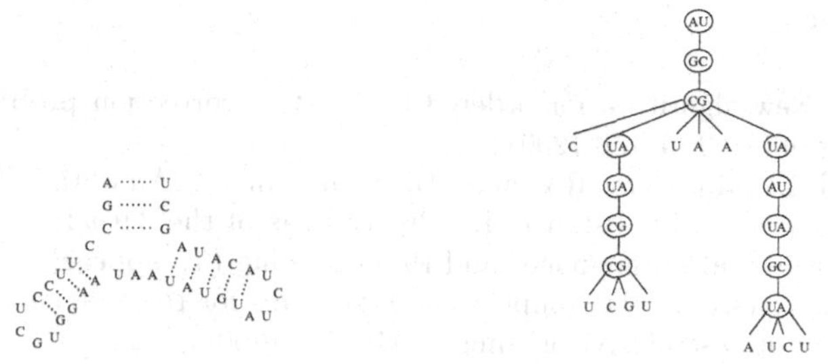

Fig. 5. Left: Secondary folding of RNA. Dotted lines represent hydrogen bonds. Right: The corresponding tree representation.

Table 1. Reduction of tree sizes for selected tRNA molecules [8]

Name	$\|T\|$	$\|\widetilde{T}\|$	Reduction (%)
Athal-chr4.trna25	52	35	33%
cb25.fpc2454.trna15	52	40	23%
CHROMOSOME_I.trna38	51	38	25%
chr1.trna1190	51	34	33%
Acinetobacter_sp_ADP1.trna45	55	38	31%
Aquifex_aeolicus.trna21	55	38	31%
Azoarcus_sp_EbN1.trna58	55	38	31%
Bacillus_subtilis.trna63	52	35	33%
Aeropyrum_pernix_K1.trna3	56	39	30%
Sulfolobus_tokodaii.trna25	53	36	32%

v-reduced tree. The last column shows the size reductions in percentage. On average, we observe a size reduction by nearly one third of the original size, which roughly translates into a one-half decrease of running time for the known cover-strategy algorithms.

5 Conclusions

We presented a new method for computing tree edit distance by incorporating structural linearity. This method can work with any existing cover-strategy based algorithm to yield a time complexity of $\mathcal{O}(n^2 + \phi(\mathcal{A}, \widetilde{n}))$ where n is the original input size and $\phi(\mathcal{A}, \widetilde{n})$ assumes the same form of the time complexity of the given algorithm $\mathcal{A}$ when it is applied to an input size $\widetilde{n}$, with $\widetilde{n} \leq n$. The magnitude of $\widetilde{n}$ is reversely related to the degree of linearity.

This result would be useful when $\widetilde{n}$ is in general substantially smaller than n. Therefore, incorporating our technique in any existing cover-strategy algorithm may yield an improved performance for such situations. One application which can readily benefit from this improvement is RNA secondary structure comparisons in computational biology.

References

1. Chen, W.: New algorithm for ordered tree-to-tree correction problem. Journal of Algorithms 40(2), 135–158 (2001)
2. Demaine, E.D., Mozes, S., Rossman, B., Weimann, O.: An optimal decomposition algorithm for tree edit distance. In: Proceedings of the 34th International Colloquium on Automata, Languages and Programming (To appear)
3. Dulucq, S., Touzet, H.: Decomposition algorithms for the tree edit distance problem. Journal of Discrete Algorithms 3, 448–471 (2005)
4. Klein, P.N.: Computing the edit-distance between unrooted ordered trees. In: Proceedings of the 6th European Symposium on Algorithms(ESA), pp. 91–102 (1998)
5. Moore, P.B.: Structural motifs in RNA. Annual review of biochemistry 68, 287–300 (1999)
6. Sleator, D.D., Tarjan, R.E.: A data structure for dynamic trees. Journal of Computer and System Sciences 26, 362–391 (1983)
7. Tai, K.: The tree-to-tree correction problem. Journal of the Association for Computing Machinery (JACM) 26(3), 422–433 (1979)
8. Genomic tRNA Database. `http://lowelab.ucsc.edu/gtrnadb/`
9. Wagner, R.A., Fischer, M.J.: The string-to-string correction problem. Journal of the ACM 21(1), 168–173 (1974)
10. Zhang, K., Shasha, D.: Simple fast algorithms for the editing distance between trees and related problems. SIAM Journal on Computing 18(6), 1245–1262 (1989)

Approximation Algorithms for Reconstructing the Duplication History of Tandem Repeats

Lusheng Wang[1,⋆], Zhanyong Wang[1], and Zhizhong Chen[2]

[1] Department of Computer Science, City University of Hong Kong, Hong Kong
cswangl@cityu.edu.hk, zhyong@cs.cityu.edu.hk
[2] Department of Mathematical Sciences, Tokyo Denki University,
Hatoyama Saitama, 350-0394, Japan
cswangl@cityu.edu.hk

Abstract. Tandem repeated regions are closely related to some genetic diseases in human beings. Once a region containing pseudo-periodic repeats is found, it is interesting to study the history of creating the repeats. It is important to reveal the relationship between repeats and genetic diseases. The duplication model has been proposed to describe the history [3,10,11]. We design a polynomial time approximation scheme (PTAS) for the case where the size of the duplication box is 1. Our PTAS is faster than the best known algorithm in [11]. For example, to reach ratio-1.5, our algorithm takes $O(n^5)$ time while the algorithm in [11] takes $O(n^{11})$ time. We also design a ratio-2 approximation algorithm for the case where the size of the duplication box is at most 2. This is the first approximation algorithm with guaranteed ratio for this case.

1 Introduction

The genomes of many species are dominated by short sequences repeated consecutively. It is estimated that over 10% of the human genome, the totality of human genetic information, consists of repeated sequences. About 10-25% of all known proteins have some form of repeated structures ranging from simple homopolymers to multiple duplications of entire globular domains. In some other species, repeated regions can even dominate the whole genome. For example, in the Kangaroo rat (Dipomys ordii) more than half of the genome consists of three patterns of repeated regions: AAG (2.4 billion repetitions), TTAGG (2.2 billion repetitions) and ACACAGCGGG (1.2 billion repetitions) [9]. Recent studies show that tandem repeats are closely related with human diseases, including neurodegenerative disorders such as fragile X syndrome, Huntington's disease and spinocerebellar ataxia, and some cancers [1,2]. These tandem repeats may occur in protein coding regions of genes or non-coding regions. Since the initial discovery of tandem repeats [12], many theories on the biological mechanisms that create and extend tandem repeats have been proposed, e.g., slipped-strand

⋆ Lusheng Wang is supported by a grant from the Research Grants Council of the Hong Kong Special Administrative Region, China [Project No. CityU 1196/03E].

G. Lin (Ed.): COCOON 2007, LNCS 4598, pp. 493–503, 2007.

mis-paring, unequal sister-chromatid exchange and unequal genetic recombination during meiosis. (See [8] for details.) The exact mechanisms responsible for tandem repeat expansion are still controversial. Thus, the study of repeated regions in biological sequences has attracted lots of attentions [4,8,11,10,7,6,5].

The Duplication Model

The model for the duplication history of tandem repeated sequences was proposed by Fitch in 1977 [3] and re-proposed by Tang et al. [10] and Jaitly et al. [11]. The model captures both evolutionary history and the observed order of sequences on a chromosome. Let $S = s_1 s_2 \ldots s_n$ be an observed string containing n segment s_i for $i = 1, 2, \ldots, n$, where each s_i has exactly m letters. Let $r_i r_{i+1} \ldots r_{i+k-1}$ be k consecutive segments in an ancestor string of S in the evolutionary history. A *duplication event* generates $2k$ consecutive segments $lc(r_i) lc(r_{i+1}) \ldots lc(r_{i+k-1})$ $rc(r_i) rc(r_{i+1}) \ldots rc(r_{i+k-1})$ by (approximately) copying the k segments $r_i r_{i+1} \ldots r_{i+k-1}$ twice, where both $lc(r_{i+j})$ and $rc(r_{i+j})$ are approximate copies of r_{i+j}. See Figure 1. Assume that the n segments $s_1, s_2, \ldots s_n$ were formed from a locus by tandem duplications. Then, the locus had grown from a single copy through a series of duplications. A duplication replaces a stretch of DNA containing several repeats with two identical and adjacent copies of itself. If the stretch contains k repeats, the duplication is called a *k-duplication.*

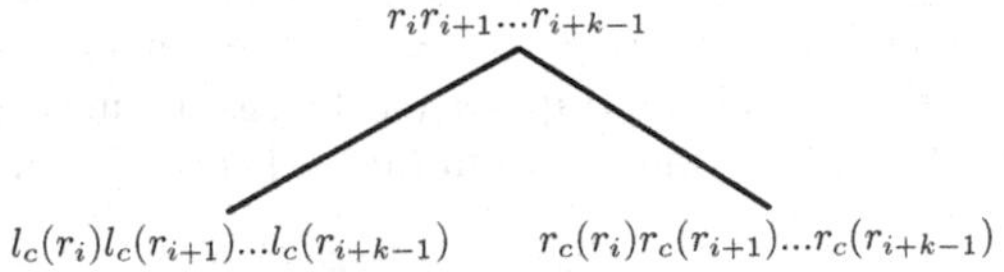

Fig. 1. A duplication event generates consecutive segments $lc(r_i)\ lc(r_{i+1}) \ldots lc(r_{i+k-1})$ $rc(r_i)\ rc(r_{i+1}) \ldots rc(r_{i+k-1})$ by (approximately) copying the k segments $r_i r_{i+1} \ldots r_{i+k}$ twice, where both $lc(r_{i+j})$ and $rc(r_{i+j})$ are approximate copies of r_{i+j}

A duplication model M for tandem repeated sequences is a directed graph that contains nodes, edges and blocks. A node in M represents a repeated segment. A directed edge (u, v) from u to v indicates that v is a child of u. A node s is an ancestor of a node t if there is a directed path from s to t. A node that has no outgoing edges is called a *leaf*; it is labeled with a given segment s_i. A non-leaf node is called an internal node; it has a left child and a right child. The root, which has only outgoing edges, represents the original copy at the locus. A block in M represents a duplication event. Each internal node appears in a unique block; no node is an ancestor of another in a block. If the block corresponds to a k-duplication, then it contains k nodes $v_1, v_2, \ldots, v_k$ from left to right. Let $lc(v_i)$ and $rc(v_i)$ be the left and right child of v_i. Then, $lc(v_1), lc(v_2), \ldots, lc(v_k), rc(v_1), rc(v_2), \ldots, rc(v_k)$ are placed from left to right in the model. Hence, for any i and j, $1 \le i < j \le k$, the edges $(v_i, rc(v_i))$ and $(v_j, lc(v_j))$ cross each other. However, no other edge crosses in the model. For simplicity, we will only draw a box for a block representing a q-duplication event for $q \geq 2$. Figure 2 gives an example. We also refer q as the size of the duplication box. The cost on each edge in the duplication model is the hamming distance between the two segments associated with the two ends of the edge. The cost of the duplication model is the total cost of all edges in the model.

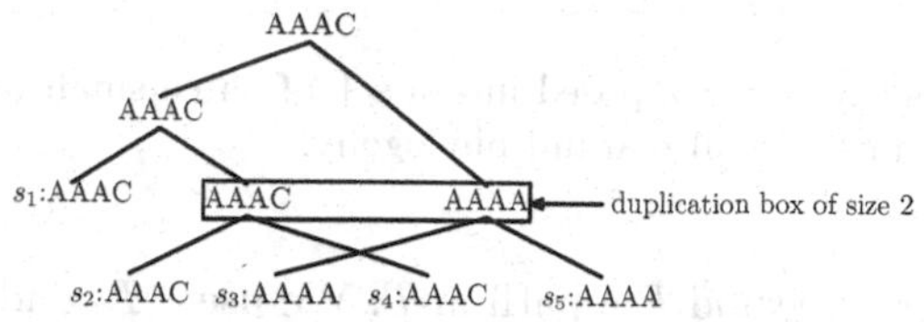

Fig. 2. A duplication model for $S = s_1s_2s_3s_4s_5$, where $s_1 = AAAC$, $s_2 = AAAC$, $s_3 = AAAA$, $s_4 = AAAC$, and $s_5 = AAAA$

Reconstructing the Duplication History: Given $S = s_1, s_2, \ldots s_n$, find a duplication model with the smallest cost.

In this paper, we only consider the cases, where the size of the duplication box is 1 or at most 2. When the size of the duplication box is 1, we design a polynomial time approximation scheme (PTAS). For ratio $1+\frac{2^t}{t2^t-2q}$, the running time is $O(n^{k+1}(f(k)+g(k)))$, where $k = 2^t - q$, and $f(k)$ and $g(k)$ are constant for any constant k. Our PTAS is faster than the best known algorithm in [11]. For example, to reach ratio-1.5, our algorithm takes $O(n^5)$ time while the algorithm in [11] takes $O(n^{11})$ time. We also design a ratio-2 approximation algorithm for the case where the size of the duplication box is at most 2. This is the first approximation algorithm with guaranteed ratio for this case.

2 The PTAS When the Size of the Duplication Box is 1

A *full phylogeny* is a phylogeny in which all internal nodes are associated with sequences. Any sequence in $\{s_1, s_2, .., s_n\}$ is called a *leaf sequence.* A sequence of length m is a *non-leaf sequence* if it is not in $\{s_1, s_2, .., s_n\}$. A *real full k-*phylogeny is a full phylogeny with k leaves, where each internal node except possibly the root is assigned a non-leaf sequence.

Consider a full phylogeny T, where some of the internal nodes are assigned some leaf sequences. The full phylogeny T can be decomposed into a set of small real full phylogenies. The roots of those (except the one on top) small real full phylogenies are leaf sequences and they are connected via identical leaf sequences. (See Figure 3.)

A phylogeny for n terminals is called a k-size phylogeny if the sizes of all its real full phylogenies are *at most* k. The basic idea of the approximation algorithm is to assign leaf sequences to some internal nodes so that the whole phylogeny is decomposed into a set of small phylogenies and we do local optimization for each small phylogeny.

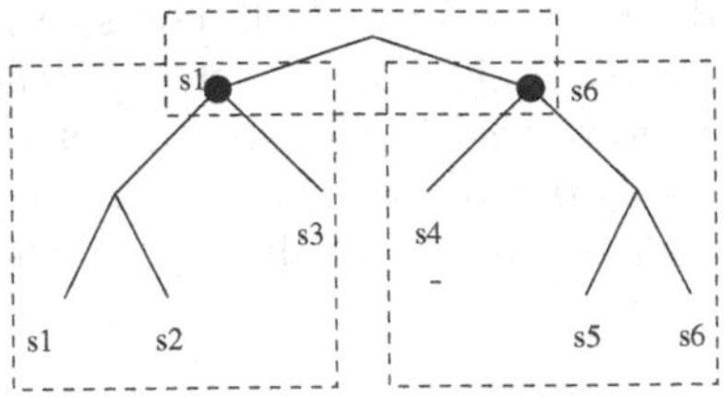

Fig. 3. A full phylogeny is decomposed into a set of three small real full phylogenies. Each rectangle contains a small real full phylogeny.

Let T be a binary tree and P a path in T. We use $c(T)$ and $c(P)$ to represent their costs. Let T_{opt} be the optimal phylogeny for $s_1, s_2, \ldots, s_n$.

2.1 The Ratio for $k = 2^t$

Now, we want to show that there exists a 2^t-size phylogeny T for $s_1, s_2, \ldots, s_n$ such that the cost $c(T)$ is at most $(1 + \frac{1}{t})c(T_{opt})$. In order to prove the theorem, we need some definitions and a preliminary lemma. Let T_{opt} be the optimal phylogeny for $s_1, s_2, \ldots, s_n$. Starting from the root, a *counter-clockwise walk* along the outside of the optimal tree T_{opt} travels through all the edges twice, one in each direction, and comes back to the root. The cost of the clockwise walk is twice of the optimal, i.e., $2c(T_{opt})$.

Consider an internal node v in T_{opt}. We use $l(v)$ and $r(v)$ to represent the left most and right most decedent leaves of v in T_{opt}, respectively. Let $in(T)$ be the set of all internal nodes other than the root T.

For each internal node $u \in in(T_{opt})$, there is a *connecting* path $P(u)$ such that it is from $r(u)$ to u in the counter-clockwise walk if u is the left child of its parent, or it is from u to $l(u)$ if u is the right child of its parent. (See Figure 4.)

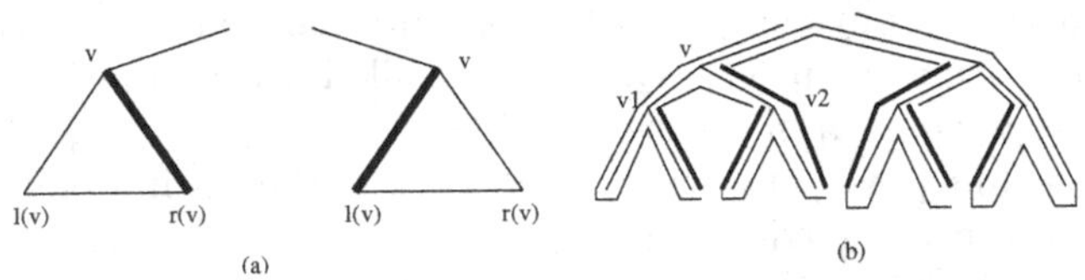

Fig. 4. (a) The connecting paths of v. (b) The walk and connecting paths.

The total length of all connecting paths in the clockwise walk is at most $c(T_{opt})$. Formally, we have

Lemma 1

$$\sum_{u \in in(T_{opt})} c(P(u)) \le c(T_{opt}).$$

Proof. The total length of the clockwise is $2c(T_{opt})$. Let v_1 and v_2 be the two children of v. The path

$$P : r(v_1), \ldots, v_1, v, v_2, \ldots, l(v_2)$$

in the clockwise walk of T_{opt} connects the two subtrees $T_{opt}(v_1)$ and $T_{opt}(v_2)$. Deleting the two edges (v_1, v) and (v, v_2), we have the two connecting paths $P(v_1)$ and $P(v_2)$, where path $P(v_1)$ in the clockwise walk is from $r(v_1)$ to v_1 and $P(v_2)$ is from v_2 to $l(v_2)$. Note that, for each node v in T_{opt}, there are two edges (v, v_1) and (v, v_2) in T_{opt}.

$$\sum_{v \in in(T_{opt}) \cup r} d(v, v_1) + d(v, v_2) = c(T_{opt}). \tag{1}$$

By (1), the total length of all the deleted edges in the clockwise walk is $c(T_{opt})$. Therefore, the total length of the connecting paths is at most $2(T_{opt}) - c(T_{opt}) \leq c(T_{opt})$. □

Theorem 1. *There exists a 2^t-size phylogeny T for s_1, s_2, ..., s_n such that the cost $c(T)$ is at most $(1 + \frac{1}{t})c(T_{opt})$.*

Proof. Let T_{opt} be an optimal tree. Let V_i be a set containing all nodes at level i in T_{opt}. We partition the nodes of T_{opt} into t groups G_0, G_1, ..., G_{t-1}, where

$$G_i = \bigcup_{i=j \ mod \ t} V_j.$$

For any node v in G_i $(i = 0, 1, \ldots, t-1)$, define $T_{opt}(v, t)$ to be a subtree of T_{opt} rooted at v containing $t + 1$ levels of nodes. If v is at level k, then all the leaves in $T_{opt}(v, t)$ are at level $k + t$. Given a subtree $T_{opt}(v, t)$, we can obtain a full 2^t-phylogeny $ST_{opt}(v, t)$, where 2^t is the number of leaves in $T_{opt}(v, t)$, as follows: (1) We replace the sequence associated with every leaf u in $T_{opt}(v, t)$ with the sequence that is on the leaf $r(u)$ if u is the left child of its parent, or on the leaf $l(u)$ if u is the right child of its parent. (Note that u is an internal node in T_{opt} and $l(u)$ and $r(u)$ are leaves in T_{opt}.) (2) any other nodes including v in $ST_{opt}(v, t)$ are assigned the sequence as in T_{opt}.

Let r be the root of T_{opt}. The set of all subtrees $ST_{opt}(r, i)$ and $ST_{opt}(v, t)$ for $v \in G_i$, forms a 2^t-size full phylogeny $ST_{opt}[i, t]$ for s_1, s_2, ..., s_n. Here each internal node $u \in G_i$ appears as a leaf of some $ST_{opt}(v, t)$ once, and appears as the root of the subtree $ST_{opt}(u, t)$ once. The full 2^t-phylogenies $ST_{opt}(v, t)$ for $v \in G_i$ are connected via the nodes associated with common sequences s_q's. Figure 5 gives an example for $t = 2$. The left to right linear order among the leaves in T_{opt} is still kept in $ST_{opt}[i, t]$.

Consider the cost $c(ST_{opt}[i, t])$ of $ST_{opt}[i, t]$.

$$c(ST_{opt}[i, t]) \leq c(T_{opt}) + \sum_{v \in G_i - r} c(P(v)),$$

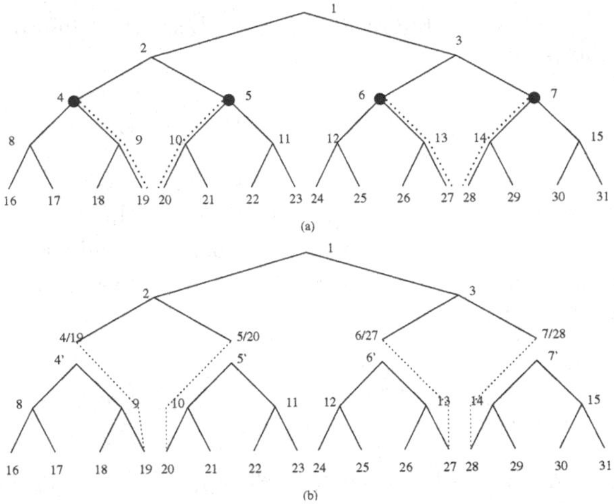

Fig. 5. (a) The tree T_{opt}. The dark nodes are in G_0. (b) The tree $ST_{opt}(i, t)$ for $t = 2$ and $i = 0$. The connection of subtrees $ST_{opt}(v, t)$ is via those nodes sharing common sequences. For example, the subtree $ST_{opt}(4, 2)$ is connected with the subtree $ST_{opt}(1, 2)$ with the sequence on node 19.

where the extra cost $\sum_{v \in G_i - r} c(P(v))$ are the costs for those connecting paths P(v) for $v \in G_i$. The total cost of the t trees $ST(0,t)$, $ST(2,t)$, ..., $ST(t-1,t)$ is

$$\sum_{i=0}^{t-1} c(ST_{opt}(i,t)) \leq t \cdot c(T_{opt}) + \sum_{i=0}^{t-1} \sum_{v \in G_i - r} c(P(v)) \leq (t+1) \cdot c(T_{opt}). \quad (2)$$

From (4), the cost of t trees is at most $(t+1) \cdot c(T_{opt})$. Thus, we know that there exists a $ST_{opt}(i, t)$ whose cost $c(ST_{opt}(i, t))$ is at most $\frac{t+1}{t} c(T_{opt}) = (1 + \frac{1}{t}) c(T_{opt})$. This completes the proof of Theorem 1. □

The Algorithm

Now, we describe our algorithm. Consider a fixed integer t. Let $D(i, j, k)$ be the cost of a minimum cost k-*tree* $T(i, j, k)$ for segment s_i, s_{i+1}, ..., s_j, where $i \leq j$, such that (1) the leaves are labeled, from left to right, by s_i, s_{i+1}, ..., s_j; and (2) the top of the tree is a k-tree.

Consider the computation of $D(i, j, k)$. The top of the tree is a k-tree that divides the consecutive segments $s_i, s_{i+1}, \ldots, s_j$, denoted as (i, j), into k parts, $(k_1, k_2 - 1)$, $(k_2, k_3 - 1)$, ..., $(k_{k-1}, k_k - 1)$, (k_k, k_{k+1}), where $k_1 = i$ and $k_{k+1} = j$.

Consider the k-tree with topology T_{top} at the top of $T(i, j, t)$. The k leaves of T_{top} are $s_{i_1}, s_{i_2}, \ldots, s_{i_k}$, where s_{i_j} is either k_j of $k_{j+1} - 1$ depending on the topology T_{top}, i.e., if leaf j is the left child of its parent in T_{top}, then $i_j = k_{j+1} - 1$; if leaf j is the right child of its parent, then $i_j = k_j$. Let $c(s_{i_1}, s_{i_2}, \ldots, s_{i_k}, T_{top})$ be the cost the k-tree at the top with topology T_{top} such that s_{i_j} is assigned to the j-th leaf of T_{top}.

$$D(i,j,k) = c(s_{i_1}, s_{i_2}, \ldots, s_{i_k}, T_{top}) + \sum_{j=1}^{k} D(k_j, k_{j+1}-1, k). \quad (3)$$

In fact, equation (3) is for the ideal case, where the top topology T_{top} is a tree F whose leaves are all at the same level and the number of leaves is k. We have to consider the degenerated cases, where the top topology T_{top} becomes a tree obtained by deleting some subtrees of F. In that case, we have to have a formula corresponding to (3), where if a subtree in F is deleted, we have to merge some consecutive regions $(k_j, k_{j+1}-1)$, ..., $(k_{jj}, k_{jj+1}-1)$ corresponding to the leaves of the subtree into one region $(s_j, s_{jj+1}-1)$. To compute $D(i,j,k)$, we have to consider all the degenerated case. The number of all degenerated cases is upper bounded by 2^{2^t-1} since there are $2^t - 1$ internal nodes in F and each node may become a leaf in the degenerated topology.

It follows that $D(1, n, 2^t)$ is at most $(1 + \frac{1}{t})$ times of the optimum.

Computing $D(i,j,k)$ needs to know $k-1$ breaking points in (i,j). Thus, for each $D(i,j,k)$ the time required is $O(n^{k-1}(f(k)+g(k)))$, where $f(k)$ is the total number of possible degenerated cases upper bounded by 2^{k-1} and $g(k)$ is the time required for computing $c(s_{i_1}, s_{i_2}, \ldots, s_{i_k}, Top)$. Since there are $O(n^2)$ $D(i,j,t)$'s for a fixed t, the total time required for the algorithm is $O(n^{k+1}(f(k)+g(k)))$.

Theorem 2. *The algorithm is a PTAS with ratio* $1 + \frac{1}{t}$ *and runs in* $O(n^{2^t+1}$ $(f(2^t) + g(2^t)))$ *time.*

For a ratio-1.5 algorithm, the running time of our approach is $O(n^5(f(2)+g(2)))$, where the old algorithm in [11] requires $O(n^{11} + n^5 g(r))$.

2.2 The Ratio for k-Size Phylogenies

For any $k > 1$, k can be written as $k = 2^t - q$ for some $t \geq 2$ and $0 \leq q \leq 2^t - 1$.

Theorem 3. *The algorithm is a PTAS with ratio* $1 + \frac{2^t}{t2^t - 2q}$ *and runs in* $O(n^{k+1}$ $(f(k) + g(k)))$ *time.*

To reach ratio 1.67, we needs $O(n^4)$ time, while the algorithm in [11] needs $O(n^7)$ time.

3 A Ratio-2 Algorithm When the Size of the Duplication Box ≤ 2

In this section, we consider the case where each block in a duplication model has size at most 2. We give a ratio-2 algorithm for this case.

Let T be a minimum cost tree fitting a duplication model for segments $s_1, s_2, \ldots, s_n$. Consider the counter-clockwise walk of T starting from s_1 and ending with s_n. In this walk, if we ignore all the internal nodes and directly connect the leaves (segments) according to the order in the walk, we get a *spanning path* for T. Obviously, if the size of all the duplication boxes is 1, then the spanning path is $s_1, s_2, \ldots s_i, s_{i+1}, \ldots, s_n$. However, if there are size-2 duplication boxes, the order will be different. An arbitrary order of the n segments may not admit a duplication model and thus may not be a spanning path. Zhang et. al. gives an algorithm to test if a given order of the segments can fit a duplication model [4].

Let T_{opt} be the optimal (minimum cost) tree for all possible duplication models. SP_{opt} is the spanning path for T_{opt}.

Lemma 2. *The cost of the spanning path SP_{opt} is at most twice of that for T_{opt}.*

From Lemma 2, if we can compute SP_{opt}, then we get a ratio-2 algorithm. However, SP_{opt} is defined from the optimal solution T_{opt} and is hard to compute. Instead, we will compute a spanning path with minimum cost among all possible duplication models for $S = s_1 s_2 \ldots s_n$. Note that if the size of all duplication boxes is 1, then for any range $[i, j]$, the set of segments s_k's with $k \in [i, j]$ forms one sub-path of the spanning path. The key idea for computing a spanning path with minimum cost is based on the observation that the spanning path for a duplication model can be decomposed into a set of ranges $[i, j]$ such that for each range $[i, j]$, all the segments s_k's with $k \in [i, j]$ form at most two sub-paths of the spanning path.

Decomposition of the Spanning Path

We give a decomposition of the spanning path that will be used for the computation of the optimal spanning path. Let T be the tree for a duplication model. In the decomposition, each internal node in T corresponds to a sub-path. A sub-path is *complete* if the sub-path contains all the segments s_k with $k \in [i, j]$, where i is the smallest index and j is the largest index in the sub-path. Two sub-paths form a *complete pair* if the two sub-paths contain all the segments s_k with $k \in [i, j]$, where i is the smallest sub-index, j is the largest sub-index in the two sub-paths and one path is from s_i to $s_{j'}$ and the other path is from $s_{i'}$ to s_j with $i < i' < j' < j$.

Let us consider the duplication boxes at the bottom of T. For a size-1 box at the bottom, it corresponds to a sub-path (actually an edge) s_i and s_{i+1}. For a size-2 box at the bottom, it corresponds to a complete pair (actually two edges), (s_i, s_{i+2}) and (s_{i+1}, s_{i+3}). For simplicity, we use $[s_i, s_j]$ to represent a complete sub-path from s_i to s_j and use $[s_i, s_j, s_k, s_l]$ to indicate a complete pair containing a sub-path from s_i to s_k and a sub-path from s_j to s_l.

Lemma 3. *Each internal node in a size-*1 *box of the duplication model corresponds to a complete path.*

Lemma 4. *Each size-*2 *box in the duplication model corresponds to a complete pair of paths.*

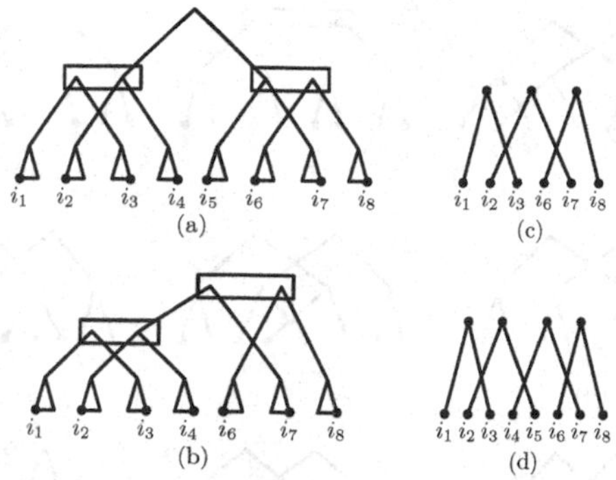

Fig. 6. Cases (a) to (d)

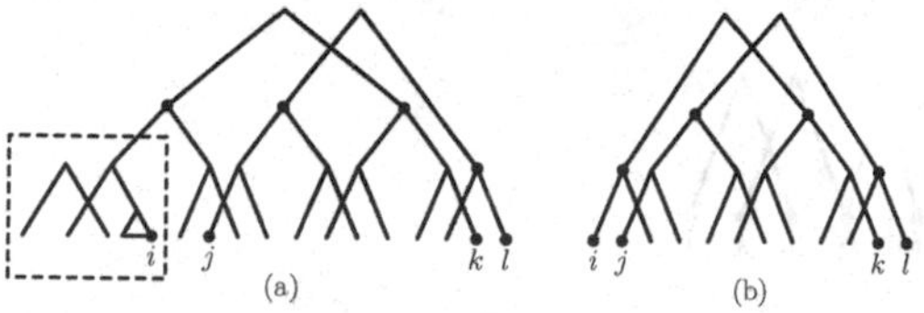

Fig. 7. Cases (a) There are more than four nodes that are co-related. (b) We 'cut' at node i. The component shown in the dashed part in (a) and the component in (b) share the common node i.

Let $E[i, j]$ be the minimum cost for all complete paths $[s_i, s_j]$. Let $D[i, i_2, i_3, j]$ denote the minimum cost for all complete pairs of paths $[s_i, s_{i_2}, s_{i_3}, s_j]$. Let $d(s_i, s_j)$ be the distance between the two segments s_i and s_j.

Lemma 5

$$E[i,i] = 0; E[i,i+1] = d(s_i, s_{i+1}). \tag{4}$$

$$E[i,j] = \min\{\min_{i<i'\leq j} d(s_{i'}, s_{i'+1}) + E[i,i'] + E[i'+1,j], \min_{i\leq i_2<i_3\leq j} D[i,i_2,i_3,j] + d(s_{i_2}, s_{i_3})\}. \tag{5}$$

Two size-2 boxes could be connected via a node and form a component containing three co-related nodes as shown in Figure 6 (a) and (b). The resulting configuration is shown in Figure 6 (c), where we use $[s_{i_1}, s_{i_2}, s_{i_3}, s_{i_6}, s_{i_7}, s_{i_8}]$ to denote the component and $F[i_1, i_2, i_3, i_6, i_7, i_8]$ to denote its cost. This case can be further extended to have four co-related nodes as shown in Figure 6 (d). We use $[s_{i_1}, s_{i_2}, s_{i_3}, s_{i_4}, s_{i_5}, s_{i_6}, s_{i_7}, s_{i_8}]$ to denote the component and $G[i_1, i_2, i_3, i_4, i_5, i_6, i_7, i_8]$ to denote its cost. There is no need to consider further extension of Figure 6 (d) since a size-2 duplication event contains at most four nodes. If there are more than four nodes that are co-related, we can 'cut' the component such that it has four co-related nodes. (See Figure 7.) Thus, in order to get Lemma 6 there are ten cases that we have to consider as shown in Figure 8.

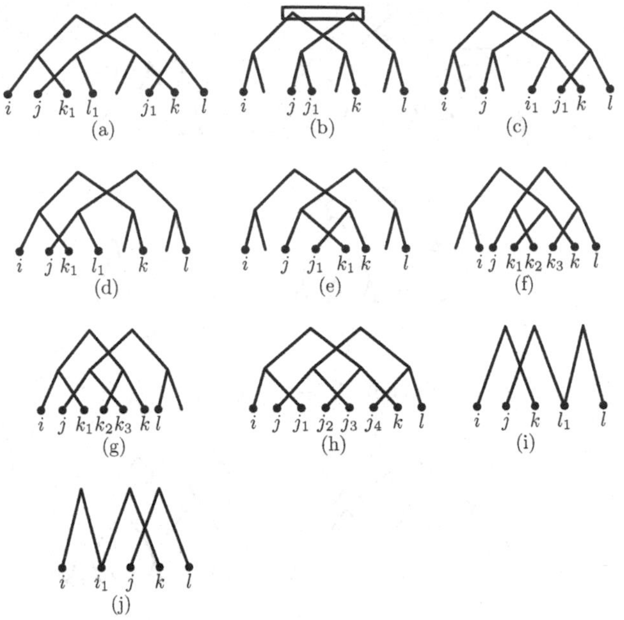

Fig. 8. Cases (a) to (j)

Besides, we can show that

$$F[i_1, i_2, i_3, , i_6, i_7, i_8] = \min_{i_3 < i_4 < i_6} \{(E[i_1, i_2, i_3, i_4] + E[i_4, i_6, i_7, i_8]), \\ (E[i_1, i_2, i_3, i_4] + E[i_4 + 1, i_6, i_7, i_8] + d(s_{i_4}, s_{i_4+1}))\} \text{ and}$$

$$G[i, j, j_1, j_2, j_3, j_4, k, l] = \min_{j_1 < l_1 < j_2} \{(E[i, j, j_1, l_1] + F[l_1, j_2, j_3, j_4, k, l]), \\ (E[i, j, j_1, l_1] + F[l_1 + 1, j_2, j_3, j_4, k, l] + d(s_{l_1}, s_{l_1+1}))\}.$$

Lemma 6. *$D[i, j, k, l]$ is the minimum of*

$$\begin{cases} \min_{j<k_1<l_1<j_1<k}(D[i,j,k_1,l_1] + D[l_1+1, j_1, k, l] + d(s_{k_1}, s_{l_1+1}) \\ \quad + d(s_{l_1}, s_{j_1})) \\ \min_{j<j_1<k}(E[i, j-1] + E[j, j_1] + E[j_1+1, k] + E[k+1, l] \\ \quad + d(s_{j-1}, s_{j_1+1}) + d(s_{j_1}, s_{k+1})) \\ \min_{j<i_1<j_1<k}(E[i, j-1] + E[j, i_1-1] + D[i_1, j_1, k, l] + d(s_{j-1}, s_{i_1}) \\ \quad + d(s_{i_1-1}, s_{j_1})) \\ \min_{j<k_1<l_1<k}(D[i, j, k_1, l_1] + E[l_1+1, k] + E[k+1, l] + d(s_{k_1}, s_{l_1-1}) + \\ \quad d(s_{l_1}, s_{k+1})) \\ \min_{j<j_1<k_1<k}(E[i, j-1] + D[j, j_1, k_1, k] + E[k+1, l] + d(s_{j-1}, s_{j_1}) \\ \quad + d(s_{k_1}, s_{k+1})) \\ \min_{i+1<k_1<k_2<k_3<k} F[i+1, k_1, k_2, k_3, k, l] + d(s_i, s_{k_1}) + d(s_{k_2}, s_{k_3}), \ (j = i+1) \\ \min_{j<k_1<k_2<k_3<k} F[i, j, k_1, k_2, k_3, k] + d(s_{k_1}, s_{k_2}) + d(s_{k_3}, s_l), \ (l = k+1) \\ \min_{j<j_1<j_2<j_3<j_4<k}(G[i, j, j_1, j_2, j_3, j_4, k, l] + d(s_{j_1}, s_{j_2}) + d(s_{j_3}, s_{j_4})). \\ \min_{k<l_1<l} D[i, j, k, l_1] + E[l_1, l]) \\ \min_{i<i_1<j} E[i, i_1] + D[i_1, j, k, l]\} \end{cases}$$

References

1. Boby, T., Patch, A.-M., Aves, S.J.: TRbase: a database relating tandem repeats to disease genes for the human genome. Bioinformatics, Advance Access published on October 12, 2004
2. Subramanian, S., Mishra, R.K., Singh, L.: Genome-wide analysis of Bkm sequences (GATA repeats): predominant association with sex chromosomes and potential role in higher order chromatin organization and function. Bioinformatics 19, 681–685 (2003)
3. Fitch, W.: Phylogenies constrained by cross-over process as illustrated by human hemoglobins in a thirteen cycle, eleven amino-acid repeat in human apolipoprotein A-I. Genetics 86, 623C644 (1977)
4. Zhang, L., Ma, B., Wang, L., Xu, Y.: Greedy method for inferring tandem duplication history. Bioinformatics 19, 1497–1504 (2003)
5. Otu, H.H., Sayood, K.: A new sequence distance measure for phylogenetic tree construction. Bioinformatics 19, 2122–2130 (2003)
6. Macas, J., Mszros, T., Nouzov, M.: PlantSat: a specialized database for plant satellite repeats. Bioinformatics 18, 28–35 (2002)
7. Lillo, F., Basile, S., Mantegna, R.N.: Comparative genomics study of inverted repeats in bacteria. Bioinformatics 18, 971–979 (2002)
8. Benson, G., Dong, L.: Reconstructing the duplication history of a tandem repeat. In: Proceedings of the Seventh International Conference on Intelligent Systems for Molecular Biology (ISMB-99), pp. 44–53 (1999)
9. Widegren, B., Arnason, U., Akusjarvi, G.: Characteristics of a conserved 1,579-bp highly repetitive component in the killer whale, Ornicus orca. Mol. Biol. Evol. 2, 411–419 (1985)
10. Tang, M., Waterman, M.S., Yooseph, S.: Zinc Finger Gene Clusters and Tandem Gene Duplication. Journal of Computational Biology 9(2), 429–446 (2002)
11. Jaitly, D., Kearney, P.E., Lin, G.-H., Ma, B.: Methods for reconstructing the history of tandem repeats and their application to the human genome. J. Comput. Syst. Sci. 65(3), 494–507 (2002)
12. Wyman, A.H., White, R.: A highly polymorphic locus in human DNA. Proc. Natl. Acad. Sci. 77, 6745–6758 (1980)

Priority Algorithms for the Subset-Sum Problem

Yuli Ye and Allan Borodin

Department of Computer Science
University of Toronto
Toronto, ON, Canada M5S 3G4
{y3ye,bor}@cs.toronto.edu

Abstract. Greedy algorithms are simple, but their relative power is not well understood. The priority framework [5] captures a key notion of "greediness" in the sense that it processes (in some locally optimal manner) one data item at a time, depending on and only on the current knowledge of the input. This algorithmic model provides a tool to assess the computational power and limitations of greedy algorithms, especially in terms of their approximability. In this paper, we study priority algorithm approximation ratios for the Subset-Sum Problem, focusing on the power of revocable decisions. We first provide a tight bound of $\alpha \approx 0.657$ for irrevocable priority algorithms. We then show that the approximation ratio of fixed order revocable priority algorithms is between $\beta \approx 0.780$ and $\gamma \approx 0.852$, and the ratio of adaptive order revocable priority algorithms is between 0.8 and $\delta \approx 0.893$.

1 Introduction

Greedy algorithms are of great interest because of their simplicity and efficiency. In many cases they produce reasonable (and sometimes optimal) solutions. Surprisingly, it is not obvious how to formalize the concept of a greedy algorithm and given such a formalism how to determine its power and limitations with regard to natural combinatorial optimization problems. Borodin, Nielson and Rackoff [5] suggested the priority model which provides a rigorous framework to analyze greedy-like algorithms. In this framework, they define fixed order and adaptive (order) priority algorithms, both of which capture a key notion of greedy algorithms in the sense that they process one data item at a time. For fixed order priority, the ordering function used to evaluate the priority of a data item is fixed before execution of the algorithm, while for adaptive priority, the ordering function can change during every iteration of the algorithm. By restricting algorithms to this framework, approximability results and limitations[1] for many problems have been obtained; for example, scheduling problems [5,18], facility

[1] We note that similar to the study of online competitive analysis, negative priority results are in some sense incomparable with hardness of approximation results as there are no explicit complexity considerations as to how a priority algorithm can choose its next item and how it decides what to do with that item. Negative results are derived from the structure of the algorithm.

G. Lin (Ed.): COCOON 2007, LNCS 4598, pp. 504–514, 2007.

location and set cover [1], job interval selection (JISP and WJISP) [12], and various graph problems [6,9]. The original priority framework specified that decisions (being made for the current input item) are irrevocable. Even within this restrictive framework, the gap between the best known algorithm and provable negative remains significant for most problems. Following [10,4], Horn [12] extended the priority framework to allow revocable acceptances when considering packing problems. That is, input items could be accepted and then later rejected, the only restriction being that a feasible solution is maintained at the end of each iteration. The revocable (decision) priority model is intuitively more powerful and almost as conceptually simple as the irrevocable model and it is perhaps surprising that it is not a more commonly used type of algorithm. Erlebach and Spieksma [10] and independently Bar-Noy et al. [4] provide a simple revocable priority approximation algorithm for the WJISP problem, and Horn [12] formalizes this model and provides an approximation upper bound[2] of $\approx 1/(1.17)$ for the special case of the weighted interval scheduling problem. Moore's [17] optimal "greedy algorithm" for the unweighted throughput maximization problem without release times (i.e. $1||\sum_j \bar{U}_j$ in Graham's scheduling notation) can be implemented as a fixed order revocable priority algorithm. It is not difficult to show that this problem cannot be solved optimally by an irrevocable priority algorithm.

The *Subset-Sum Problem* (SSP) is one the most fundamental NP-complete problems [11], and perhaps the simplest of its kind. Approximation algorithms for SSP have been studied extensively in the literature. The first FPTAS (for the more general knapsack problem) is due to Ibarra and Kim [13], and the best current approximation algorithm is due to Kellerer et al. [15], having time and space complexity $O(min\{\frac{n}{\varepsilon}, n+\frac{1}{\varepsilon^2}\log\frac{1}{\varepsilon}\})$ and $O(n+\frac{1}{\varepsilon})$ respectively. Greedy-like approximation algorithms have also been studied for SSP; an algorithm called greedy but using multiple passes, has approximation ratio 0.75, see [16]. In this paper, we study priority algorithms for SSP. Although in some sense one may consider SSP to be a "solved problem", the problem still presents an interesting challenge for the study of greedy algorithms. We believe the ideas employed for SSP will be applicable to the study of simple algorithms for other (say scheduling) problems which are not well understood, such as the throughput maximization problem (with release times) and some of its more tractable subcases. In particular, can we derive priority approximation algorithms for throughput maximization when all jobs have a fixed processing time (i.e. $1|r_j, p_j = p|\sum_j w_j\bar{U}_j$)? (We note that Horn's [12] $1/(1.17)$ bound also applies to this problem.) Baptiste [3] optimally solves this special case of throughput maximization using a dynamic programming algorithm with time complexity $O(n^7)$. (See also Chuzhoy et al. [8] and Chrobak et al. [7] for additional throughput maximization results.)

In spite of the conceptual simplicity of the SSP problem and the priority framework, there is still a great deal of flexibility in how one can design algorithms, both in terms of the ordering and in terms of which items to accept and

[2] As we are considering maximization problems in this paper, all approximation ratios will be ≤ 1 so that negative results become upper bounds on the ratio.

(for the revocable model) which items to discard in order to fit in a new item. We give a tight bound of $\alpha \approx 0.657$ for irrevocable priority algorithms showing that in this case adaptive ordering does not help. For fixed order revocable algorithms, we can show that the best approximation ratio is between $\beta \approx 0.780$ and $\gamma \approx 0.852$; for adaptive revocable priority algorithms, the best approximation ratio is between 0.8 and $\delta \approx 0.893$. All omitted proofs can be found in [19].

2 Definitions and Notation

We use **bold** font letters to denote sets of data items. For a given set $\mathbf{R}$ of data items, we use $|\mathbf{R}|$ to denote its cardinality and $\|\mathbf{R}\|$, its total weight. For a data item u, we use u to represent the singleton set $\{u\}$ and $2u$, the multi-set $\{u, u\}$; we also use u to represent the weight of u since it is the only attribute. The term u here is an overloaded term, but the meaning will become clear in the actual context. For set operations, we use $\oplus$ to denote set addition, and use $\ominus$ to denote set subtraction.

2.1 The Subset-Sum Problem

Given a set of n data items with positive weights and a capacity c, the *maximization version* of SSP is to find a subset such that the corresponding total weight is maximized without exceeding the capacity c. Without loss of generality, we make two assumptions. First of all, the weights are all scaled to their relative values to the capacity; hence we can use 1 instead of c for the capacity. Secondly, we assume each data item has weight $\in (0, 1]$. An instance of SSP is a set $\mathbf{I} = \{u_1, u_2, \ldots, u_n\}$ of n data items, where the set $\mathbf{I}$ is the *input set*, and $u_1, u_2, \ldots, u_n$ are the *data items*. A *feasible* solution is a subset $\mathbf{B}$ of $\mathbf{I}$ such that $\|\mathbf{B}\| \leq 1$. An *optimal* solution is a feasible solution with maximum weight. Let $\mathcal{A}$ be an algorithm for SSP, we denote **ALG** the solution achieved by $\mathcal{A}$ and **OPT**, the optimal solution, then the approximation ratio of $\mathcal{A}$ on that instance is denoted by $\rho = \frac{\|\mathbf{ALG}\|}{\|\mathbf{OPT}\|}$. The *approximation ratio* of $\mathcal{A}$ is the infimum of the set of ratios achieved by $\mathcal{A}$ over all instances of SSP.

2.2 Priority Model

We base our terminology and model on that of [5], and start with the class of fixed order irrevocable priority algorithms for SSP. For a given instance, a *fixed order irrevocable* priority algorithm maintains a feasible solution $\mathbf{B}$ throughout the algorithm. The structure of the algorithm[3] is as follows:

[3] We formalize the allowable (fixed) orderings by saying that the algorithm specifies a total ordering on all possible input items. The items that constitute the actual input set $\mathbf{I}$ will then inherit this ordering. That is, the priority model insists that the ordering satisfies Arrow's Independence of Irrelevant Attributes IIA Axion [2]. For adaptive orderings the algorithm can construct a new IIA ordering based on all the items that it has already seen as well as those items it can deduce are not in the input set.

FIXED ORDER IRREVOCABLE PRIORITY
Ordering: Determine a total ordering of all possible data items
while **I** is not empty
  next := index of the data item in **I** that comes first in the ordering
  Decision: Decide whether or not to add u_{next} to **B**, and then remove u_{next} from **I**
end while

An *adaptive irrevocable* priority algorithm is similar to a fixed order one, but instead of looking at a data item according to some pre-determined ordering, the algorithm is allowed to reorder the remaining data items in **I** at each iteration. This gives the algorithm an advantage since now it can take into account the information that has been revealed so far to determine which is the best data item to consider next. The structure of an adaptive irrevocable priority algorithm is described as follows:

ADAPTIVE IRREVOCABLE PRIORITY
while **I** is not empty
  Ordering: Determine a total ordering of all possible (remaining) data items
  next := index of the data item in **I** that comes first in the ordering
  Decision: Decide whether or not to add u_{next} to **B**, and then remove u_{next} from **I**
end while

The above defined priority algorithms are "irrevocable" in the sense that once a data item is admitted to the solution it cannot be removed. We can extend our notion of "fixed order" and "adaptive" to the class of revocable priority algorithms, where revocable decisions on accepted data items are allowed. Accordingly, those algorithms are called *fixed order revocable* and *adaptive revocable* priority algorithms respectively. The extension[4] to revocable acceptances provides additional power; for example, as shown in [14], *online irrevocable* algorithms for SSP cannot achieve any constant bound approximation ratio, while *online revocable* algorithms can achieve a tight approximation ratio of $\frac{\sqrt{5}-1}{2} \approx 0.618$.

2.3 Adversarial Strategy

We utilize an adversary in proving approximation bounds. For a given priority algorithm, we run the adversary against the algorithm in the following scheme. At the beginning of the algorithm, the adversary first presents a set of data items to the algorithm, possibly with some data items having the same weight. Furthermore, our adversary promises that the actual input is contained in this set[5]. Since weight is the only input parameter, the algorithm give the same pri-

[4] This extension applies to priority algorithms for packing problems.
[5] This assumption is optional. The approximation bounds clearly hold for a stronger adversary.

ority to all items having the same weight[6]. At each step, the adversary asks the algorithm to select one data item in the remaining set and make a decision on that data item. Once the algorithm makes a decision on the selected item, the adversary then has the power to remove any number of data items in the remaining set; this repeats until the remaining set is empty, which then terminates the algorithm.

For convenience, we often use a diagram to illustrate an adversarial strategy. A diagram of an adversarial strategy is an acyclic directed graph, where each node represents a possible state of the strategy, and each arc indicates a possible transition. Each state of the strategy contains two boxes. The first box indicates the current solution maintained by the algorithm, the second box indicates the remaining set of data items maintained by the adversary. A state can be either terminal or non-terminal. A state is *terminal* if and only if it is a sink, in the sense that the adversary no longer need perform any additional action; we indicate a terminal state using **bold** boxes. Each transition also contains two lines of actions. The first line indicates the action taken by the algorithm and the second line indicates the action taken by the adversary. Sometimes the algorithm may need to reject certain data items in order to accept a new one, so an action may contain multiple operations which occur at the same time; we use $\oslash$ if there is no action. Note that to calculate a bound for the approximation ratio of an algorithm, it is sufficient to consider the approximation ratios achieved in all terminal states. We will see such diagrams in Sect. 3.

3 Priority Algorithms and Approximation Bounds

We first define four constants that will be used for our results. Let α, β, γ and δ be the real roots (respectively) of the equations $2x^3+x^2-1=0$, $2x^2+x-2=0$, $10x^2-5x-3=0$ and $6x^2-2x-3=0$ between 0 and 1. The corresponding values are shown in Table 1.

Table 1. Corresponding values

name	α	β	γ	δ
value	≈ 0.657	≈ 0.780	≈ 0.852	≈ 0.893

We now give a tight bound for irrevocable priority algorithms. It is interesting that there is no approximability difference between fixed order and adaptive irrevocable priority algorithms.

Theorem 1. *There is a fixed order irrevocable priority algorithm for SSP with approximation ratio α, and every irrevocable priority algorithm for SSP has approximation ratio at most α.*

[6] Technically we can use an item number identifier to further distinguish items, but by providing sufficiently many copies of an item the adversary can effectively achieve what the statement claims.

The case for revocable priority algorithms is more interesting. The ability to make revocable acceptances gives the algorithm a certain flexibility to "regret". The data items admitted into the solution are never "safe" until the termination of the algorithm. Therefore, if there is enough "room", it never hurts to accept a data item no matter how "bad" it is, as we can always reject it later at any time and with no cost. For the rest of the paper, we assume our algorithms will take advantage of this property. We start with fixed order revocable priority algorithm by giving two tight bounds for non-increasing order (i.e. items are ordered so that $u_1 \geq u_2 \ldots \geq u_n$) and non-decreasing order revocable priority algorithms.

Theorem 2. *There is a non-increasing order revocable priority algorithm for SSP that has approximation ratio α, and every such algorithm has approximation ratio at most α. (Note that the simple ordering here is different from the fixed order irrevocable algorithm of Theorem 1.)*

Theorem 3. *There is a non-decreasing order revocable priority algorithm for SSP that has approximation ratio β, and every such algorithm has approximation ratio at most β.*

The improvement using non-decreasing order is perhaps counter-intuitive [7] as one might think it is more flexible to fill in with small items at the end. Next, we give a approximation bound for any fixed order revocable priority algorithm; this exhibits the first approximation gap we are unable to close. The technique used in the proof is based on a chain of possible item priorities. It turns out, in order to achieve certain approximation ratio, some data items must be placed before some other data items. This order relation is transitive and therefore, has to be acyclic.

Theorem 4. *No fixed order revocable priority algorithm of SSP can achieve approximation ratio better than γ.*

Proof. Let $u_1 = 0.2$, $u_2 = \frac{1}{2}\gamma - \frac{1}{10} \approx 0.326$, $u_3 = 0.5$, and $u_4 = 0.8$. For a data item u, we denote by $rank(u)$ its priority. There are four cases:

1. If $rank(u_4) > rank(u_3)$, then the adversarial strategy is shown in Fig. 1. If the algorithm terminates via state $\mathbf{s}_1$, then

$$\rho = \frac{\|\mathbf{ALG}\|}{\|\mathbf{OPT}\|} = \frac{u_3}{u_4} < \gamma.$$

   If the algorithm terminates via state $\mathbf{s}_2$, then

$$\rho = \frac{\|\mathbf{ALG}\|}{\|\mathbf{OPT}\|} \leq \frac{u_4}{2u_3} = u_4 < \gamma.$$

[7] As another example, in the identical machines makespan problem, it is provably advantageous to consider the largest items first.

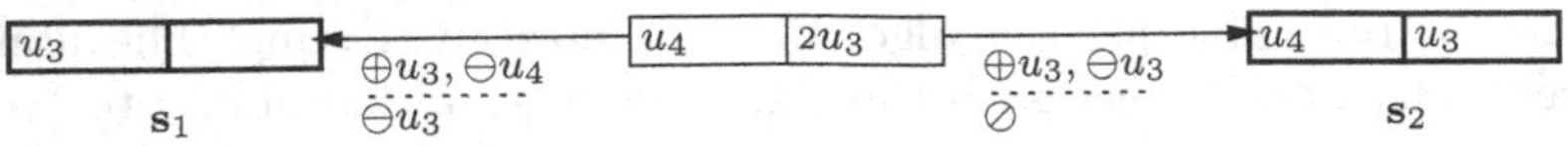

Fig. 1. Adversarial strategy for $rank(u_4) > rank(u_3)$

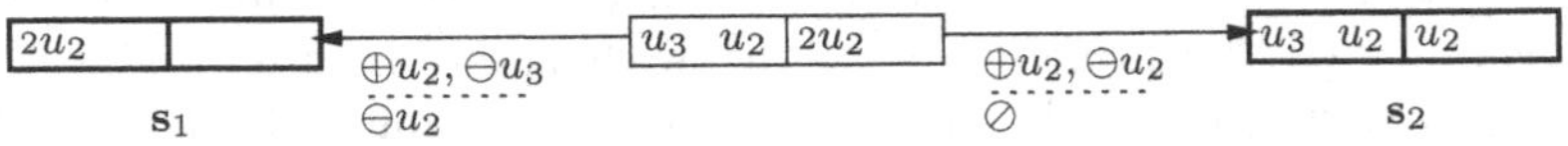

Fig. 2. Adversarial strategy for $rank(u_3) > rank(u_2)$

2. If $rank(u_3) > rank(u_2)$, then the adversarial strategy is shown in Fig. 2. If the algorithm terminates via state $\mathbf{s}_1$, then

$$\rho = \frac{\|\mathbf{ALG}\|}{\|\mathbf{OPT}\|} = \frac{2u_2}{u_2 + u_3} = \frac{\gamma - \frac{1}{5}}{\frac{1}{2} + \frac{1}{2}\gamma - \frac{1}{10}} = \frac{10\gamma - 2}{5\gamma + 4} < \gamma.$$

If the algorithm terminates via state $\mathbf{s}_2$, then

$$\rho = \frac{\|\mathbf{ALG}\|}{\|\mathbf{OPT}\|} \leq \frac{u_2 + u_3}{3u_2} = \frac{\frac{1}{2} + \frac{1}{2}\gamma - \frac{1}{10}}{\frac{3}{2}\gamma - \frac{3}{10}} = \frac{5\gamma + 4}{15\gamma - 3} < \gamma.$$

3. If $rank(u_2) > rank(u_1)$, then the adversarial strategy is shown in Fig. 3.

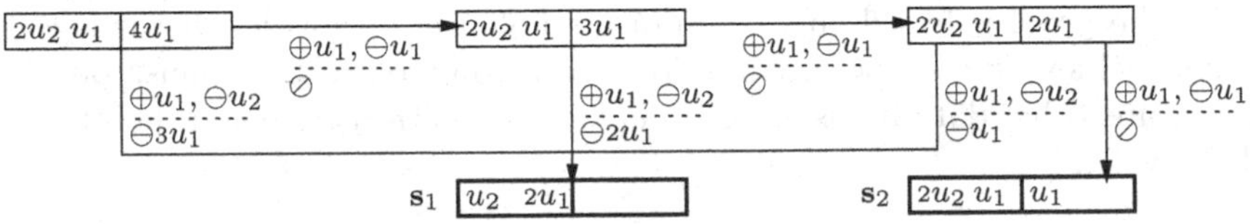

Fig. 3. Adversarial strategy for $rank(u_2) > rank(u_1)$

If the algorithm terminates via state $\mathbf{s}_1$, then

$$\rho = \frac{\|\mathbf{ALG}\|}{\|\mathbf{OPT}\|} \leq \frac{2u_1 + u_2}{u_1 + 2u_2} = \frac{\frac{2}{5} + \frac{1}{2}\gamma - \frac{1}{10}}{\frac{1}{5} + \gamma - \frac{1}{5}} = \frac{\frac{1}{2}\gamma + \frac{3}{10}}{\gamma} = \frac{5\gamma + 3}{10\gamma} = \gamma.$$

If the algorithm terminates via state $\mathbf{s}_2$, then

$$\rho = \frac{\|\mathbf{ALG}\|}{\|\mathbf{OPT}\|} \leq \frac{u_1 + 2u_2}{5u_1} = u_1 + 2u_2 = \frac{1}{5} + \gamma - \frac{1}{5} = \gamma.$$

4. If $rank(u_1) > rank(u_2) > rank(u_3) > rank(u_4)$, then the adversarial strategy is shown in Fig. 4.
If the algorithm terminates via state $\mathbf{s}_1$, then

$$\rho = \frac{\|\mathbf{ALG}\|}{\|\mathbf{OPT}\|} = \frac{u_1 + u_3}{u_2 + u_3} = \frac{\frac{1}{5} + \frac{1}{2}}{\frac{1}{2}\gamma - \frac{1}{10} + \frac{1}{2}} = \frac{7}{5\gamma + 4} < \gamma.$$

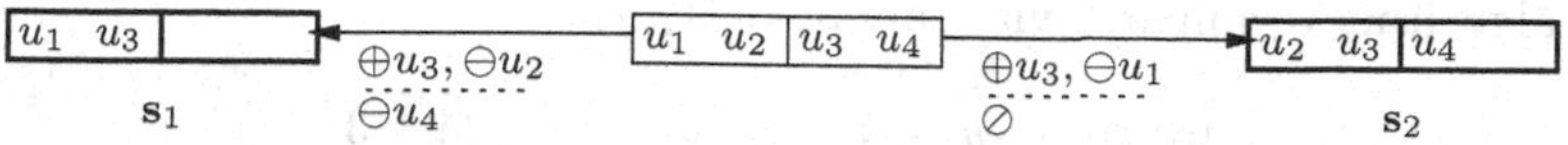

Fig. 4. Adversarial strategy for $rank(u_1) > rank(u_2) > rank(u_3) > rank(u_4)$

If the algorithm terminates via state $\mathbf{s}_2$, then

$$\rho = \frac{\|\mathbf{ALG}\|}{\|\mathbf{OPT}\|} \leq \frac{u_2 + u_3}{u_1 + u_4} = u_2 + u_3 = \frac{1}{2}\gamma - \frac{1}{10} + \frac{1}{2} < \gamma.$$

As a conclusion, no fixed order revocable priority algorithm of SSP can achieve approximation ratio better than γ. This completes the proof. □

Finally, we study adaptive revocable priority algorithms. This is the strongest class of algorithms studied in this paper and arguably represents the ultimate approximation power of greedy algorithms (for packing problems). We show that no such algorithm can achieve an approximation ratio better than δ, and then we develop a relatively subtle algorithm having approximation ratio 0.8 in Theorem 6, thus leaving another gap in what is provably the best approximation ratio possible.

Theorem 5. *No adaptive revocable priority algorithm of SSP can achieve approximation ratio better than* δ.

Proof. Let $u_1 = \frac{1}{3}\delta \approx 0.298$ and $u_2 = 0.5$. For a given algorithm, we utilize the following adversary strategy shown in Fig. 5. If the algorithm terminates via state $\mathbf{s}_1$ or $\mathbf{s}_2$, then

$$\rho = \frac{\|\mathbf{ALG}\|}{\|\mathbf{OPT}\|} \leq \frac{3u_1}{2u_2} = 3u_1 = \delta.$$

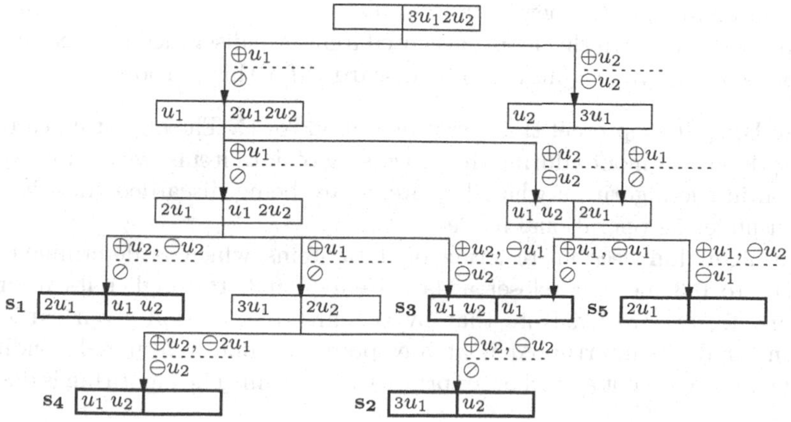

Fig. 5. Adversarial strategy for adaptive revocable priority algorithms

If the algorithm terminates via state $\mathbf{s}_3$ or $\mathbf{s}_4$, then

$$\rho = \frac{\|\mathbf{ALG}\|}{\|\mathbf{OPT}\|} \leq \frac{u_1 + u_2}{3u_1} = \frac{\frac{1}{3}\delta + \frac{1}{2}}{\delta} = \frac{2\delta + 3}{6\delta} = \delta.$$

If the algorithm terminates via state $\mathbf{s}_5$, then

$$\rho = \frac{\|\mathbf{ALG}\|}{\|\mathbf{OPT}\|} = \frac{2u_1}{u_1 + u_2} = \frac{\frac{2}{3}\delta}{\frac{1}{3}\delta + \frac{1}{2}} = \frac{4\delta}{2\delta + 3} < \delta.$$

In all three cases, the adversary forces the algorithm to have approximation ratio no better than δ; this completes the proof. □

Our 0.8 approximation adaptive priority algorithm is facilitated by a few simplifying observations. First of all, since we are targeting a ratio of 0.8, the algorithm can terminate whenever it detects an item u in the remaining input such that a subset of $(\mathbf{B} \oplus u)$ has total weight ≥ 0.8; such an item is *adaptively* given the highest priority and we call this a *terminal condition*. In our algorithm description, it is understood that the algorithm first *adaptively* checks for such a condition, and terminates if it is satisfied. Note that the running time of checking such condition is bounded by a constant. Secondly, data items outside $(0.2, 0.8)$ are not needed for the analysis of the algorithm. That is, items in $[0.8, 1]$ trivialize the problem and items in $(0, .2]$ can be considered at the end and added to $\mathbf{B}$ (if necessary) to achieve the desired bound. Finally, we assume the current set of accepted items $\mathbf{B}$ operates in one of the following four modes:

1. `Queue Mode`: In this mode, accepted items are discarded in the FIFO order to accommodate the new data item u.
2. `Queue_1 Mode`: In this mode, the first accepted item is never discarded, the rest data items are discarded in the FIFO order to accommodate the new data item u.
3. `Stack Mode`: In this mode, accepted items are discarded in the FILO order to accommodate the new data item u.
4. `Optimum Mode`: In this mode, accepted items are discarded to maximize $\|\mathbf{B}\|$; the new data item u may also be discarded for this purpose.

We use $\mathbf{B}_{\text{mode}}$ to represent the operational mode of $\mathbf{B}$. The algorithm can switch among these four modes during the processing of data items; we do not explicitly mention in the algorithm what data items are being discarded since it is well-defined under its operational mode.

The algorithm uses an ordering of data items which is determined by its distance to 0.3, i.e., the closer a data item to 0.3, the higher its priority is, breaking tie arbitrarily. Note that by the first observation given earlier, this ordering may be interrupted if at any point of time a terminal condition is satisfied, so this is not a fixed order priority algorithm. The algorithm is described below.

Algorithm ADAPTIVE

```
B := ∅;
if the first data item is in (0.2, 0.35) then
    B_mode := Queue;
else
    B_mode := Queue_1;
end if
while I contains a data item ∈ (0.2, 0.4] do
    let u be the next data item in I;
    I := I ⊖ u;
    accept u;
end while
if B contains exactly three data items and all are ∈ (0.2, 0.3] then
    B_mode := Stack;
else
    B_mode := Optimum;
end if
while I contains a data item ∈ (0.4, 0.8) do
    let u be the next data item in I;
    I := I ⊖ u;
    accept u;
end while
```

Theorem 6. *Algorithm* ADAPTIVE *achieves approximation ratio 0.8 for SSP.*

It is seemingly a small step from a 0.78 algorithm to a 0.8 algorithm, but the latter algorithm requires a substantially more refined approach and detailed analysis. The detailed analysis can be found in the full paper. The merit, we believe, in studying such a class of "simple algorithms" is that the simplicity of the structure suggests algorithmic ideas and allows a careful analysis of such algorithms. Limiting ourselves to simple algorithmic forms and exploiting the flexibility within such forms may very well give us a better understanding of the structure of a given problem and a better chance to derive new algorithms.

4 Conclusion

We analyze different types of priority algorithms for SSP leaving open two approximability gaps, one for fixed order and one for adaptive revocable priority algorithms. It is interesting that such gaps and non-trivial algorithms exist for such a simple class of algorithms and such a basic problem as SSP. We optimistically believe that surprisingly good algorithms can be designed within the revocable priority framework for problems which are currently not well understood.

References

1. Angelopoulos, S., Borodin, A.: On the power of priority algorithms for facility location and set cover. Algorithmica 40, 271–291 (2004)
2. Arrow, K.: Social Choice and Individual Values. Wiley, Chichester (1951)
3. Baptiste, P.: Polynomial time algorithms for minimizing the weighted number of late jobs on a single machine with equal processing times. Journal of Scheduling 2, 245–252 (1999)
4. Bar-Noy, A., Guha, S., Naor, J., Schieber, B.: Approximating throughput in real-time scheduling. SIAM Journal of Computing 31, 331–352 (2001)
5. Borodin, A., Nielsen, M., Rackoff, C.: (Incremental) priority algorithms. Algorithmica 37, 295–326 (2003)
6. Borodin, A., Boyar, J., Larsen, K.: Priority algorithms for graph optimization problems. In: Persiano, G., Solis-Oba, R. (eds.) WAOA 2004. LNCS, vol. 3351, pp. 126–139. Springer, Heidelberg (2005)
7. Chrobak, M., Durr, C., Jawor, W., Kowalik, L., Kurowski, M.: On scheduling equal length jobs to maximize throughput. To appear in Journal of Scheduling.
8. Chuzhoy, J., Ostrovsky, R., Rabani, Y.: Approximation algorithms for the job interval scheduling problem and related scheduling problems. In: Proceedings of 42nd Annual IEEE Symposium of Foundations of Computer Science, pp. 348–356. IEEE Computer Society Press, Los Alamitos (2001)
9. Davis, S., Impagliazzo, R.: Models of greedy algorithms for graph problems. In: Proceedings of the 15th Annual ACM-SIAM Symposium on Discrete Algorithms, pp. 381–390. ACM Press, New York (2004)
10. Erlebach, T., Spieksma, F.: Interval selection: Applications, algorithms, and lower bounds. Journal of Algorithms 46, 27–53 (2003)
11. Garey, M., Johnson, D.: Computers and Intractability: A Guide to the Theory of NP-completeness. Freeman, San Francisco (1979)
12. Horn, S.: One-pass algorithms with revocable acceptances for job interval selection. Master's thesis, University of Toronto (2004)
13. Ibarra, O., Kim, C.: Fast approximation algorithms for the knapsack and sum of subset problem. Journal of the ACM 22, 463–468 (1975)
14. Iwama, K., Taketomi, S.: Removable online knapsack problems. In: Widmayer, P., Triguero, F., Morales, R., Hennessy, M., Eidenbenz, S., Conejo, R. (eds.) ICALP 2002. LNCS, vol. 2380, pp. 293–305. Springer, Heidelberg (2002)
15. Kellerer, H., Mansini, R., Pferschy, U., Speranza, M.: An efficient fully polynomial approximation scheme for the subset-sum problem. Journal of Computer and System Science 66, 349–370 (2003)
16. Martello, S., Toth, P.: Knapsack Problems: Algorithms and Computer Implementations. John Wiley and Sons, Chichester (1990)
17. Moore, J.: An n-job, one machine sequencing algorithm algorithm for minimizing the number of late jobs. Management Science 15, 102–109 (1968)
18. Regev, O.: Priority algorithms for makespan minimization in the subset model. Information Processing Letters 84, 153–157 (2002)
19. Ye, Y., Borodin, A.: Priority algorithms for the subset-sum problem. Technical Report, University of Toronto (2007) `http://www.cs.toronto.edu/~bor`

Distributed Approximation Algorithms for Weighted Problems in Minor-Closed Families*

A. Czygrinow[1] and M. Hańćkowiak[2]

[1] Department of Mathematics and Statistics
Arizona State University
Tempe, AZ 85287-1804, USA
andrzej@math.la.asu.edu

[2] Faculty of Mathematics and Computer Science
Adam Mickiewicz University, Poznań, Poland
mhanckow@amu.edu.pl

Abstract. We give efficient distributed approximation algorithms for weighted versions of the maximum matching problem and the minimum dominating set problem for graphs from minor-closed families. To complement these results we indicate that no efficient distributed algorithm for the minimum weight connected dominating set exists.

1 Introduction

Efficient distributed algorithms for only very few graph-theoretic problems are known. At the same time there has been much more success in designing efficient distributed algorithms in case the underlying topology of the network has additional properties. For example, many problems can be solved efficiently in constant maximum degree graphs and some problems admit rather easy distributed approximations in graphs of bounded arboricity (for example in planar graphs). In this paper, we will study distributed complexity of three fundamental problems in proper minor-closed families of graphs. We will show that the maximum-weight matching problem and the minimum-weight dominating set problem admit efficient distributed approximations but the minimum-weight connected dominating set problem does not. This extends and complements the results from [3] where distributed complexity of unweighted versions of the above problems is analyzed. Note however that algorithms for weighted problems are significantly different than the ones from [3]. In fact, even the distributed complexity of weighted and unweighted problems can be different. For example, in [3] we proved that the minimum-connected dominating set problems admits an efficient distributed approximation in connected graphs which come from minor-closed families. This is not the case for the weighted analog as we indicate in the last section of this paper. The algorithms for weighted versions of the maximum matching and the minimum dominating set problems are in turn based on a completely new and provably more powerful partitioning algorithm than the corresponding clustering procedure in [3].

* This work was supported by grant N206 017 32/2452 for years 2007-2010.

G. Lin (Ed.): COCOON 2007, LNCS 4598, pp. 515–525, 2007.

1.1 Terminology and Notation

We will consider the message-passing distributed model (see Linial [10]). In this model, network is represented by an undirected graph with vertices corresponding to processors, and edges corresponding to communication links between processors. The network is synchronized and computations proceed in discrete rounds. In a single round a vertex can send and receive messages from its neighbors, and can perform some local computations. Neither the amount of local computations nor the lengths of messages are restricted in any way. We will also assume that nodes in the network have unique identifiers which are positive integers from $\{1,\dots,|G|\}$ where $|G|$ is the order of G and is globally known. Although different possible measures of efficiency of a distributed algorithm can be assumed, following [10] we call a distributed algorithm *efficient* if it runs in a poly-logarithmic (in the order of the graph) number of rounds. Consequently, if the diameter of the underlying network is poly-logarithmic then any of the above problems admits a trivial efficient solution. In this paper, we shall focus on distributed approximation algorithms for minor-closed families. All graphs are simple and in the graph-theoretic terminology we will follow [5]. A graph H is called a *minor* of G if it can be obtained from a subgraph of G by a series of edge contractions. A family $\mathcal{C}$ is called *minor-closed* if for any graph $G \in \mathcal{C}$ every minor of G is also in $\mathcal{C}$. A family $\mathcal{C}$ is proper if there is a graph G which is not in $\mathcal{C}$ and is non-trivial if it contains a graph with at least one edge. We will always assume that our minor-closed family is both proper and non-trivial. Although the most important example of a minor-closed family is certainly the class of planar graphs, algorithmic questions for different minor-closed classes of graphs, like the family of graphs with a bounded tree-width or a bounded genus, have recently attracted attention. Let $\mathcal{C}$ be a minor-closed family and let $\rho := \sup_{G\in\mathcal{C}} \frac{||G||}{|G|}$ be the edge density of $\mathcal{C}$. It is known (see for example [12]) that ρ is finite if and only if $\mathcal{C}$ is a proper minor-closed family. We will write $\mathcal{C}_\rho$ for a minor-closed family with edge density ρ and assume that ρ is known by an algorithm. A *matching* in graph G is a subset M of edges of G with no two edges from M sharing a vertex. For an edge-weighted graph $(G,\bar{\omega})$ with $\bar{\omega} : E(G) \to R^+ \cup \{0\}$ we denote by $\beta(G)$ the maximum weight of a matching in G, that is $\beta(G) = \max_M \sum_{e\in M} \bar{\omega}(e)$. A *dominating set* in a graph G is a subset D of vertices such that for every vertex $v \notin D$ a neighbor of v is in D. For a vertex-weighted graph (G,ω) with $\omega : V(G) \to R^+ \cup \{0\}$ we denote by $\gamma(G)$ the minimum weight of a dominating set in G, that is $\gamma(G) = \min_D \sum_{v\in D} \omega(v)$. Finally, a dominating set D is called a *connected dominating set* in G if it is a dominating set and the subgraph of G induced by D is connected.

We will denote by $|G|$ the order of G, that is the number of vertices of G and by $||G||$ the size of G, that is the number of edges of G. As already noted we assume that each vertex v has a unique identifier $ID(v)$ and $ID : V(G) \to \{1,\dots,|G|\}$. Since our partitioning algorithm will be applied to an auxiliary graph of G and it will be important to distinguish between the range of identifiers in the auxiliary graph and the order of the graph we denote by $ID(H) = \bigcup_{v\in V(H)}\{ID(v)\}$.

1.2 Results

We give distributed approximation algorithms for the maximum-weight matching problem and for the minimum-weight dominating set problem for graphs from a minor-closed family $\mathcal{C}_\rho$. In the case of the maximum-weight matching problem we will give a distributed algorithm which given a positive integer d finds in an edge-weighted graph $(G, \bar{\omega})$ with $G \in \mathcal{C}_\rho$ a matching M of weight $\bar{\omega}(M) \geq \left(1 - \frac{1}{\log^d |G|}\right) \beta(G)$. The algorithm runs in a poly-logarithmic number of rounds. (Theorem 4.) For the minimum-weight dominating set problem, we will prove that there is a distributed algorithm which given a positive integer d finds in vertex-weighted graph (G, ω) with $G \in \mathcal{C}_\rho$ a dominating set D such that $\omega(D) \leq \left(1 + \frac{1}{\log^d |G|}\right) \gamma(G)$. This algorithm again runs in a poly-logarithmic number of rounds. (Theorem 5.) For the minimum-weight connected dominating set problem we indicate that to accomplish any finite multiplicative approximation error, $\Omega(|G|)$ rounds are needed. Both algorithms use a vertex partitioning procedure which partitions the vertex set of a graph G into sets $V_1, \ldots, V_l$ so that each $G[V_i]$ has a poly-logarithmic diameter and the weight of the border vertices is small with respect to the total weight of G (see Corollary 3 for a precise statement).

1.3 Related Work

We will briefly put our results in a more general context. The reader is directed to Elkin's survey [6] for a more comprehensive overview of distributed approximation algorithms. Let us first note that efficient distributed algorithms that find exact solutions to the above problems do not exist (even for unweighted analogs). For example, the minimum dominating set problem and the maximum matching problem when restricted to a cycle G cannot be found in $o(|G|)$ rounds ([10]). In addition, to achieve a poly-logarithmic approximation ratio for minimum dominating set at least $\max\{\Omega(\sqrt{\log |G| / \log\log |G|})\Omega(\log \Delta / \log\log \Delta)\}$ rounds are required ([8]).

Distributed approximation algorithms for planar graphs were studied in [2] and [4]. In particular, [2] contains an efficient distributed approximation for the maximum-weight independent set problem in planar graphs and [3] contains efficient distributed algorithms for unweighted versions of the three problems in minor-closed families of graphs.

1.4 Organization

In the rest of the paper we will first discuss vertex partitioning problems in weighted graphs and give our main auxiliary procedure (Section 2). Then, in Section 3, we give our approximation algorithms.

2 Partitioning of Vertex-Weighted Graphs

We will start with fixing some general graph-theoretic terminology. For a graph G, $V(G)$ will denote the vertex set of G and $E(G)$ will denote the edge set of

G. If U, U' are two disjoint subsets of $V(G)$ then $E_G(U, U')$ denotes the set of edges with one endpoint in U, another in U'. For $v \in V(G)$, $N(v)$ denotes the set of neighbors of v in G and if $U \subseteq V(G)$ then $N_G(U) := \bigcup_{u \in U} N(u) \setminus U$. For two vertices u, u', $dist_G(u, u')$ is the length of the shortest path between u and u', the diameter of G, $diam_G$, is the maximum of $dist_G(u, u')$ over all pair (u, u'), and for sets U, U', we set $dist_G(U, U') := \min_{u \in U, u' \in U'} dist_G(u, u')$. In addition for a subgraph H of G we will consider two different diameters of H. The strong diameter of H, $SDiam_G(H)$, will be defined as as $diam_H$ and the weak diameter of H, $WDiam_G(H)$, will be defined as $max_{u,u \in V(H)} dist_G(u, u')$. Clearly $WDiam_G(H) \leq SDiam_G(H)$.

Let $\mathcal{C}_\rho$ be a minor-closed family of graphs G with the edge-density ρ, that is

$$\rho = \sup \frac{||G||}{|G|}$$

where the supremum is taken over all graphs from $\mathcal{C}_\rho$. It is known (see [12] for this and many other results) that ρ is finite if and only if $\mathcal{C}_\rho$ is a proper minor-closed family. In addition, in the paper, we will always assume that $\mathcal{C}_\rho$ is proper and $\rho > 0$.

Although vast majority of the paper is concerned with vertex-weighted graphs, we will start with a brief discussion that shows a connection between distributed partitioning problems for vertex-weighted and edge-weighted graphs (Section 2.1). In Section 2.2, we will give an efficient distributed partitioning algorithm for vertex-weighted graphs from $\mathcal{C}_\rho$. The distributed algorithm is deterministic but we assume that both $|G|$ and ρ are known to all vertices of G.

2.1 Weighted Graphs

For a graph $G \in \mathcal{C}_\rho$ we will consider two types of weight functions on G. Pair (G, ω) with $\omega : V(G) \to R^+ \cup \{0\}$ will be called *vertex-weighted graph* G and the pair $(G, \bar{\omega})$ with $\bar{\omega} : E(G) \to R^+ \cup \{0\}$ will be called *edge-weighted graph* G. We need some more notation and terminology. Let (G, ω) be a vertex-weighted graph. For a set $S \subseteq V(G)$ we define $\omega(S) := \sum_{v \in S} \omega(v)$. A vertex of S is called a *border vertex* in S if it has a neighbor in $V(G) \setminus S$. The set of all border vertices in S is denoted by $\partial(S)$ and for a partition $P = (V_1, V_2, \ldots, V_l)$ of $V(G)$ we set $\partial(P) := \bigcup_{i=1}^{l} \partial(V_i)$. In the case of an edge-weighted graph $(G, \bar{\omega})$ we define $\bar{\partial}(S)$ to be the set of all edges with one endpoint in S and another in $V(G) \setminus S$. Then for a partition $P = (V_1, V_2, \ldots, V_l)$ of $V(G)$, $\bar{\partial}(P) := \bigcup_{i=1}^{l} \bar{\partial}(V_i)$.

Definition 1. *Let (G, ω) be a vertex-weighted graph and let $a(\cdot), b(\cdot)$ be functions to R. A partition $P = (V_1, V_2, \ldots, V_l)$ of $V(G)$ is called an (a, b)-vertex-weight partition if the following two conditions are satisfied:*

- *For $i = 1, \ldots, l$, $G[V_i]$ is connected and $WDiam_G(G[V_i]) \leq a(|G|)$.*
- *$\omega(\partial(P)) \leq \omega(V(G))/b(|G|)$.*

Similarly we define an (a, b)-edge-weight partition of $(G, \bar{\omega})$. We will be almost exclusively interested in cases when both a and b are poly-logarithmic functions.

In [2], a distributed algorithm that finds a $(\log |G|, \log |G|)$- edge-weight partition of an edge-weighted planar graph G is given. The algorithm runs in a poly-logarithmic number of rounds. This edge-weight partition can be used to give distributed approximation algorithms for the maximum-weight independent set problem. In addition, a similar procedure can be used to give distributed approximations for the unweighted versions of the maximum matching problem or the minimum dominating set problem in graphs G which come from a fixed minor-closed family. On the other hand, the edge-weight partition property is not strong enough to yield approximations for weighted analogs of the maximum matching problem or the minimum dominating set problem. As we will show in the next section, vertex-weight partition can be found by a distributed algorithm efficiently and can be used to design approximations for the weighted versions of the above two problems. Let us first note however the vertex-weight partition is indeed stronger than an edge-weight partition.

2.2 Partitioning Algorithm

We will now give an algorithm which finds an (a, b)-vertex-weight partition. Let us start by fixing some additional terminology. Let G be a graph from $\mathcal{C}_\rho$ and let $\omega : V(G) \to R^+ \cup \{0\}$. A small modification (change in the number of iterations) of the algorithms CLUSTERING and WISPlanar from [2] yields the following two facts.

Lemma 1. *Let $\mathcal{C}_\rho$ be a minor-closed family of graphs. Let $G \in \mathcal{C}_\rho$ and let $(G, \bar{\omega})$ be an edge-weighted graph. There exists a distributed algorithm which given a constant $d > 1$ finds in $O(\log |G| \log^* |G|)$ rounds a (D_ρ, d)-edge-weight partition for some constant $D_\rho = D_\rho(d)$.*

Lemma 2. *Let $\mathcal{C}_\rho$ be a minor-closed family of graphs. Let $G \in \mathcal{C}_\rho$ and let (G, ω) be a vertex-weighted graph. There exists a distributed algorithm which given a constant $d > 2\rho$ finds in $O(\log |G| \log^* |G|)$ rounds a maximal independent set I in G with*

$$\omega(I) \geq \omega(V(G))/d.$$

Our first procedure, HEAVY SUBSET, finds a subset of vertices of a large weight which induces subgraphs of small weak diameter. As the procedure is a bit technical we will divide it into two phases.

HEAVY SUBSET PHASE 1. Use the algorithm DECOMPOSITION from [3] to find a partition $(V_1, \ldots, V_k)$ of $V(G)$ with properties that each V_i is an independent set, $k = O(\log |G|)$, and for every i, if $v \in V_i$ then $|N(v) \cap \bigcup_{j>i} V_j| \leq 3\rho$. Give the orientation (u, v) (from u to v) to every edge $\{u, v\}$ with $u \in V_i$ and $v \in V_j$ whenever $i < j$ and define the weight of (u, v) by setting $\bar{\omega}(u, v) := \omega(u)$. Note that the out-degree of this directed graph is at most 3ρ. Let $d := 3 \cdot \rho / \left(1 - \frac{1}{2\rho+1}\right)$ and let D_ρ denote the constant from Lemma 1. Find a (D_ρ, d)-edge-weight partition $(V_1, \ldots, V_k)$ of $(G, \bar{\omega})$. Consider two sets of vertices: B (black) and W (white). Set initially $B := V(G)$ and $W := \emptyset$. For every vertex u, in parallel, if $u \in V_i$ and there is a vertex $v \in V \setminus V_i$ such that (u, v) is an arc, then change the color of u to white. We will end the phase one here. First note the following fact.

Fact 1. *After* HEAVY SUBSET PHASE 1 *all edges with endpoints in different* V_i*'s have at least one endpoint in* W.

In addition, we have the following easy lemma.

Lemma 3. *Let* B *be the set of black vertices in* G *after* HEAVY SUBSET PHASE 1. *We have*

$$\omega(B) \geq \frac{\omega(V(G))}{2\rho + 1}.$$

HEAVY SUBSET PHASE 2. After the execution of phase one $(V_1, \ldots, V_k)$ is a partition of $V(G)$ and every edge with endpoints in different V_i's has at least one endpoint in W. Consequently some of the border vertices of each V_i can be white. A vertex w is called a *troubler* if $w \in W$ and for some $v_i \in V_i \cap B$ and $v_j \in V_j \cap B$ with $i \neq j$, $v_i w v_j$ is a path (of length two) in G. In other word a troubler is a white vertex which is connected by an edge with two black vertices in different V_i's. Clearly only a border vertex can be a troubler. Recall that in phase one we gave an orientation to all edges of G. We shall now define an auxiliary hypergraph. For each troubler w, if w is in V_i and has more than one out-neighbor in $B \cap (V \setminus V_i)$ then consider the hyper-edge f_w consisting of these out-neighbors and let H be the hypergraph $\mathcal{H} := (B, \bigcup\{f_w\})$. Note that as $\mathcal{H}$ is on B, there can be many isolated vertices in $\mathcal{H}$. In addition $|f_w| \leq 3\rho$ for any troubler w as the out-degree is at most 3ρ.

Next task is to find a "heavy" maximal independent set I in $\mathcal{H}$. This is done by consider the graph G' with $V(G') := B$ and the edge set $E(G')$ obtained in the following way. Every troubler w selects two distinct vertices $u, v \in f_w$ and adds the edge $\{u, v\}$ to $E(G')$. Then every edge in G' corresponds to a path uwv in G with $w \in W$ and different paths contain different w's. Therefore G' is a topological minor of G and so $G' \in \mathcal{C}_\rho$. Use Lemma 2 to find a maximal independent set I in G' with $\omega(I) \geq \omega(B)/(2\rho + 1) \geq \omega(V(G))/(2\rho + 1)^2$ and repaint vertices from $B \setminus I$ with the white color. Repeat the process after updating sets f_w and the hypergraph $\mathcal{H}$. Note that in each round of the above procedure, the size of f_w drops by at least one and so after $3\rho - 1$ rounds $|f_w| \leq 1$ for every troubler w. Consequently, the last instance of I from the loop above is an independent set in the initial hypergraph $\mathcal{H}$. In addition we see that I has a large weight.

Fact 2. *Set* I *of vertices in* G *is an independent set in the hypergraph* $\mathcal{H}$ *and* $\omega(I) \geq \omega(V(G))/(2\rho + 1)^{3\rho}$.

Now for every troubler w with $|f_w| = 1$ let u_w denote the vertex in f_w and let $S = \{\{w, u_w\} : |f_w| = 1\}$. Consider the subgraph $\tilde{G}_i$ of $G[V_i]$ induced by edges which have at least one endpoint in $B \cap V_i$. In addition, consider another auxiliary graph G'': For every $i = 1, \ldots, k$, contract every component U of the subgraph $\tilde{G}_i$ to a vertex and let v_U denote the vertex obtained from set U. Put an edge between two vertices $v_U, v_{U'}$ whenever $E_G(U,U') \cap S \neq \emptyset$. In addition, set $\omega(v_U) := \sum_{w \in B \cap U} \omega(w)$. Finally, note that the graph G'' is in $\mathcal{C}_\rho$ and so by Lemma 2, we can find an independent set I in G'' of weight which is at least $\omega(V(G''))/(2\rho+1)$ which by Fact 2 is at least $\omega(V(G))/(2\rho+1)^{3\rho+1}$. Now repaint all vertices of B which are not in a set U with $v_U \in I$ with the white color. Finally consider the subgraph $\tilde{G}$ of G induced by edges from $E(G)$ which have at least one endpoint in B and return the components of $\tilde{G}$. This is the end of phase two.

Lemma 4. *Let B be the set obtained in Phase 2 and let $L_1, L_2, \ldots, L_p$ be the components of $\tilde{G}$ which are returned in the end of phase two. We have:*

- $\omega(B) \geq \omega(V(G))/(2\rho+1)^{3\rho+1}$.
- *For* $i = 1, \ldots, p$, $WDiam_G(L_i) \leq D_\rho + 2$.

Proof. We have already proved the first part. Recall that I denotes the independent set in G'' obtained in HEAVYSUBSET PHASE 2 and let $v_U, v_{U'} \in I$. We will first show that $dist_G(U \cap B, U' \cap B) \geq 3$. To that end assume first that $U \subset V_i$ and $U' \subset V_j$ with $i \neq j$ and suppose that there is a path of length at most two with one endpoint in $U \cap B$ and another in $U' \cap B$. Clearly the path cannot have length one as every edge from $E_G(V_i, V_j)$ has one endpoint in W (Fact 1). Consequently the path has length two and has the form $v_i w v_j$ with $w \in W$. Thus w is a troubler after all of the iterations in $\mathcal{H}$ and either $v_i = u_w$ or $v_j = u_w$ which yields an edge between $v_U, v_{U'}$ in G'' and contradicts the fact that I is independent.

Now suppose that $U, U' \subset V_i$. Then the graphs induced by U, U' are components in $\tilde{G}_i$ and so the distance between $U \cap B$ and $U' \cap B$ in $G[V_i]$ is at least three. In addition, if there is vertex $w \in V_j$ for $j \neq i$ with a neighbor in $U \cap B$ and $U' \cap B$ then $w \in W$ and $|f_w| \geq 2$ which is not possible. Now take a component L of $\tilde{G}$ and let $v_U \in I$ be such that L and U intersect in a black vertex. Also let i be such that $U \subseteq V_i$. If $N_{\tilde{G}}(U \cap B) \subseteq V_i$ then the vertex set of L is a subset of V_i and so the diameter of L is at most D_ρ. Otherwise take a vertex $u \in N_{\tilde{G}}(U \cap B) \setminus V_i$. Then u is white and so u is troubler. As there are no edges in $\tilde{G}$ with both endpoints white, u has at most two neighbors in $\tilde{G}$ and both of them are black. If one of them is in $U' \neq U$ then $dist_G(B \cap U, B \cap U') \leq 2$. Consequently u has one neighbor in $\tilde{G}$ from $B \cap U$. Therefore $V(L)$ is a subset of $N_G(U \cap B) \cup U \subseteq N_G(V_i) \cup V_i$. Since $WDiam_G(G[V_i]) \leq D_\rho$ we have $WDiam_G(L) \leq D_\rho + 2$. □

Now we can describe our main partitioning algorithm.

VERTEX-WEIGHT PARTITION. Given is a vertex-weighted graph (G, ω) with $G \in \mathcal{C}_\rho$ and a positive integer t which can depend on $|G|$. Iterate with i from 1 to t. In the ith iteration, invoke HEAVY SUBSET to find the set of black vertices B and the components $L_1, L_2, \ldots, L_p$ from Lemma 4. Obtain the minor G_i of G_{i-1} by contracting each of the $L_1, L_2, \ldots, L_p$ to a single vertex and set the weight of the vertex obtained from L_i to be equal to the total weight of white vertices in L_i. Let G_t be the graph obtained from G after all t iterations. For each vertex Q

in G_t consider the set V_Q of all vertices in G which have been contracted to Q in the above iterations and return the partition $P = (V_Q | Q \in V(G_t))$.

Lemma 5. *Let $P = (V_1, V_2, \ldots, V_k)$ be the partition of graph $G \in \mathcal{C}_\rho$ obtained by* Vertex-weight Partition *with given parameter t.*

(a) Let D_ρ be the constant from Lemma 1 obtained by setting $d = 3 \cdot \rho / \left(1 - \frac{1}{2\rho+1}\right)$. Then

$$WDiam_G(G[V_i]) \leq (D_\rho + 3)^t.$$

(b) $\omega(\partial(P)) \leq \left(1 - \frac{1}{(2\rho+1)^{3\rho+1}}\right)^t \omega(V(G))$.

Proof. Let G_i denote the graph obtained after the ith iteration of Vertex-Weight Partition and let $G_0 := G$. To show (a), let $diam_i$ be the maximum of $WDiam_G(G[V'])$ over subsets $V' \subset V(G)$ that are contracted to single vertices in G_i. Clearly $diam_0 = 0$ and by Lemma 4 (part two) $diam_{i+1} \leq (D_\rho + 3) \cdot diam_i + D_\rho + 2$. Consequently $diam_t \leq (D_\rho + 3)^t$. To verify part (b), we consider the sequence of partitions $\{P_i\}$ where P_i is the partition of $V(G)$ obtained be creating a partition class for each vertex Q in G_i consisting of vertices that has been contracted to Q in iterations $1, \ldots, i$. Let $\partial_i = \partial(P_i)$ with $\partial_0 := V(G)$. Note that after the coloring of $V(G_i)$ performed by Heavy Subset only white vertices in a component L have neighbors in $V(G_i) \backslash L$. Therefore, $\omega(\partial_i)$ is smaller than or equal to the weight of white vertices in G_i. By definition of the weights in Vertex-Weight Partition, $\omega(V(G_{i+1}))$ is equal to the weight of white vertices in G_i and so $\omega(V(G_i)) \leq \left(1 - \frac{1}{(2\rho+1)^{3\rho+1}}\right)^{i-1} \omega(V(G))$ which in view of Lemma 4 (part one) gives

$$\omega(\partial_i) \leq \left(1 - \frac{1}{(2\rho+1)^{3\rho+1}}\right)^i \omega(V(G)).$$

□

For the next corollary, recall that $ID : V(G) \to \{1, \ldots, m\}$ is a function with $ID(v)$ equal to the identifier of vertex v. Although, as mentioned in the introduction, in the original graph G, $ID(V(G))$ is assumed to be equal to $V(G)$, in applications we will partition auxiliary graphs and it will be important to distinguish between $|G|$ and the order of the auxiliary graph.

Corollary 3

(a) There is a distributed algorithm which given $0 < \epsilon < 1$ finds in a vertex-weighted graph (G, ω) with $G \in \mathcal{C}_\rho$ an (a, b)-vertex-weight partition $P = (V_1, \ldots, V_k)$ with $b \geq 1/\epsilon$ and $a \leq D(\epsilon)$ for some constant $D(\epsilon)$. The algorithm runs in $O(\log |G| \log^* |G|)$ rounds.

(b) There is a distributed algorithm which given a positive integer p finds in a vertex-weighted graph (G, ω) with $G \in \mathcal{C}_\rho$ and $ID(v) \leq m$ for every $v \in V(G)$ an (a, b)-vertex-weight partition $P = (V_1, \ldots, V_k)$ with $b = \log^p m$ and $a = polylog(m)$. The algorithm runs in a pol-logarithmic (in m) number of rounds.

3 Applications

We will now show how to use the vertex-weight partition to design distributed approximations for the minimum-weight dominating set problem and the maximum-weight matching problem.

3.1 Matchings

Let us start with the maximum-weight matching problem. Let $(G, \bar{\omega})$ be an edge-weighted graph with $G \in \mathcal{C}_\rho$. In the algorithm, we first find a subgraph of G and use it to define the vertex-weighted graph (G, ω). Then we apply the partitioning procedure from Corollary 3. The procedure takes a positive integer d as an input.

ApproxMWM. Use the algorithm Decomposition from [3] to find a partition of $V(G)$ into k sets $V_1, \ldots, V_k$ so that each V_i is an independent set, $k = O(\log |G|)$, and for every i, if $v \in V_i$ then $|N(v) \cap \bigcup_{j>i} V_j| \leq 3\rho$. For every vertex v if $v \in V_j$ then v properly colors all edges in $E(\{v\}, \bigcup_{j>i} V_j)$ using colors from $\{1, \ldots, 3\rho\}$. Let F_i be the subgraph of G induced by edges of color i. Then F_i is a forest every component of which has diameter $O(\log |G|)$. Find in each F_i a maximum weight matching N_i and let $Q := \bigcup N_i$. Now for every vertex v set $\omega(v) := \bar{\omega}(e)$ where e is an edge in Q of maximum weight which is incident to v. If no such edge exists set $\omega(v) := 0$. Use the algorithm from Corollary 3 (b) with $p = d+1$ and $m = |G|$ to obtain a vertex-weight partition $P = (V_1, \ldots, V_k)$ of (G, ω). Find a maximum weight matching M_i in each of $(G[V_i], \bar{\omega})$ and return $\bigcup M_i$.

Theorem 4. *Let $(G, \bar{\omega})$ be an edge-weighted graph with $G \in \mathcal{C}_\rho$. There is a distributed algorithm which given a positive integer d finds a matching M in G with*

$$\bar{\omega}(M) \geq \left(1 - \frac{1}{\log^d |G|}\right) \beta(G)$$

where $\beta(G)$ is the weight of a maximum-weight matching in G. The algorithm runs in a poly-logarithmic number of rounds.

Proof. We use ApproxMWM. Note that $\bar{\omega}(Q) \leq 3\rho\beta(G)$ and so the total vertex weight of G satisfies $\omega(V(G)) \leq 3\rho\beta(G)$. Moreover we have that, for every edge $\{u, v\} \in E(G)$, $\bar{\omega}(\{u, v\}) \leq \omega(u) + \omega(v)$. Indeed if $\{u, v\} \in Q$ then this is clear. If $\{u, v\}$ is not in Q and $\{u, v\}$ is in F_i then there exist at most two edges $e_1 = \{u, w\}, e_2 = \{v, z\}$ in M_i such that $\bar{\omega}(\{u, v\}) \leq \bar{\omega}(e_1) + \bar{\omega}(e_2)$. Consequently $\bar{\omega}(\{u, v\}) \leq \omega(u) + \omega(v)$. We have

$$\omega(\partial(P)) \leq \omega(V(G)) / \log^{d+1} |G| \leq \frac{3\rho\beta(G)}{\log^{d+1} |G|} \leq \frac{\beta(G)}{\log^d |G|}. \quad (1)$$

Every matching in G contains two types of edges: edges with both endpoints in some V_i and edges that are incident to $\partial(P)$. The total weight of the latter is at most $\frac{\beta(G)}{\log^d |G|}$ by (1) and so the matching M returned by ApproxMWM satisfies

$$\beta(G) \leq \omega(M) + \frac{\beta(G)}{\log^d |G|}.$$

□

3.2 Dominating and Connected Dominating Sets

In this section we discuss dominating set problems. Due to space limitations, we will not be able to provide full details (the full version is available from authors' web sites). Let (G, ω) be a vertex-weighted graph. Recall that for any $D \subseteq V(G)$ we have $\omega(D) := \sum_{v \in D} \omega(v)$. We will denote by $\gamma(G) = \min \omega(D)$ where the minimum is taken over all dominating sets in graph G. For a vertex v, recall that $N(v)$ denotes the set of neighbors of v and $N[v] := N(v) \cup \{v\}$. Pick one vertex in $N[v]$, $s(v)$, with $\omega(s(v)) := \min_{w \in N[v]} \omega(w)$ and set $\bar{D} := \bigcup\{s(v)\}$.

Lemma 6. *Let (G, ω) be a vertex-weighted graph and let $\bar{D} := \bigcup\{s(v)\}$. Then $\bar{D}$ is a dominating set and $\omega(\bar{D}) \leq |G|\gamma(G)$.*

Our approximation algorithm proceeds in two main phases. First we find a constant approximation of $\gamma(G)$ and next we find a more accurate approximation. For a dominating set D in G, every vertex $v \in V(G) \setminus D$ selects one vertex w in $N(v) \cap D$ and joints group U_w. Let G_D be obtained from G by contracting $U_w \cup \{w\}$ to a single vertex u_w with $\omega(u_w) := \omega(w)$. Then, clearly, $\omega(V(G_D)) = \omega(D)$. Our algorithm ApproxMWDS is given a positive integer d which will be used in the second phase of the procedure.

ApproxMWDS Phase 1. Let $D := \bar{D}$. We iterate with i from 1 to $\log_2 |G|$. In the ith iteration, we consider (G_D, ω) and use Corollary 3 (a) with $\epsilon = 1/2$ to find a vertex-weight partition $(V_1', \ldots, V_k')$ of G_D. This gives a partition P of G by setting V_j to be the the union of U_w's with $u_w \in V_j'$. In each $G[V_j]$ we find a dominating set D_i with $\omega(D_j) = \gamma(G[V_j])$ and set $D := \bigcup_{j=1}^k D_j$. Then we have the following fact.

Lemma 7. *Let (G, ω) be a vertex-weighted graph with $G \in \mathcal{C}_\rho$ and let D be the set obtained by* ApproxMWDS Phase 1. *Then $\omega(D) \leq \gamma(G)/2$.*

ApproxMWDS Phase 2. Let D be the dominating set obtained from ApproxMWDS Phase 1. Consider G_D and use the algorithm from Corollary 3 (b) with $p := d + 1$ and $m = |G|$ to find a vertex-weight partition $(V_1', \ldots, V_k')$ of G_D. This gives a partition $P = (V_1, \ldots, V_k)$ of G as in phase one and we again find an optimal solution in each of $G[V_i]$'s and return the union.

Theorem 5. *Let $\mathcal{C}_\rho$ be a minor-closed family. There exists a distributed algorithm which given a positive integer d finds in vertex-weighted graph (G, ω) with $G \in \mathcal{C}_\rho$ a dominating set D with*

$$\omega(D) \leq \left(1 + \frac{1}{\log^d |G|}\right) \gamma(G).$$

The algorithm runs in a poly-logarithmic number of rounds.

In the case of the weighted connected dominating set problem, one can show that no non-trivial approximation factor can be obtained in $o(|G|)$ rounds even when G is a cycle.

References

1. Awerbuch, B., Goldberg, A.V., Luby, M., Plotkin, S.A.: Network Decomposition and Locality in Distributed Computation. In: Proc. 30th IEEE Symp. on Foundations of Computer Science (FOCS), pp. 364–369. IEEE Computer Society Press, Los Alamitos (1989)
2. Czygrinow, A., Hańćkowiak, M.: Distributed algorithms for weighted problems in sparse graphs. Journal of Discrete Algorithms 4(4), 588–607 (2006)
3. Czygrinow, A., Hańćkowiak, M.: Distributed almost exact approximations for minor-closed families. In: Azar, Y., Erlebach, T. (eds.) ESA 2006. LNCS, vol. 4168, pp. 244–255. Springer, Heidelberg (2006)
4. Czygrinow, A., Hańćkowiak, M., Szymańska, E.: Distributed approximation algorithms in planar graphs, 6th Conference on Algorithms and Complexity (CIAC). In: Calamoneri, T., Finocchi, I., Italiano, G.F. (eds.) CIAC 2006. LNCS, vol. 3998, pp. 296–307. Springer, Heidelberg (2006)
5. Diestel, R.: Graph Theory, 3rd edn. Springer, Heidelberg (2005)
6. Elkin, M.: An Overview of Distributed Approximation. In ACM SIGACT News Distributed Computing Column. 35(4,132), 40–57 (2004)
7. Kuhn, F., Wattenhofer, R.: Constant-Time Distributed Dominating Set Approximation. 22nd ACM Symposium on the Principles of Distributed Computing (PODC), pp. 25–32. ACM Press, New York (2003)
8. Kuhn, F., Moscibroda, T., Wattenhofer, R.: What Cannot Be Computed Locally! In: Proceedings of 23rd ACM Symposium on the Principles of Distributed Computing (PODC), pp. 300–309. ACM Press, New York (2004)
9. Kutten, S., Peleg, D.: Fast distributed construction of k-dominating sets and applications. In: Proceedings of the 14th ACM symposium on Principles of Distributed Computing (PODC), pp. 238–251. ACM Press, New York (1995)
10. Linial, N.: Locality in distributed graph algorithms. SIAM Journal on Computing 21(1), 193–201 (1992)
11. Luby, M.: A simple parallel algorithm for the maximal independent set problem. SIAM Journal on Computing 15(4), 1036–1053 (1986)
12. Nesetril, J., de Mendez, P.O.: Colorings and homomorphisms of minor closed classes. In: Aronov, B., Basu, S., Pach, J., Sharir, M. (eds.) Discrete and Computational Geometry, The Goodman-Pollack Festschrift. Algorithms and Combinatorics, vol. 25, pp. 651–664. Springer, Heidelberg (2003)

A 1-Local 13/9-Competitive Algorithm for Multicoloring Hexagonal Graphs

Francis Y.L. Chin[1,⋆], Yong Zhang[1], and Hong Zhu[2,⋆⋆]

[1] Department of Computer Science, The University of Hong Kong, Hong Kong
{chin,yzhang}@cs.hku.hk
[2] Institute of Theoretical Computing, East China Normal University, China
hzhu@sei.ecnu.edu.cn

Abstract. In the frequency allocation problem, we are given a mobile telephone network, whose geographical coverage area is divided into cells, wherein phone calls are serviced by assigning frequencies to them so that no two calls emanating from the same or neighboring cells are assigned the same frequency. The problem is to use the frequencies efficiently, i.e., minimize the span of frequencies used. The frequency allocation problem can be regarded as a multicoloring problem on a weighted hexagonal graph. In this paper, we give a 1-local 4/3-competitive distributed algorithm for multicoloring a triangle-free hexagonal graph, which is a special case. Based on this result, we then propose a 1-local 13/9-competitive algorithm for multicoloring the (general-case) hexagonal graph, thereby improving the previous 1-local 3/2-competitive algorithm.

1 Introduction

Wireless communication based on Frequency Division Multiplexing (FDM) technology is widely used in the area of mobile computing today. In such FDM networks, a geographic area is divided into small cellular regions or *cells*, each containing one base station. Base stations communicate with each other via a high-speed wired network. Calls between any two clients (even within the same cell) must be established through base stations. When a call arrives, the nearest base station must allocate a frequency from the available spectrum to the call without causing any interference to other calls. In practice, when the same frequency is assigned to two different calls emanating from cells that are geographically close to each other, interference may occur which distorts the radio signals. To avoid interference, the temptation is to use many frequencies. However, spectrum is a scarce resource and efficient utilization of the available spectrum is essential for FDM networks.

The *frequency allocation problem* has been extensively studied [6,9,12,13,1,10,2,3,4]. Both the off-line and online versions of the problem have

⋆ This research was supported by Hong Kong RGC Grant HKU-7113/07E.

⋆⋆ This research was supported in part by National Natural Science Fund (grant no. 60496321).

G. Lin (Ed.): COCOON 2007, LNCS 4598, pp. 526–536, 2007.

been studied. For the off-line problem on cellular networks (where cells are hexagonal regions as shown in Fig. 1 and the calls to be serviced are known *a priori*), McDiarmid and Reed [12] have shown that the problem of minimizing the span of frequencies to satisfy all the call requests is NP-hard, and 4/3-approximation algorithms were given in [12,14].

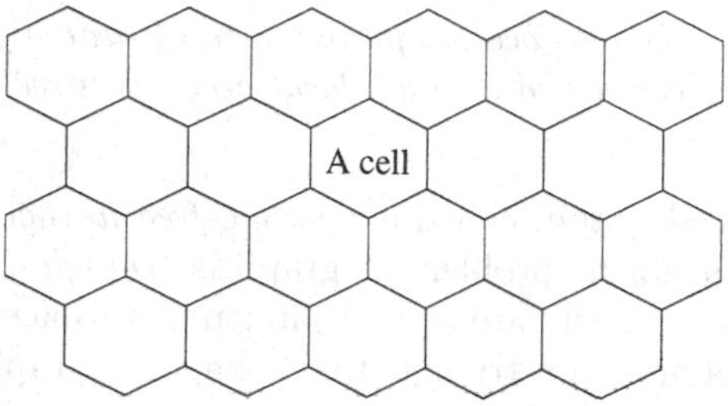

Fig. 1. Example of a cellular network (with hexagonal cells)

For the online version, there are mainly three strategies, which have been introduced: the *fixed allocation assignment* (FAA) [11], the *greedy algorithm* (Greedy) [5] and the *hybrid algorithm* [4]. FAA partitions cells into independent sets which are each assigned a separate set of frequencies. It is easy to see that FAA is 3-competitive as cellular networks are 3-colorable. Greedy assigns the minimum available number (frequency) to a new call so that the call does not interfere with calls of the same or adjacent cells. Caragiannis et al. [5] proved that the competitive ratio of Greedy for FAC is at least 17/7 and at most 2.5. Chan et al. [3] gave a tighter analysis to show that Greedy is 17/7-competitive. Furthermore, Chan et al. [4] proposed a 2-competitive hybrid algorithm, which combines the FAA and Greedy approach and which is optimal since the lower bound of this problem is also 2.

The frequency allocation problem in cellular network can be abstracted to the problem of *multicoloring a weighted hexagonal graph*, in which each vertex has a weight which specifies how many different colors have to be assigned to the vertex. Given the constraint that the same color cannot be assigned to adjacent vertices, the target is minimize the number of assigned colors.

In frequency allocation problem, the size of cellular network is very large, when handling a call request, the computation will be very complex if all information needed. In reality, each server in the cells knows its position before processing the sequence of calls; when satisfying call requests, each server only know its local information, i.e., some information within a fixed distance. In this paper, we focus on *distributed algorithms* for the multicoloring, i.e., each vertex is an independent server, which runs the algorithm to assign multicolors to the vertex based on what is known as *k-local* information. The concept of *k-local* distributed algorithms was introduced by Janssen et al [8], where an algorithm is k-local if the computation at a vertex depends only on the information of the neighboring vertices of at most k distance away (suppose each edge has unit distance). We can also assume that in multicoloring problem, each vertex is also know its position in the graph.

In [8], Janssen et al. proved (the next lemma) that a k-local c-approximate off-line algorithm can be easily converted to a k-local c-competitive online algorithm. Thus, to design a k-local online algorithm, we need only to focus on the off-line problem.

Lemma 1. *[8] Let A be a k-local c-approximate off-line algorithm for multicoloring. Then A can be converted into a k-local c-competitive online algorithm for multicoloring.*

An interesting induced graph, called a *triangle-free hexagonal graph*, has been studied for the multicoloring problem. A graph is *triangle-free* if there are no 3-cliques in the graph, i.e., there are no three mutually-adjacent vertices with positive weights. An example of a triangle-free hexagonal graph is shown in Fig. 2.

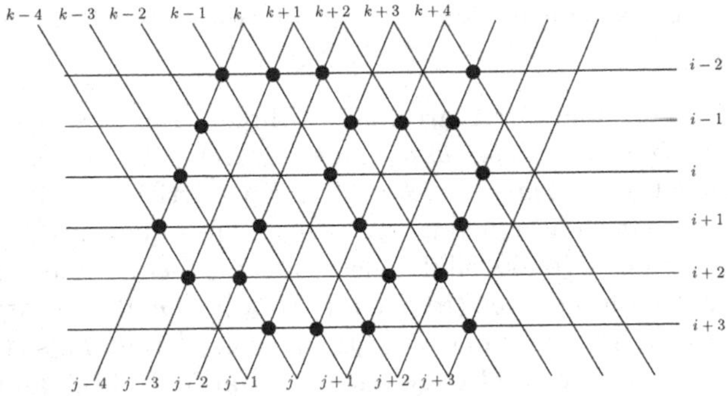

Fig. 2. An example of a triangle-free hexagonal graph, where solid circles are vertices with positive weights

The best known competitive ratios for 0-, 1-, 2- and 4-local distributed algorithms for multicoloring on (general) hexagonal graphs are 3, 3/2, 4/3 and 4/3, respectively [8,17]. It is possible to do better for triangle-free hexagonal graphs. For example, in [16], a 2-local 5/4 competitive algorithm was given, and an inductive proof for the 7/6 ratio was reported in [7].

In this paper, we first give a 1-local 4/3-competitive algorithm for multicoloring a triangle-free hexagonal graph in Section 3. Based on this result, we then propose, in Section 4, a 1-local 13/9-competitive algorithm for the multicoloring problem in hexagonal graph, which improves the previous 3/2-competitive result. Section 2 introduces some preliminary terminology to be used in this paper, and Section 5 concludes.

2 Preliminary Terminology

Suppose the hexagonal graph has been three-colored with each vertex colored with one of three *base colors*, without loss of generality, say *Red*, *Green* and

Blue. We assume a transitive order on these three base colors: namely, *Red* < *Green* < *Blue*.

We use a 3-coordinate system to represent each vertex. In particular, referring to the lines shown in Fig. 2, each vertex can be represented by coordinate (i, j, k) where i is the coordinate for the horizontal line, j for the up-sloping line and k for the down-sloping line. (In fact, two coordinates are enough to represent vertices since coordinate k is redundant, but we find it convenient to refer to three coordinates.) For example, a vertex with coordinate (i, j, k) and its six neighboring vertices, denoted by UL, L, DL, UR, R and DR, are represented as shown in Fig. 3.

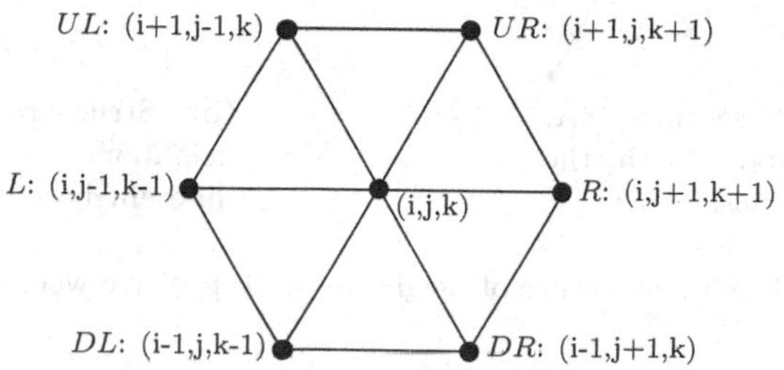

Fig. 3. Coordinates of a vertex (i, j, k) and its neighboring vertices

Next, we use the definition in [17] to define the *parity* of a vertex with respect to its various neighbors. We say that the parity of a vertex v with coordinate (i, j, k) is:

1. ***odd*** (alternatively, ***even***) with respect to its L or R neighbor if $j \equiv 1 \mod 2$ (correspondingly, $j \equiv 0 \mod 2$);
2. ***odd*** (alternatively, ***even***) with respect to its UL or DR neighbor if $i \equiv 1 \mod 2$ (correspondingly, $i \equiv 0 \mod 2$);
3. ***odd*** (alternatively, ***even***) with respect to its DL or UR neighbor if $k \equiv 1 \mod 2$ (correspondingly, $k \equiv 0 \mod 2$).

Let w_v be the weight of vertex v, which corresponds to the number of colors needed to multicolor v. After the multicoloring assignment, each vertex v will be assigned a *set* F_v *of colors*, such that $F_v \subset Z^+$ and $|F_v| = w_v$, where, for any two adjacent vertices u and v, $F_u \cap F_v = \phi$.

3 Multicoloring in Triangle-Free Hexagonal Graphs

In this section, we shall study the problem of multicoloring a special type of hexagonal graph. Finding a good solution for this problem will lead to an algorithm for finding good solutions for general hexagonal graphs.

A graph is *triangle-free* if no three mutually-adjacent vertices have positive weights. For a given vertex u with positive weight w_u, from this definition of triangle-free graph, only two possible configurations may exist for the structure of neighbors with positive weights, which are shown in Fig. 4. There exist a simple structure in triangle-free graph, i.e., a vertex has only one neighbor, we can regard this structure as the case in Fig. 4(b).

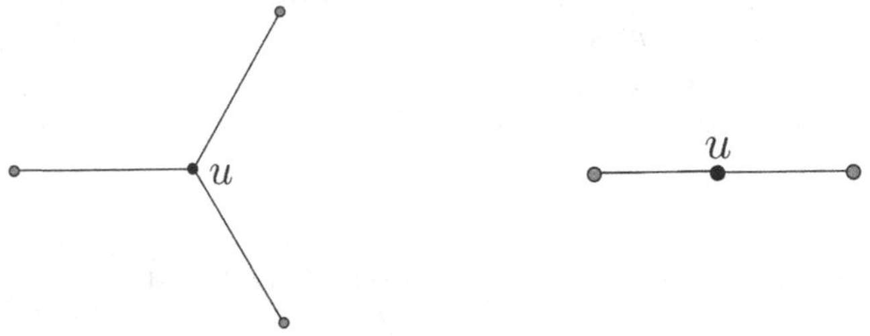

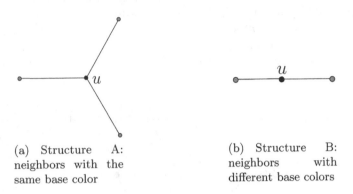

(a) Structure A: neighbors with the same base color

(b) Structure B: neighbors with different base colors

Fig. 4. Structure of neighbors with positive weight

Consider vertex u with positive weight w_u. Compute $c_u = w_u + \max\{w_v \mid v$ is u's neighbor$\}$. c_u is the weight of the maximum 2-clique adjacent to u, which also gives the minimum number of colors needed for multicoloring a triangle-free hexagonal graph. From the definition of c_u, any feasible coloring of vertex u and its neighbors requires at least c_u colors.

Let $d_u = \lceil c_u/3 \rceil$. For each vertex u, we define four color sets, each of size d_u:

1. $colorset_u(Red) = \{j \in \{1, \ldots, 4d_u\} \mid j = 1 \mod 4\}$,
2. $colorset_u(Green) = \{j \in \{1, \ldots, 4d_u\} \mid j = 2 \mod 4\}$,
3. $colorset_u(Blue) = \{j \in \{1, \ldots, 4d_u\} \mid j = 3 \mod 4\}$, and
4. $extraset_u = \{j \in \{1, \ldots, 4d_u\} \mid j = 0 \mod 4\}$.

We will give a strategy to multicolor any vertex u with weight w_u by assigning w_u colors from the above four sets so that no adjacent vertices are assigned the same color. The assignment strategy assigns multicolors to u according to its base color and neighboring structure and can be described as follows.

Assume vertex u with base color X has neighboring structure A, i.e. all its neighbors have the same base color $Y \neq X$. Let the third base color be Z where $Z \neq X$ and $Z \neq Y$. In this case, our strategy would be to assign multicolors to vertex u first from $colorset_u(X)$, then $colorset_u(Z)$ and finally $colorset_u(Y)$.

On the other hand, if vertex u with base color X has neighboring structure B, then all three base colors will be used by u and its neighbors. The strategy will first assign multicolors from $colorset_u(X)$, then from the extra color set $extraset_u$ and finally from $colorset_u(Y)$ where base color $Y > Z \neq X$.

Note that in both cases, the colors in each color set may be assigned either from bottom to top or from top to bottom, depending on the base color or the parity of the vertex so as to avoid any conflicts.

THE STRATEGY

1. If vertex u has no neighbors, just assign w_u colors from 1 to w_u.
2. If vertex u with base color X has neighboring structure A (Fig. 4(a)), let Y be the base color of u's neighbors and Z be the other third color. Assign w_u multicolors to vertex u as follows:
   (a) Assign colors from $colorset_u(X)$ in bottom-to-top order.
   (b) If not enough, assign colors from $colorset_u(Z)$ in bottom-to-top order if $X < Y$; top-to-bottom otherwise.
   (c) If still not enough, assign colors from $colorset_u(Y)$ in top-to-bottom order.
3. If vertex u with base color X has neighboring structure B (Fig. 4(b)), let Y and Z be the base colors of the left neighbor and the right neighbor, respectively. Without loss of generality, assume $Y > Z$. Assign w_u multicolors to vertex u as follows:
   (a) Assign colors from $colorset_u(X)$ in bottom-to-top order.
   (b) If not enough, assign colors from $extraset_u$ in bottom-to-top order if u is *odd* with respect to its left or right neighbor; top-to-bottom order otherwise.
   (c) If still not enough, assign colors from $colorset_u(Y)$ in top-to-bottom order.

Theorem 1. *The above strategy is 1-local and can solve the multicoloring problem in triangle-free hexagonal graphs with performance ratio 4/3.*

Proof. From the description of the strategy, it is clear that the colors assigned to any vertex depend only on neighboring information within distance 1, and thus, the strategy is 1-local.

To prove that the above strategy solves the multicoloring problem, we must prove that the colors assigned to any two adjacent vertices u and v are all different. As it turns out, we need to analyze the three structures shown in Fig. 5. X and Y denote the two respective different base colors of u and v. It is easy to see that different kinds of color sets of u and v have no common colors. For example, $colorset_u(Red) \bigcap extraset_v = \emptyset$.

From the definition of c_u, we have $c_u \geq w_u + w_v$ and, since $d_u = \lceil c_u/3 \rceil$, $c_u \leq 3d_u$.

For Case A, the strategy would assign colors to u from $colorset_u(X)$, then $colorset_u(Z)$ and then $colorset_u(Y)$, and would assign colors to v from $colorset_v(Y)$, then $colorset_v(Z)$ and then $colorset_v(X)$. There are three subcases to consider:

(A-1) Case where u is assigned colors from $colorset_u(X)$ and v is assigned colors from $colorset_v(X)$ after exhausting all colors in $colorset_v(Y)$ and $colorset_v(Z)$. Then, the weight w_v of vertex v should be very large since all the colors in $colorset_v(Y)$ and $colorset_v(Z)$ must be used up. Since $w_u + w_v \leq c_u \leq 3d_u$, $w_u + w_v \leq c_v \leq 3d_v$, and since u and v use $colorset(X) = colorset_u(X) \cup colorset_v(X)$ from opposite directions (depending on whether $X < Y$ or otherwise), u and v will not be assigned the same color.

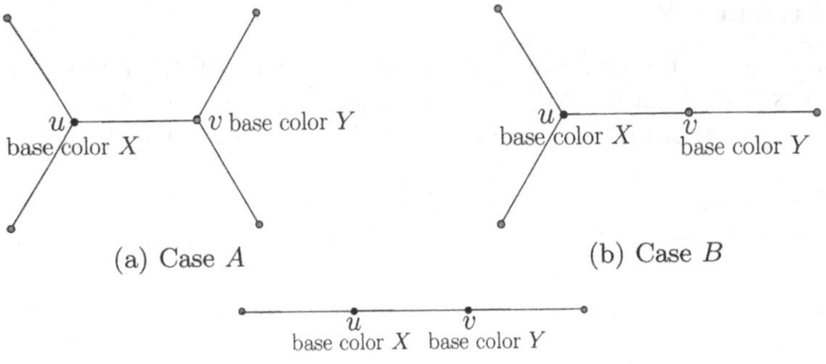

(a) Case A (b) Case B (c) Case C

Fig. 5. The local structure of vertices u and v

(A-2) Case where u is assigned colors from $colorset_u(Z)$ and v is assigned colors from $colorset_v(Z)$. Then, all the colors in $colorset_u(X)$ are assigned to u and all the colors in $colorset_v(Y)$ are assigned to v. Since $w_u + w_v \leq 3d_u, 3d_v$, and since u and v use $colorset(Z) = colorset_u(Z) \cup colorset_v(Z)$ from opposite directions, u and v will not be assigned the same color.

(A-3) Case where u is assigned colors from $colorset_u(Y)$ and v is assigned colors from $colorset_v(Y)$. By similar analysis as in case (A-1), we can say that u and v will not be assigned the same color.

In Case B, the strategy would assign colors to u from $colorset_u(X)$, then $colorset_u(Z)$ and then $colorset_u(Y)$. Also, the strategy would assign colors to v from $colorset_v(Y)$, then $extraset_v$ and then $colorset_v(X)$ or $colorset_v(Z)$ depending whether $X < Z$ or otherwise. Without loss of generality, we assume $colorset_v(X)$ is used. There are two subcases to consider:

(B-1) Case where u is assigned colors from $colorset_u(X)$ and v is assigned colors from $colorset_v(X)$ after exhausting all colors in $colorset_v(Y)$ and $extraset_v$. This means the weight w_v of v is very large. Since $w_u + w_v \leq 3d_u, 3d_v$, and since u and v use $colorset(X) = colorset_u(X) \cup colorset_v(X)$ from opposite directions, u and v will not be assigned the same color.

(B-2) Case where u is assigned colors from $colorset_u(Y)$ and v is assigned colors from $colorset_v(Y)$. By similar analysis as in case (B-1), we can say that u and v will not be assigned the same color.

In Case C, the strategy would assign colors to u from $colorset_u(X)$, then $extraset_u$ and then $colorset_u(Y)$, and would assign colors to v from $colorset_v(Y)$, then $extraset_v$ and then $colorset_v(X)$. There are three subcases to consider:

(C-1) Case where u is assigned colors from $colorset_u(X)$ and v is assigned colors from $colorset_v(X)$ after exhausting all colors in $colorset_v(Y)$ and $extraset_v$. This means the weight w_v of v is very large. Since

$w_u + w_v \leq 3d_u, 3d_v$, and since u and v use $colorset(X) = colorset_u(X) \cup colorset_v(X)$ from opposite directions, u and v will not be assigned the same color.

(C-2) Case where u is assigned colors from $extraset_u$ and v is assigned colors from $extraset_v$. This means all the colors in $colorset_u(X)$ and $colorset_v(Y)$ have been assigned to u and v, respectively. Since $w_u+w_v \leq 3d_u, 3d_v$, and since u and v use $extraset = extraset_u \cup extraset_v$ from opposite directions (the parities of u and v are different), u and v will not be assigned the same color.

(C-3) Case where u is assigned colors from $colorset_u(Y)$ and v is assigned colors from $colorset_v(Y)$. By similar analysis as in case (C-1), we can say that u and v will not be assigned the same color.

For the whole triangle-free hexagonal graph, the maximal weight clique (2-clique) $c = \max_u\{c_u\}$ is a lower bound on the optimal value, and our algorithm uses at most $4\max_u\{\lceil c_u/3 \rceil\}$ colors. Thus, the above strategy has a performance ratio of 4/3. □

From Theorem 1, we can easily have a 1-local 4/3-competitive online algorithm for frequency allocation in triangle-free cellular networks.

4 Multicoloring in Hexagonal Graphs

In this section, we consider multicoloring hexagonal graphs. Our strategy works in two stages. In the first stage, each vertex assigns colors using local information on the weights of this vertex and its neighboring vertices. After the first stage, some vertices may be unsatisfied, i.e. not all of the necessary colors have been assigned, and the unsatisfied vertices, along with the edges connecting them, form a triangle-free graph. Applying the algorithm in the previous section, each vertex can be assigned colors, to satisfy all the remaining unsatisfied vertices, by using 1-local information. Combining these two stages, we have a 1-local algorithm for multicoloring hexagonal graphs.

We now describe the first stage, which is similar to the first stage in [12]. In [12], the algorithm needs to have the global information about the maximum weights of "all" 3-cliques in the graph (stage 1) so as to have an acyclic graph of the remaining unsatisfied vertices (for stage 2). As for an algorithm which is 1-local, only the maximal weights of the local 3-cliques will be available (stage 1) and a triangle-free hexagonal graph (which can be cyclic) will result (for stage 2). Consider vertex u with base color X. Let C_u be the maximal weights among the 3-cliques including u, and let $k_u = \lceil C_u/3 \rceil$. For the three base colors *Red*, *Green* and *Blue*, we define a cyclic order among them as *Red* $\rightarrow$ *Green*, *Green* $\rightarrow$ *Blue* and *Blue* $\rightarrow$ *Red*. If $X \rightarrow Y$, let m_u be the maximal weight of the neighboring vertices with color Y. We define color sets: $colorset_u(Red) = \{j \in \{1, \ldots, 3k_u\} \mid j \equiv 1 \mod 3\}$, $colorset_u(Green) = \{j \in \{1, \ldots, 3k_u\} \mid j \equiv 2 \mod 3\}$ and $colorset_u(Blue) = \{j \in \{1, \ldots, 3k_u\} \mid j \equiv 0 \mod 3\}$. In the first stage, vertex u with base color X and weight w_u is assigned colors from these sets using the strategy described as follows:

1. Vertex u is assigned colors from $colorset_u(X)$ in bottom-to-top order.
2. If not enough and $m_u < k_u$, vertex u is assigned the upper $\min\{k_u - m_u, w_u - k_u\}$ colors from $colorset_u(Y)$.

After the first stage, each vertex has been assigned with some colors. The remaining graph contains only those vertices whose calls have not been totally satisfied, i.e., the number of assigned colors in vertex u is less than its weight.

Lemma 2. *The remaining graph is triangle-free, i.e., contains no 3-clique.*

Proof. If some vertex u is still unsatisfied, it must be that $w_u > \max\{k_u, 2k_u - m_u\}$. Thus, the remaining unsatisfied weight in vertex u is $w'_u = w_u - \max\{k_u, 2k_u - m_u\}$. For any three mutually-adjacent vertices (3-clique) u, v and t, since $\min\{C_u, C_v, C_t\} \geq w_u + w_v + w_t$, $\min\{k_u, k_v, k_t\} \geq \min\{w_u, w_v, w_t\}$ and at most two of $\{w'_u, w'_v, w'_t\}$ are strictly positive, at least one of the vertices (u, v and t) has all its required colors totally assigned in the first stage. Therefore, the remaining graph contains no 3-clique, i.e., is a triangle-free hexagonal graph. □

Lemma 3. *The total weight of two neighboring vertices u and v in the remaining graph is at most* $\max\{C_u, C_v\}/3$.

Proof. For the remaining unsatisfied vertices, since $w'_u = w_u - \max\{k_u, 2k_u - m_u\}$, we have $w'_u \leq w_u - (2k_u - m_u) = w_u + m_u - 2k_u \leq C_u - 2k_u$. For any two adjacent unsatisfied vertices u and v, we can also get $w'_u + w'_v \leq w_u - k_u + w_v - k_v \leq C_v/3$. If $C_u \geq C_v$, which implies $k_u \geq k_v$, then we have $w'_u + w'_v \leq C_v - 2k_v$ as $C_v \geq w_u + w_v$. Similarly, if $C_u \leq C_v$, which implies $k_u \leq k_v$, then we have $w'_u + w'_v \leq C_u - 2k_u \leq C_u/3$. Thus, the total remaining weight of any two adjacent unsatisfied vertices is at most $\max\{C_u, C_v\}/3$. □

From Lemma 2, the remaining graph is triangle-free, so in the second stage, we can use the algorithm in Section 3 to process the remaining unsatisfied vertices. Each vertex gets the remaining weight information from its adjacent vertices and the total number of colors used in this stage is at most $4\max_u \lceil \frac{C_u}{9} \rceil$ (Theorem 1 and Lemma 3).

Combining these two stages, we use at most $\max_u(3k_u + 4\lceil \frac{C_u}{9} \rceil) \leq \max_u(3\lceil \frac{C_u}{3} \rceil + 4\lceil \frac{C_u}{9} \rceil) \leq \frac{13}{9}C_u + 7$ colors. Since C_u is a lower bound on the optimal solution, the performance ratio for our strategy is 13/9.

Thus, we have the following theorem.

Theorem 2. *For the multicoloring problem in hexagonal graphs, a 1-local 13/9-competitive algorithm can be achieved.*

5 Conclusion

We have given a 13/9-approximation algorithm for multicoloring hexagonal graphs. This implies a 13/9-competitive solution for the online frequency allocation problem, which involves servicing calls in each cell in a cellular network. The distributed algorithm is practical in the sense that frequency allocation can be done based on information about its neighbors and itself only. We note that, in fact, when calls are requested or released in a cell, a constant number of frequencies might have to be reassigned so as to actually achieve the 13/9-competitive bound.

Acknowledgements. The authors thank Dr. Bethany M.Y. Chan for her efforts in making this paper more readable.

References

1. Aardal, K.I., van Hoesel, S.P.M., Koster, A.M.C.A., Mannino, C., Sassano, A.: Models and solution techniques for frequency assignment problems. Quarterly Journal of the Belgian, French and Italian Operations Research Societies (4OR) 1(4), 261–317 (2003)
2. Chan, W.-T., Chin, F.Y.L., Ye, D., Zhang, Y., Zhu, H.: Frequency Allocation Problem for Linear Cellular Networks. In: Asano, T. (ed.) ISAAC 2006. LNCS, vol. 4288, pp. 61–70. Springer, Heidelberg (2006)
3. Chan, W.-T., Chin, F.Y.L., Ye, D., Zhang, Y., Zhu, H.: Greedy Online Frequency Allocation in Cellular Networks. Information Processing Letters 102, 55–61 (2007)
4. Chan, W.-T., Chin, F.Y.L., Ye, D., Zhang, Y.: Online Frequency Allocation in Cellular Networks. To appear in Proc. of the 19th ACM Symposium on Parallelism in Algorithms and Architectures (SPAA '07)
5. Caragiannis, I., Kaklamanis, C., Papaioannou, E.: Efficient on-line frequency allocation and call control in cellular networks. Theory Comput. Syst. 35(5), 521–543 (2002) A preliminary version of the paper appeared in SPAA 2000
6. Hale, W.: Frequency assignment: Theory and applications. Proceedings of the IEEE 68(12), 1497–1514 (1980)
7. Havet, F.: Channel assignment and multicoloring of the induced subgraphs of the triangular lattice. Discrete Math. 233, 219C231 (2001)
8. Janssen, J., Krizanc, D., Narayanan, L., Shende, S.M.: Distributed online frequency assignment in cellular networks. J. Algorithms 36(2), 119–151 (2000)
9. Jaumard, B., Marcotte, O., Meyer, C.: Mathematical models and exact methods for channel assignment in cellular networks. In: Sansò, B., Soriano, P. (eds.) Telecommunications Network Planning, pp. 239–255. Kluwer Academic Publishers, Dordrecht (1999)
10. Katzela, I., Naghshineh, M.: Channel assignment schemes for cellular mobile telecommunication systems: A comprehensive survey. IEEE Personal Communications 3(3), 10–31 (1996)
11. MacDonald, V.: Advanced mobile phone service: The cellular concept. Bell Systems Technical Journal, vol. 58(1) (1979)
12. McDiarmid, C., Reed, B.A.: Channel assignment and weighted coloring. Networks 36(2), 114–117 (2000)
13. Narayanan, L.: Channel assignment and graph multicoloring. In: Stojmenović, I. (ed.) Handbook of Wireless Networks and Mobile Computing, pp. 71–94. John Wiley & Sons, Chichester (2002)
14. Narayanan, L., Shende, S.M.: Static frequency assignment in cellular networks. Algorithmica 29(3), 396–409 (2001)

15. Narayanan, L., Tang, Y.: Worst-case analysis of a dynamic channel assignment strategy. Discrete Applied Mathematics 140(1-3), 115–141 (2004)
16. Sparl, P., Zerovnik, J.: 2-local 5/4-competitive algorithm for multicoloring trianglr-free hexagonal graphs. Information Processing Letters 90, 239–246 (2004)
17. Sparl, P., Zerovnik, J.: 2-local 4/3-competitive algorithm for multicoloring hexagonal graphs. J. Algorithms 55(1), 29–41 (2005)

Improved Algorithms for Weighted and Unweighted Set Splitting Problems*

Jianer Chen and Songjian Lu

Department of Computer Science
Texas A&M University
College Station, TX 77843-3112, USA
{chen,sjlu}@cs.tamu.edu

Abstract. In this paper, we study parameterized algorithms for the SET SPLITTING problem, for both weighted and unweighted versions. First, we develop a new and effective technique based on a probabilistic method that allows us to develop a simpler and more efficient (deterministic) kernelization algorithm for the unweighted SET SPLITTING problem. We then propose a randomized algorithm for the weighted SET SPLITTING problem that is based on a new subset partition technique and has its running time bounded by $O^*(2^k)$, which even significantly improves the previously known upper bound for the unweigthed SET SPLITTING problem. We also show that our algorithm can be de-randomized, thus derive the first fixed parameter tractable algorithm for the weighted SET SPLITTING problem.

1 Introduction

Let X be a set. A *partition* of X is a pair of subsets (X_1, X_2) of X such that $X_1 \cup X_2 = X$ and $X_1 \cap X_2 = \emptyset$. We say that a subset S of X is *split* by the partition (X_1, X_2) of X if S intersects with both X_1 and X_2. The SET SPLITTING problem is defined as follows: given a collection $\mathcal{F}$ of subsets of a ground set X, construct a partition of X that maximizes the number of split subsets in $\mathcal{F}$.

A more generalized version of the SET SPLITTING problem is the *weighted* SET SPLITTING problem, in which each subset in the collection $\mathcal{F}$ is associated with a weight that is a real number, and the objective is to construct a partition of the ground set that maximizes the sum of the weights of the split subsets.

The SET SPLITTING problem is an important NP-complete problem [10]. A number of well-known NP-complete problems are related to the SET SPLITTING problem, including the HITTING SET problem that is to find a small subset of the ground X that intersects all subsets in a collection $\mathcal{F}$, and the SET PACKING problem that is to find a large sub-collection $\mathcal{F}'$ of the collection $\mathcal{F}$ of subsets such that the subsets in $\mathcal{F}'$ are all pairwise disjoint.

In terms of approximability, the SET SPLITTING problem is APX-complete [3]. Andersson and Engebretsen [2] gave an approximation algorithm for the

* This work was supported in part by the National Science Foundation under the Grants CCR-0311590 and CCF-0430683.

G. Lin (Ed.): COCOON 2007, LNCS 4598, pp. 537–547, 2007.

problem that has an approximation ratio bounded by 0.724. Zhang and Ling [16] presented an improved approximation algorithm of approximation ratio 0.7499 for the problem. Better approximation algorithms can be achieved if we further restrict the number of elements in each subset in the input [11,16,17,18].

On the need of applications, such as the analysis of micro-array data, people have studied the parameterized version of the SET SPLITTING problem, by associating each input with a parameter k, which is in general small. Formally, an instance of a *parameterized weighted* SET SPLITTING problem consists of a collection $\mathcal{F}$ of subsets of a ground set X, in which each subset has a weight, and a parameter k. The objective is to construct a partition of the ground set X that maximizes the weight sum of k split subsets in $\mathcal{F}$, or report that no partition of X can split k subsets in $\mathcal{F}$. A restricted version of the parameterized weighted SET SPLITTING problem is the *parameterized unweighted* SET SPLITTING problem in which all subsets in the collection $\mathcal{F}$ has weight 1.

In this paper, we are mainly concerned with *fixed parameter tractable algorithms* [8] for the parameterized SET SPLITTING problems, where the algorithms run in time $f(k)n^{O(1)}$, with $f(k)$ being a function that only depends on the parameter k. In particular, for small values of the parameter k, a fixed parameter tractable algorithm for the parameterized SET SPLITTING problem may become effective in practice. Since we will be only considering parameterized versions of the SET SPLITTING problems, we will drop the word "parameterized" when we refer to the problems.

The unweighted SET SPLITTING problem has been studied in the literature. Dehne, Fellows, and Rosamond [6] were the first to study the problem and provided a fixed parameter tractable algorithm of running time $O^*(72^k)$ for the problem[1]. In the same paper, the authors also proved that the unweighted SET SPLITTING problem has a kernel of size bounded by $2k$: that is, there is a polynomial time algorithm that on a given instance $(X, \mathcal{F}, k)$ of unweighted SET SPLITTING, produces another instance $(X', \mathcal{F}', k')$ for the problem such that $|X'| \leq |X|$, $|\mathcal{F}'| < 2k$, $k' \leq k$, and that the set X has a partition that splits k subsets in the collection $\mathcal{F}$ if and only if the set X' has a partition that splits k' subsets in the collection $\mathcal{F}'$. Later, Dehne, Fellows, Rosamond, and Shaw [7] developed an improved algorithm of running time $O^*(8^k)$ for the problem. The improved algorithm was obtained by combining the recently developed techniques *greedy localization* and *modeled crown reduction* in the study of parameterized algorithms. The current best algorithm for the unweighted SET SPLITTING problem is developed by Lokshtanov and Sloper [14], where they used Chen and Kanj's result for MAX-SAT problem [4] and reached a time complexity of $O^*(2.65^k)$.

No fixed parameter tractable algorithms have been known for the weighted SET SPLITTING problem. In fact, none of the techniques developed previously for the unweighted SET SPLITTING problem, such as those in [6,7,14], seems to be extendable to the weighted case.

---

[1] Following the recent convention, by the notation $O^*(c^k)$, where $c > 1$ is a constant, we refer to a function of order $O(c^k n^{O(1)} g(k))$, where $g(k) = c^{o(k)}$.

In this paper, we develop new techniques in dealing with the SET SPLITTING problems. First, we develop a new and effective technique based on a probabilistic method that allows us to develop a (deterministic) kernelization algorithm for the unweighted SET SPLITTING problem. The new kernelization algorithm is simpler and more efficient compared to the previous algorithm given in [6]. We then propose a randomized algorithm for the weighted SET SPLITTING problem that is based on a new subset partition technique and has its running time bounded by $O^*(2^k)$. This even significantly improves the previous best upper bound $O^*(2.65^k)$ given in [14] for the (simpler) unweighted SET SPLITTING problem. We also show that, using the subset partition family proposed by Naor, Schulman, and Srinivasan [15], we can de-randomize our randomized algorithm, which gives the first fixed parameter tractable algorithm for the weighted SET SPLITTING problem.

2 A New Kernelization Algorithm for the Unweighted SET SPLITTING Problem

In this section, we focus on the unweighted SET SPLITTING problem. By a *kernelization algorithm* for unweighted SET SPLITTING, we mean a polynomial time algorithm that, on an instance $(X, \mathcal{F}, k)$ of unweighted SET SPLITTING, produces another instance $(X', \mathcal{F}', k')$ for the problem such that the size of the instance $(X', \mathcal{F}', k')$ only depends on the parameter k. The instance $(X', \mathcal{F}', k')$ will be called a *kernel* for the instance $(X, \mathcal{F}, k)$. Dehne, Fellows, Rosamond and Shaw [6] developed a kernelization algorithm by which the kernel $(X', \mathcal{F}', k')$ satisfies the conditions $|\mathcal{F}'| < 2k$ and that each subset in $\mathcal{F}'$ has at most $2k$ elements. Lokshtanov and Sloper [14] used the crown decomposition and obtained a kernel that both $|\mathcal{F}'|$ and $|X'|$ are less than $2k$. We introduce a new method to find the kernel for the unweighted SET SPLITTING problem. What is interesting in our method is that we use a probabilistic method to derive a deterministic kernelization algorithm. In particular, our method is simpler, has lower time complexity, and can also get a better kernel in term of the number of subsets in $\mathcal{F}'$ if there are many subsets that have more than two elements.

Lemma 1. *Given an instance $(X, \mathcal{F}, k)$ of the unweighted* SET SPLITTING *problem, let m_1 be the number of subsets in $\mathcal{F}$ that have only one element. If $|\mathcal{F}| - m_1 \geq 2k$, then a partition of X exists that splits at least k subsets in $\mathcal{F}$.*

Proof. For each subset $S \in \mathcal{F}$, if S has at least two elements, we pick any two elements from S. Let V be the set of all these elements picked from the subsets in $\mathcal{F}$ that have more than one element. Note that for two subsets S_1 and S_2 in $\mathcal{F}$ that have more than one element, the two elements in S_1 and the two elements in S_2 may not be disjoint.

Suppose $|V| = t$. We randomly partition V into two subsets V_l and V_r, such that $|V_l| = \lfloor t/2 \rfloor$, $|V_r| = t - |V_l|$, i.e. we randomly pick $\lfloor t/2 \rfloor$ elements of V into V_l and let the remaining $t - \lfloor t/2 \rfloor$ elements of V in V_r. Thus, for any subset S in $\mathcal{F}$:

$$Pr(\text{S is split}) \begin{cases} \geq \frac{2\binom{t-2}{\lfloor t/2 \rfloor - 1}}{\binom{t}{\lfloor t/2 \rfloor}} = \frac{2\lfloor t/2 \rfloor (t - \lfloor t/2 \rfloor)}{t(t-1)} > \frac{1}{2}, & \text{if } S \text{ has more than one element} \\ =0, & \text{otherwise.} \end{cases}$$

If we let:

$$X_S = \begin{cases} 1, & \text{if } S \text{ is split,} \\ 0, & \text{otherwise,} \end{cases}$$

then the expectation of the number of split subsets in $\mathcal{F}$ satisfies

$$E\left(\sum_{S \in \mathcal{F}} X_S\right) \geq \frac{1}{2}(|\mathcal{F}| - m_1),$$

Therefore, if $|\mathcal{F}| - m_1 \geq 2k$, there must exist a partition of the ground set X such that the number of split subsets in $\mathcal{F}$ is at least k. This completes the proof of the lemma. □

The following lemma shows that we can directly include subsets of at least k elements in our split subsets while we are solving the unweighted SET SPLITTING problem.

Lemma 2. *Let $(X, \mathcal{F}, k)$ be an instance of the unweighted* SET SPLITTING *problem, and let S be a subset in $\mathcal{F}$ that contains at least k elements. Then there is a partition of X that splits k subsets in $\mathcal{F}$ if and only if there is a partition of X that splits $k-1$ subsets in $\mathcal{F} - \{S\}$.*

Proof. Suppose that there is a partition (X_l, X_r) of X that splits k subsets in $\mathcal{F}$. Then it is obvious that (X_l, X_r) splits (at least) $k-1$ subsets in $\mathcal{F} - \{S\}$.

On the other hand, suppose that there is a partition (X_l, X_r) of X that splits $k-1$ subsets $S_1, \ldots, S_{k-1}$ in $\mathcal{F} - \{S\}$. Let $l_i, r_i \in S_i$, $l_i \in X_l$, and $r_i \in X_r$, for all $1 \leq i \leq k-1$. Since S has at least k elements, there are at least two elements l and r in S such that $l \notin \{r_1, \ldots, r_{k-1}\}$ and $r \notin \{l_1, \ldots, l_{k-1}\}$. Therefore, if we modify the partition (X_l, X_r) to enforce l in X_l and r in X_r (note that this modification still keeps l_i in X_l and r_i in X_r for all $1 \leq i \leq k-1$), then the new partition splits the subset S, as well as the $k-1$ subsets $S_1, \ldots, S_{k-1}$ in $\mathcal{F} - \{S\}$. In consequence, the new partition of the ground set X splits (at least) k subsets in the collection $\mathcal{F}$. □

Now we are ready to state our first kernelization result.

Theorem 1. *Given an instance $(X, \mathcal{F}, k)$ of the unweighted* SET SPLITTING *problem, we can construct in time $O(N + 2k^2)$, where N is the input size in terms of $(X, \mathcal{F}, k)$, a kernel $(X', \mathcal{F}', k')$ such that $|\mathcal{F}'| < 2k$, $k' \leq k$, $|X'| < 2k^2$, and that each subset in $\mathcal{F}'$ has at most $k-1$ elements.*

Proof. We use the following procedure to find the kernel: given an instance $(X, \mathcal{F}, k)$ of the unweighted SET SPLITTING problem, we delete each subset that has only one element, also delete each subset that has at least k elements and decrease k by 1. Let the resulting instance be $(X', \mathcal{F}', k')$. If $|\mathcal{F}| \geq 2k$, then by Lemma 1 (note that $\mathcal{F}'$ contains no subsets of one element), we know that the given instance is a "Yes" instance; otherwise we have $|\mathcal{F}'| < 2k$, $k' \leq k$. Moreover, since we have removed all subsets of at least k elements, every subset in $\mathcal{F}'$ contains at most $k-1$ elements. In consequence, $|X'| < 2k^2$. The time to find this kernel is obviously $O(N + 2k^2)$. □

Theorem 1 improves the time complexity of the kernelization algorithm given in [6], which takes time $O(N + n^4)$, as well as that given in [14], which takes time $O(N + n^2)$, where n is a number satisfying $|\mathcal{F}| = O(n)$ and $|X| = O(n)$.

From intuition, when we randomly partition X into $X_l \cup X_r$ such that each element in X has a probability of 1/2 to be assigned to X_l and a probability of 1/2 to be assigned to X_r, a big subset has more chance to be split. This is true, if many subsets in $\mathcal{F}$ have many elements, we can obtain a better kernel, or a kernel that has fewer subsets in $\mathcal{F}'$.

Lemma 3. *Let $(X, \mathcal{F}, k)$ be an instance of the unweighted* SET SPLITTING *problem. Suppose the number of subsets that have i elements is m_i for $1 \leq i \leq k-1$ and the number of subsets that have at least k elements is m'_k. If $\sum_{i=2}^{k-1} \frac{2^i-2}{2^i} m_i + m'_k \geq k$, then a partition of X exists that splits at least k subsets in $\mathcal{F}$.*

Proof. Let $S_1, \ldots, S_{m'_k}$ be the subsets in $\mathcal{F}$ that have at least k elements and let $\mathcal{F}_{<k} = \mathcal{F} - \{S_1, \ldots, S_{m'_k}\}$.

We use a randomized process to partition X into (X_l, X_r) and let each element in X go to X_l with a probability of 1/2 and go to X_r with a probability of 1/2, then for any subset $S \in \mathcal{F}_{<k}$ that has i elements:

$$Pr(S \text{ is split}) = \frac{2^i - 2}{2^i}.$$

If we let:

$$X_S = \begin{cases} 1, \text{ if } S \text{ is split,} \\ 0, \text{ otherwise.} \end{cases}$$

then the expectation of the number of split subsets in $\mathcal{F}_{<k}$ satisfies

$$E\left(\sum_{S \in \mathcal{F}_{<k}} X_S\right) = \sum_{i=1}^{k-1} \sum_{|S|=i} E(S \text{ is split}) = \sum_{i=1}^{k-1} \frac{2^i - 2}{2^i} m_i.$$

So there exists a partition of X such that the number of subsets in $\mathcal{F}_{<k}$ that are split is at least $\sum_{i=1}^{k-1} \frac{2^i-2}{2^i} m_i$. Hence if $\sum_{i=1}^{k-1} \frac{2^i-2}{2^i} m_i \geq k - m'_k$, there must exist a partition of X such that $k - m'_k$ subsets in $\mathcal{F}_{<k}$ are split. By repeatedly using Lemma 2, we conclude that there is a partition of X that splits $k - m'_k + 1$ subsets in $\mathcal{F}_{<k} \cup \{S_1\}$; there is a partition of X that splits $k - m'_k + 2$ subsets in $\mathcal{F}_{<k} \cup \{S_1, S_2\}$; and so on. In consequence, there is a partition of X that splits k subsets in $\mathcal{F}_{<k} \cup \{S_1, \ldots, S_{m'_k}\} = \mathcal{F}$. □

Using the procedure that is similar to Theorem 1, but counting the number of subsets in $\mathcal{F}$ of different size and using the result of Lemma 3, we have the following Theorem that is stronger than Theorem 1.

Theorem 2. *Given an instance $(X, \mathcal{F}, k)$ of* SET SPLITTING *problem, suppose the number of subsets in $\mathcal{F}$ that have i elements is m_i for $1 \leq i \leq k-1$ and the number of subsets that have at least k elements is m'_k. Then in time $O(N+2k^2)$, where N is the input size in terms of $(X, \mathcal{F}, k)$, we can find a kernel $(X', \mathcal{F}', k')$ such that $|\mathcal{F}'| < 2k - \sum_{i=3}^{k-1} \frac{2^{i-1}-2}{2^{i-1}} m_i - 2m'_k$, that $k' \leq k$, that each subset in $\mathcal{F}'$ has at most $k-1$ elements, and that $|X'| < 2k^2$.*

Proof. We use the procedure that is similar to Theorem 1 to find the kernel: given an instance $(X, \mathcal{F}, k)$ of the unweighted SET SPLITTING problem, we delete each subset that has only one element, also delete each subset that has more than $k-1$ elements and decrease k by 1. In this procedure, we also obtain m_i for $1 \leq i \leq k-1$ and m'_k. By Lemma 3, if $\sum_{i=2}^{k-1} \frac{2^i-2}{2^i} m_i + m'_k \geq k$, we know the given instance is a "Yes" instance; otherwise $\sum_{i=2}^{k-1} \frac{2^i-2}{2^i} m_i + m'_k < k$, i.e. $|\mathcal{F}'| = \sum_{i=2}^{k-1} m_i < 2k - \sum_{i=3}^{k-1} \frac{2^{i-1}-2}{2^{i-1}} m_i - 2m'_k$. So we find a kernel $(X', \mathcal{F}', k')$ that $|\mathcal{F}'| < 2k - \sum_{i=3}^{k-1} \frac{2^{i-1}-2}{2^{i-1}} m_i - 2m'_k$, that $k' \leq k$, that each subset in $\mathcal{F}'$ has at most $k-1$ elements, and that $|X'| < 4k^2$. The time needed to find the kernel is $O(N + 2k^2)$. □

3 A Randomized Algorithm for the Weighted SET SPLITTING Problem

For the unweighted SET SPLITTING problem, Lokshtanov and Sloper [14] have currently the best parameterized algorithm, whose running time is bounded by $O^*(2.65^k)$. Unfortunately, their method does not seem to be extendable to the weighted case, neither do the methods presented in [6,7] for the unweighted case. In fact, no previous work is known that gives a fixed parameter tractable algorithm for the weighted SET SPLITTING problem.

In this section, we present a randomized algorithm to solve the weighted SET SPLITTING problem. Our basic idea is that if a given instance $(X, \mathcal{F}, k)$ of the weighted SET SPLITTING problem has a partition of the ground set X that splits k subsets in the collection $\mathcal{F}$, then there exists a subset X' of at most $2k$ elements in X such that a proper partition of the elements in X' can split at least k subsets in $\mathcal{F}$. If we use a randomized process to partition X into (X_l, X_r) and let each element in X go to X_l with a probability of $1/2$ and go to X_r with a probability of $1/2$, then the probability that the elements in X' are partitioned properly is at least $2/2^{2k}$. Thus, if we try $O(4^k)$ times, we have a good chance to find a proper partition if it exists. In fact, we can do better than this in a randomized algorithm.

Theorem 3. *The weighted* SET SPLITTING *problem can be solved by a randomized algorithm of running time $O(2^k N)$, where N is the input size in terms of $(X, \mathcal{F}, k)$.*

Proof. Let $(X, \mathcal{F}, k)$ be an instance of the weighted SET SPLITTING problem. Suppose that there is a partition of the ground set X that splits at least k subsets

Algorithm-1 SetSplitting($X, \mathcal{F}, k$)
input: A ground set X, a collection $\mathcal{F}$ of subsets of X, and an integer k
output: A partition (X_l, X_r) of X and k subsets in $\mathcal{F}$ that are split by (X_l, X_r), or report "no partition of X splits k subsets in $\mathcal{F}$".

1. $\mathcal{Q}_0 = \emptyset$;
2. **for** $i = 1$ **to** $10 \cdot 2^k$ **do**
2.1. randomly partition X into X_l and X_r such that each element in X has a probability 1/2 in X_l and a probability 1/2 in X_r;
2.2. let $\mathcal{Q}$ be the collection of subsets in $\mathcal{F}$ that are split by (X_l, X_r);
2.3. **if** $\mathcal{Q}$ contains at least k subsets **then** delete all but the k subsets of maximum weight in $\mathcal{Q}$;
2.4. **if** the weight sum of subsets in $\mathcal{Q}$ is larger than that in $\mathcal{Q}_0$ **then** $\mathcal{Q}_0 = \mathcal{Q}$;
3. return $\mathcal{Q}_0$.

Fig. 1. Randomized algorithm for WEIGHTED SET SPLITTING problem

in the collection $\mathcal{F}$. Let (X_l, X_r) be a partition of the ground set X and let S_1, ..., S_k be k subsets in the collection $\mathcal{F}$ that are split by the partition (X_l, X_r), such that the weight sum of S_1, ..., S_k is the maximum over all collections of k subsets in $\mathcal{F}$ that can be split by a partition of X. More specifically, let (l_1, r_1), ..., (l_k, r_k) be k pairs of elements in the ground set X such that $l_i, r_i \in S_i$, $l_i \in X_l$, and $r_i \in X_r$ for all $1 \leq i \leq k$. Note that it is possible that $l_i = l_j$ or $r_i = r_j$ for some $i \neq j$. In consequence, each of the sets $\{l_1, \ldots, l_k\}$ and $\{r_1, \ldots, r_k\}$ may contain fewer than k elements.

We construct a graph $G = (V, E)$, where $V = \{l_1, l_2, \ldots, l_k\} \cup \{r_1, r_2, \ldots, r_k\}$ and $E = \{(l_i, r_i) \mid 1 \leq i \leq k\}$. It is obvious that G is a bipartite graph with the left vertex set $L = \{l_1, l_2, \ldots, l_k\}$ and the right vertex set $R = \{r_1, r_2, \ldots, r_k\}$. Suppose that the graph G has t connected components $C_1, \cdots, C_t$, where $C_i = (V_i, E_i)$, with $n_i = |V_i|$ and $m_i = |E_i|$, for $1 \leq i \leq t$. Then $n_i \leq m_i + 1$ for $1 \leq i \leq t$ and $\sum_{i=1}^{t} m_i = k$. If we use a randomized process to partition X into (X_l, X_r) and let each element in X go to X_l with a probability of 1/2 and go to X_r with a probability of 1/2, then for each connected component C_i of the graph G, the probability that the vertex set V_i of C_i is properly partitioned, i.e., either $L \cap V_i \subseteq X_l$ and $R \cap V_i \subseteq X_R$, or $R \cap V_i \subseteq X_l$ and $L \cap V_i \subseteq X_R$, is $2/2^{n_i}$. Therefore, the total probability that the vertex set V_i for every connected component C_i is properly partitioned, i.e., that the pair (l_i, r_i) intersects with both X_l and X_r for all $1 \leq i \leq k$, is not less than

$$\frac{2}{2^{n_1}} \cdot \frac{2}{2^{n_2}} \cdot \cdots \cdot \frac{2}{2^{n_t}} \geq \frac{2}{2^{m_1+1}} \cdot \frac{2}{2^{m_2+1}} \cdot \cdots \cdot \frac{2}{2^{m_t+1}} = \frac{2^t}{2^{\sum_{i=1}^{t} m_i + t}} = \frac{1}{2^k}.$$

The algorithm in Figure 1 implements the above idea. By the above discussion, each random partition (X_l, X_r) constructed in step 2.1 has a probability of at least $1/2^k$ to split the k subsets S_1, ..., S_k (recall that S_1, ..., S_k are the k subsets in $\mathcal{F}$ whose weight sum is the maximum over all collections of k subsets in $\mathcal{F}$ that are split by a partition of X). Since step 2 loops $10 \cdot 2^k$ times, with a probability of at least

$$1 - \left(1 - \frac{1}{2^k}\right)^{10 \cdot 2^k} \geq 99.99\%,$$

one partition (X_l, X_r) constructed by step 2.1 splits the k subsets $S_1, \ldots, S_k$. For this partition (X_l, X_r), steps 2.2-2.4 produces a collection $\mathcal{Q}$ of k subsets in $\mathcal{F}$ whose weight sum is the maximum over all collections of k subsets in $\mathcal{F}$ that are split by a partition of the ground set X.

Since each execution of steps 2.1-2.4 obviously takes time $O(N)$, we conclude that the running time of the algorithm **SetSplitting** is bounded by $O(2^k N)$. □

Obviously, the algorithm **SetSplitting** of running time $O(2^k N)$ can be directly used to solve the unweighted SET SPLITTING problem, and its running time significantly improves the previous best algorithm [14] for the unweighted SET SPLITTING problem. Moreover, the algorithm **SetSplitting** is much simpler than the one presented in [14].

4 Derandomization

For many parameterized NP-Complete problems, such as the k-PATH, 3-SET PACKING, and 3D-MATCHING problems, the solution is a subset of size $O(k)$. If we have a way to partition this subset properly, we only need to deal with several subproblems of smaller solution sizes, such as two $(\frac{k}{2})$-path problems. For a given subset of size k, or a k-subset, we have 2^k ways of partitioning it into two subsets. If we know this k-subset, we can enumerate all possible partitions in time $O(2^k)$. But usually, we do not know this k-subset. In this section, we present a set partition method such that given any set V of size n and an integer k, we can get a partition family $\mathcal{F}(V, k)$ to partition any k-subset S of V in all 2^k partition ways. The method was described in an extended abstract by Noar, Schulman, and Srinivasan [15], with many details omitted. In this section, we provide further details and concrete constructions for the parts related to our problems, and show how these techniques can be used to derive efficient parameterized algorithms for the weighted SET SPLITTING problem.

Lemma 4 **([1]).** *Suppose $n = 2^d - 1$ and $k = 2t + 1 \leq n$. Then there exists a uniform probability space Ω of size $2(n+1)^t$ and k-wise independent random variables $\xi_1, \ldots, \xi_n$ over Ω each of which takes the values 0 and 1 with probability 1/2.*

The Ω in lemma 4 is a $n \times 2(n+1)^{\lfloor k/2 \rfloor}$ matrix that takes only value 0 and 1. In our case, we can see each row as an element in a set V of size n and each column as a partition of V. By the lemma 4, for each k-subset S of V and each partition of S, there are $\frac{1}{2^k} 2(n+1)^{\lfloor k/2 \rfloor}$ columns to partition S this way. This result was used by paper [15] to prove Lemma 7.

Lemma 5. **a)** *Given two sets A and B and a relation $R \subset A \times B$, if $(a, b) \in R$, we say element a in set A relates to element b in set B. If each element in set A relates to at least fraction p of elements in set B, then there is an element b' in set B such that there are at least fraction p of elements in A that relate to it.*

b) *For any given n and k and $k \leq n$, if $t > ek2^k(\log n + 1)$, then $\binom{n}{k}2^k(1 - \frac{1}{2^k})^t < 1$.*

Proof. **a)** Each element in set A relates to $p|B|$ elements in set B, then there are $p|B||A|$ elements in set $R \subset A \times B$. So there is at least one element in set B that there exist $p|A|$ elements in set A that relate to it.

b) Because $\binom{n}{k}2^k(1 - \frac{1}{2^k})^t < n^k 2^k(1 - \frac{1}{2^k})^t$, just let $t = ek2^k(\log n + 1)$, you can verify the inequality: $\binom{n}{k}2^k(1 - \frac{1}{2^k})^t < 1$ for $t > ek2^k(\log n + 1)$. □

Lemma 6. [9] *Given $U = \{1, 2, \ldots, n\}$ with $p = n + 1$ a prime number, and given $W \subset U$ with $|W| = k$, then there exists a $k' \in U$, such that mapping $x \to (k'x \mod p) \mod k^2$ is one-to-one from W to $\{0, 1, 2, \ldots, k^2 - 1\}$.*

Lemma 6 was proposed by Fredman, Komlos, and Szemeredi. It was used in Theorem 4 to reduce the size of ground set V from n to k^2. Now we give the most important Lemma in the construction of partition family $\mathcal{F}(V, k)$.

Lemma 7. [15] *For any given set V of size n and $k \leq n$, there is a partition function family $\mathcal{F}(V, k)$ of V such that for any k-subset V' of V, any partition of V' can be done by a partition function in $\mathcal{F}(V, k)$. Furthermore the cardinality of $\mathcal{F}(V, k)$ is $O(k2^k \log n)$. And $\mathcal{F}(V, k)$ can be constructed deterministically in time $O(\binom{n}{k}k2^{2k}n^{\lfloor \frac{k}{2} \rfloor})$.*

Proof. For each k-subset S of V and each partition p of S, we make a pair (S, p) and let A be a set of all such pairs; thus $|A| = \binom{n}{k}2^k$. Let $B = \Omega$, where Ω is the probability space as in Lemma 4. Any $(S, p) \in A$ and $p' \in B$, if p' agree the partition p in subset S (We say partition p' covers pair (S, p)), let $((S, p), p') \in R \subset A \times B$.

By Lemma 4, each $(S, p) \in A$, if we randomly choose $p' \in B$, the probability that p' covers (S, p) is $\frac{1}{2^k}$, i.e. there are $|B|\frac{1}{2^k}$ elements in B that cover (S, p). So by Lemma 5-*a*), there is a partition p' in B such that p' covers $|A|\frac{1}{2^k}$ pairs in A. We put this p' into $\mathcal{F}(V, k)$ and delete the pairs that are covered by p'. The number of pairs remaining in A is $|A|(1 - \frac{1}{2^k})$. If we do this process again, the number of pairs remaining in A becomes $|A|(1 - \frac{1}{2^k})^2$. By Lemma 5-*b*), if we repeat this process $ek2^k(\log n + 1) + 1$ times, the number of pairs remaining in A is less than 1. This means that every pair in A is covered by some partitions in $\mathcal{F}(V, k)$ and the cardinality of $\mathcal{F}(V, k)$ is $O(k2^k \log n)$.

Because $|A| = \binom{n}{k}2^k$, $|B| = 2n^{\lfloor k/2 \rfloor}$, the time to construct $\mathcal{F}(V, k)$ is:

$$\sum_{i=0}^{O(k2^k \log n)} |A||B| \left(1 - \frac{1}{2^k}\right)^i < O\left(\binom{n}{k} k2^{2k} n^{\lfloor \frac{k}{2} \rfloor}\right).$$

□

Lemma 7 gave a way to construct partition family $\mathcal{F}(V,k)$ of cardinality $O(k2^k \log n)$, but the time complexity is too large. In fact, Lemma 7 is only a middle step result, the final construction is in the theorem 4.

Theorem 4. [15] *For any given set V of size n and $k \leq n$, there is a partition function family $\mathcal{F}(V,k)$ of V such that for any k-subset V' of V, any partition of V' can be done by a partition function in $\mathcal{F}(V,k)$ applying to V'. The family $\mathcal{F}(V,k)$ contains $O(n2^{k+12\log^2 k-4\log k})$ partitions of V, and can be constructed in time $O(n2^{k+12\log^2 k-4\log k})$.*

Proof. The construction of $\mathcal{F}(V,k)$ includes three steps:

First step: Mapping V to $\{0,1,\ldots,k^2-1\}$ by function $f_i(x) = (ix \mod p) \mod k^2$, where p is a prime number between n and $2n$. From number theory, this prime number p must exist. By Lemma 6, for a fixed k-subset V' of V, there is $1 \leq i < 2n$ such that every element in V' is mapped to different value in $\{0,1,\ldots,k^2-1\}$ by $f_i(x)$. This step has $O(n)$ branches.

Second step: Partitioning $\{0,1,\ldots,k^2-1\}$ into $4\log k$ subsets $V_1, V_2, \ldots, V_{4\log k}$ such that each subset has $\frac{k}{4\log k}$ elements of the fixed k-subset V'. This step has $O(k^{8(\log k-1)})$ branches.

Third step: Enumerating all possible combinations of $\mathcal{F}(V_i, \frac{k}{4\log k})$ that obtained by using the method from Lemma 7. The number of branches in this step is $O((\frac{k}{4\log k})^{4\log k}(2\log k)^{4\log k}2^k)$.

Because the time to construct all $\mathcal{F}(V_i, \frac{k}{4\log k})$ is $O(k2^{\frac{3k}{4}+\frac{k}{2\log k}})$, both the cardinality of $\mathcal{F}(V,k)$ and the time to construct $\mathcal{F}(v,k)$ are $O(n2^{k+12\log^2 k-4\log k})$. □

Using the result of Theorem 4, we derive the first fixed parameter tractable algorithm for the weighted SET SPLITTING problem, as given below.

Theorem 5. *The WEIGHTED SET SPLITTING problem can be solved determinately by an algorithm of running time $O((|\mathcal{F}|k+n))n2^{2k+12\log^2 k})$.*

Proof. Given an instance of WEIGHTED SET SPLITTING problem, if it has at least k subsets in $\mathcal{F}$ that can be split by a partition of X, then there exist two sets $\{l_1, l_2, \ldots, l_k\}$ and $\{r_1, r_2, \ldots, r_k\}$. If we can partition $\{l_1, l_2, \ldots, l_k\}$ and $\{r_1, r_2, \ldots, r_k\}$ into two different groups, then there are at least k subsets in $\mathcal{F}$ that are split by this partition. Using the partition family $\mathcal{F}(V,2k)$ that is constructed as in Theorem 4, we can do it. So we replace line 1 in Algorithm-1 by looping all partitions in this $\mathcal{F}(V,2k)$, then we can find a partition to split at least k subsets in $\mathcal{F}$ determinately if this partition exists. □

Finally, we remark that the partition function family $\mathcal{F}(V,k)$ can also be used to derandomize the algorithms for k-PATH, 3D-MATCHING and 3-SET PACKING problems that are based on the divide-and-conquer techniques [5,12].

References

1. Alon, N., Babai, L., Itai, A.: A fast and simple randomized parallel algorithm for the maximal independent set problem. Journal of Algorithms 7, 567–683 (1986)
2. Andersson, G., Engebretsen, L.: Better approximation algorithms and tighter analysis for set splitting and not-all-equal sat. In: ECCCTR: Electronic colloquium on computational complexity (1997)
3. Ausiello, G., Crescenzi, P., Gambosi, G., Kann, V., Marchetti-Spaccamela, A., Protasi, M.: Complexity and Approximation: Combinatorial Optimization Problems and Their Approximability Properties. Springer, Heidelberg (1999)
4. Chen, J., Kanj, I.: Improved Exact Algorithms for Max-Sat. Discrete Applied Mathematics 142, 17–27 (2004)
5. Chen, J., Lu, S., Sze, S., Zhang, F.: Improved algorithms for path, matching, and packing problems. In: SODA, pp. 298–307 (2007)
6. Dehne, F., Fellows, M., Rosamond, F.: An FPT Algorithm for Set Splitting. In: Bodlaender, H.L. (ed.) WG 2003. LNCS, vol. 2880, pp. 180–191. Springer, Heidelberg (2003)
7. Dehne, F., Fellows, M., Rosamond, F., Shaw, P.: Greedy localization, iterative compression, modeled crown reductions: new FPT techniques, and improved algorithm for set splitting, and a novel $2k$ kernelization of vertex cover. In: Downey, R.G., Fellows, M.R., Dehne, F. (eds.) IWPEC 2004. LNCS, vol. 3162, pp. 127–137. Springer, Heidelberg (2004)
8. Downey, R., Fellows, M.: Parameterized Complexity. Springer, Heidelberg (1999)
9. Fredman, M., Komlos, J., Szemeredi, E.: Storing a sparse table with $O(1)$ worst case access time. Journal of the ACM 31, 538–544 (1984)
10. Garey, M., Johnson, D.: Computers and Intractability: A Guide to the Theory of NP-Completeness. Freeman, San Francisco (1979)
11. Kann, V., Lagergren, J., Panconesi, A.: Approximability of maximum splitting of k-sets and some other apx-complete problems. Information Processing Letters 58(3), 105–110 (1996)
12. Kneis, J., Molle, D., Richter, S., Rossmanith, P.: Divide-and-color. In: Fomin, F.V. (ed.) WG 2006. LNCS, vol. 4271, pp. 58–67. Springer, Heidelberg (2006)
13. Liu, Y., Lu, S., Chen, J., Sze, S.: Greedy localization and color-coding: Improved Matching and Packing Algorithms. In: Bodlaender, H.L., Langston, M.A. (eds.) IWPEC 2006. LNCS, vol. 4169, pp. 84–95. Springer, Heidelberg (2006)
14. Lokshtanov, D., Sloper, C.: Fixed parameter set splitting, linear kernel and improved running time. Algorithms and Complexity in Durham 2005, King's College Press, Texts in Algorithmics 4, 105–113 (2005)
15. Naor, M., Schulman, L., Srinivasan, A.: Splitters and near-optimal derandomization. In: FOCS, pp. 182–190 (1995)
16. Zhang, H., Ling, C.: An improved learning algorithm for augmented naive bayes. In: Cheung, D., Williams, G.J., Li, Q. (eds.) PAKDD 2001. LNCS (LNAI), vol. 2035, pp. 581–586. Springer, Heidelberg (2001)
17. Zwick, U.: Approximation algorithms for constraint satisfaction problems involving at most three variables per constraint. In: SODA, pp. 201–220 (1998)
18. Zwick, U.: Outward rotations: A tool for rounding solutions of semidefinite programming relaxation, with applications to max cut and other problem. In: STOC, pp. 679–687 (1999)

An $\frac{8}{5}$-Approximation Algorithm for a Hard Variant of Stable Marriage

Robert W. Irving* and David F. Manlove*

Department of Computing Science, University of Glasgow, Glasgow G12 8QQ, UK
rwi@dcs.gla.ac.uk, davidm@dcs.gla.ac.uk

Abstract. When ties and incomplete preference lists are permitted in the Stable Marriage problem, stable matchings can have different sizes. The problem of finding a maximum cardinality stable matching in this context is known to be NP-hard, even under very severe restrictions on the number, size and position of ties. In this paper, we describe a polynomial-time $\frac{8}{5}$-approximation algorithm for a variant in which ties are on one side only and at the end of the preference lists. The particular variant is motivated by important applications in large scale centralized matching schemes.

1 Introduction

Background

An instance of the Stable Marriage problem with Ties and Incomplete Lists (SMTI) comprises a set of n_1 men $m_1, \ldots, m_{n_1}$ and a set of n_2 women $w_1, \ldots$, w_{n_2}. Each person has a *preference list* consisting of a subset of the members of the opposite sex, his or her *acceptable partners*, listed in order of preference, with ties, consisting of two or more persons of equal preference, permitted. If man m and woman w appear on each other's preference list then (m, w) is called an *acceptable pair*. If w precedes w' on m's list then m is said to *prefer* w to w', while if w and w' appear together in a tie on m's list then m is said to be *indifferent* between w and w'.

A *matching* is a set M of acceptable pairs so that each person appears in at most one pair of M. If M is a matching and $(m, w) \in M$ we write $w = M(m)$ and $m = M(w)$, and we say that m and w are *partners* in M. A pair (m, w) is a *blocking pair* for M, or *blocks* M, if m is either unmatched in M or prefers w to $M(m)$, and simultaneously w is either unmatched in M or prefers m to $M(w)$. A matching for which there is no blocking pair is said to be *stable*.

SMTI is an extension of the classical Stable Marriage problem (SM) introduced by Gale and Shapley [2]. In the classical case, the numbers of men and women are equal, all preference lists are complete, i.e., they contain all members of the opposite sex, and ties are not permitted, i.e., all preferences are strict. Gale and Shapley proved that, for every instance of SM, there is at least one

* Supported by EPSRC research grant EP/E011993/1.

G. Lin (Ed.): COCOON 2007, LNCS 4598, pp. 548–558, 2007.

stable matching, and they described an $O(n^2)$ time algorithm to find such a matching; this has come to be known as the *Gale-Shapley algorithm.*

This algorithm is easily extended to the case in which the numbers of men and women differ and preference lists are incomplete (SMI – Stable Marriage with Incomplete lists); it has complexity $O(a)$ in this case, where a is the number of acceptable pairs [4]. In the case of SMI, not everyone need be matched in a stable matching. In general, for a given instance of SMI, there may be many stable matchings – exponentially many in extreme cases – but all stable matchings have the same size and match exactly the same sets of men and women [15,3].

The Gale-Shapley algorithm may be applied from either the men's side or the women's side, and in general these two applications will produce different stable matchings. When applied from the men's side, the *man-optimal* stable matching is found; in this, every man has the best partner that he can have in any stable matching, and every woman the worst. When the algorithm is applied from the women's side, the *woman-optimal* stable matching results, with analogous properties. Exceptionally, the man-optimal and woman-optimal stable matchings may coincide, in which case this is the unique stable matching, but in general there may be other stable matchings – possibly exponentially many – between these two extremes. However, for a given instance of SMI, all stable matchings have the same size and match exactly the same sets of men and women [15,3].

The situation for SMTI is dramatically different. Again, at least one stable matching exists for every instance, and can be found in $O(a)$ time by breaking all ties in an arbitrary way to give an instance of SMI, and applying the Gale-Shapley algorithm to that instance. However, the ways in which ties are broken can significantly affect the outcome. In particular, not all stable matchings need be of the same size, and in the most extreme case, there may be two stable matchings M and M' with $|M| = 2|M'|$. Furthermore, the problem of finding a stable matching of maximum cardinality for an instance of SMTI – problem MAX-SMTI – is NP-hard [13]. This hardness result holds even under severe restrictions, for example, if the ties are on one side only, each list contains at most one tie, and that tie, if present, is at the end of the list.

Practical Applications

The practical importance of stable matching problems arises from their application in the assignment of applicants to positions in various job markets. The many-one version of the problem has come to be known as the Hospitals/Residents problem (HR) because of its widespread application in the medical employment domain [15,9,14,1,16]. In an instance of HR, each resident has a preference list of acceptable hospitals, while each hospital has a preference list of acceptable residents together with a quota of positions. A matching M is a set of acceptable resident-hospital pairs such that each resident is in at most one pair, and each hospital is in a number of pairs that is bounded by its quota. If pair (r, h) is in M, we write $h = M(r)$ and $r \in M(h)$, so that $M(h)$ is a *set* of residents for each hospital h. A matching is stable if it admits no blocking pair, i.e., an acceptable pair (r, h) such that r is either unmatched in M or prefers h

Men's preferences	Women's preferences
$m_1 : w_1 \ \ w_2$	$w_1 : (\ m_1 \ \ m_2\ )$
$m_2 : w_1$	$w_2 : m_1$

Fig. 1. An instance of SSMTI with stable matchings of sizes 1 and 2.

to $M(r)$, and simultaneously h is either under quota or prefers r to at least one member of $M(h)$.

As in the case of SMI, all stable matchings for an instance of HR have the same size, and so-called *resident-optimal* and *hospital-optimal* stable matchings can be found by applying an extended version of the Gale-Shapley algorithm from the residents' side or the hospitals' side respectively. But if ties are allowed in the preference lists - the Hospitals/Residents problem with Ties (HRT) – then as in the case of SMTI, stable matchings can have different sizes, and it is NP-hard to find a stable matching of maximum size, even under severe restrictions on the number, size, and position of ties.

Variants of the extended Gale-Shapley algorithm are routinely used in a number of countries, including the United States [14], Canada [1] and Scotland [16], to allocate graduating medical students to hospital posts, and in a variety of other countries and contexts. In large scale matching schemes of this kind, participants, particularly large popular hospitals, may not be able to provide a genuine strict preference order over what may be a very large number of applicants, so that HRT is a more appropriate model than HR. If artificial tie-breaking is carried out, either by the participant, because a strictly ordered list is required by the matching scheme, or by the administrators of the scheme, prior to running an algorithm that requires strict preferences, then the size of the resulting stable matching is likely to be affected. Breaking ties in different ways will typically yield stable matchings of different sizes; what would be ideal would be to find a way of breaking the ties that maximizes the size of the resulting stable matching, but the NP-hardness of this problem makes this an objective that is unlikely to be feasible.

A special case of HRT arises if residents are required to strictly rank their chosen hospitals but hospitals are asked to rank only as many of their applicants as they reasonably can, and then place the remainder in a single tie at the end. For example this variant has been employed in the Scottish Foundation Allocation Scheme (SFAS). The correspondingly restricted version of SMTI, where all men's lists are strict and women's lists may contain one tie at the end, is the special case in which all quotas are equal to one. We refer to these restricted versions of SMTI and HRT as *Special SMTI/HRT* (SSMTI/SHRT), and use the terms MAX-SSMTI and MAX-SHRT for the problems of finding maximum cardinality stable matchings in these cases, which remain NP-hard problems [13].

Figure 1 shows an example of SSMTI in which there are two stable matchings, $M_1 = \{(m_1, w_2), (m_2, w_1)\}$ of size 2 and $M_2 = \{(m_1, w_1)\}$ of size 1. A tie in the preference lists is indicated by parentheses.

Related Results

It is trivial to establish that there can be at most a factor of two difference between the sizes of a minimum and maximum cardinality stable matching for an instance of SMTI, and as a consequence, breaking ties arbitrarily and applying the Gale-Shapley algorithm gives a 2-approximation algorithm for MAX-SMTI. A number of improved approximation algorithms for versions of SMTI have recently been proposed.

For the general case, Iwama et al [11] gave an algorithm with a performance guarantee of $(2 - c/\sqrt{n})$, (for the case of n men and n women), for a constant c. Very recently, Iwama et al [12] gave the first approximation algorithm for the general case with a constant performance guarantee better than 2, namely $\frac{15}{8}$. From the inapproximability point of view, Halldórsson et al showed the problem to be APX-complete [5], and gave a lower bound of $\frac{21}{19}$ on any polynomial-time approximation algorithm (assuming $P \neq NP$) [6]. This lower bound applies even to MAX-SSMTI.

As far as special cases are concerned, Halldórsson et al [6] gave a $(2/(1+L^{-2}))$-approximation algorithm for the case where all ties are on one side, and are of length at most L – so, for example, this gives a bound of $\frac{8}{5}$ when all ties are of length 2. If ties are on both sides and restricted to be of length 2, a bound of $\frac{13}{7}$ is shown in [6]. Halldórsson et al [7] also described a randomized algorithm with an expected performance guarantee of $\frac{10}{7}$ for the same special case under the additional restriction that there is at most one tie per list.

The Contribution of This Paper

In this paper, we focus on the problem MAX-SSMTI described above, and give a polynomial-time $\frac{8}{5}$-approximation algorithm for this case. The algorithm is relatively easy to extend to MAX-HRT, and the same $\frac{8}{5}$ performance guarantee holds in this more general setting. We also show that this performance guarantee is the best that can be proved for the algorithm by providing an example for which this bound is realised.

2 The Algorithm

In what follows, we assume that each man's preference list is strict, and each woman's preference list is strict except for a tie (of length ≥ 1) at the end. The algorithm consists of three phases. The first phase is a variant of the Gale-Shapley algorithm for the classical stable marriage problem, applied from the women's side, but with women proposing only as far as the tie (if any) in their list. This results in a provisional matching involving precisely those men who received proposals. The second phase adds to this provisional matching a maximal set of acceptable pairs from among the remaining men and women. Finally, in Phase 3 all ties are broken, favouring unmatched men over matched men, and the standard Gale-Shapley algorithm is run to completion on the resulting instance of SMI.

```
assign each person to be free;
while (some woman w is free) and (w has a non-empty list)
    and (w has an untied man m at the head of her list) {
    w proposes, and becomes engaged to m;
    for each successor w' of w on m's list {
        if w' is engaged to m
            break the engagement, so that w' becomes free;
        delete the pair (m, w') from the preference lists;
    }
}
```

Fig. 2. Phase 1 of Algorithm SSMTI-APPROX

Phase 1 of Algorithm SSMTI-APPROX

The first phase of the algorithm is a variant of the Gale-Shapley algorithm for the classical stable marriage problem, applied from the women's side. During this phase, zero or more deletions are made from the preference lists – by the deletion of the pair (m, w), we mean the removal of w from m's list and the removal of m from that of w. Initially, everyone is free. During execution of the algorithm, a woman may alternate between being free and being engaged, but once a man becomes engaged, he remains in that state, though the identity of his partner may change over time. A free woman w who still has an untied man on her current list proposes to the first such man and becomes (at least temporarily) engaged to that man. When a man m receives a proposal from woman w, he rejects his current partner (if any), setting her free, and all pairs (m, w') such that m prefers w to w' are deleted. This phase of the algorithm is summarised in Figure 2.

When Phase 1 of the algorithm terminates on a given instance I, a woman w's preference list must be in one of three possible states – it may be empty, it may consist of a single tie, or it may have a unique untied man m at its head. In the latter case, it is clear that m cannot be the unique man at the head of any other woman's list, and that w is the last entry in m's list.

Lemma 1. *On termination of Phase 1 of Algorithm SSMTI-APPROX,*
(i) no deleted pair can belong to a stable matching;
(ii) if man m is the unique man at the head of some woman's list then m is matched in every stable matching.

Proof. (i) Suppose that (m, w) is a deleted pair that belongs to a stable matching M, and that (m, w) was the first such pair to be deleted in an execution of Phase 1 of the algorithm. This must have happened because m received a proposal from some woman w' whom he prefers to w. Woman w' is either unmatched in M or prefers m to $M(w)$, because any pair (m', w') such that w' prefers m' to m must have been previously deleted, and by our assumption, this pair cannot be in a stable matching. Hence (m, w') blocks M, a contradiction.

$V = Y_1 \cup Q_1$;
$E = \{(m, w) \in Y_1 \times Q_1 : (m, w)$ is a Phase 1 acceptable pair$\}$;
construct the bipartite graph $G = (V, E)$;
K = a maximum cardinality matching in G;
for each pair $(m, w) \in K$
    promote m from the tie to the head of w's list;
re-activate the proposal sequence of Phase 1;

Fig. 3. Phase 2 of Algorithm SSMTI-APPROX

(ii) Suppose that m is the unique man at the head of woman w's list, and that M is a stable matching in which m is unmatched. Then, by part (i), w is either unmatched in M or prefers m to $M(w)$, so that (m, w) blocks M, a contradiction. □

We refer to the men who appear untied at the head of some woman's list after Phase 1 of the algorithm as the *Phase 1 X-men* and the other men as the *Phase 1 Y-men*, and we denote these sets by X_1 and Y_1 respectively. Likewise, the women who have an untied man at the head of their list are the *Phase 1 P-women* and the others are the *Phase 1 Q-women*, denoted by P_1 and Q_1. So the engaged pairs at the end of Phase 1 constitute a perfect matching between X_1 and P_1, and the essence of Lemma 1(ii) is that each member of X_1 is matched in every stable matching. We call the preference lists that remain after Phase 1 the *Phase 1 lists*, and if man m and woman w are in each other's Phase 1 lists, we say that (m, w) is a *Phase 1 acceptable pair.*

Phase 2 of Algorithm SSMTI-APPROX

In Phase 2 of the algorithm, we seek to increase the number of men who are guaranteed to be matched. To this end, we find a maximum cardinality matching K of Y_1 to Q_1, where a pair (m, w) can be in this matching only if $m \in Y_1$, $w \in Q_1$, and (m, w) is a Phase 1 acceptable pair. For each pair (m, w) in K, we break the tie in w's Phase-1 list by promoting m to the head of that list (and leaving the rest of the tie intact). We then re-activate the proposal sequence of Phase 1, which will lead to a single proposal corresponding to each pair in K, and which may result in some further deletions from the preference lists, but no rejections and no other proposals. This produces an instance I' of SSMTI that is a refinement of the original instance I – or more properly, a refinement of the variant of I that results from application of Phase 1; clearly any matching that is stable for I' is also stable for I, but not necessarily vice-versa. Phase 2 of the algorithm is summarised in Figure 3.

Lemma 2. *Every man who is untied at the head of some woman's list on termination of Phase 2 of Algorithm SSMTI-APPROX is matched in every stable matching for I'.*

Proof. The proof is completely analogous to that of Lemma 1(ii). □

Note that, while Lemma 2 can be expected, in many cases, to give a stronger lower bound on the size of a stable matching than is given by Lemma 1, this need not be the case. It is perfectly possible that the Phase 1 Q-women have only Phase 1 X-men in their preference lists, and that, as a consequence, K is the empty matching. However, we now extend the set of X-men and P-women to include those who became engaged during Phase 2. Henceforth, we use the term *X-men* to refer to those men who appear untied at the head of some woman's list, and the *P-women* are the women who have an X-man at the head of their list, after Phase 2 of the algorithm. Let x be the number of X-men and P-women, and suppose that these sets are $X = \{m_1, \ldots, m_x\}$ and $P = \{w_1, \ldots, w_x\}$ respectively. We also define $Y = \{m_{x+1}, \ldots, m_{n_1}\}$ and $Q = \{w_{x+1}, \ldots, w_{n_2}\}$, and refer to these sets as the *Y-men* and *Q-women* respectively.

Lemma 3. *Let A be a matching that is stable for I', and let M be a maximum cardinality stable matching for I. Then*
(i) A Y-man who is matched in M must be matched in M with a P-woman.
(ii) $|M| \leq |A| + x$;
(iii) $|M| \leq 2x$.

Proof. (i) Suppose that m is a Y-man and that $(m, w) \in M$. Then if w were a Q-woman, she must be a Phase 1 Q-woman who failed to be matched during Phase 2, and therefore the matching found in Phase 2 could have been extended by adding the pair (m, w), contradicting its maximality.
(ii) By Lemma 2, all of the X-men are matched in A. So the only men who can be matched in M but not in A are Y-men. By (i), such a man must be matched in M with a P-woman. The inequality follows, as there are just x P-women.
(iii) Men matched in M are either X-men, and there are x of these, or Y-men matched with P-women (by (i)), and there are x of the latter, hence at most $2x$ such men in total. □

Phase 3 of Algorithm SSMTI-APPROX

Phase 3 of the algorithm involves completely breaking the remaining ties and then applying to the resulting instance of SMI the standard Gale-Shapley algorithm (or at least the extended version of that algorithm that deletes redundant entries from the preference lists - see [4]). The algorithm may be applied from either the men's or women's side; as is well known, the size of the resulting matching will be the same in each case. Tie-breaking is carried out according to just one restriction, namely, for each tie, the Y-men are given priority over the X-men. In other words, each tie is resolved by listing the Y-men that it contains, in arbitrary order, followed by the X-men that it contains, again in arbitrary order. It is immediate that the algorithm produces a matching that is stable for the original instance of SMTI. For an instance I of SMTI, we denote by I'' an instance of SMI obtained by application of Phases 1 and 2 of Algorithm SSMTI-APPROX, followed by tie-breaking according to this rule. Again

```
for each woman w
      break the tie (if any) in w's list, placing the Y-men ahead of the X-men;
/* Now apply the standard Gale-Shapley algorithm */
assign each person to be free;
while (some man m is free) and (m has a non-empty list) {
      w = the first woman on m's list;
      m proposes, and becomes engaged to w;
      for each successor m' of m on w's list {
            if m' is engaged to w
                  break the engagement, so that m' becomes free;
            delete the pair (m', w) from the preference lists;
      }
}
return the set A of engaged pairs;
```

Fig. 4. Phase 3 of Algorithm SSMTI-APPROX

it is immediate that a matching that is stable for I'' is also stable for I. Phase 3 of the algorithm is summarised in Figure 4.

The Performance Guarantee

Let A be a matching produced by application of Algorithm SSMTI-APPROX, and let M be a maximum cardinality stable matching for the original instance I of SSMTI. As previously established, all of the X-men are matched in A. Suppose that exactly r of the Y-men, say $m_{x+1}, \ldots, m_{x+r}$, are matched in M but not in A. Let us call these men the *extra men* (for M), and their partners in M the *extra women.*

Lemma 4. *(i) Each extra woman is matched in A.*
(ii) An extra woman is either matched in A to a Y-man or strictly prefers her A-partner to her M-partner.

Proof. Let w be an extra woman, and let m be her partner in M. Recall that M is stable for the original instance I, while A is stable for the refined instance I'' (of SMI), and therefore also for the instances I' and I (of SSMTI).
(i) By definition, m is an extra man and therefore is not matched in A, so that if w is not matched in A it is immediate that the pair (m, w) blocks A.
(ii) Let a be w's A-partner. If w strictly prefers m to a then, since m is unmatched in A, the pair (m, w) blocks A in I, a contradiction. If m and a are tied in w's list, and a is an X-man then, when that tie was broken to form I'', m, being a Y-man, must have preceded a in the resulting strict preference list. Hence, again since m is unmatched in A, the pair (m, w) blocks A in I'', a contradiction. □

We partition M's extra men into two sets U and V; those in U have an M-partner who is matched in A to an X-man, and those in V have an M-partner who is matched in A to a Y-man. Suppose, without loss of generality, that

$U = \{m_{x+1}, \ldots, m_{x+s}\}$ and $V = \{m_{x+s+1}, \ldots, m_{x+r}\}$, i.e., $|U| = s$, $|V| = r - s$. Let $M(U)$ denote the set of women who are matched in M to a man in U. Suppose that, among the men who are matched in A with women in $M(U)$, exactly t ($\leq s$) are unmatched in M. (These are all X-men, by definition of U, but some X-men – those who became so during Phase 2 of the algorithm, need not be matched in M.)

Our next lemma gives us certain inequalities involving the sizes of matchings M and A that will enable us to establish the claimed performance guarantee for Algorithm SSMTI-APPROX.

Lemma 5. *(i)* $|M| \leq |A| + r - t$.
(ii) $|A| \geq x + r - s$.
(iii) $|A| \geq r + s - t$.

Proof. (i) All the X-men are matched in A, but at least t of them are not matched in M, and the Y-men who are matched in M but not in A number exactly r.
(ii) Consider the set V. Each woman w who is the partner in M of a man in V is matched in A to a Y-man, and there are $r - s$ such women w. This gives us $r - s$ of the Y-men who are matched in A, and together with all x of the X-men who, by Lemma 1(ii), are all matched in A, we have a total of $x + r - s$ distinct men who are matched in A.
(iii) Consider the set U, and suppose that (m_{x+j}, w_{i_j}) is in M for $j = 1, \ldots, s$. By definition of U, each w_{i_j} has an X-man as her partner in A; without loss of generality, suppose that (m_j, w_{i_j}) is in A, for $j = 1, \ldots, s$. By Lemma 4, w_{i_j} strictly prefers her A-partner m_j to her M-partner m_{x+j}. Each m_j is an X-man and $s - t$ is the number of these men who are matched in M; suppose, without loss of generality, that (m_j, w_{k_j}) is in M, for $j = 1, \ldots, s - t$. Then none of these w_{k_j} can be an extra woman, for the M-partners of the latter are Y-men. Also, each of the men m_j prefers w_{k_j} to w_{i_j}, for otherwise (m_j, w_{i_j}) would block M. Furthermore, each w_{k_j} must be matched in A. For if not, the pair (m_j, w_{k_j}) would block A. It follows that we have a total of $r + s - t$ women who must be matched in A, namely the r extra women, by Lemma 4, and the $s - t$ women $w_{k_1}, \ldots, w_{k_{s-t}}$. □

We are now in a position to establish our main theorem.

Theorem 1. *For a given instance of SSMTI, let M be a maximum cardinality stable matching and let A be a stable matching returned by Algorithm SSMTI-APPROX. Then $|M| \leq 8|A|/5$.*

Proof. By Lemma 5 we have $|A| \geq \max(x + r - s, r + s - t) \geq \frac{1}{2}((x + r - s) + (r + s - t)) = x/2 + r - t/2$. So, by Lemma 3(iii), $|A| \geq |M|/4 + r - t/2$. Hence, by Lemma 5(i), $|A| \geq |M|/4 + |M| - |A| + t - t/2$, and so $2|A| \geq 5|M|/4 + t/2$, from which the claimed bound follows. □

Complexity of the Algorithm

The worst-case complexity of Algorithm SSMTI-APPROX is dominated by the maximum cardinality matching step in Phase 2. Using the Hopcroft-Karp algorithm [8], this can be achieved in $O(\sqrt{n}a)$ time, where n is the total number of men and women, and a is the sum of the lengths of the preference lists. However, there is a variant of the algorithm that achieves the same performance guarantee but with $O(a)$ complexity. This is obtained by observing that, in Phase 2, it suffices to find a *maximal* matching – i.e., a matching that cannot be extended to a larger matching by adding further pairs – rather than a maximum cardinality matching, of men in Y_1 to women in Q_1. The only place in the subsequent argument where the relevant property of this matching is needed is in the proof of Lemma 3(i), and it is indeed merely maximality that is required. A maximal matching can be found in $O(a)$ time, and all other parts of the algorithm are merely variants of the Gale-Shapley algorithm. It is not hard to show that these can also be implemented to run in $O(a)$ time (see [4]).

Tightness of the Approximation Guarantee

This is the tightest bound that can be established for Algorithm SSMTI-APPROX. Figure 5 shows an example where the ratio of $|M|$ to $|A|$ is $\frac{8}{5}$. The matching $M = \{(m_1, w_5), (m_2, w_6), (m_3, w_7), (m_4, w_8), (m_5, w_1), (m_6, w_2), (m_7, w_3), (m_8, w_4)\}$ is a maximum cardinality stable matching of size 8, whereas if ties are broken simply by removing the parentheses, the algorithm returns matching $A = \{(m_1, w_2), (m_2, w_3), (m_3, w_5), (m_4, w_6), (m_8, w_1)\}$, of size 5. By duplicating this pattern, we can obtain arbitrarily large instances realising the $\frac{8}{5}$ ratio.

Men's preferences	Women's preferences
$m_1 : w_5 \ \ w_2$	$w_1 : m_3 \ \ (\ m_8 \ \ m_5 \ )$
$m_2 : w_6 \ \ w_3$	$w_2 : m_1 \ \ m_6$
$m_3 : w_5 \ \ w_7 \ \ w_8 \ \ w_1$	$w_3 : m_2 \ \ m_7$
$m_4 : w_6 \ \ w_8 \ \ w_7 \ \ w_4$	$w_4 : m_4 \ \ m_8$
$m_5 : w_1$	$w_5 : (\ m_3 \ \ m_1 \ )$
$m_6 : w_2$	$w_6 : (\ m_4 \ \ m_2 \ )$
$m_7 : w_3$	$w_7 : (\ m_4 \ \ m_3 \ )$
$m_8 : w_1 \ \ w_4$	$w_8 : (\ m_3 \ \ m_4 \ )$

Fig. 5. An instance of SSMTI with ratio $\frac{8}{5}$

Extension to Special HRT

In view of the fact that our study was motivated by practical applications of the HRT problem, it is important to note that we can obtain exactly the same $\frac{8}{5}$ performance guarantee for an analogous algorithm for the special case of HRT in which each hospital's preference list has a tie of length ≥ 1 at the end. Full details of the extended algorithm and a correctness proof can be found in [10].

3 Summary and Open Problems

We have described a polynomial-time approximation algorithm with a performance guarantee of $\frac{8}{5}$ for a maximum cardinality stable matching in NP-hard variants of the Stable Marriage and Hospitals/Residents problems that are of significant practical interest. We have also shown that this performance guarantee is the best that can be proved for the algorithm.

The most obvious open question to pursue is whether this or a similar approach can yield useful performance guarantees for more general versions of SMTI and HRT, for example when there can be a single tie at the end of the lists on both sides, or when the lists on one side can contain arbitrary ties.

References

1. (Canadian Resident Matching Service) `http://www.carms.ca/jsp/main.jsp`
2. Gale, D., Shapley, L.S.: College admissions and the stability of marriage. American Mathematical Monthly 69, 9–15 (1962)
3. Gale, D., Sotomayor, M.: Some remarks on the stable matching problem. Discrete Applied Mathematics 11, 223–232 (1985)
4. Gusfield, D., Irving, R.W.: The Stable Marriage Problem: Structure and Algorithms. MIT Press, Cambridge (1989)
5. Halldórsson, M., Irving, R.W., Iwama, K., Manlove, D.F., Miyazaki, S., Morita, Y., Scott, S.: Approximability results for stable marriage problems with ties. Theoretical Computer Science 306(1-3), 431–447 (2003)
6. Halldórsson, M., Iwama, K., Miyazaki, S., Yanagisawa, H.: Improved approximation of the stable marriage problem. In: Di Battista, G., Zwick, U. (eds.) ESA 2003. LNCS, vol. 2832, pp. 266–277. Springer, Heidelberg (2003)
7. Halldórsson, M.M., Iwama, K., Miyazaki, S., Yanagisawa, H.: Randomized approximation of the stable marriage problem. Theoretical Computer Science 325(3), 439–465 (2004)
8. Hopcroft, J.E., Karp, R.M.: A $n^{5/2}$ algorithm for maximum matchings in bipartite graphs. SIAM Journal on Computing 2, 225–231 (1973)
9. Irving, R.W.: Matching medical students to pairs of hospitals: a new variation on a well-known theme. In: Bilardi, G., Pietracaprina, A., Italiano, G.F., Pucci, G. (eds.) ESA 1998. LNCS, vol. 1461, pp. 381–392. Springer, Heidelberg (1998)
10. Irving, R.W., Manlove, D.F.: 8/5-approximation algorithms for hard variants of the stable marriage and hospitals/residents problems. Technical Report TR-2007-232, University of Glasgow, Department of Computing Science (February 2007)
11. Iwama, K., Miyazaki, S., Yamauchi, N.: A $\left(2 - c\frac{1}{\sqrt{n}}\right)$ approximation algorithm for the stable marriage problem. In: Deng, X., Du, D.-Z. (eds.) ISAAC 2005. LNCS, vol. 3827, pp. 902–914. Springer, Heidelberg (2005)
12. Iwama, K., Miyazaki, S., Yamauchi, N.: A 1.875–approximation algorithm for the stable marriage problem. In: Proceedings of SODA 2007, pp. 288–297 (2007)
13. Manlove, D.F., Irving, R.W., Iwama, K., Miyazaki, S., Morita, Y.: Hard variants of stable marriage. Theoretical Computer Science 276(1-2), 261–279 (2002)
14. (National Resident Matching Program) `http://www.nrmp.org/about_nrmp/`
15. Roth, A.E.: The evolution of the labor market for medical interns and residents: a case study in game theory. Journal of Political Economy 92(6), 991–1016 (1984)
16. (Scottish Foundation Allocation Scheme) `http://www.nes.scot.nhs.uk/sfas/`

Approximation Algorithms for the Black and White Traveling Salesman Problem

Binay Bhattacharya[1,*], Yuzhuang Hu[2], and Alexander Kononov[3]

[1] School of Computing Science, Simon Fraser University, Burnaby, Canada, V5A 1S6
{binay,yhu1}@cs.sfu.ca
[2] Laboratory "Mathematical Models of Decision Making", Sobolev Institute of Mathematics, Acad. Koptyug Avenue, 630090 Novosibirsk, Russia
alvenko@math.nsc.ru

Abstract. The black and white traveling salesman problem (BWTSP) is to find the minimum cost hamiltonian tour of an undirected complete graph G, containing black and white vertices, subject to two restrictions: the number of white vertices, and the cost of the subtour between two consecutive black vertices are bounded. This paper focuses on designing approximation algorithms for the BWTSP in a graph satisfying the triangle inequality. We show that approximating the tour which satisfies the length constraint is NP-hard. We then show that the BWTSP can be approximated with tour cost $(4 - \frac{3}{2Q})$ times the optimal cost, when at most Q white vertices appear between two consecutive black vertices. When exactly Q white vertices appear between two consecutive black vertices, the approximation bound can be slightly improved to $(4 - \frac{15}{8Q})$. This approximation bound is further improved to 2.5 when $Q = 2$.

1 Introduction

In this paper, we consider an extension of the classical traveling salesman problem (TSP). The problem is defined on an undirected graph, $G = (V, E)$, where a vertex set, $V = V_B \cup V_W$, is partitioned into a set of *black vertices*, V_B, and a set of *white vertices*, V_W, and an edge set, E, with edge costs $w(e)$ for all $e \in E$ satisfying the triangle inequality. The black and white traveling salesman problem (BWTSP) is to determine a minimum cost hamiltonian tour of G subject to the following restrictions:

1. *Cardinality constraint* in which the number of white vertices on "black to black" paths is bounded above by a positive integer constant Q, and
2. *Length constraint* in which the cost of any path between two consecutive black vertices is bounded above by a positive value L.

Clearly, BWTSP reduces to the classical TSP when $L = Q = \infty$, and is therefore NP-hard. An application of the directed BWTSP arises in short-haul airline operations ([14,13]). The flight leg between two stations p and q is determined by a white vertex v_{pq} and a maintenance station s corresponds to a

* Research was partially supported by MITACS and NSERC.

G. Lin (Ed.): COCOON 2007, LNCS 4598, pp. 559–567, 2007.

black vertex v_s. An arc represents a leg-leg, leg-maintenance, or maintenance-leg sequence. The problem is to determine a flying sequence such that the number of takeoffs and landings, as well as the total operating cost between any two maintenance sequences are bounded as above. The undirected case has applications in telecommunications ([6,15]). Cosares et al. [6] and Wasem [15] describe an application of the undirected BWTSP arising in the design of telecommunication ring networks, in which black vertices are "ring offices" and white vertices are "hubs". In order to achieve a survivable synchronous optical network (SONET) architecture, any two consecutive ring offices on the network must be separated by at most Q hubs and a length not exceeding L. Another particular case of the BWTSP is the vehicle routing problem (VRP) where each client has unit demand, the vehicle has capacity Q, and maximal route length of the vehicle is at most L.

Attempts have been made to optimally solve BWTSP for small size problems ([2,15]). Ghiani, Laporte and Semet recently developed an exact branch-and-cut algorithm for the undirected case ([10]). Mak and Boland [13] have proposed a simulated annealing algorithm for the directed BWTSP and have applied it to instances involving 36 vertices. Bourgeois, Laporte and Samet [2] proposed five heuristic algorithms for the BWTSP, along with extensive computational computational comparisons.

In this paper we are interested in designing efficient approximation algorithms with guaranteed performances for the BWTSP. We show that BWTSP cannot be approximated if the length constraint is specified. However, approximation algorithms with guaranteed performances can be designed when only the cardinality constraint is specified. BWTSP with the cardinality constraint $Q = 1$ occurs in routing papers with different names: bipartite traveling salesperson problem or k-delivery problem where $k = 1$. Anily and Hassin [1] have shown a 2.5-approximation algorithm for another generalization of this problem, known as the swapping problem. Their algorithm finds a perfect matching M, consisting of edges that connect black and white vertices, and it uses the Christofides-Serdyukov heuristic [4] to find a tour, T, of the black vertices. The final route consists of visiting the black vertices in the sequence specified by the tour T, using the matching edges in M. Later, Chalasani and Motwani [3] developed a 2-approximation algorithm for 1-delivery problem using some combinatorial properties of bipartite spanning trees and matroid intersection. We expand the idea proposed in [1] and present a $(4 - \frac{3}{2Q})$-approximation algorithm, when the number of white vertices between two consecutive black vertices is bounded above by Q. The bound can be slightly improved to $(4 - \frac{15}{8Q})$, if the number of white vertices is exactly $Q \cdot |V_B|$. When $|V_W| = 2 \cdot |V_B|$, the bound can be improved to 2.5.

The organization of this paper is as follows: In section 2, we show that the BWTSP is NP-hard when the length constraint is specified. Section 3 deals with the BWTSP when only the cardinality constraint is specified. Various approximation algorithms are provided for different variants of the cardinality constraint. Section 4 discusses our conclusions.

2 BWTSP with Length Constraint

We first show that the following problem is NP-complete. Given a complete weighted graph G, with black and white vertices, satisfying the triangle inequality, determine whether G has a BWTSP route wherein the cost of the path between two consecutive black vertices in the cycle is no more than L. The above result then implies that the problem of designing approximation algorithms for the BWTSP is NP-hard, if the length constraint is specified.

Let us consider an instance of the hamiltonian path problem. Let $G = (V, E)$ be the input graph with $|V| = n$. Consider the following graph G'. G' has n black vertices V_B and n copies of V (say, $V_1, V_2, \ldots, V_n$) which are all white. Suppose $L = n + 1$ and $Q = \infty$. The cost of the edge between u and v is fixed as follows:

(i) if $u \in V_B$ and $v \in V_B$, $w(u,v) = 2$,
(ii) if $u \in V_B$ and $v \in V_i$, for any i, $w(u,v) = 1$,
(iii) if $u \in V_i$ and $v \in V_j$, for any $i \neq j$, $w(u,v) = 2$,
(iv) if $u \in V_i$, $v \in V_i$ and $(u, v) \in E$, $w(u,v) = 1$, and
(v) if $u \in V_i$, $v \in V_i$ and $(u, v) \notin E$, $w(u,v) = 2$.

G' is a complete weighted graph satisfying the triangle inequality. Clearly, G' has a BWTSP route satisfying the length constraint, if and only if, the graph G has a hamiltonian path.

3 BWTSP with Only the Cardinality Constraint Specified

The result from the previous section implies the impossibility of finding approximate solutions to BWTSP when the length constraint is specified. Therefore, we consider the case where only the cardinality constraint is satisfied. In other words, Given a graph $G = (V, E)$ where $V = V_B \cup V_W$, $V_B \cap V_W = \emptyset$, with the edges satisfying the triangle inequality, determine a minimum cost traveling salesman tour such that the number of white vertices between two consecutive black vertices in the tour is at most a given integer Q.

Let $|V_B| = n$ and $|V_W| = m$. Without any loss of generality, we assume that $m \leq Q \cdot n$, otherwise an instance does not have a feasible solution. Also note that if $m \leq Q$, any Hamiltonian cycle satisfies the cardinality constraint and we get the classical traveling salesman problem. So in our paper, we assume that $Q < m \leq Q \cdot n$.

3.1 Lower Bounds

Let L^* be the length of the optimal tour of the BWTSP in $G = (V, E)$, satisfying the cardinality constraint. The fact that the given graph satisfies the triangle inequality implies that the cost of the optimal traveling salesman tour that visits only a subset of vertices is a lower bound of optimal tour of the BWTSP. Let L^*_B and L^*_W denote the lengths of the optimal traveling salesman tour of the black and white vertices respectively. Hence $L^* \geq L^*_B$ and $L^* \geq L^*_W$.

We define a Q-factor of G as a set of edges $E_Q \subseteq E$, such that for each black vertex $v \in V_B$, $\sigma(v) \leq Q$, and for each white vertex, $v \in V_W$, $\sigma(v) = 1$, where $\sigma(v)$ is the number of edges of E_Q incident on v.

Given a tour T_{BW} of black and white vertices, a white vertex w is said to be closer to a black vertex u than to a black vertex v in T_{BW} if the number of vertices between w and u in T_{BW} is less than the number of vertices between w and v in T_{BW}. Suppose in the optimal tour we connect each white vertex to the closest black vertex. If black to black path in the optimal tour has an odd number of white vertices, the middle white vertex can be connected to either of the black vertices. Our strategy of connection is such that each black vertex is allowed to be connected to one such middle vertex. This way each black vertex is connected to at most Q white vertices. Thus, the obtained set of edges is a Q-factor.

Let $L^*_{E_Q}$ be the total cost of edges connecting the black and white vertices, using the above rule on the optimal tour of the BWTSP. We can estimate the cost of each edge between black and white vertices by using the triangle inequality. When Q is even, clearly $L^*_{E_Q} \leq \frac{Q}{2} L^*$. A similar inequality results when Q is odd.

3.2 Approximation Algorithm when $Q < m \leq Q \cdot n$

We describe our approximation algorithm below. In the following we assume that in any tour of G there are at most Q white vertices between two consecutive black vertices. Each step is followed by a brief discussion and implementation details if needed.

Algorithm $BWTSP(n, m)$

Step 1: Construct a complete bipartite graph K in the following way. One part contains n black vertices, V_B, and the other part contains m white vertices, V_W.

Step 2: Find a minimum cost Q-factor E_Q of K.
We solve this problem in the following way. We add $Q - 1$ copies of each black vertex to the first part of bipartite graph. We also copy the edges incident on the black vertex. We then find a minimum cost matching M in the augmented complete bipartite graph K [7]. Note that the total cost of the edges in E_Q, denoted by$\|E_Q\|$, is at most $L^*_{E_Q}$.

Step 3: Transform graph G to graph $\hat{G}$ as follows. Let h_v be the degree of black vertex v in the induced graph (V, E_Q). For each black vertex v add $Q - h_v$ "dummy" white vertices, and associate them with black vertex v in the following way. Each dummy white vertex is connected to v with edge cost zero. The dummy vertices are connected to other black and white vertices with edge costs being the same as the edge costs with v. This way we get $Q \cdot n$ white vertices in total. It is easy to show that the triangle inequality is still satisfied in $\hat{G}$.

Consider any tour where the black vertices $v_1, v_2, \ldots, v_n$ are ordered, and the number of white vertices between consecutive black vertices v_i and v_{i+1} is $t = h_{v_i}$. Let these white vertices be in order $b_1, b_2, \ldots, b_t$. We now augment the path $v_i, b_1, b_2, \ldots, b_t, v_{i+1}$ to $v_i, a_1, a_2, \ldots, a_s, b_1, b_2, \ldots, b_t, v_{i+1}$, for $s+t = Q$. Here $a_1, a_2, \ldots, a_s$ are the dummy white vertices associated with v_i. This means that the cost of the edge between v_i and a_1 is 0, and the cost of the edge between a_s and b_1 is the same as the cost of the edge between v_i and b_1. Therefore the cost of the augmented black and white tour is the same as that of the original tour. Let $\hat{G}_W = (\hat{V}_W, \hat{E}_W)$ be the graph induced by the white vertices.

Step 4: Find a near optimal hamiltonian tour $\hat{T}_W$ in graph $\hat{G}_W = (\hat{V}_W, \hat{E}_W)$. This tour fixes the order of the white vertices in the proposed tour of the BWTSP. We use Christofides-Serdyukov algorithm [4] to obtain $\hat{T}_W$. Let $L_{\hat{T}_W}$ denote the length of $\hat{T}_W$. From the discussions in step 3, the optimal TSP tour $\hat{T}^*_W$ involving all the white vertices of $\hat{G}_W$ has the same cost as the optimal TSP tour of G, and therefore the cost of $\hat{T}^*_W$ is less than L^*. So we have $L_{\hat{T}_W} \leq 1.5L^*$.

Step 5: Partition the tour $\hat{T}_W$ into paths P_i on Q vertices, $i = 1, 2, \ldots n$ of minimum cost.

Let $P_i = (u_{i1}, u_{i2}, \ldots, u_{iQ})$, $i = 1, 2, \ldots, n$ be the minimum cost paths. Since there exist Q different ways to partition tour $\hat{T}_W$, the total cost of the paths in $P_i, i = 1, 2, \ldots, n$ is at most $\frac{Q-1}{Q} L(\hat{T}_W)$.

Step 6: Construct a bipartite multigraph H in the following way. One part contains the vertices V_B and the other part contains n vertices $y_1, y_2, \ldots, y_n$ where the element y_i represents path P_i computed in step 5. Now (u, y), $u \in V_B$ and $y \in \{y_1, y_2, \ldots, y_n\}$, is an edge in H, if and only if there exists an edge (u, v') in E_Q (computed in step 2) such that $u \in V_B$ and v' is a vertex of the path represented by y.

Thus H is a bipartite multigraph and each vertex of H is of degree Q.

Step 7: Find a proper edge coloring of H in Q colors.

Each vertex in H has degree Q. According to König [11], the chromatic index of a bipartite multigraph with maximum degree h is h. It is also shown in [11] that in a bipartite multigraph, there exists a matching that saturates all the vertices with the maximum degree. Therefore, the edges of H can be colored using Q colors, and the set of edges of the same color covers the vertices of H. Let $C_1, C_2, \ldots, C_Q$ be the partition of the set of edges of E_Q where C_i contains all the i-colored edges. Note here that each C_i determines an assignment between black vertices and paths $P_1, P_2, \ldots, P_n$.

Step 8: Select the set C_q from $C_1, C_2, \ldots, C_Q$ with minimum length.

Clearly, $\|C_q\| \leq \frac{1}{Q}\|E_Q\|$. Therefore, $\|C_q\| \leq \frac{1}{2}L^*$.

Step 9: Let v_i be the black vertex assigned to P_i. Construct two hamiltonian tours $R_1 = (v_1, u_{11}, u_{12}, \ldots, u_{1Q},\ v_2, u_{21}, u_{22}, \ldots, u_{2Q}, \ldots, v_i, u_{i1}, u_{i2}, \ldots, u_{iQ}, v_{i+1}, \ldots, v_n, u_{n1}, u_{n2}, \ldots, u_{nQ})$ and $R_2 = (u_{11}, u_{12}, \ldots, u_{1Q}, v_1, u_{21}, u_{22}, \ldots, u_{2Q}, v_2, \ldots, u_{i1}, u_{i2}, \ldots, u_{iQ}, v_i, \ldots, u_{n1}, u_{n2}, \ldots, u_{nQ}, v_n)$. Remove the dummies from R_1 and R_2, and take the tour, say R, with the minimal cost as a BWTSP tour of G.

3.3 Performance Analysis

We can estimate the total cost of each tour by separately estimating the cost of the edges between the white vertices and between the black and white vertices. The total length of the edges between the white vertices is the total cost of n paths obtained in step 4 and is at most $\frac{Q-1}{Q} L_{\hat{T}_W}$.

We now estimate the total edge cost of the edges between black and white vertices in tour R. Let us first consider the total cost of edges connected to the black vertex v_i in routes R_1 and R_2. For route R_1, suppose $(v_i, u_{ik}) \in C_q$ for some k, $1 \le k \le Q$, then

$$w(u_{i-1,Q}, v_i) + w(v_i, u_{i1}) \le w(u_{i-1Q}, u_{i1}) + L(u_{i1}, u_{i2}, \ldots, u_{ik}) + w(u_{ik}, v_i) \\ + L(u_{i1}, u_{i2}, \ldots, u_{ik}) + w(v_i, u_{ik}).$$

Here $L(P_i)$ indicates the cost of path P_i.

For tour R_2 we have

$$w(u_{iQ}, v_i) + w(v_i, u_{i+1,1}) \le w(u_{ik}, v_i) + L(u_{ik}, \ldots, u_{iQ}) + w(u_{ik}, v_i) \\ + L(u_{ik}, \ldots, u_{iQ}) + w(u_{iQ}, u_{i+1,1}).$$

Since $\min\{L_{R_1}, L_{R_2}\} \le \frac{L_{R_1}+L_{R_2}}{2}$ we can now write

$$\begin{aligned} \min\{L_{R_1}, L_{R_2}\} &\le (4\|C_Q\| + 2\frac{Q-1}{Q} L_{\hat{T}_W} + 2L_{\hat{T}_W})/2 \\ &\le 2\|C_Q\| + \frac{Q-1}{Q} L_{\hat{T}_W} + L_{\hat{T}_W} \\ &\le (4 - \frac{3}{2Q})L^*. \end{aligned}$$

Theorem 1: The black and white traveling salesman problem with only the cardinality constraint can be approximated to within $(4-\frac{3}{2Q})$, where Q is the maximum number of consecutive white vertices that can appear in the route.

Running Time. The following computations dominate the running time of the algorithm.

1. Computing near optimal hamiltonian tour $\hat{T}_W$ of $\hat{G}_W = (V_W, \hat{E}_W)$ (Step 4). The running time of Christofides-Serdyukov's algorithm [4] to compute $\hat{T}_W$ is dominated by the perfect matching problem in a subgraph of $\hat{G}_W$ which, in the worst case, takes $O(|V_W|^3)$ time[12]. Since $|V_W| = Q \cdot n$, it takes $O(Q^3 n^3)$ time to compute $\hat{T}_W$.
2. Computing a Q-factor E_Q (Step 2).
   This problem has been shown to be equivalent to a matching problem in a complete bipartite graph K involving $O(Q \cdot n)$ vertices. Therefore, we can find the minimum cost Q-factor of K in $O(Q^3 n^3)$ time.
3. Finding a proper edge coloring of bipartite multigraph H (Step 6).
   We can use the algorithm proposed by Cole and Hopcroft [5] to find an edge coloring of the bipartite multigraph in $O(Q^2 n^2 \log n)$ time.

Thus, the approximation algorithm proposed to solve cardinality constrained *BWTSP* in a graph with n black vertices and m ($Q < m \leq Q \cdot n$) white vertices takes $O(Q^3 n^3)$ time to compute.

3.4 Approximation Algorithms When $m = Q \cdot n$

In this section we discuss ways of improving the performance bounds in the case when $m = Q \cdot n$. Let E^*_{BW} be the $2n$ edges of the optimal BWTSP solution L^* connecting the black and white edges. The cost of the Q-factor, described in section 3, can be alternately bounded by $\|E^*_{BW}\| + \frac{Q-2}{2} * L^*$. The argument for this new bound is almost the same once the edges of E^*_{BW} are separated. Also note that G and $\hat{G}$ are the same in Step 3 of $BWTSP(n, m)$. Let $\alpha = \frac{\|E^*_{BW}\|}{L^*}$, then we can write

$$\min\{L_{R_1}, L_{R_2}\} \leq 2\|C_Q\| + \frac{Q-1}{Q} L_{\hat{T}_W} + L_{\hat{T}_W} \leq (4 - \frac{7}{2Q} + \frac{2\alpha}{Q})L^*.$$

We describe below another algorithm, called $BWTSP2(n, Q \cdot n)$, where the number of white vertices between two consecutive black nodes is exactly Q. Most of the steps of Algorithm $BWTSP(n, m)$ are the same in the new algorithm. Steps 3, 4 and 5 are replaced by step 3-4-5 and steps 8 and 9 are replaced by steps 8-9(a) and 8-9(b). As before, each new step is followed by a brief discussion and comments, if needed.

Algorithm $BWTSP2(n, Q \cdot n)$

Step 3-4-5: Partition the white vertices into a set of paths, each containing exactly Q vertices, using the algorithm of Goemans and Williamson [9].

As described in [9], the exact path partitioning problem (partitioning $G_W = (V_W, E_W)$ into disjoint paths, each path containing exactly Q vertices) can be approximated to within $4(1 - \frac{1}{Q})(1 - \frac{1}{|V|})$, i.e. within $4(1 - \frac{1}{Q})$. Thus the set of paths we found in this step has a total cost of less than $4(1 - \frac{1}{Q})(1 - \alpha)L^*$. This step can be performed in time $O(n^2 \log n)$ [9] which was later improved to $O(n^2)$ [8]. Let P_W be the set of paths.

Step 8-9(a): Find a near-optimal tour T_B in $G_B = (V_B, E_B)$.

Step 8(b): Construct a tour involving the edges of C_q, P_W and T_B.

This tour uses the edges of C_q and the edges of the paths in P_W, found in step 3-4-5, twice and the edges of T_B once. Thus we can get a feasible solution of BWTSP with a cost less than $(1.5 + 8(1 - \frac{1}{Q})(1 - \alpha) + \frac{2}{Q}(\alpha + \frac{Q-2}{2}))L^*$, i.e. less than $(10.5 - \frac{10}{Q} - (8 - \frac{10}{Q})\alpha)L^*$.

We now have two solutions with the costs of $(4 - \frac{7}{2Q} + \frac{2\alpha}{Q})L^*$ and $(10.5 - \frac{10}{Q} - (8 - \frac{10}{Q})\alpha)L^*$ respectively. We can choose the one with the smaller cost to be our final solution. It is interesting to verify that the two cost functions have the same value for all Q when $\alpha = \frac{13}{16}$. Substituting $\frac{13}{16}$ for α, the approximation ratio is then $4 - \frac{15}{8Q}$. The gain of $\frac{3}{8Q}$ from our previous ratio of $4 - \frac{3}{2Q}$ is meaningful for small Q.

Further Improvement when $m = 2n$. In the following we show that for $m = 2n$ ($Q = 2$), the approximation bound can be improved to 2.5. This is due to the fact that both the white-white edges (denoted by E^*_{WW}) and the black-white edges (denoted by E^*_{BW}) of an optimal BWTSP solution L^*_{BW} can be efficiently approximated. Note that there are n edges in E^*_{WW} and $2n$ edges in E^*_{BW}. The algorithm is formally described below.

Algorithm $BWTSP3(n, 2 \cdot n)$

Step 1: Construct a bipartite graph K in the following way. One part contains n black vertices of V_B and the other part contains all $2n$ white vertices of V_W. Only the edges of G connecting the black and white vertices are present in K.

Step 2: Find a minimum cost 2-factor E_2 of K.
Since E^*_{BW} is a 2-factor, then $\|E_2\| \leq \|E^*_{BW}\|$.

Step 3: Find a minimum cost perfect matching M of $G_W = (V_W, E_W)$.
Clearly $\|M\| \leq \|E^*_{WW}\|$. Consider the induced graph $(V, M \cup E_2)$. This graph is a collection of cycles. Each cycle involving black and white vertices is a tour (on a subset of vertices) satisfying the cardinality constraint. We represent each cycle by $CYCLE(v)$ where v is an arbitrary black vertex in the cycle. Let V_A be the set of arbitrary black vertices chosen to represent the cycles. We note that $\|M \cup E_2\| = \|M\| + \|E_2\| \leq \|E^*_{BW}\| + \|E^*_{WW}\| = L^*$.

Step 4: Find a near-optimal TSP tour T_A of G_A, the subgraph of G induced by the vertices of V_A.

Step 5: Starting from an arbitrary black vertex $v \in V_A$ we make the round of all vertices of $CYCLE(v)$. After that we move to the next black vertex in tour T_A. The order in which the vertices appear in this walk define a tour T.
The cost of T is bounded by the total cost of the edges of $M \cup E_2$ and edges in tour T_A. So we have $L_T = L_{T_A} + \|M\| + \|E_2\| \leq \frac{5}{2}L^*$ and therefore, the algorithm $BWTSP3(n, 2n)$ is a 2.5-approximation algorithm.

4 Conclusions

We have designed approximation algorithms with guaranteed performances for the black and white traveling salesman problem. We have shown that BWTSP cannot be approximated if the length constraint is specified. However, approximation algorithms with guaranteed performances can be designed when the cardinality constraint Q is only specified. For arbitrary Q, the designed algorithm has an approximation ratio of less than 4. This ratio is slightly improved for smaller values of Q if the number of white vertices is exactly $Q \cdot n$ where n is the number of black vertices. The approximation bound of 2.5 is possible for the BWTSP when the number of white vertices is exactly $2n$.

References

1. Anily, S., Hassin, R.: The swapping problem. Networks 22, 419–433 (1992)
2. Bourgeois, M., Laporte, G., Samet, F.: Heuristics for the black and white traveling salesman problem. Computers and Operations Research 30, 75–85 (2003)
3. Chalasani, P., Motwani, R.: Approximating capacitated routing and delivery problems. SIAM Journal on Computing 28, 2133–2149 (1999)
4. Christofides, N.: The traveling salesman problem. In: Christofides, N., Mingozzi, A., Toth, P., Sandi, C. (eds.) Combinatorial Optimization, pp. 315–318 (1979)
5. Cole, R., Hopcroft, J.: On edge coloring bipartite graphs. SIAM Journal on Computing 11, 540–546 (1982)
6. Cosares, S., Deutsch, D.N., Saniee, I., Wasem, O.J.: SONET Toolkit: A decision support system for designing robust and cost effective fibre-optic networks. Interfaces 25, 20–40 (1995)
7. Dinitz, D.: The solution of two assignment problems. In: Fridman, A.A. (ed.) Russian; Studies in Discrete Optimization, Nauka, Moscow, pp. 333–348 (1976)
8. Gabow, H.N., Pettie, S.: The dynamic vertex minimum problem and its application to clustering-type approximation algorithms. In: Penttonen, M., Schmidt, E.M. (eds.) SWAT 2002. LNCS, vol. 2368, pp. 190–199. Springer, Heidelberg (2002)
9. Goemans, M.X., Williamson, D.P.: A general approximation technique for constrained forest problems. SIAM Journal on Computing 24(2), 296–317 (1995)
10. Ghiani, G., Laporte, G., Semet, F.: The black and white traveling salesman problem. Operations Research 54, 366–378 (2006)
11. König, D.: Über Graphen und ihre Anwendungen. Math. Annalen 77, 453–465 (1916)
12. Lawler, E.L.: Combinatorial Optimization: Networks and Matroids. Holt. Rinehart and Winston, New York (1976)
13. Mak, V., Boland, N.: Heuristic approaches to the asymmetric traveling salesman problem with replenishment arcs. International Transactions in Operations Research 7, 431–437 (2000)
14. Talluri, K.T.: The four-day aircraft maintenance routing problem. Transportation Science 32, 43–53 (1998)
15. Wasem, O.J.: An algorithm for designing rings in survivable fibre networks. IEEE Transactions on Reliability 40, 428–432 (1991)

Author Index

Lecture Notes in Computer Science

For information about Vols. 1–4499

please contact your bookseller or Springer

Vol. 4605: D. Papadias, D. Zhang, G. Kollios (Eds.), Advances in Spatial and Temporal Databases. X, 479 pages. 2007.

Vol. 4602: S. Barker, G.-J. Ahn (Eds.), Data and Applications Security XXI. X, 291 pages. 2007.

Vol. 4600: H. Comon-Lundh, C. Kirchner, H. Kirchner, Rewriting, Computation and Proof. XVI, 273 pages. 2007.

Vol. 4598: G. Lin (Ed.), Computing and Combinatorics. XII, 570 pages. 2007.

Vol. 4597: P. Perner (Ed.), Advances in Data Mining. XI, 353 pages. 2007. (Sublibrary LNAI).

Vol. 4596: L. Arge, C. Cachin, T. Jurdzinski, A. Tarlecki (Eds.), Automata, Languages and Programming. XVII, 953 pages. 2007.

Vol. 4595: D. Bošnački, S. Edelkamp (Eds.), Model Checking Software. X, 285 pages. 2007.

Vol. 4592: Z. Kedad, N. Lammari, E. Métais, F. Meziane, Y. Rezgui (Eds.), Natural Language Processing and Information Systems. XIV, 442 pages. 2007.

Vol. 4591: J. Davies, J. Gibbons (Eds.), Integrated Formal Methods. IX, 660 pages. 2007.

Vol. 4590: W. Damm, H. Hermanns (Eds.), Computer Aided Verification. XV, 562 pages. 2007.

Vol. 4589: J. Münch, P. Abrahamsson (Eds.), Product-Focused Software Process Improvement. XII, 414 pages. 2007.

Vol. 4588: T. Harju, J. Karhumäki, A. Lepistö (Eds.), Developments in Language Theory. XI, 423 pages. 2007.

Vol. 4587: R. Cooper, J. Kennedy (Eds.), Data Management. XIII, 259 pages. 2007.

Vol. 4586: J. Pieprzyk, H. Ghodosi, E. Dawson (Eds.), Information Security and Privacy. XIV, 476 pages. 2007.

Vol. 4585: M. Kryszkiewicz, J.F. Peters, H. Rybinski, A. Skowron (Eds.), Rough Sets and Intelligent Systems Paradigms. XIX, 836 pages. 2007. (Sublibrary LNAI).

Vol. 4584: N. Karssemeijer, B. Lelieveldt (Eds.), Information Processing in Medical Imaging. XX, 777 pages. 2007.

Vol. 4583: S.R. Della Rocca (Ed.), Typed Lambda Calculi and Applications. X, 397 pages. 2007.

Vol. 4582: J. Lopez, P. Samarati, J.L. Ferrer (Eds.), Public Key Infrastructure. XI, 375 pages. 2007.

Vol. 4581: A. Petrenko, M. Veanes, J. Tretmans, W. Grieskamp (Eds.), Testing of Software and Communicating Systems. XII, 379 pages. 2007.

Vol. 4580: B. Ma, K. Zhang (Eds.), Combinatorial Pattern Matching. XII, 366 pages. 2007.

Vol. 4578: F. Masulli, S. Mitra, G. Pasi (Eds.), Applications of Fuzzy Sets Theory. XVIII, 693 pages. 2007. (Sublibrary LNAI).

Vol. 4577: N. Sebe, Y. Liu, Y. Zhuang (Eds.), Multimedia Content Analysis and Mining. XIII, 513 pages. 2007.

Vol. 4576: D. Leivant, R. de Queiroz (Eds.), Logic, Language, Information, and Computation. X, 363 pages. 2007.

Vol. 4575: T. Takagi, T. Okamoto, E. Okamoto, T. Okamoto (Eds.), Pairing-Based Cryptography. XI, 408 pages. 2007.

Vol. 4574: J. Derrick, J. Vain (Eds.), Formal Techniques for Networked and Distributed Systems – FORTE 2007. XI, 375 pages. 2007.

Vol. 4573: M. Kauers, M. Kerber, R. Miner, W. Windsteiger (Eds.), Towards Mechanized Mathematical Assistants. XIII, 407 pages. 2007. (Sublibrary LNAI).

Vol. 4572: F. Stajano, C. Meadows, S. Capkun, T. Moore (Eds.), Security and Privacy in Ad-hoc and Sensor Networks. X, 247 pages. 2007.

Vol. 4570: H.G. Okuno, M. Ali (Eds.), New Trends in Applied Artificial Intelligence. XXI, 1194 pages. 2007. (Sublibrary LNAI).

Vol. 4569: A. Butz, B. Fisher, A. Krüger, P. Olivier, S. Owada (Eds.), Smart Graphics. IX, 237 pages. 2007.

Vol. 4566: M.J. Dainoff (Ed.), Ergonomics and Health Aspects of Work with Computers. XVIII, 390 pages. 2007.

Vol. 4565: D.D. Schmorrow, L.M. Reeves (Eds.), Foundations of Augmented Cognition. XIX, 450 pages. 2007. (Sublibrary LNAI).

Vol. 4564: D. Schuler (Ed.), Online Communities and Social Computing. XVII, 520 pages. 2007.

Vol. 4563: R. Shumaker (Ed.), Virtual Reality. XXII, 762 pages. 2007.

Vol. 4562: D. Harris (Ed.), Engineering Psychology and Cognitive Ergonomics. XXIII, 879 pages. 2007. (Sublibrary LNAI).

Vol. 4561: V.G. Duffy (Ed.), Digital Human Modeling. XXIII, 1068 pages. 2007.

Vol. 4560: N. Aykin (Ed.), Usability and Internationalization, Part II. XVIII, 576 pages. 2007.

Vol. 4559: N. Aykin (Ed.), Usability and Internationalization, Part I. XVIII, 661 pages. 2007.

Vol. 4551: J.A. Jacko (Ed.), Human-Computer Interaction, Part II. XXIII, 1253 pages. 2007.

Vol. 4550: J.A. Jacko (Ed.), Human-Computer Interaction, Part I. XXIII, 1240 pages. 2007.